THE WILEY BICENTENNIAL–KNOWLEDGE FOR GENERATIONS

Each generation has its unique needs and aspirations. When Charles Wiley first opened his small printing shop in lower Manhattan in 1807, it was a generation of boundless potential searching for an identity. And we were there, helping to define a new American literary tradition. Over half a century later, in the midst of the Second Industrial Revolution, it was a generation focused on building the future. Once again, we were there, supplying the critical scientific, technical, and engineering knowledge that helped frame the world. Throughout the 20th Century, and into the new millennium, nations began to reach out beyond their own borders and a new international community was born. Wiley was there, expanding its operations around the world to enable a global exchange of ideas, opinions, and know-how.

For 200 years, Wiley has been an integral part of each generation's journey, enabling the flow of information and understanding necessary to meet their needs and fulfill their aspirations. Today, bold new technologies are changing the way we live and learn. Wiley will be there, providing you the must-have knowledge you need to imagine new worlds, new possibilities, and new opportunities.

Generations come and go, but you can always count on Wiley to provide you the knowledge you need, when and where you need it!

WILLIAM J. PESCE
PRESIDENT AND CHIEF EXECUTIVE OFFICER

PETER BOOTH WILEY
CHAIRMAN OF THE BOARD

COMPREHENSIVE ORGANIC REACTIONS IN AQUEOUS MEDIA

COMPREHENSIVE ORGANIC REACTIONS IN AQUEOUS MEDIA

Second Edition

CHAO-JUN LI
McGill University, Montreal, Canada

TAK-HANG CHAN
McGill University, Montreal, Canada
Hong Kong Polytechnic University, China

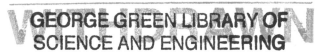

WILEY-INTERSCIENCE
A John Wiley & Sons, Inc., Publication

Published by John Wiley & Sons, Inc., Hoboken, New Jersey.
Published simultaneously in Canada. *10 06539960*

For general information on our other products and services or for technical support, please contact our Customer Care Department within the United States at (800) 762-2974, outside the United States at (317) 572-3993 or fax (317) 572-4002.

Wiley also publishes its books in a variety of electronic formats. Some content that appears in print may not be available in electronic formats. For more information about Wiley products, visit our web site at www.wiley.com.

Wiley Bicentennial Logo: Richard J. Pacifico

Library of Congress Cataloging-in-Publication Data is available.

ISBN: 978-0-471-76129-7

Printed in the United States of America

10 9 8 7 6 5 4 3 2 1

Dedicated to the authors' wives and children

CONTENTS

3 ALKENES   25

PREFACE TO THE SECOND EDITION

Nearly a decade ago, we published the first monograph on the subject of organic chemistry in water with the title of *Organic Reactions in Aqueous Media*. The book was divided into eight chapters, including introduction, pericyclic reactions, nucleophilic additions and substitutions, metal-mediated reactions, transition-metal catalyzed reactions, oxidations and reductions, and industrial applications. At that time, only a limited number of references existed in the literature on this subject, and such a division appears clear and concise, yet quite comprehensive. The book has served as a general guide for subsequent research efforts in the field. Since then, however, there has been an explosion of research activities in this important area. Now, organic reactions in water have become some of the most exciting research endeavors. The types of organic reactions reported in water are as diverse as those in organic solvents in classical organic chemistry. Completely new and unique reactivities have also been discovered in water. It has become difficult for researchers to follow the field and for students to be familiar with the overall picture of the area. As the subject is so broad at present, we have decided to discuss the topics on the basis of functional groups, which is parallel to most classical organic chemistry textbooks. Consequently, this new edition includes 12 chapters: introduction, alkanes, alkenes, alkynes, alcohols and ethers, organic halides, aromatics, aldehydes and ketones, carboxylic acid and derivatives, conjugated carbonyl compounds, amines, and pericyclic reactions. We hope that this comprehensive second edition will provide researchers and students with a tool in their endeavors. We would like to thank Charlene Keh, Liang Chen, Xiaoquan Yao, Zhiping Li, Weishi Miao, Xun He, Liang Zhao, Yuhua Zhang, Jianqing Feng, and Stephanie Boluk for their technical assistance.

CHAO-JUN LI
Montreal, Canada
TAK-HANG CHAN
Hong Kong, China

CHAPTER 1

INTRODUCTION

Of all the inorganic substances acting in their own proper nature, and without assistance or combination, water is the most wonderful. If we think of it as the source of all the changefulness and beauty which we have seen in clouds; then as the instrument by which the earth we have contemplated was modeled into symmetry, and its crags chiseled into grace; then as, in the form of snow, it robes the mountains it has made, with what transcendent light which we could not have conceived if we had not seen; then as it exists in the form of torrent, in the iris which spans it, in the morning mist which rises from it, in the deep crystalline pools which mirror its hanging shore, in the broad lake and glancing river; finally, in that which is to all human minds the best emblem of unwearied, unconquerable power, the wild, various, fantastic, tameless unity of the sea; . . . It is like trying to paint a soul.[1]

Aqueous chemistry is one of the oldest forces of change in the solar system. It started less than 20 million years after the gases of the solar nebula began to coalesce into solid objects.[2] Water is also the most abundant volatile molecule in comets. On the earth, the oceans alone contain about 1.4×10^{21} kilograms or 320,000,000 cubic miles of water. Another 0.8×10^{21} kilogram is held within the rocks of the earth's crust, existing in the form of water of hydration. The human

Comprehensive Organic Reactions in Aqueous Media, Second Edition, by Chao-Jun Li and Tak-Hang Chan
Copyright © 2007 John Wiley & Sons, Inc.

1

body is roughly 65% water by weight. Some organs like the brain and lungs are composed of nearly 80% water.[3]

Water is the basis and bearer of life. For millions of years, water had been at work to prepare the earth for the evolution of life. It is the solvent in which numerous biochemical organic reactions (and inorganic reactions) take place. All these reactions affecting the living system have inevitably occurred in an aqueous medium. On the other hand, modern organic chemistry has been developed almost on the basis that organic reactions are often to be carried out in organic solvents. It is only within the last two decades or so that people have again focused their attention on carrying out organic reactions in water. This development is in large part due to the study by Breslow on the Diels-Alder reactions.[4] Since then, many organic reactions that are traditionally carried out exclusively in organic solvents, such as the Barbier-Grignard-type reaction, have been successfully performed in an aqueous medium. Furthermore, novel reactions have been discovered, for which the use of water as solvent is critical.

Why should we consider using water as a solvent for organic reactions? There are many potential advantages:

- *Cost.* Water is the cheapest solvent available on earth; using water as a solvent can make many chemical processes more economical.
- *Safety.* Many organic solvents are flammable, potentially explosive, mutagenic, and/or carcinogenic. Water, on the other hand, is none of these.
- *Synthetic efficiency.* In many organic syntheses, it may be possible to eliminate the need for the protection and deprotection of functional groups, and save many synthetic steps. Water-soluble substrates can be used directly. This will be especially useful in carbohydrate and protein chemistry.
- *Simple operation.* In large industrial processes, isolation of the organic products can be performed by simple phase-separation. It is also easier to control the reaction temperature, since water has the largest heat capacities of all substances.
- *Environmental benefits.* It may alleviate the problem of pollution by organic solvents since water can be recycled readily and is benign when released into the environment (when no harmful residue is present).
- *Potential for new synthetic methodologies.* Compared to reactions in organic solvents, the use of water as a reaction solvent has

been explored much less in organic chemistry. There are many opportunities to develop novel synthetic methodologies that have not been discovered before.

This chapter will briefly survey some basic physical and chemical properties of water as well as the possible relevance of these properties to aqueous organic chemistry in terms of reactivity, selectivity (chemo-, regio-, and stereo-), and phase-separations.

1.1 THE STRUCTURE AND FORMS OF WATER[5]

In the 1780s, Cavendish and Lavoisier established that water is composed of hydrogen and oxygen. Gay-Lussac and Humboldt discovered in 1805 that the ratio of hydrogen and oxygen in a water molecule is two to one. And in 1842, Dumas found that the ratio of the combining weights of hydrogen and oxygen is very close to 2 to 16 in the molecule.

Water has two σ bonds, two lone pairs of electrons on oxygen, and a bond angle of 104.5° at oxygen.

Water exists in three basic forms: vapor, liquid, and solid. The relationship among the three forms of water is described by the pressure-volume-temperature phase diagram (Figure 1.1).

The structure of water in its liquid state is very complicated and is still a topic of current research. The structure of liquid water, with its molecules connected together by hydrogen bonds, gives rise to several anomalies when compared with other liquids.[6]

In its solid state, however, the basic structural features of ordinary hexagonal ice (ice I) are well established. In this structure (Figure 1.2), each water molecule is hydrogen bonded to four others in nearly perfect tetrahedral coordination. This arrangement leads to an open lattice in which intermolecular cohesion is large.

In its gaseous state, water molecules become dissociated from each other and can aggregate easily into small clusters. Increasing temperature and decreasing pressure help the dissociation; whereas increasing both the temperature and pressure increases the gasification of the liquid

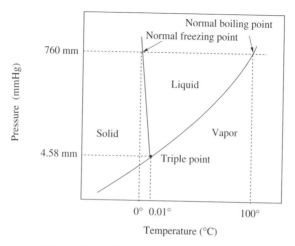

Figure 1.1 Phase diagram for water.

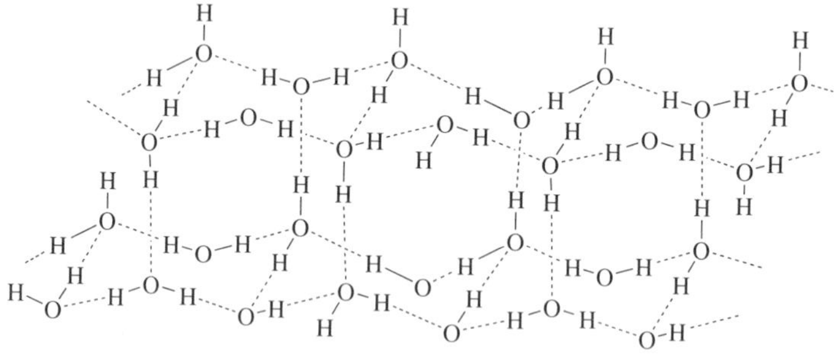

Figure 1.2 The structure of hexagonal ice.

phase as well as the aggregation of the gaseous phase. Under extremely high temperature and high pressure, water reaches a supercritical state in which the distinct gaseous phase and liquid phase no longer exist.

1.2 PROPERTIES OF WATER

The principal physical properties of water are shown in Table 1.1.[7]

Water has its highest density at 3.98°C and decreases as the temperature falls to 0°C. For this reason, ice is lighter than water and floats, which insulates deeper water from the cold and prevents it from freezing. This property has fundamental importance in nature. The density

TABLE 1.1   Principal Physical Properties of Water

Density, g/ml (3.98°C)	1.00
Melting point, °C (at 760 mm Hg)	0.00
Boiling point, °C (at 760 mm Hg)	100.00
Temperature at maximum density, °C	3.98
Heat of melting, (cal/g °C)	79.71
J/g °C	333.75
Heat of vaporization, cal/(g °C)(0°C)	595.40
(J/g °C)	2260.00
Specific heat, cal/(g °C)(15°C)	1.00
(J/(g °C)	4.19
Surface tension, mN/m (20°C)	72.75
Dynamic viscosity, mN.s/m² (20°C)	1.00
Specific electrical conductivity, s/m (25°C)	5.10^{-6}
Critical temperature, °C	374.00
Critical density, g/cm³	0.322
Critical specific volume, cm³/g	3.110
Dielectric constant (20°C)	80.20

of water also decreases with the increase of temperature above 3.98°C. It reaches the density of ice at about 70°C.

The viscosity of water also changes with temperature. It decreases with an increase in temperature because of the reduction in the number of hydrogen bonds binding the molecules together. The viscosity of water has an influence on the movement of solutes in water and on the sedimentation rate of suspended solids.

Water has the highest value for the specific heat of all substances. The specific heat of water is the standard against which the values for all the other substances are determined. The high value for the specific heat of water is because of the great heat capacity of the water mass. This means that rapid changes in ambient temperature result in only slow changes in water temperature. Such an effect is important for aquatic organisms. It is also an important advantage in using water as solvent to control the temperature for both endo- and exothermic reactions, especially in large-scale industrial processes.

If the surface of a liquid is regarded as an elastic membrane, then the surface tension is the breaking force of this membrane. Water has one of the highest surface tensions of all liquids. For example, the surface tension of ethanol at 20°C is 22 mN/m, while that of water is 72.75 mN/m. The surface tension of water decreases with temperature.

The presence of surface active agents, such as detergents, also decreases the surface tension of water.

1.3 SOLVATION

Water is a very good solvent for many substances, which is of fundamental importance in nature. The solubility of a chemical substance is dependent upon the temperature. The solubility of gases such as oxygen, nitrogen, and carbon dioxide in water usually decreases with a rise in temperature. However, there are some gases whose solubility increases with an increase in temperature. An example of such is helium. Similarly, there is variation in the relation between the solubility of solids in water and the temperature. For example, the solubility of $AgNO_3$ rapidly increases with an increase in temperature, but for NaCl there is only a slight increase in solubility with a rise in temperature. On the other hand, when temperature increases there is a decrease in solubility for Li_2CO_3. The influence of the temperature on the solubility of substances is dictated by the heat of solution of the substance, which is the heat emitted or absorbed during the dissolution of one mole of a substance in one liter of water.

Metal ions in aqueous solution exist as complexes with water. The solubility of organic compounds in water depends primarily on their polarity and their ability to form hydrogen bonds with water. Organic compounds with a large part of polar components such as acetic acid, "dissolve" in water without limit. In such cases, the polar part dominates. By contrast, soaps and detergents have a polar "end" attached to a relatively large nonpolar part of the molecule. They have limited solubility and the molecules tend to coalesce to form micelles.

1.4 HYDROPHOBIC EFFECT

Polar compounds and compounds that ionize can dissolve readily in water. These compounds are said to be hydrophilic. In contrast to hydrophilic substances, hydrocarbons and other nonpolar substances have very low solubility in water because it is energetically more favorable for water molecules to interact with other water molecules rather than with nonpolar molecules. As a result, water molecules tend to exclude nonpolar substances, forcing them to associate with themselves in forming drops, thereby minimizing the contact area between water and the organic substance. This phenomenon of repulsion of nonpolar substances by water is referred to as the hydrophobic effect.[8] The hydrophobic effect plays a critical role in biological systems. For example, it is the primary force in determining the folding patterns of proteins and the self-assembly of biological membranes.

Hydrophobic interaction, the association of a relatively nonpolar molecule or group in water with other nonpolar molecules, is not due to the mutual attraction of the nonpolar molecules. Rather, it is due to the large *cohesive energy density* (c.e.d.) of water. Cohesive energy density of a liquid is the energy of vaporization in calories per cubic centimeter of the liquid, which reflects the van der Waals forces holding the molecules of the liquid together, whereas the internal pressure of a liquid describes the sensitivity of energy to a change in volume and reflects the strength of intermolecular forces. The large cohesive energy density causes the polar water molecules surrounding the nonpolar compounds to associate with each other. Internal pressure of water, however, may also play a role in hydrophobic interaction. Table 1.2 contains the internal pressure and cohesive energy density for some common solvents at 25°C.

In thermodynamic terms, solutes can be divided into two classes. For hydrophobic solutes in dilute solution in water, the partial Gibbs free energy of solution is positive. This is because water molecules that surround a less polar molecule in solution are more restricted in

TABLE 1.2   Internal Pressures and Cohesive Energy Densities for Some Common Solvents (25°C)[9]

Solvent	c.e.d. (cal cm^{-3})	Internal pressure (cal cm^{-3})
Water	550.2	41.0
Formamide	376.4	131.0
Methanol	208.8	70.9
Dimethyl sulfoxide	168.6	123.7
Dimethylformamide	139.2	114
Acetonitrile	139.2	96
Acetone	94.3	79.5
Benzene	83.7	88.4
Carbon tetrachloride	73.6	80.6
Diethyl ether	59.9	63.0
Hexane	52.4	57.1

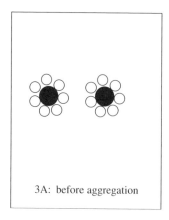 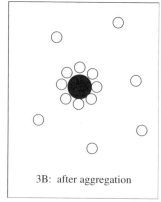

3A: before aggregation   3B: after aggregation

Figure 1.3   (● nonpolar solute; ○ water).

their interactions with other water molecules, and these restricted water molecules are relatively immobile and ordered (3A, Figure 1.3). However, water molecules in the bulk solvent phase are much more mobile and disordered. Thermodynamically, there is a net gain in the combined entropy of the solvent and the nonpolar solute when the nonpolar groups aggregate and water is freed from its ordered state surrounding the nonpolar groups (3B, Figure 1.3).

Hydrophobic interaction, however, is a relatively weak interaction. For example, the energy required to transfer a $-CH_2-$ unit from a hydrophobic to an aqueous environment is about 3 kJ mol^{-1}.

Thus, the transfer of a hydrophobic molecule from a pure state to an aqueous solution is an unfavorable process due to the large decrease of entropy resulting from the reorganization of the water molecules surrounding the solutes. By contrast, the partial Gibbs free energy for dissolving a hydrophilic solute in water is negative. This is because hydrophilic solutes can bind water molecules through hydrogen bonding.

1.5   SALT EFFECT

The effect of dissolved hydrophilic electrolytes on the interaction between organic solutes and water can be described by the salting-in and salting-out effects. Dissolved electrolytes usually increase the internal pressure in water, through a volume-reducing process that involves polarization and attraction of solvent molecules around the ionic species (electrostriction). For example, a 3 M aqueous solution of sodium bromide has an internal pressure of around 75 cal cm^{-3}, whereas the internal pressure of water at 25°C is only 41 cal cm^{-3}.

When the dissolved salt increases the internal pressure of aqueous solution to a certain extent, the nonelectrolyte is squeezed out (salting out). On the other hand, when the dissolved salt reduces the internal pressure of the solution, more of the nonelectrolyte is able to dissolve (salting in). All the electrolytes except perchloric acid increase the internal pressure of water and cause a salting out of organic species. For example, saturated sodium chloride is used to separate organic compounds from water.

Water's internal pressure acts on the volume of activation ($\Delta V^{\neq}$) of a reaction in the same way as an externally applied pressure does. Thus, the internal pressure of water influences the rates of nonpolar reactions in water in the same direction as external pressures. Nonpolar reactions with a negative volume of activation will thus be accelerated by the internal pressure of water, whereas nonpolar reactions with a positive volume of activation will be slowed by the internal pressure. For example, at 20°C the rate of Diels-Alder reaction between cyclopentadiene and butenone, which is known to have a negative volume of activation, in a 4.86 M LiCl solution is about twice as that of the reaction in water alone (Eq. 1.1).[4]

$$\tag{1.1}$$

Solvent	$k_2 \times 10^5$, (M^{-1}s^{-1})
H$_2$O	4400 ± 70
H$_2$O + LiCl (4.86 M)	10800

1.6 WATER UNDER EXTREME CONDITIONS[3,10]

Ordinary water behaves very differently under high temperature and high pressure. Early studies of aqueous solutions under high pressure showed a unique anomaly that was not observed with any other solvent.[11] The electrolytic conductance of aqueous solutions increases with an increase in pressure. The effect is more pronounced at lower temperatures. For all other solvents, the electrical conductivity of solutions decreases with increase in pressure. This unusual behavior of water is due to its peculiar associative properties.[12]

Thermal expansion causes liquid water to become less dense as the temperature increases. At the same time, the liquid vapor becomes more dense as the pressure rises. For example, the density of water varies from 1.0 g/cm^3 at room temperature to 0.7 g/cm^3 at 306°C. At the critical point, the densities of the two phases become identical and they become a single fluid called *supercritical fluid*. Its density at this point is only about 0.3 g/cm^3 (Figures 1.4 and 1.5).

In the region of supercritical point, most properties of supercritical water vary widely. The most prominent of these is the heat capacity at constant pressure, which approaches infinity at the critical point. Even 25°C above T_c, at 80 bar away from P_c, the heat capacity of water is an order of magnitude greater than its value at higher or lower pressure.

The dielectric constant of dense, supercritical water can range from 5 to 20 simply upon variation of the applied pressure.

As the temperature increases from ambient to the critical point, the electrolytic conductance of water rises sharply and is almost independent of the pressure. Macroscopically, this is due to the decrease in water viscosity over this range. The primary cause for the fall in viscosity is a disintegration of water clusters.

However, the conductances (being as much as tenfold greater than at room temperature) begin to drop off after entering the supercritical

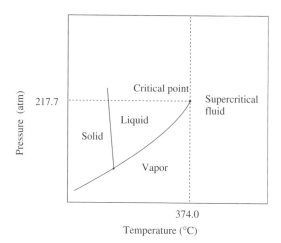

Figure 1.4   Phase diagram of water around the supercritical region.

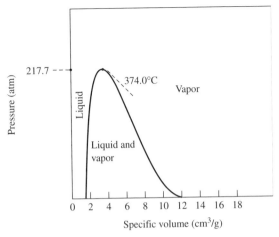

Figure 1.5 Pressure-volume diagram for water around the supercritical region.

region. The degree of drop is dependent on the pressure. Low pressures bring about a sharp drop, while at high pressures the decline is much less severe. Dissolved salts associate with themselves and behave as weak electrolytes.

On the other hand, water, itself an extremely weak electrolyte at room temperature, dissociates to a greater extent as the temperature rises (Eq. 1.2). For example, K_w increases to about 10^{-6} at $1000°C$ and a density of 1.0 g/ml.

$$2\,H_2O \rightleftharpoons H_3O^+ + {}^-OH \tag{1.2}$$

The increased dissociation of water in conjunction with the increased association of the electrolyte in the supercritical region has a fundamental influence on chemical reactions. Some reactions such as hydrolysis become faster in supercritical water. For example, there are at least eight species (KCl, KOH, HCl, HOH, K^+, Cl^-, H^+, and OH^-) for potassium chloride in supercritical water.

On the other hand, the increase in temperature decreases the intermolecular interaction (hydrogen bonding) between water molecules, which lessens the squeezing-out effect for nonpolar solutes. At the supercritcal state, water exhibits an "antiaqueous property." For example, water at high temperatures exhibits considerable, and sometimes complete, miscibility with nonpolar compounds.

Because oxygen, carbon dioxide, methane, and other alkanes are completely miscible with dense supercritical water, combustion can occur in this fluid phase. Both flameless oxidation and flaming combustion can take place. This leads to an important application in the treatment of organic hazardous wastes. Nonpolar organic wastes such as polychlorinated biphenyls (PCBs) are miscible in all proportions in supercritical water and, in the presence of an oxidizer, react to produce primarily carbon dioxide, water, chloride salts, and other small molecules. The products can be selectively removed from solution by dropping the pressure or by cooling. Oxidation in supercritical water can transform more than 99.9 percent of hazardous organic materials into environmentally acceptable forms in just a few minutes. A supercritical water reactor is a closed system that has no emissions into the atmosphere, which is different from an incinerator.

Quantum chemistry calculations suggest that supercritical water can provide new reaction pathways by forming structures with the reacting molecule that lower the activation energies for bond breakage and formation.[10b] The calculation shows that the more water molecules participate in the reaction, the lower the activation energy of the reaction is. For example, calculation on the following gas-shift reactions (Eq. 1.3) shows that the activation energy for the first step of the reaction is 61.7 kcal/mol if no additional water molecules participate in the reaction ($n = 0$); whereas the participation of an additional water molecule ($n = 1$) would lower the activation energy to half of its original value (35.6 kcal/mol). A similar decrease in activation energy by the participation of an additional water molecule was found for the second step, and the participation of more water molecules in the reaction would further reduce the activation energy.

$$CO + (n + 1)H_2O \longrightarrow HCOOH + nH_2O \longrightarrow CO_2 + H_2 + nH_2O \qquad (1.3)$$

$$\text{Step 1} \qquad\qquad\qquad \text{Step 2}$$

In this case water is effectively acting as a catalyst for the reaction by lowering the energy of activation. These catalytic water molecules are more likely to participate in the reaction under supercritical conditions because their high compressibility promotes the formation of solute-solvent clusters.

REFERENCES

1. Ruskin, J., cited in Coles-Finch, W. and Hawks, E. *Water in Nature*. T. C. & E. C. Jack Ltd., London, 1933.

2. Endress, M., Bischoff, A., and Zinner, E., *Nature* **1996**, *379*, 701.

3. Eisenberg, D., Kauzmann, W., *The Structure and Properties of Water*. Oxford University Press, 1969.

4. Rideout, D. C., and Breslow, R., *J. Am. Chem. Soc.* **1980**, *102*, 7816.

5. *Water and Aqueous Solutions* Horne, R. A., ed. Wiley-Interscience, New York, 1972.

6. Dojlido, J. R., Best, G. A., *Chemistry of Water and Water Pollution*. Ellis Horwood, 1993.

7. See *CRC Handbook of Chemistry and Physics*, 75th ed. CRC Press, 1994: also ref. 6.

8. Tanford, C., *The Hydrophobic Effect: Formation of Micelles and Biological Membranes,* 2nd ed., John Wiley & Sons, New York, 1980; Blokzijl, W., Engberts, J. B. F. N., *Angew. Chem., Int. Ed. Engl.* **1993**, *32*, 1545.

9. Dack, M. R. J., *Chem. Soc. Rev.* **1975**, *4*, 211; Gordon, J. E., *J. Phy. Chem.* **1966**, *70*, 2413.

10. Brummer, S. B., Gancy, A. B., in *Water and Aqueous Solutions: Structure, Thermodynamics, and Transport Processes*, Horne, R. A., ed. Wiley-Interscience, New York, 1972; Shaw, R. W., Brill, T. B., Clifford, A. A., Eckert, C. A., Franck, E. U., *C & EN*, Dec. 23, **1991**, 26.

11. Cohen, E., *Piezochemie, Kondensierter Systeme, Akademische Verlagsgesellschaft*. Leipzig, 1919.

12. Kavanau, J. L., *Water and Solute-Water Interactions*. Holden-Day, San Francisco, 1964.

CHAPTER 2

ALKANES

Because simple alkanes consist of only nonactivated C–H bonds and C–C bonds, these compounds are generally considered as nonreactive except under vigorous conditions such as high temperature and radical halogenations. On the other hand, there has been great interest in developing methods that can functionalize alkanes efficiently, selectively, and under mild conditions. As petroleum and natural gas are the prime feedstocks for the chemical industry and energy supplies globally at the present time, the success of such methods has potential and fundamental implications in a variety of fields including chemicals, energy, medicine, and the environment. Within the past two decades, significant progress has been made in the *activation* of alkanes under milder conditions.[1] Such activations can be performed even in aqueous conditions.[2]

2.1 OXYGENATION OF ALKANES

The most extensive studies of alkane reactions in aqueous media are on the oxygenation reaction. In fact, nature has used monooxygenase (found in mammalian tissue) and other enzymes to catalyze the oxidation of alkanes to give alcohols in aqueous environments at ambient

Comprehensive Organic Reactions in Aqueous Media, Second Edition, by Chao-Jun Li and Tak-Hang Chan
Copyright © 2007 John Wiley & Sons, Inc.

conditions.[3] Nature has used both aerobic (involving molecular oxygen) and anaerobic oxidation for alkane hydroxylation. The enzymes isolated from methane oxidizing bacteria are called in general terms *methane monooxygenase* (MMO). The MMOs studied in most detail are the water-soluble ones from *Methylococcus capsulatus*. Methane shows the highest activity among alkanes with such MMOs.[4] More detailed discussion on such enzymatic reactions is beyond the scope of this book.

On the other hand, there has been a long-held interest in developing the chemists' equivalent of enzymatic hydroxylation of C–H bonds in alkanes by using transition-metals and other reagents under mild conditions. The reaction of hydrocarbons with OH radicals, produced by the photolysis of H_2O_2 or HNO_3, at 17.5°C in aqueous solution, gives various oxidation products.[5] Some of the most successful progress has been made using water. For example, hydrophobic alkanes in aqueous media were functionalized using a biomimetic methane monooxygenase (MMO) enzyme assembly, including the active site and a hydrophobic pocket in a derivatized surface silica system using *tert*-Bu hydroperoxide/O_2 as the oxidants (Eq. 2.1).[6]

$$\text{(2.1)}$$

PEO = poly(ethylene oxide)
PPO = poly(propylene oxide
TPA = *tris*[(2-pyridyl)methyl]amine)

For transition-metal catalyzed hydroxylation of alkane C–H bonds, the reactions of alkanes with platinum(II) complexes were the most successful. In an aqueous solution of hexachloroplatinic acid and Na_2PtCl_4, alkanes were converted into a mixture of isomeric alkyl chlorides, alcohols, and ketones, and the platinum(IV) is reduced to platinum(II).[7] The kinetics of the reaction with methane as the alkane have been described in detail.[8]

Two main mechanisms may be proposed for the first step of the alkane interaction with platinum(II) complexes: (1) oxidative addition followed by reversible proton elimination; and (2) electrophilic substitution with simultaneous proton abstraction. Isotopic experiments with enriched ^{195}Pt complexes showed[9] that the oxidation process involves an electron transfer to generate σ-alkylplatinum(IV) complex, which reacts with nucleophile H_2O to afford an alcohol ROH (Eq. 2.2).

$$R\text{-}Pt(II)Cl_3{}^{2-} + {}^*Pt(IV)Cl_6{}^{-2} \longrightarrow R\text{-}Pt(IV)Cl_5{}^{2-} + {}^*Pt(II)Cl_4{}^{-2}$$

$$\downarrow {}_{H_2O}$$

$$R\text{-}OH$$

(2.2)

The direct use of dioxygen in the catalytic oxidation of nonactivated C–H bonds under mild conditions without using a coreductant was reported by using a $K_2PtO_4/CuCl_2$ catalytic system to give the selective functionalization of α-amino acids in water.[10] The aqueous medium avoided the need for protection of functional groups (Eq. 2.3).

$$\underset{NH_2}{\text{COOH}} \xrightarrow[\substack{K_2PtO_4/CuCl_2 \\ H_2O,\ 100°C}]{O_2} \underset{NH_2}{\overset{O}{\longrightarrow}}$$

(2.3)

35% yield

In addition to platinum-catalyzed oxidation of alkane C–H bonds, other oxidation systems have also been developed, some of which were performed under aqueous conditions. The most well known is the Gif system developed by Barton and co-workers[11] for oxidation and oxidative functionalization of alkanes under mild conditions. An early version of the Gif-type system consisted of an Fe(II/III) complex, a reducing agent (Fe, Zn, or H_2S), and dioxygen (or air), operating in a pyridine/acetic acid medium often with a small amount of water (about 6.6%).[11a] More recent versions of the Gif reagents have been represented by Fe(III)/H_2O_2 or Fe(III)/t-BuOOH, often together with the presence of 2-picolinic acid.[11b] While the mechanism of the Gif chemistry is still a matter of continuous debate,[11c,d] Gif chemistry attracted considerable interest for industrial applications, especially in relation to the conversion of cyclohexane to cyclohexanol/cyclohexanone (Eq. 2.4), which is an important feedstock for the Nylon manufacturing process. The possible role of water, either as an additive to the media or as part of the 30% H_2O_2 used, has not been examined carefully.

$$\text{Gif chemistry}$$

(2.4)

The reaction of alkanes by the aqueous Fenton-type oxidation with ruthenium complexes,[12] iron salts,[13] and chromium oxides[14] generates a mixture of alcohols and ketones (and formaldehyde for methane) in water or a water/CH_2Cl_2 mixture. Efficient and highly selective conversion of cyclooctane into cyclooctanone is obtained under pure biphasic conditions through *tert*-BuOOH activation by the in-situ formation of colloidal ruthenium species arising from $RuCl_3 \cdot 3H_2O$.[15] Aerobic photooxygenation of alkanes by using polyoxotungstate $W_{10}O_{32}{}^{4-}$ in water was also developed.[16] These methods often generate a mixture of oxidation products. Aqueous solutions of permanganate oxidize methane into carbon dioxide as the sole product at 40–100°C.[17] Recently, polyoxometalates have been used to oxidize alkanes effectively in water. Manganese substituted polyoxometalates were effective catalysts for the oxidation of alkanes to ketones with ozone in an aqueous reaction media (Table 2.1). A reactive manganese ozonide species was proposed as the intermediate.[18]

Recently, Breslow[19] reported a regioselective and stereoselective hydroxylation of steroid substrates catalyzed by manganese porphyrin

TABLE 2.1   Oxidation of Alkanes with Ozone Catalyzed by $Li_{12}[Mn^{II}_2\text{-}ZnW(ZnW_9O_{34})_2]$ in 40 percent t-BuOH-Water[a]

Substrate	Product, mol%	Conversion, mol%
Ethylbenzene	Acetophenone, 85; (1-Phenylethanol, 15)	82
Diphenylmethane	Benzophenone, >98	62
Tetrahydronaphthalene	α- and β-Tetralone, >98	56
Cumene	Acetophenone, >98	38
Cyclohexane	Cyclohexanone, >98	41
Cyclooctane	Cyclooctanone, >98	38
n-Decane	2-, 3-, 4-, and 5-Decanone, >98	28

[a]Reaction conditions: 1 mmol substrate, 10 μmol $Li_{12}[Mn^{II}_2\text{-}ZnW(ZnW_9O_{34})_2]$, 2.5 mL 40% t-BuOH-water, 45 min, 2°C.

systems (cytochrome P-450 mimic) carrying cyclodextrin binding groups, using iodosobenzene as oxidant. Hydrogen peroxide and other simple oxidants such as sodium hypochlorite are not effective in water. When thiol ligands were added to the catalyst, either covalently attached or hydrophobically bound, hydrogen peroxide became an effective oxidant for the remote hydroxylation (Eq. 2.5).

$$(2.5)$$

In addition to metal catalyzed oxygenation of nonactivated alkane C–H bonds, oxofunctionalization of C–H bonds can also occur in water by using dioxiranes.[20] Alkylketones and alkylketoesters could be regioselectively oxidized at the δ-position of the aliphatic chain by dioxiranes generated in situ by oxone in a mixture of H_2O/MeCN

(1/1.5 v/v) (Eq. 2.6).[21] The reaction was proposed to proceed via a concerted nonradical mechanism involving the formation of a δ-hydroxy ketone intermediate and a subsequent cyclization to give a hemiketal. This hemiketal prevents further oxidation at the δ site.

$$\text{(2.6)}$$

86% yield

2.2   HALOGENATION OF ALKANES

Halogenation of alkanes is an important reaction to generate halogenated compounds. Such reactions are often carried out via a radical process at high temperature. However, with the use of transition metal together with light, it is possible to carry out halogenation at room temperature.[22] Under phase-transfer conditions, adamantane has been brominated by using a combination of CBr_4/NaOH (Eq. 2.7).[23] The reaction is proposed to involve the single-electron oxidation of HO^- by CBr_4 to generate $HO\cdot$ radical as the initial step (Scheme 2.1). In the presence of halide ions, the Shilov-type oxidation of alkanes in aqueous media generates alkyl halides. For example, the reaction of methane with chlorine in water at 125°C in the presence of platinum chloride affords methyl chloride that is partially hydrolyzed to methanol in situ.[24]

$$\text{(2.7)}$$

2.3   FORMATION OF CARBON–CARBON BONDS

Radical ion $SO_4^{-\cdot}$ generated in situ from $S_2O_8^{2-}$ in aqueous solution at 100°C reacts with methane and ethane to generate alkyl radicals. These radicals react with another $SO_4^{-\cdot}$ to afford $ROSO_3^-$. Under an atmosphere of CO, the corresponding acid, RCOOH, can be obtained.[25] Recently, it has been shown that it is possible to couple methane with

$$CBr_4 + HO^- \quad \rightleftharpoons \quad CBr_4^{\overline{\bullet}} + HO\bullet$$

$$CBr_4^{\overline{\bullet}} \quad \rightleftharpoons \quad \bullet CBr_3 + Br^-$$

$$RH + \bullet CBr_3 \quad \rightleftharpoons \quad R\bullet + HCBr_3$$

$$R\bullet + CBr_4 \quad \rightleftharpoons \quad RBr + Br_3C\bullet$$

Scheme 2.1

CO to generate acetic acid in aqueous conditions by means of several catalysts (Table 2.2).[26] $RhCl_3$ catalyzed the direct formation of methanol and acetic acid from methane, CO, and O_2 in a mixture of perfluorobutyric acid and water with a turnover rate at approximately $2.9 \ h^{-1}$ based on Rh at $80–85°C$.[27] Under similar conditions, ethane was more active and gave ethanol, acetic acid, and methanol.

2.4 D/H EXCHANGE OF ALKANES IN WATER

The reaction of alkanes at approximately $100°C$ in a sealed tube in a D_2O-CD_3COOD solution of $PtCl_4^{2-}$ generates the H–D exchange product.[28] The process can occur in pure D_2O; however, the addition of acetic acid increases the rate of reaction by a factor of 30. H–D exchange in cyclohexane also occurs when anions X^- are added to the system. The sequence for the platinum(II) ligands on the rate of the H–D exchange is as follows: $PPh_3 = py < DMSO < CN^- < NO^{2-} < NH_3 < I^- < Br^- < Cl^- < F^- = H_2O$. Unbranched alkanes are the most reactive and branched hydrocarbons show a reactivity

TABLE 2.2 Methane Carboxylation in Aqueous Solution

				Products (concentration/10^3 mol dm^{-3})		
CH_4	$\xrightarrow[\text{aqueous solution, pH} = 7.3]{\text{CO, Air, NaVO}_3, 80°C}$			$MeCO_2H$	$MeOH$	$HCHO$
		t/h	5	0.3	0.2	0.03
			15	0.6	0.4	0.1
			25	1.0	0.6	0.5

$1° > 2° > 3°$, which was attributed to steric factors. Multiple H–D exchanges often occur, which suggests that the alkane molecule can exchange several hydrogen atoms for deuteriums without leaving the coordination sphere of the metal complex.

The air-stable complex $Cp^*(PMe_3)IrCl_2$ efficiently catalyzes the exchange of deuterium from D_2O into both activated and nonactivated C–H bonds of organic molecules without added acid or stabilizers. Selectivity is observed in many cases, with activation of primary C–H bonds occurring preferentially (Eq. 2.8).[29]

$$R\text{-}H \xrightarrow[135°C, D_2O]{5 \text{ mol}\% \text{ Ir}_{cat}} R\text{-}D$$

(2.8)

$$Ir_{cat} = $$

REFERENCES

1. For representative reviews, see: Shilov, A. E., Shul'pin, G. B., *Reactions of Saturated Hydrocarbons in the Presence of Metal Complexes*, Kluwer Academic, New York, 2002; Shilov, A. E., Shul'pin, G. B., *Chem. Rev.* **1997**, *97*, 2879; Ritleng, V., Sirlin, C., Pfeffer, M., *Chem. Rev.* **2002**, *102*, 1731; Crabtree, R. H., *Chem. Rev.* **1985**, *85*, 245; Fujiwara, Y., Tabaki, K., Taniguchi, Y., *Synlett.* **1996**, 591; Sen, A., Benvenuto, M. A., Lin, M, Hutson, A. C., Basickes, N., *J. Am. Chem. Soc.* **1994**, *116*, 998; Periana, R. A., Taube, D. J., Gamble, S., Taube, H., Satoh, T., Fujii, H., *Science* **1998**, *280*, 560; Dyker, G. *Angew. Chem. Int. Ed.* **1999**, *38*, 1698; Arndsten, B. A., Bergman, R. G., Mobley, T. A., Peterson, T. H., *Acc. Chem. Res.* **1995**, *28*, 154.

2. For examples, see: Gol'dshleger, N. F., Tyabin, M. B., Shilov, A. E., Shteinman, A. A., *Zh. Fiz. Khim.* **1969**, *43*, 2174; Periana, R. A., Taube, D. J., Gamble, S., Taube, H., Satoh, T., Fujii, H., *Science* **1998**, *280*, 560; Mylvaganam, K., Bacskay, G. B., Hush, N. S., *J. Am. Chem. Soc.* **2000**, *122*, 2041; Balzarek, C., Weakley, T. J. R., Tyler, D. R., *J. Am. Chem. Soc.* **2000**, *122*, 9427; Jere, F. T., Miller, D. J., Jackson, J. E., *Org. Lett.* **2003**, *5*, 527; Klei, S. R., Tilley, T. D., Bergman, R. G., *Organometallics* **2002**, *21*, 4905; Sen, A., Lin, M., *J. C. S. Chem. Commun.* **1992**, 508.

3. For examples, see: Vincent, J. B., Olivier-Lilley, G. L., Averill, B. A., *Chem. Rev.* **1990**, *90*, 1447; Wilkins, R. G., *Chem. Soc. Rev.* **1992**, *21*, 171; Que, L., Jr., Dong, Y., *Acc. Chem. Res.* **1996**, *29*, 190.

4. Shimoda, M., Okura, I., *J. Chem. Soc., Chem. Commun.* **1990**, 533.

5. Berces, T., Trotman-Dickenson, A. F., *J. Chem. Soc.* **1961**, 4281.

6. Neimann, K., Neumann, R., Rabion, A., Buchanan, R. M., Fish, R. H., *Inorg. Chem.* **1999**, *38*, 3575.

7. Gol'dshleger, N. F., Es'kova, V. V., Shilov, A. E., Shteinman, A. A., *Zh. Fiz. Khim.* **1972**, *46*, 1353.

8. Kushch, L. A., Lavrushko, V. V., Misharin, Y. S., Moravskii, A. P., Shilov, A. E., *Nouv. J. Chim.* **1983**, *7*, 729.

9. Luinstra, G. A., Wang, L., Stahl, S. S., Labinger, J. A., Bercaw, J. E., *Organometallics* **1994**, *13*, 755; Luinstra, G. A., Wang, L., Stahl, S. S., Labinger, J. A., Bercaw, J. E., *J. Organomet. Chem.* **1995**, *504*, 75.

10. Dangel, B. D., Johnson, J. A., Sames, D., *J. Am. Chem. Soc.* **2001**, *123*, 8149.

11. Barton, D. H. R., Gastiger, M. J., Motherwell, W. B., *J. Chem. Soc., Chem. Commun.* **1983**, 41; Kiani, S., Tapper, A., Staples, R. J., Stavropoulos, P., *J. Am. Chem. Soc.* **2000**, *122*, 7503; Barton, D. H. R. *Tetrahedron*, **1998**, *54*, 5805; Stavropoulos, P., Celenligil-Cetin, R., Tapper, A. E., *Acc. Chem. Res.* **2001**, *34*, 745.

12. Goldstein, A. S., Drago, R. S., *J. Chem. Soc., Chem. Commun.* **1991**, 21.

13. Briffaud, T., Larpent, C., Patin, H., *J. Chem. Soc., Chem. Commun.* **1990**, 1193.

14. Druzhinina, A. N., Nizova, G. V., Shul'pin, G. B., *Bull. Acad. Sci. USSR, Div. Chem. Sci.* **1990**, *39*, 194.

15. Launay, F., Roucoux, A., Patin, H., *Tetrahedron Lett.* **1998**, 39, 1353.

16. Muradov, N. Z., Rustamov, M. I., *Kinet. Katal.* **1989**, *30*, 248.

17. Belavin, B. V., Kresova, E. I., Moravskii, A. P., Shilov, A. E., *Kinet. Katal.* **1990**, *31*, 764.

18. Neumann, R., Khenkin, A. M., *Chem. Commun.* **1998**, 1967.

19. Fang, Z., Breslow, R., *Bioorg. Med. Chem. Lett.* **2005**, *15*, 5463.

20. For some reviews, see: Adam, W., *Peroxide Chemistry: Mechanistic and Preparative Aspect of Oxygen Transfer*. Wiley-VCH; Weinheim, 2000; Adam, W., Saha-Moller, C. R., Ganeshpure, P. A., *Chem. Rev.* **2001**, *101*, 3499; Shi, Y., *Acc. Chem. Res.* **2004**, *37*, 488; Curci, R., D'Accolti, L., Fusco, C., *Acc. Chem. Res.* **2006**, *39*, 1.

21. Yang, D., Wong, M.–K., Wang, X.–C., Tang, Y.–C., *J. Am. Chem. Soc.* **1998**, *120*, 6611.

22. Shul'pin, G. B., Lederer, P., Geletiy, Y. V., *J. Gen. Chem. USSR* **1987**, *57*, 543.

23. Schreiner, P. R., Lauenstein, O., Kolomitsyn, I. V., Nadi, S., Fokin, A. A., *Angew. Chem. Int. Ed. Engl.* **1998**, *37*, 1895.

24. Horvath, I. T., Cook, R. A., Millar, J. M., Kiss, G., *Organometallics* **1993**, *12*, 8.

25. Lin, M., Sen, A., *J. Chem. Soc. Chem. Commun.* **1992**, 892.

26. Lin, M., Sen, A., *Nature* **1994**, *368*, 613; Nizova, G. V., Shul'pin, G. B., Süss-Fink, G., Stanislas, S., *Chem. Commun.* **1998**, 1885; Asadullah, M., Taniguchi, Y., Kitamura, T., Fujiwara, Y., *Appl. Organomet. Chem.* **1998**, *12*, 277.

27. Lin, M., Hogan, T. E., Sen, A., *J. Am. Chem. Soc.* **1996**, *118*, 4574.

28. Gol'dshleger, N. F., Tyabin, M. B., Shilov, A. E., Shteinman, A. A., *Zh. Fiz. Khim.* **1969**, *43*, 2174

29. Klei, S. R., Golden, J. T., Tilley, T. D., Bergman, R. G., *J. Am. Chem. Soc.* **2002**, *124*, 2092.

CHAPTER 3

ALKENES

3.1 REDUCTION

3.1.1 Hydrogenation

Because of the electron-richness of simple carbon–carbon double bonds, their reduction to C–C single bonds is usually difficult with common electron-rich reducing reagents such as metal hydrides. On the other hand, the use of transition-metal can change the electronic nature of C=C into electron-demanding through coordination with the transition-metals. The carbon–carbon double bond thus could be reduced with various reducing reagents. It should be noted that hydrogenation is an important process in biochemistry to adjust the saturated/unsaturated lipid ratio within the cell membranes.[1] The most widely studied chemical method for alkene reduction is the transition-metal catalyzed hydrogenation of alkenes. Because there is no "free" reducing hydridic ion involved in such reductions, it is possible to carry out catalytic hydrogenations in water. The catalytic hydrogenations in aqueous solution by using a variety of catalysts have been reported since the 1960s.[2] Water-soluble substrates were hydrogenated directly. Cyclohexene was hydrogenated in a two-phase system with [RhCl$_3$ · 3H$_2$O] in the presence of an excess of

---

Comprehensive Organic Reactions in Aqueous Media, Second Edition, by Chao-Jun Li and Tak-Hang Chan
Copyright © 2007 John Wiley & Sons, Inc.

Ph_2PPhSO_3Na (TPPMS), in connection with the use of a cosolvent.[3] The cosolvent has an effect on the reactivity of the catalyst, which increases in the order of dimethylacetamide < dimethoxyethane < ethanol < methanol. Borowski et al.[4] used [RhCl(TPPMS)$_3$] and [RuHCl(TPPMS)$_3$] for hydrogenation without the use of a cosolvent. Some double bond migration was observed. On the other hand, Larpent et al. used the more water-soluble complex, [RhCl(TPPTS)$_3$], {TPPTS = $P(PhSO_3H)_3$}, prepared in situ by mixing [RhCl$_3$·3H$_2$O] and the ligand, for hydrogenation.[5] Many olefins were hydrogenated with 100 percent conversion and complete selectivity on the C–C double bond. Various functional groups survived the reaction condition. Through ^{31}P NMR study, the active catalyst was found to involve the phosphine oxide $OP(PhSO_3Na)_3$. The observation has been confirmed by further experiments in which no hydrogenation was observed when the amount of phosphine oxide was not sufficient.[6] However, the hydrogenation of α,β-unsaturated aldehydes to saturated aldehydes proceeds without the involvement of TPPTS oxide.[7] A rhodium amphos complex, [Rh(NBD)(amphos)$_2$]$^{3+}$, was found to be more air-stable and could be recycled easily.[8] An observation that is useful for synthetic application is the transition in selectivity when going from organic solvent to water for the hydrogenation of a diene acid (Eq. 3.1).[9] This unusual selectivity could be attributed to the coordination between the carboxylic group and the metal center.

$$
\begin{array}{c}
\text{\~\~\~\~CO}_2\text{H} \\
\uparrow \quad \text{PhH, 66\%} \\
\xrightarrow[\text{RhCl[P(}p\text{-tolyl)}_3]_3]{\text{H}_2} \\
\downarrow \quad \text{PhH/H}_2\text{O, 81\%} \\
\text{\~\~\~\~CO}_2\text{H}
\end{array}
\tag{3.1}
$$

More recently, catalytic hydrogenations of alkenes by other catalysts in water have been explored. For example, water-soluble ruthenium complex $RuCl_2(TPPTS)_3$ has been used for the catalytic hydrogenation of unsaturated alkenes (and benzene).[10] Hydrogenation of nonactivated alkenes catalyzed by water-soluble ruthenium carbonyl clusters was reported in a biphasic system.[11] The tri-nuclear clusters undergo transformation during reaction but can be reused repeatedly without loss of activity. The organometallic aqua complex [Cp*Ir^{III}(H$_2$O)$_3$]$^{2+}$

acts as a catalyst precursor for hydrogenation of water-soluble alkenes in water under acidic conditions under H_2 at a pressure of 0.1–0.7 MPa at 25°C.[12] Water-soluble analogues of Vaska's complex, *trans*-[IrCl(CO)(PPh$_3$)$_2$], have been shown to be an active catalyst for the hydrogenation of olefinic double bonds in short-chain unsaturated acids in aqueous solution.[13] Nanopalladium particles supported on an amphiphilic polystyrene-poly(ethylene glycol) resin catalyzed hydrogenation of olefins under aqueous conditions.[14] Pd on C in subcritical water employing sodium formate as a hydrogen source provides a green procedure for the catalytic reduction or hydrogenation of olefins (and acetylenes).[15] Size-selective hydrogenation of olefins can be performed by dendrimer-encapsulated palladium nanoparticles (Eq. 3.2).[16] These dendrimer-encapsulated catalysts (DECs) catalyzed the hydrogenation of allyl alcohols in a 4:1 methanol/water mixture. The results indicate that steric crowding on the dendrimer periphery, which increases with dendrimer generation, can act as an adjustable-mesh nanofilter. Catalytic hydrogenation of olefins also takes place effectively in supercritical CO_2 with Pd(0) nanoparticles dispersed in the fluid phase using a water-in-CO_2 microemulsion consisting of water, sodium *bis*(2-ethylhexyl) sulfosuccinate (AOT) as a surfactant, and 1-octanol as a cosolvent.[17] The catalyst can be reused for the next runs of the hydrogenations.

$$\text{Ar} \diagup\!\!\!\diagdown \xrightarrow[\text{H}_2 \text{ (1 atm), H}_2\text{O, 25}^\circ\text{C, 24 h}]{\substack{\text{nanopalladium catalyst} \\ \text{(5 mol\% Pd)}}} \text{Ar}\diagup\diagdown \qquad (3.2)$$

The hydrogenation of alkenes and alkynes in water can also use silanes as hydrogen sources. Tour reported that by using palladium acetate as catalyst, triethoxysilane reduced C–C unsaturated bonds to saturation in a mixture of THF and water.[18] The reaction showed excellent chemo- and stereoselectivity. Water was essential to the reaction. In the absence of water, 95% of the starting alkene remained unchanged (Eq. 3.3).

2.5 equiv (EtO)$_3$SiH;
0.05 equiv Pd(OAc)$_2$

THF/H$_2$O (5:1)

96% yield,
cis/trans = 3:1

$$(3.3)$$

3.2 ELECTROPHILIC ADDITIONS

3.2.1 Reaction with Halogen

Electrophilic addition of halogen to alkenes is a classical reaction in organic chemistry. For decades, such additions have been carried out in aqueous media both with electron-rich and electron-deficient alkenes.[19] Halohydrin formation is a common competing reaction.[20] There is a dramatic micellar effect for alkene bromination in water attributed to a change in the equilibrium between Br_2 and Br_3^-: In CTAB micelles, the equilibrium favors the less reactive Br_3^- and the overall reaction is therefore slower than in nonionic micelles of Brij-35 or in anionic micelles of sodium dodecylsulfate. The addition of Bu_4NBr to anionic or nonionic micelles perturbs their surface, which assists in the binding of anions (e.g., Br^- or Br_3^-) and slows reaction. The addition of chlorine to olefins also occurs in water.[21] Other halogens and halogen equivalents react similarly. For example, the addition of BrCl (I) to $MeCH=CH_2$ (II) in aqueous HCl gives $BrCH_2CHClMe$ (III) and $ClCH_2CHBrMe$ (IV) in the proportion 54:46.[22] Direct addition of bromine chloride (BrCl) to unsaturated carboxylic acids in aqueous medium gave the corresponding 2-bromo-3-chlorocarboxylic acids.[23] Reactions of alkenes with elemental iodine activated by 30% aqueous hydrogen peroxide in the presence of methanol or water as the external nucleophile source result in the formation of vicinal iodoalkoxy- or iodohydroxy-substituted alkanes (Eq. 3.4).[24]

$$
\begin{array}{c}
Ph \\
\diagup \\
Ph
\end{array}
\quad
\xrightarrow[\substack{aq\ CH_3CN \\ 96\%}]{I_2/30\%\ aq\ H_2O_2}
\quad
Ph\text{---}\underset{OR}{\overset{Ph}{\underset{|}{\overset{|}{C}}}}\text{---}CH_2I
\qquad (3.4)
$$

3.2.2 Reaction with Hydrogen Halides

The addition of hydrogen halide to alkene is another classical electrophilic addition of alkene. Although normally such reactions are carried out under anhydrous conditions, occasionally aqueous conditions have been used.[25] However, some difference in regioselectivity (Markovnikov and anti-Markovnikov addition) was observed. The addition product formed in an organic solvent with dry HBr gives exclusively the 1-Br derivative; whereas with aq. HBr, 2-Br derivative is formed. The difference in the products formed by the two methods is believed to be due primarily to the difference in the solvents and not to the presence of any peroxide in the olefin.[26]

3.2.3 Addition of Water

The hydration of an alkene double bond under strongly acidic conditions is again a classical reaction that involves a carbocation intermediate, which often leads to various competing reaction products.[27] The regiochemistry of the water addition follows the Markovnikov rule.[28]

Recently, there has been interest in developing various solid acid-catalyzed[29] and transition-metal catalyzed hydration reactions of alkenes both to establish milder conditions for synthetic purposes as well as for the potential to control regio- and stereochemistry. Direct catalytic hydration of terminal alkenes to primary alcohol would be an inexpensive route for the preparation of alcohols useful to industry. The reaction between *trans*-PtHCl(PMe$_3$)$_2$ and NaOH in a 1:1 mixture of water and 1-hexene yields a species that, at 60°C and in the presence of the phase-transfer catalyst PhCH$_2$NEt$_3$Cl, catalyzes selective hydration to give n-hexanol at a rate of 6.9 ± 0.2 turnovers/h (Eq. 3.5).[30] Labeling experiments with *trans*-PtDCl(PMe$_3$)$_2$ show that the hydration involves reductive elimination of a C–H bond. The proposed mechanism involves the attack of hydroxide to the coordinated alkene.

$$H_2O \ + \ RHC{=}CH_2 \ \xrightarrow[\substack{R = CH_3(CH_2)_3, \ HO(CH_2)_2; \ 60°C \\ R = CH_3(CH_2)_9; \ 100°C}]{\substack{PtHCl(PMe_3)_2/NaOH, \\ NEt_3(CH_2Ph)Cl}} \ RCH_2{-}CH_2OH \qquad (3.5)$$

The hydration of propylene with sulfuric acid catalyst in high-temperature water was investigated using a flow reaction system.[31] The major product is isopropanol. A biopolymer-metal complex, wool-supported palladium-iron complex (wool-Pd-Fe), has been found to be a highly active catalyst for the hydration of some alkenes to the corresponding alcohols. The yield is greatly affected by the Pd/Fe molar ratio in the wool-Pd-Fe complex catalyst and the catalyst can be reused several times without remarkable change in the catalytic activity.[32]

3.2.4 Oxymercuration/Oxymetalation

Oxymercuration/demercuration provides a milder alternative for the conventional acid-catalyzed hydration of alkenes. The reaction also provides the Markovnikov regiochemistry for unsymmetrical alkenes.[33] Interestingly, an enantioselective/inverse phase-transfer catalysis (IPTC) reaction for the Markovnikov hydration of double bonds by an oxymercuration-demercuration reaction with cyclodextrins as catalysts was recently reported.[34] Relative to the more common phase-transfer

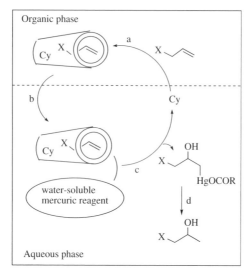

Figure 3.1 Simplified mechanism of an IPTC process mediated by cyclo-dextrins.

catalysis, inverse phase transfer catalysis (IPTC) reactions—in which a lipophilic reactant is transported to the aqueous phase by the cata-lyst—are relatively rare. In this example, cyclodextrin (Cy, α or β) was used as the catalyst and the substrate (an allylic ether or an amine) was transferred by cyclodextrin from the organic hexane phase to the aque-ous phase (Figure 3.1, steps a and b). Water-soluble mercury reagent, $Hg(OCOCF_3)_2$, then reacted with the substrate, presumably still within the cyclodextrin cavity, to give the oxymercuration product (step c). The intermediate was not isolated but was demercurated with alkaline borohydride to give the hydration product (step d). Moderate ee (up to 32%) and yields (14–60%) were obtained for allylic amines and protected allylic alcohol as starting materials.

Oxytelluration of olefins with $PhTeBr_3$ or Ph_2Te_2-Br_2 in aqueous THF generates (β-hydroxyalkyl)aryltellurium dihalides. The reaction is *trans*-stereospecific in the case of *cis*-2-butene and *cis*- and *trans*-4-octenes and regiospecific in the cases of all terminal olefins. Their reactions with reducing agents in aqueous sodium hydroxide generate the Markovnikov hydration products.[35]

3.2.5 Epoxidation

In the natural environment of water and air, biological oxygenations of organic compounds are catalyzed by cytochrome P-450. One of the most important oxygenations by P-450 is the epoxidation of alkenes. Recently, extensive studies have been carried out in synthesizing water-soluble metalloporphyrins as epoxidation and other oxygenation catalysts in aqueous media to mimic the properties of P-450. By using these water-soluble metalloporphyrins, epoxidation of alkenes can be accomplished with a variety of oxidizing reagents such as PhIO, NaClO, O_2, H_2O_2, ROOH, $KHSO_5$, and so forth. This area has been reviewed in detail.[36] For example, carbamazepine reacted with $KHSO_5$ catalyzed by water-soluble iron and manganese porphyrins (see structures **3.1** and **3.2**) in water to give the corresponding expoxide (Eq. 3.6).[37] Mn^{III}-salen complexes bearing perfluoroalkyl substituents have been organized as a Langmuir film on an aqueous subphase containing a urea/hydrogen peroxide adduct (UHP, the oxidant) and cinnamyl alcohol (the substrate). The catalytic activity of the monolayer for the epoxidation of the alkene dissolved in water was investigated.[38] The reaction rate exhibits first-order dependence on oxidant concentration and zero-order dependence on alkene concentration, in agreement with the reaction orders reported for Mn^{III}-salen-catalyzed epoxidation reactions carried out in solution.

$$\text{(3.6)}$$

An efficient metal-catalyzed epoxidation of alkenes using aqueous hydrogen peroxide was described by Venturello, who used a tungstate catalyst under phase transfer conditions. However, chlorinated solvents were normally required to give high reaction rates and yields of epoxides.[39] Noyori and co-workers developed the use of sodium tungstate dihydrate with 30% hydrogen peroxide in toluene as an epoxidation system for alkenes.[40] The use of a quaternary ammonium hydrogen sulfate as a phase transfer catalyst was also required and the reaction had to be conducted at 90°C (Eq. 3.7).

$$\text{(3.7)}$$

An alternative, milder oxidation method has recently been developed by Richardson and co-workers.[41] It was found that hydrogen peroxide together with sodium hydrogen carbonate can efficiently epoxidize

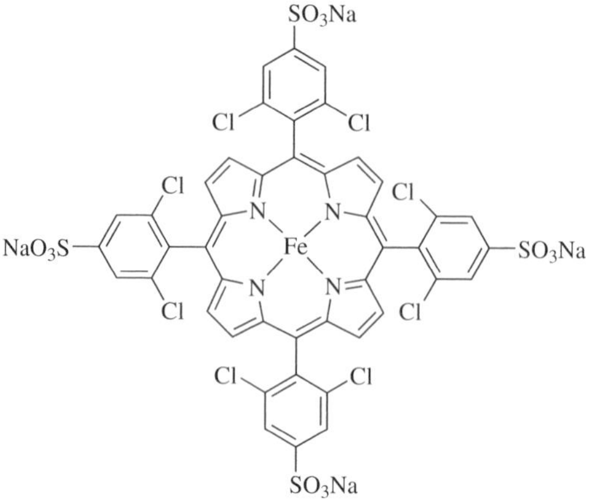

3.1 FeTDCPPS

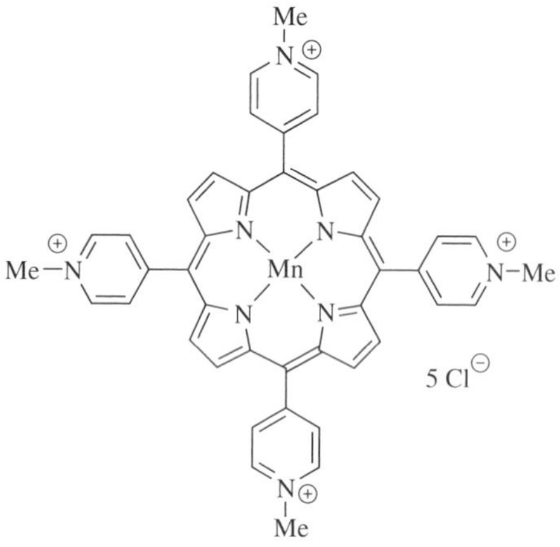

3.2 MnTMPyP

alkenes in aqueous media. The reaction was presumed to proceed via a peroxymonocarbonate, HCO_4^-, intermediate. The rate of epoxidation was relatively slow, however, and the yield of the epoxide was moderate. An improvement has been found by Burgess, who showed that the addition of simple manganese salts such as manganese sulfate greatly accelerated the reaction.[42] For substituted alkenes, the reactions were fast at room temperature and the yields of epoxides high (Eq. 3.8). Since the reagents and metal catalyst used are both inexpensive, the method is considered to be an inexpensive, scalable, and environmentally benign method of alkene epoxidation. However, the oxidation system could not be used for the epoxidation of terminal alkenes and, for water-insoluble alkenes, an organic cosolvent such as t-butanol or DMF was required.

$$R\diagdown\joinrel=\joinrel R' \xrightarrow[\text{H}_2\text{O}/t\text{-butanol or DMF, rt}]{\text{NaHCO}_3/\text{H}_2\text{O}_2/\text{MnSO}_4} R\diagup\!\!\diagdown_{\displaystyle O}\!\!R' \qquad (3.8)$$

Water/solvent/CO_2 ternary systems have also been exploited to perform homogeneous oxidation of organic substrates using water-soluble catalysts and oxidants.[43] By employing CO_2-expanded $CH_3CN/H_2O_2/H_2O$ mixture, a variety of olefins were oxidized homogeneously with high (>85%) epoxidation selectivities. The addition of pyridine to the homogeneous system greatly enhanced the reactivity by an order of magnitude. The reaction was proposed to involve the formation of monoperoxycarbonic acid as a catalyst. Epoxidation of geraniol in water generates the corresponding 2,3-epoxide selectively in good yields in micellar systems composed of novel amphiphilic hydroperoxides, α-alkoxyalkyl hydroperoxides, with a catalytic amount of $MoO_2(acac)_2$ at 30°C.[44] Hydrogen peroxide, generated in situ by the enzymatic oxidation of glucose using glucose oxidase, is coupled to a catalytic system of sodium bicarbonate/manganese sulfate to epoxidize alkenes in aqueous media (Figure 3.2). $PhCH=CH_2$ was converted to the epoxide in 90% yield by this method, whereas cyclooctene oxide was generated in >99% yield from the parent alkene.[45]

Methyltrioxorhenium, supported on silica functionalized with polyether tethers, catalyzed the epoxidation of alkenes with 30% aq H_2O_2 in high selectivity compared to the ring opening products observed in homogeneous media in the absence of an organic solvent.[46]

Alkenes can also be epoxidized directly with a variety of organic peroxy acids or related reagents such as peroxy carboximidic acid, RC(NH)OOH, which is readily available through an in situ reaction of

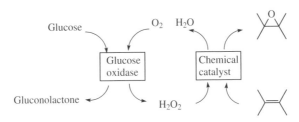

Figure 3.2

a nitrile with hydrogen peroxide. Thus, reactions of alkenes with m-chloroperoxybenzoic acid in water at room temperature give the epoxides in high yields.[47] By using monoperoxyphthalic acid together with cetyltrimethylammonium hydroxide (CTAOH) as base to control the pH of the aqueous medium, highly regioselective expoxidation of allyl alcohols in the presence of other C=C bonds is possible (Eq. 3.9).[48]

$$
\underset{92\%}{\xrightarrow[\text{H}_2\text{O/CTAOH}]{\text{MPPA}}}
$$

(3.9)

Epoxidation of compounds in which the double bond is conjugated to electron-withdrawing groups occurs only very slowly, or not at all with peroxy acids or alkyl peroxides. On the other hand, epoxidation with hydrogen peroxide under basic biphase conditions, known as the Weitz-Scheffer epoxidation (Scheme 3.1),[49] is an efficient method for the conversion into epoxides. This reaction has been applied to many α,β-unsaturated aldehydes, ketones, nitriles, esters and sulfones, and so on. The reaction is first order in both unsaturated ketone and $^-$O$_2$H through a Michael-type addition of the hydrogen peroxide anion to the conjugated system followed by the ring closure of the intermediate enolate with the expulsion of $^-$OH. The epoxidation of electron-deficient olefins can also be performed with hydrogen peroxide in the presence of sodium tungstate as catalyst.[50] Studies on the conversion of olefins into epoxides, bromohydrins, and dibromides with sodium bromide in water–organic solvent electrolysis systems have also been reported.[51]

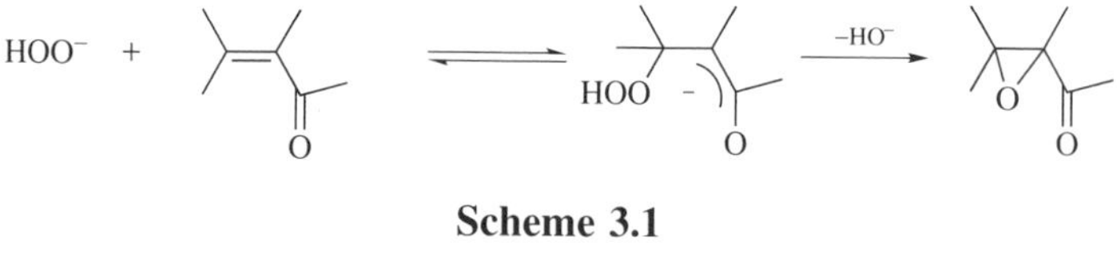

$$
\text{HOO}^- + \qquad \rightleftharpoons \qquad \xrightarrow{-\text{HO}^-} \qquad
$$

Scheme 3.1

Epoxidation of cyclooctene and other alkenes with Oxone (KHSO$_5$) was promoted effectively in an aqueous micellar solution of an amphiphilic ketone (**3.3**).[52] The amphiphilic ketone can be easily derived from hepta(ethylene glycol) monodecyl ether.

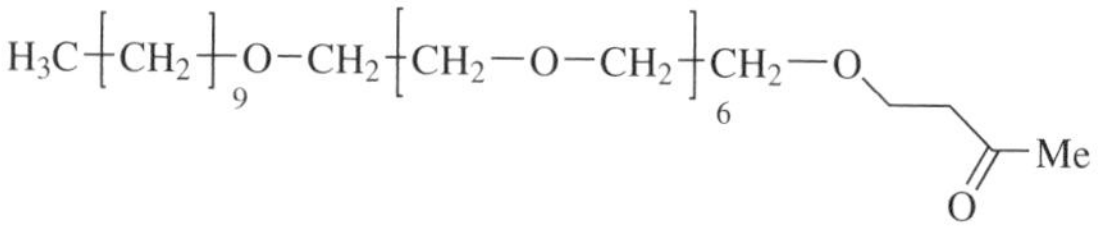

$$
\text{H}_3\text{C}\!\left[\text{CH}_2\right]_9\!\!\text{O}\!-\!\text{CH}_2\!\!\left[\text{CH}_2\!-\!\text{O}\!-\!\text{CH}_2\right]_6\!\!\text{CH}_2\!-\!\text{O} \qquad \text{Me} \qquad \textbf{3.3}
$$

Recently, exceptional progress has been made in the development of chiral ketones (via dioxirane intermediates) based on asymmetric epoxidations (Eq. 3.10). Although the first such type of asymmetric epoxidation was carried out by Curci in 1984,[53] it is only in the last decade that excellent enantioselectivity of such epoxidations has been achieved. Two of the most prevalent workers in the area are Shi[54] (by using chiral sugar-based ketone, **3.4**) and Yang (chiral binapthalene derivative, **3.5**).[55] Often, these reactions are performed by using Oxone in an aqueous environment. Many other chiral ketones have also been developed and these methods have been used in various syntheses. This subject has been reviewed by many authors.[50,51]

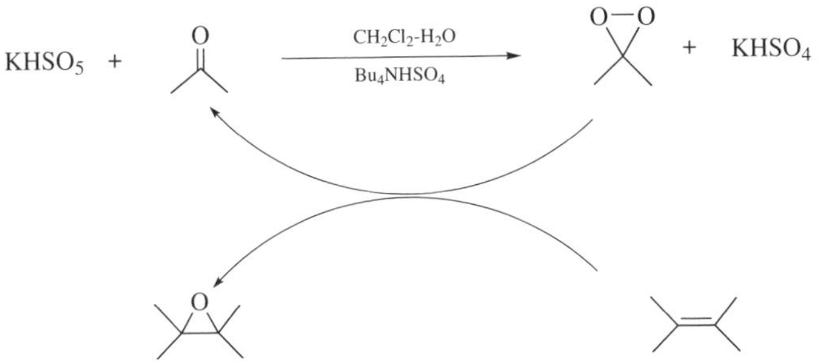

(3.10)

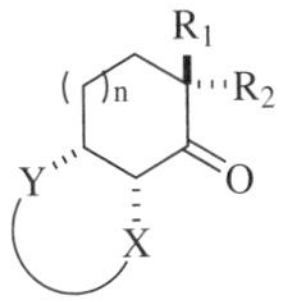

3.4 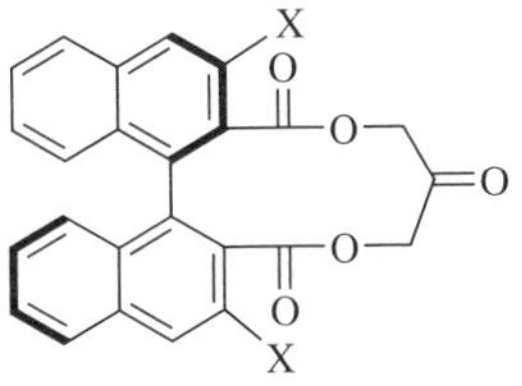   **3.5**

3.2.6   Dihydroxylation and Hydroxylamination

3.2.6.1   *Syn-Dihydroxylation.*   When the reaction was first discovered, the syn-dihydroxylation of alkenes was carried out by using a stoichiometric amount of osmium tetroxide in dry organic solvent.[56] Hoffman made the observation that alkenes could react with chlorate salts as the primary oxidants together with a catalytic quantity of osmium tetroxide, yielding syn-vicinal diols (Eq. 3.11). This catalytic reaction is usually carried out in an aqueous and tetrahydrofuran solvent mixture, and silver or barium chlorate generally give better yields.[57]

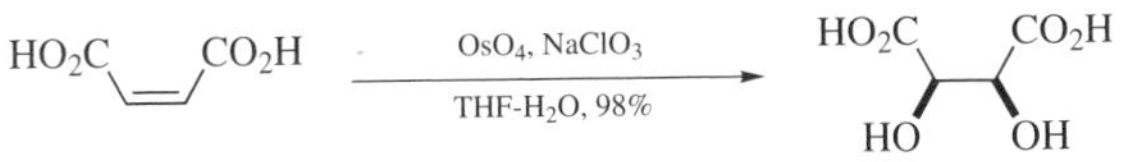

(3.11)

Syn-hydroxylation of alkenes is also effected by a catalytic amount of osmium tetroxide in the presence of hydrogen peroxide. Originally developed by Milas, the reaction can be performed with aqueous hydrogen peroxide in solvents such as acetone or diethyl ether.[58] Allyl alcohol is quantitatively hydroxylated in water (Eq. 3.12).[59]

$$\text{\ \ \ } \xrightarrow[\text{H}_2\text{O, }100\%]{\text{OsO}_4,\ \text{H}_2\text{O}_2} \text{\ \ \ } \qquad (3.12)$$

Catalytic dihydroxylations using the osmium tetroxide-*tert*-butyl hydroperoxide system are largely due to Sharpless and co-workers. The aqueous 70%-*t*-butyl hydroperoxide is commercially available and ideal for direct use in this dihydroxylation process.[60]

A very effective way of carrying out syn-dihydroxylation of alkenes is by using an osmium tetroxide-tertiary amine N-oxide system. This dihydroxylation is usually carried out in aqueous acetone in either one- or two-phase systems, but other solvents may be required to overcome problems of substrate solubility.[61]

Potassium permanganate, usually in alkaline conditions, using aqueous or aqueous-organic solvents, is a widely used oxidant for effecting syn-vicinal hydroxylation of alkenes (Eq. 3.13). However, overoxidation or alternative oxidation pathways may pose a problem, and the conditions must be carefully controlled.[62]

$$\xrightarrow[\substack{-10°\text{C}/3\text{-}5\ \text{min}\\40\%}]{\substack{\text{KMnO}_4\\ \text{H}_2\text{O}/t\text{-BuOH}/\text{NaOH}}} \qquad (3.13)$$

A photo-induced dihydroxylation of methacryamide by chromium (VI) reagent in aqueous solution was recently reported and may have potential synthetic applications in the syn-dihydroxylation of electron-deficient olefins.[63] Recently, Minato et al. demonstrated that $K_3Fe(CN)_6$ in the presence of K_2CO_3 in aqueous *tert*-butyl alcohol provides a powerful system for the osmium-catalyzed dihydroxylation of olefins.[64] This combination overcomes the disadvantages of overoxidation and low reactivity on hindered olefins related to previous processes (Eq. 3.14).

$$\xrightarrow[\substack{\text{K}_2\text{CO}_3/\text{aq. }t\text{-BuOH}\\88\%}]{\text{cat. OsO}_4/\text{K}_3\text{Fe(CN)}_6} \qquad (3.14)$$

3.2.6.2 *Anti-Dihydroxylation.*

Hydrogen peroxide in the presence of some oxides, notably tungstic oxide (WO_3) or selenium dioxide (SeO_2), reacts with alkenes to give *anti*-dihydroxlyation products (Eq. 3.15).[65] The WO_3 process is best performed at elevated temperatures (50–70°C) and in aqueous solution. Aqueous/organic mixed solvents have also been used successfully to increase the solubility of the alkene in the reaction medium. It has been proposed that peroxytungstic acid (H_2WO_5) is involved in the dihydroxylation of alkenes by the hydrogen-peroxide-tungstic oxide reagent.[65]

$$\underset{R_3}{\overset{R_1}{\diagdown}}=\underset{R_4}{\overset{R_2}{\diagup}} \quad \xrightarrow[H_2O_2]{WO_3 \text{ or } SeO_2} \quad \underset{R_1\,\, R_3}{\overset{HO \quad R_2\, R_4}{\diagdown}}\,OH \qquad (3.15)$$

3.2.6.3 *Asymmetric Dihydroxylation.*[66]

Initially, asymmetric dihydroxylation by osmium tetroxide was carried out stoichiometrically with chiral diamine ligands.[67] Early attempts to effect dihydroxylation catalytically resulted in low enantiomeric excesses of the products (lower than the stoichiometric methods).[68] The lower optical yield was attributed to the presence of a second catalytic cycle that exhibits only low or no enantioselectivity. Sharpless discovered that under aqueous-organic (two-phase) conditions with $K_3Fe(CN)_6$ as the stoichiometric reoxidant and using derivatives of cinchona alkaloids as the chiral ligands, the undesired catalytic pathway can be eliminated, thus resulting in high enantiomeric excess (ee) for the desired dihydroxy products.[69] Under such conditions, there is no oxidant other than OsO_4 in the organic phase. The addition of 1 equivalent of $MeSO_2NH_2$ to the reaction mixture significantly increases the rate of the reaction.[70] The use of two independent cinchona alkaloid units attached to a heterocyclic spacer led to further increases of the enantioselectivity and scope of the reaction.[71] A ready-made mixture containing all reagents has been commercialized under the name AD-mix.

3.6 DHQD   **3.7** DHQ

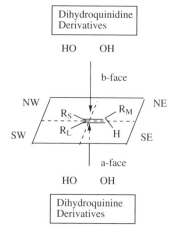

Figure 3.3 Rationale for predicting the enantiofacial selectivity in Sharpless's dihydroxylation.

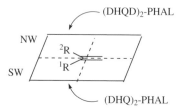

Tendency for a substituent to occupy the SW quadrant:
Aryl > Alkyl > Me BnOCH$_2$, PivOCH$_2$, R$_3$SiOCH$_2$

Figure 3.4 Rationale for enantioselectivity of 1,1-disubstituted olefins.

A model for the stereochemical outcome of the dihydroxylation has been proposed (Figure 3.3).[72] An olefin positioned according to this model will be attacked either from the top face with DHQD (**3.6**) ligands or from the bottom face with DHQ (**3.7**) ligands.

For 1,1-disubstituted olefins, the prediction of the stereochemical outcome is based on a modified model (Figure 3.4).[73]

By using the Sharpless dihydroxylation, a variety of compounds have been transformed to diols with high enantiomeric excesses. The asymmetric dihydroxylation has a wide range of synthetic applications. As an illustration, the dihydroxylation was used as the key step in the synthesis of squalestatin 1 (**3.8**) (Scheme 3.2).[74]

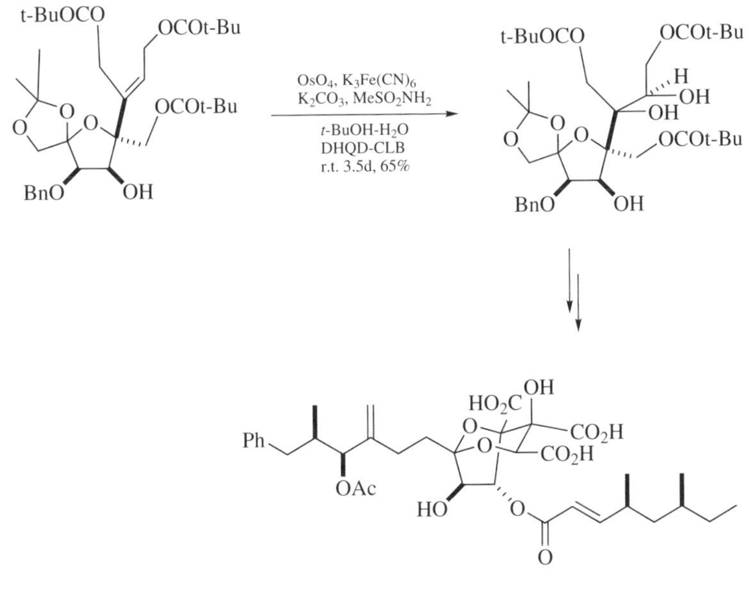

3.8 squalestatin 1

Scheme 3.2

Mono-, di-, and trisubstituted olefins undergo osmium-catalyzed enantioselective dihydroxylation in the presence of the (R)-proline-substituted hydroquinidine **3.9** to give diols in 67–95% yields and in 78–99% ee.[75] Using potassium osmate(VI) as the catalyst and potassium carbonate as the base in a *tert*-butanol/water mixture as the solvent, olefins are dihydroxylated stereo- and enantioselectively in the presence of **3.9** and potassium ferricyanide with sodium chlorite as the stoichiometric oxidant; the yields and enantiomeric excesses of the

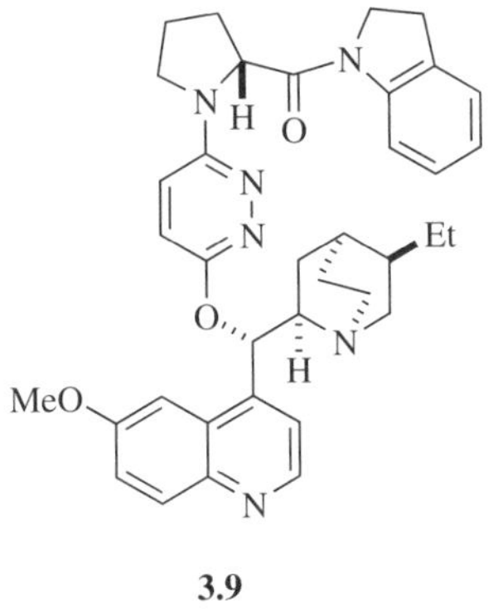

3.9

diols formed are similar to those obtained using potassium ferricyanide as the stoichiometric oxidant.

3.2.6.4 Amino-Hydroxylation.

A related reaction to asymmetric dihydroxylation is the asymmetric amino-hydroxylation of olefins, forming *vic*-aminoalcohols. The *vic*-hydroxyamino group is found in many biologically important molecules, such as the β-amino acid **3.10** (the side-chain of taxol). In the mid-1970s, Sharpless[76] reported that the trihydrate of N-chloro-*p*-toluenesulfonamide sodium salt (chloramine-T) reacts with olefins in the presence of a catalytic amount of osmium tetroxide to produce vicinal hydroxyl *p*-toluenesulfonamides (Eq. 3.16). Aminohydroxylation was also promoted by palladium.[77]

3.10

$$\text{TsNClNa·3H}_2\text{O} \quad + \quad \underset{\substack{\text{R} \\ \text{R}}}{\bigparallel} \quad \xrightarrow[\substack{t\text{-BuOH, 60°C}}]{1\%\ \text{OsO}_4} \quad \underset{\substack{\text{TsHN} \\ \text{R}}}{\overset{\text{HO} \quad \text{R}}{\bigwedge}} \quad (3.16)$$

Subsequently, stoichiometric asymmetric aminohydroxylation was reported.[78] Recently, it was found by Sharpless[79] that through the combination of chloramine-T/OsO$_4$ catalyst with phthalazine ligands used in the asymmetric dihydroxylation reaction, catalytic asymmetric aminohydroxylation of olefins was realized in aqueous acetonitrile or *tert*-butanol (Scheme 3.3). The use of aqueous *tert*-butanol is advantageous when the reaction product is not soluble. In this case, essentially pure products can be isolated by a simple filtration and the toluenesulfonamide byproduct remains in the mother liquor. A variety of olefins can be aminohydroxylated in this way (Table 3.1). The reaction is not only performed in aqueous medium but it is also not sensitive to oxygen. Electron-deficient olefins such as fumarate reacted similarly with high ee values.

3.2.7 Wacker's Oxidation

The oxidation of ethylene to acetaldehyde by palladium chloride in water has been known since the nineteenth century.[80] However, the reaction requires the use of a stoichiometric amount of PdCl$_2$, resulting in Pd(0) deposit. Anderson, in 1934, observed a similar reaction (but

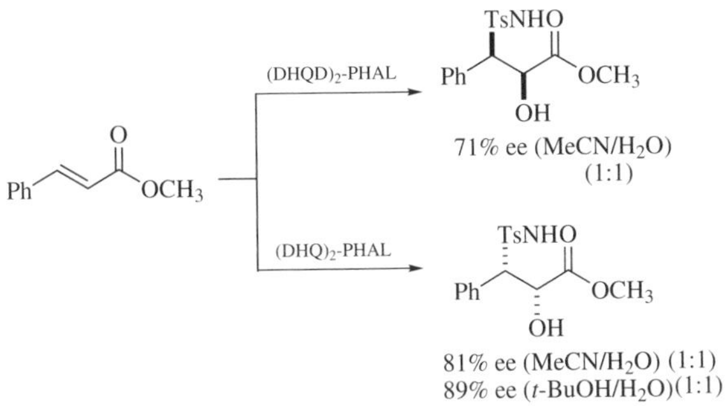

Scheme 3.3

TABLE 3.1   Catalytic Asymmetric Aminohydroxylation

Substrate	Product	ee (%, DHQ)	ee (%, DHQD)
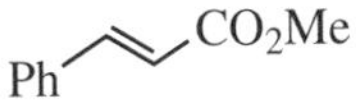	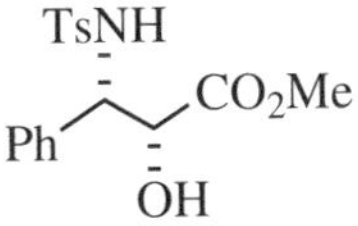	81 (MeCN/H_2O) 82 (t-BuOH/H_2O)	71
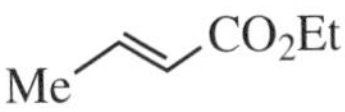	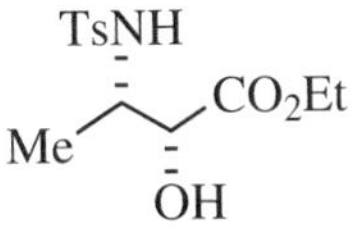	74	60
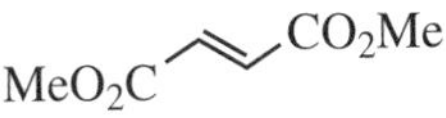	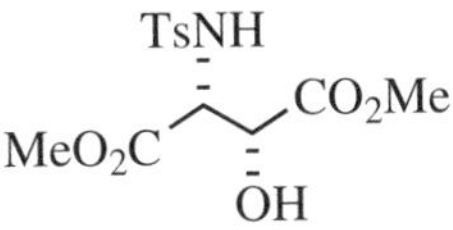	77	53
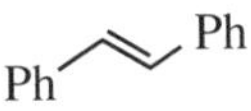	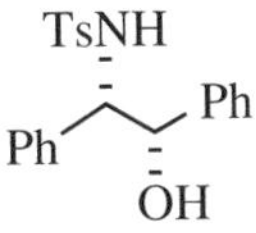	62 (MeCN/H_2O) 78 (t-BuOH/H_2O)	50
	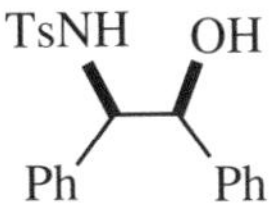	33 (MeCN/H_2O) 50 (t-BuOH/H_2O)	48
	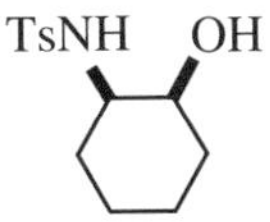	45	36

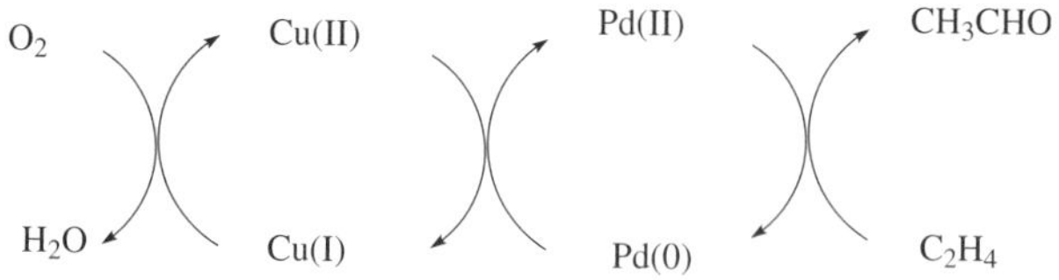

Scheme 3.4

much less efficient) with Zeise's salt in water at a higher temperature.[81] In the late 1950s, Smidt[82] of Wacker Chemie discovered that by using $CuCl_2$, Pd(0) can be oxidized back to Pd(II) before it deposits out. The $CuCl_2$ itself is reduced to cuprous chloride, which is air-sensitive and readily reoxidized back to Cu(II) (Scheme 3.4). The reaction, referred to as Wacker oxidation, is now one of the most important transition-metal catalyzed reactions in industry.

The mechanism for the product formation was originally proposed to involve a *cis* transfer of OH to a metal-bound alkene.[83] However, stereochemical studies carried out by Bäckvall[84] and by Stille[85] proved otherwise, demonstrating that a *trans* addition is involved (Eq. 3.17).

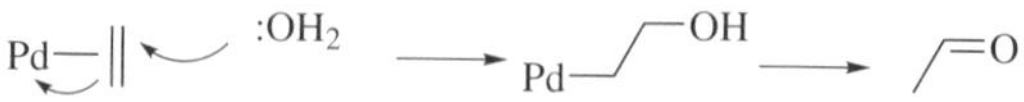

$$ (3.17) $$

For internal olefins, the Wacker oxidation is sometimes surprisingly regioselective. By using aqueous dioxane or THF, oxidation of β,γ–unsaturated esters can be achieved selectively to generate γ-keto-esters (Eq. 3.18).[86] Under appropriate conditions, Wacker oxidation can be used very efficiently in transforming an olefin to a carbonyl compound. Thus, olefins become masked ketones. An example is its application in the synthesis of (+)-19-nortestosterone (**3.11**) carried out by Tsuji (Scheme 3.5).[87]

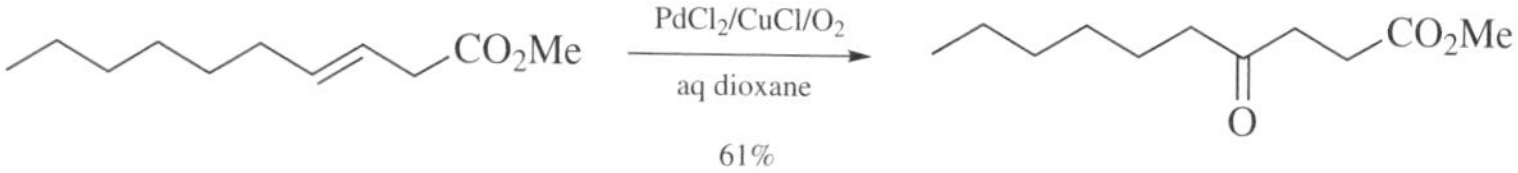

$$ (3.18) $$

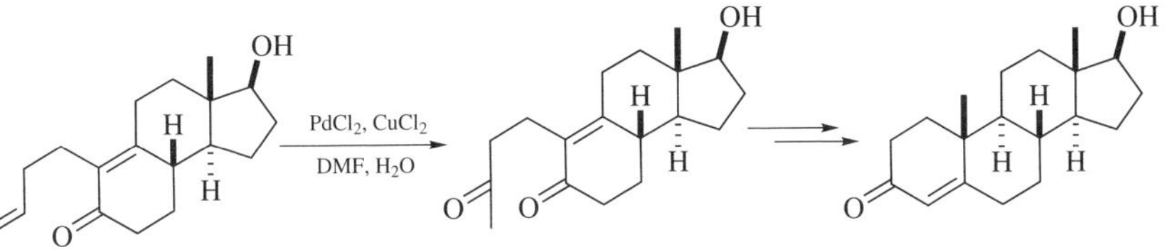

3.11

Scheme 3.5

Besides Wacker oxidation, other transition-metal catalyzed oxidations have also been carried out in aqueous medium. For example, methyl groups can be selectively hydroxylated by platinum salts in water.[88] In this way, p-toluenesulfonic acid was oxidized to benzylic alcohol, which was subsequently oxidized into the aldehyde (Eq. 3.19).[89]

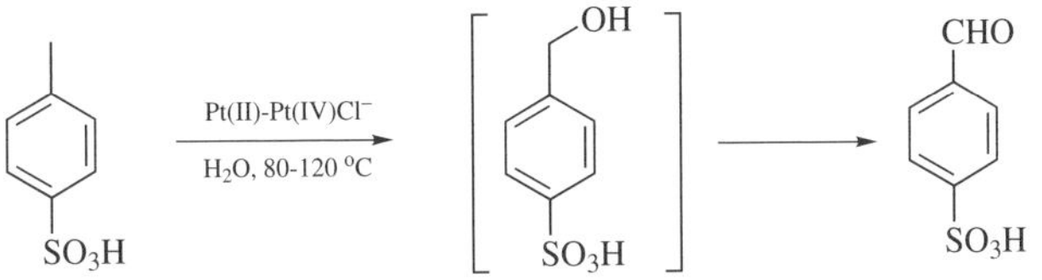

$$(3.19)$$

3.2.8 Oxidative C=C Bond Cleavage

The oxidative cleavage of C=C bond is a common type of reaction encountered in organic synthesis and has played a historical role in the structural elucidation of organic compounds. There are two main conventional methods to oxidatively cleave a C=C bond: (1) via ozonolysis and (2) via oxidation with high-valent transition-metal oxidizing reagents. A more recent method developed is via the osmium oxide catalyzed periodate oxidative cleavage of alkenes. All these methods can occur under aqueous conditions.

The cleavage of alkenes by ozone (ozonolysis) resulting in carbonyl compounds is a synthetically useful reaction.[90] Even though there is still disagreement on the exact mechanism, it is generally accepted that two heterocyclic intermediates, 1,2,3-trioxolane and 1,2,4-trioxolane, are involved. The formation of the former could be regarded as a 1,3-dipolar cyclization between ozone and the alkene and the latter as a retrocleavage followed by a recombination of the fragments (Eq. 3.20).

Depending on the synthetic purpose, the 1,2,4-trioxolane can be trans-formed to alcohols, aldehydes (and ketones), or carboxylic acids.

$$(3.20)$$

The reaction is usually performed at low temperatures, and some-times water has been used as solvent. For example, cyclooctene is ozonized in the presence of an emulsifier (polyoxyethylated lauryl alcohol) with aqueous alkaline hydrogen peroxide to give α,ω-alkanedi-carboxylic acid in one pot (Eq. 3.21).[91]

$$(3.21)$$

Thus, an alkene can be considered as a masked ketone that can be unmasked by ozonolysis. This is illustrated by the synthesis of N-acetylneuraminic acid via the ozonolysis of the precursor acrylic acid in THF/aqueous media (Eq. 3.22). The use of aqueous media facilitated the solubility of the polyhydroxyacids.[92]

$$(3.22)$$

N-acetylneuraminic acid

Ozonolysis of organic compounds in water also has biological and environmental[93] interest. Ozone preferentially attacks the base moi-ety of pyrimidine nucleotides in water.[94] For example, the reaction of ozone with uracil in water, having no substitutent at 1-position, gave the ozonolysis products in Scheme 3.6.[95] The reactions of DNA and RNA with O_3 in an aqueous environment are linked to the damage of

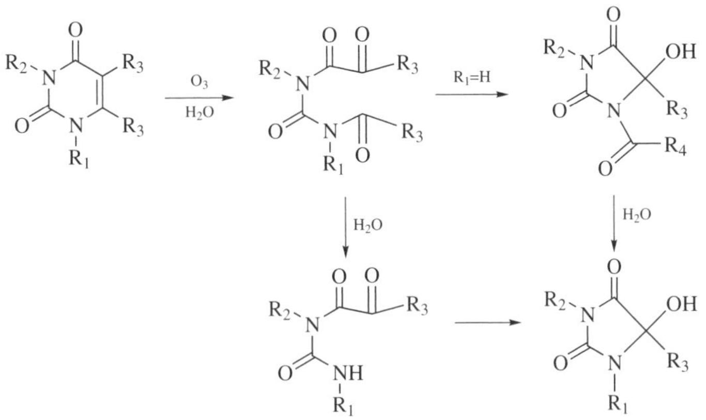

Scheme 3.6

biological systems by ozone.[96] Electrochemical ozonolysis by oxidation of water to ozone at the lead dioxide electrode leads to carboxylic acids as cleavage products in high material yield but low current yield.[97] Emulsion ozonization of methyl linoleate and methyl linolenate in aqueous alkaline hydrogen peroxide generates various acid derivatives.[98] Hydroxy acids and amino acids have been synthesized by ozonolysis of olefins in aqueous emulsions of HCN (cyanozonolysis), followed by hydrolysis of the nitriles.[99]

Oxidative cleavage of alkenes using sodium periodate proceeds effectively in a monophasic solution of acetic acid, water, and THF with very low osmium content or osmium-free. The orders of reactivity of alkenes are as follows: monosubstituted $\gg$ trisubstituted $>$ 1,2 disubstituted $>$ 1,1-disubstituted $>$ tetrasubstituted $\gg$ alkynes.[100] Cleavage with polymer-supported OsO_4 catalyst combined with $NaIO_4$ allows the reuse of the catalyst.[101]

Oxidative cleavage of alkenes can be performed with a variety of high-valent transition-metal oxidizing reagents in aqueous media. The most frequently used one is potassium permanganate. For example, by using excess $KMnO_4$, cleavage of alkenes occurred at all pH levels but was more complete at low pH levels in oil-in-water emulsion.[102] It is also possible to use a catalytic amount of permanganate together with a stoichiometric amount of an oxidizing reagent. Thus, olefinic double bonds are readily oxidized in an aqueous solution of periodate that contains only catalytic amounts of permanganate at pH $7-10$.[103] Dimethyl carbonate/water was found to be an effective system for RuO_4-catalyzed oxidation of various compounds for alkenes and others in the presence of an oxidizing reagent such as $NaIO_4$.[104] Recently, a gold(I) catalyzed oxidative cleavage of the C=C bond to afford ketone or aldehyde products with *tert*-butyl hydroperoxide as the oxidant have been reported with neocuproine in water (Eq. 3.23). Substituted styrenes yielded the corresponding aromatic ketones and aldehydes.[105] Oxidative cleavage of DNA was reported by using water-soluble iron porphyrin complex as catalyst and potassium monopersulfate as oxidizing reagent.[106]

$$(3.23)$$

3.2.9 C–C Bond Formations

Cationic polymerization of alkenes and alkene derivatives has been carried out frequently in aqueous media.[107] On the other hand, the reaction of simple olefins with aldehydes in the presence of an acid catalyst is referred to as the Prins reaction.[108] The reaction can be carried out by using an aqueous solution of the aldehyde, often resulting in a mixture of carbon–carbon bond formation products.[109] Recently, Li and co-workers reported a direct formation of tetrahydropyranol derivatives in water using a cerium-salt catalyzed cyclization in aqueous ionic liquids (Eq. 3.24).[110]

$$(3.24)$$

A further improvement on the tetrahydropyranol formation was made by using the Amberlite® IR-120 Plus resin—an acidic resin with a sulfonic acid moiety, in which a mixture of an aldehyde and homoallyl alcohol in water, in the presence of the resin and under sonication, yielded the desired tetrahydropyranol derivatives.[111]

Cho et al. reported a indium trichloride–catalyzed intramolecular Prins-type reactions of compounds having both the functionalities of homoallyl alcohol and acetal moiety. The intramolecular Prins cyclizations were performed using indium trichloride in chloroform or 25%

aqueous THF. Both 9-oxabicyclo[3.3.1]nonane and 3,9-dioxabicyclo-[3.3.1]nonane compounds were successfully obtained in moderate yields (Eq. 3.25).[112]

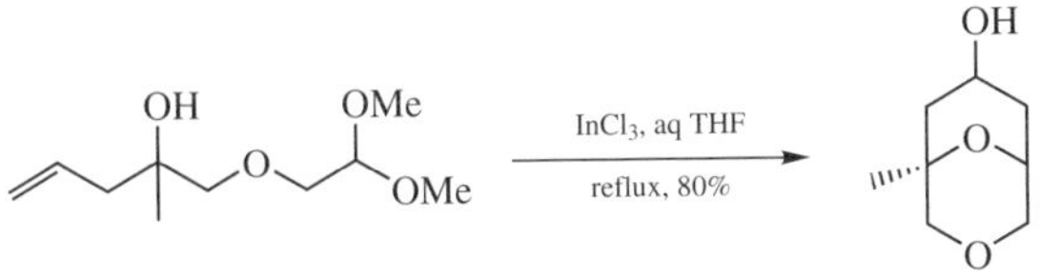

$$
\text{(3.25)}
$$

Aubele et al. studied the aqueous Prins cyclization using cyclic unsaturated acetals as oxocarbenium ion progenitors and allylsilanes are used as nucleophiles. Cyclizations proceed efficiently inside Lewis acidic micelles (of cerium salt) in water. A variety of vinyl- and aryl-substituted tetrahydropyrans with excellent stereocontrol was obtained (Eq. 3.26).[113]

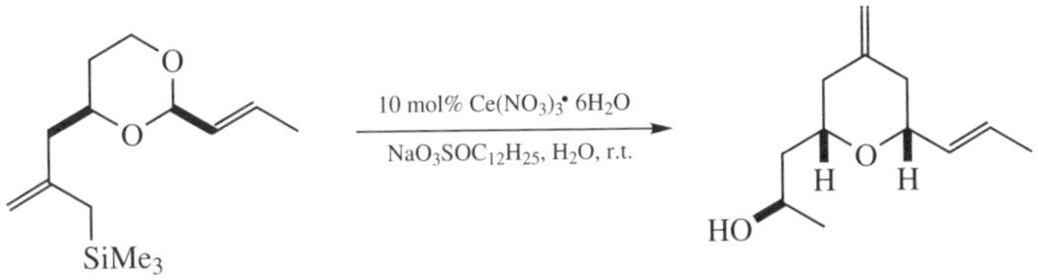

$$
\text{(3.26)}
$$

A reaction related to alkene-aldehyde coupling is the alkene-imine coupling. A one-pot cyclization involving such a reaction (Eq. 3.27) proceeds smoothly in a mixture of water-THF. The reaction has been used in the asymmetric synthesis of pipecolic acid derivatives.[114]

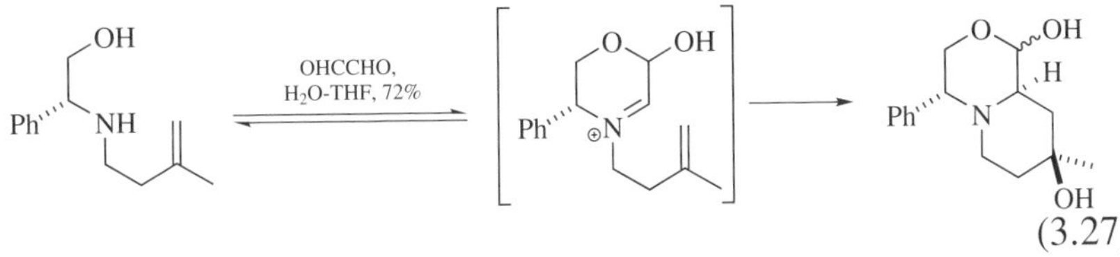

$$
\text{(3.27)}
$$

3.3 RADICAL REACTIONS OF ALKENES

3.3.1 Radical Polymerization of Alkenes

Free-radical polymerization of alkenes has been carried out in aqueous conditions.[115] Aqueous emulsion and suspension polymerization is carried out today on a large scale by free-radical routes. Polymer latexes can be obtained as products (i.e., stable aqueous dispersions

of polymer particles). Such latexes possess a unique property pro-
file and most studies on this subject are in the patent literature.[116]
Atom-transfer radical addition (ATRA) of carbon tetrachloride or chloroform to unsaturated compounds including styrene and 1-octene has
been investigated using ruthenium indenylidene catalysts. The reaction
was extended to atom-transfer radical polymerization (ATRP) by changing the monomer/halide ratio and could take place in aqueous media
(Eq. 3.28).[117]

$$ n \quad \underset{R}{\overset{R'}{\bigvee}} \quad + \quad R''X \quad \xrightarrow{[Ru]} \quad R'' \underset{R}{\overset{R'}{\left(\bigwedge\right)}}_n X \qquad (3.28) $$

3.3.2 Radical Additions

The addition of carbon-based radicals to alkenes has been shown to
be successful in water. Thus, radical addition of 2-iodoalkanamide or
2-iodoalkanoic acid to alkenols using a water-soluble radical initiator
in water generated γ-lactones (Eq. 3.29).[118]

$$ (3.29) $$

The addition of perfluoroalkyl iodides to simple olefins has been
quite successful under aqueous conditions to synthesize fluorinated
hydrocarbons.[119] In addition to carbon-based radicals, other radicals
such as sulfur-based radicals, generated from RSH-type precursors (R =
alkyl, acyl) with AIBN, also smoothly add to α-allylglycines protected at none, one, or both of the amino acid functions (NH_2 and/or
CO_2H). Optimal results were obtained when both the unsaturated amino

acid and RSH dissolved completely in the medium (dioxane/water or methanol/water are good solvent systems) (Eq. 3.30).[120]

$$\text{(3.30)}$$

3.3.3 Radical Cyclization

Radical addition to alkenes has been used in cyclizations in aqueous media. Oshima and co-worker studied triethylborane-induced atom-transfer radical cyclization of iodoacetals and iodoacetates in water.[121] Radical cyclization of the iodoacetal proceeded smoothly both in aqueous methanol and in water. Atom-transfer radical cyclization of allyl iodoacetate is much more efficient in water than in benzene or hexane. For instance, treatment of allyl iodoacetate with triethylborane in benzene or hexane at room temperature did not yield the desired lactone. In contrast, the compound cyclized much more smoothly in water and yielded the corresponding γ-lactone in high yield (Eq. 3.31).

$$\text{(3.31)}$$

Water as a reaction solvent also markedly promoted the cyclization reaction of large-membered rings. Stirring a solution of 3,6-dioxa-8-nonenyl iodoacetate in water in the presence of triethylborane at 25°C for 10 h provided the 12-membered ring product, 4-iodo-6,9-dioxa-11-undecanolide in 84% yield, whereas the cyclization in benzene afforded the lactone in only 22% yield (Eq. 3.32). *Ab initio* calculation on the cyclization indicated that the large dielectrical constant of water lowers the barrier not only of the rotation from the Z-rotamer to the E-rotamer that can cyclize but also to the cyclization constructing the γ-lactone framework. Moreover, the high cohesive energy of water also affects acceleration of the cyclization because water forces a decrease in the volume of the reactant.

$$\text{(3.32)}$$

in 20 mL benzene; 22% yield
in 20 mL water; 56% yield
in 100 mL water; 84% yield

3.4   CARBENE REACTIONS

Structurally, a carbene is the smallest member of the alkene family. As carbenes have no charge, they are expected to have a certain stability toward water. In fact, in some of the earliest work, carbenes were generated in aqueous medium under biphasic conditions via the reaction of chloroform with a strong base such as NaOH.

3.4.1   The Generation of Carbenes in Aqueous Medium

Carbenes can be generated in a variety of ways in aqueous conditions, most commonly through the thermo-[122] and photodecomposition[123] of diazo compounds and through the treatment of organic halides[124] with bases. Carbenes have also been generated in aqueous conditions from imidazolium salts.[125] Photolysis[126] and electrolysis[127] of organic compounds also generated transient carbene intermediates that undergo further reactions in aqueous conditions.

3.4.2   The Stability of Carbenes

The stability of an unsubstituted carbene is quite low in water. Highly correlated *ab initio* MO calculations have been used to study the energetics and mechanism governing the reaction between the radical $^1CH_2$ and H_2O in the gas phase and in solution, and it was found that methylene reacts in a barrierless fashion to produce the ylide-like intermediate methyleneoxonium, $H_2C\text{-}OH_2$, which in turn undergoes a 1,2-hydrogen shift to produce CH_3OH.[128] The presence of substituents appears to stabilize carbenes toward water.[129]

Carbenes are most effectively stabilized by coordinating with transition-metals and many of these complexes are stable in water. The physical and chemical properties of various transition-metal carbene complexes in aqueous media have been studied extensively by Bernasconi[130] and others. The preparation of water-stable transition-metal carbene complexes has been carried out in several ways (Scheme 3.7): (a) by reacting with alkynes (e.g., water-soluble ruthenium carbene complexes),[131] (b) by opening unstable rings,[132] (c) by conversion of complexes through transformation of the ligands,[133] and (d) by reacting with N-heterocyclic carbenes.[134] Transition-metal heterocyclic carbene complexes showed outstanding stability toward water and air and improved catalytic activities.

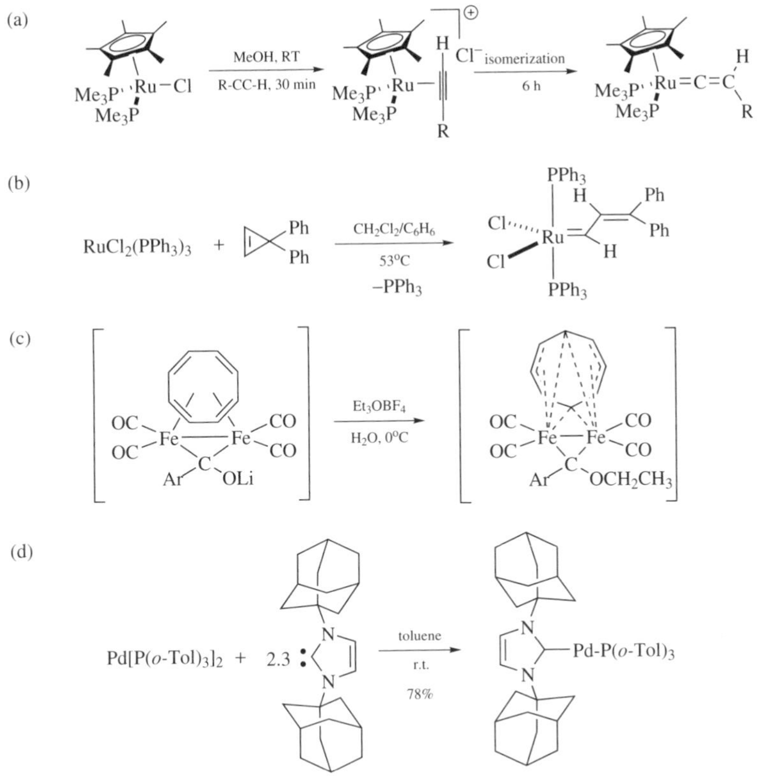

(a)

(b)

(c)

(d)

Scheme 3.7 Various methods for generating water-stable transition-metal carbene complexes.

3.4.3 The Reaction of Carbenes with Alkenes in Aqueous Medium

Cyclopropanation reactions involving ethyl diazoacetate and olefins proceed with high efficiency in aqueous media using Rh(II) carboxylates. Nishiyama's Ru(II) Py-box and Katsuki's Co(II) salen complexes that allow for highly enantioselective cyclopropanations in organic solvents can also be applied to aqueous cyclopropanations with similar results. In-situ generation of ethyl diazoacetate and cyclopropanation also proceeds efficiently (Eq. 3.33).[135]

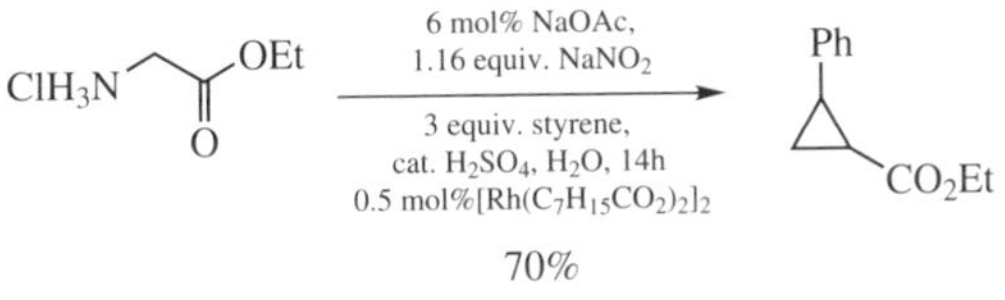

$$(3.33)$$

70%

Insertions of dichlorocarbene into tertiary C–H bonds were observed even if these were not activated by neighboring phenyl or ether groups. Yields are 3–29% under the conditions used and this insertion does not require a thermoexcitation of the dichlorocarbene as was assumed earlier.[136]

3.5 ALKENE ISOMERIZATION

In the presence of transition-metal complexes, organic compounds that are unsaturated or strained often rearrange themselves. One syntheti- cally useful transition-metal catalyzed isomerization is the olefin migra- tion reaction. Two general mechanisms have been proposed for olefin migrations, depending on the type of catalyst employed (A and B) (Scheme 3.8).[137]

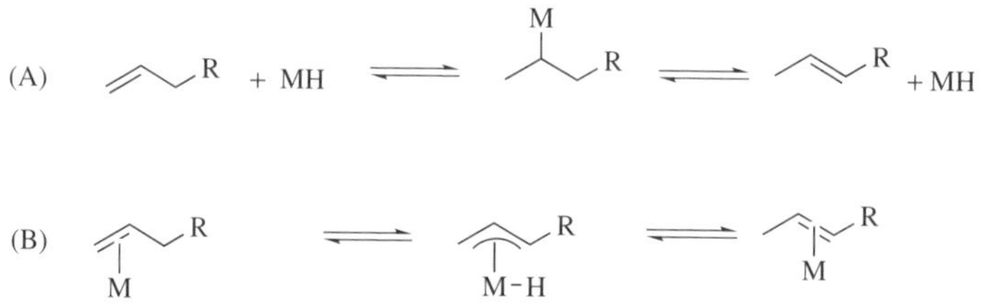

Scheme 3.8

Recently, Grubbs[138] demonstrated that olefin isomerization of allyl- lic ethers and alcohols is catalyzed by Ru(II)(H$_2$O)$_6$(tos)$_2$ (tos = p- toluenesulfonate) in aqueous medium. The olefin migration products, enols, and enol ethers thus generated are unstable and are hydrolyzed instantly to yield the corresponding carbonyl compounds (Eq. 3.34).

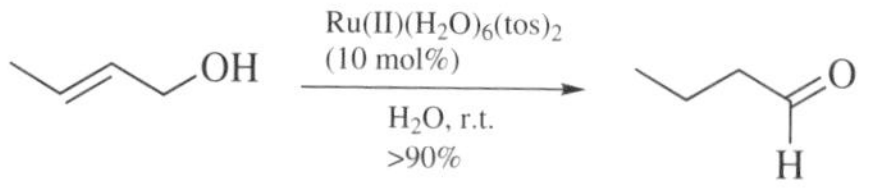

(3.34)

Li et al.[139] reported that in the presence of a catalytic amount of RuCl$_2$(PPh$_3$)$_3$, homoallylic alcohols undergo structural reorganization in which both the hydroxyl group and the olefin have been reshuffled in water (Scheme 3.9). The reaction can be conceived of as an olefin migration followed by an allylic rearrangement. Thus, allyl alcohols are rearranged similarly. The use of water is critical to the success of this reaction. Nickel-catalyzed isomerization of olefins in a two-phase system with a suitably selected Brönsted acid shows relatively stable and high catalytic activity.[140]

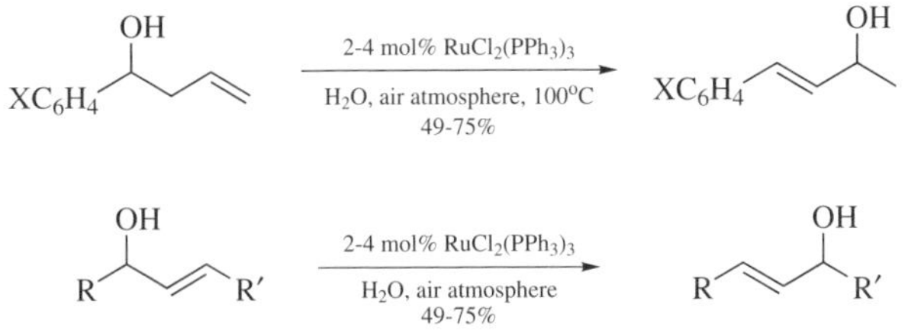

Scheme 3.9

3.6 TRANSITION-METAL CATALYZED C–C FORMATION REACTIONS

Stable transition-metal-alkene complexes can be obtained readily from their salts and alkenes in water.[141]

3.6.1 Polymerizations

Transition-metal catalyzed polymerizations of alkenes in aqueous conditions has become a well-established field over the past two decades.[142] Among other advantages, the use of water as a dispersing medium is particularly environmentally friendly. A variety of high-molecular-weight polymers ranging from amorphous or semicrystalline polyolefins to polar-substituted hydrophilic materials have now been prepared by catalytic polymerization of olefinic monomers in water. A recent example is the ethylene polymerization catalyzed by water-soluble salicylaldiminato Ni(II)–methyl complexes in water. Extremely small nanoparticles of high-molecular-weight polyethylene are formed under organic solvent-free aqueous conditions (Eq. 3.35).[142c]

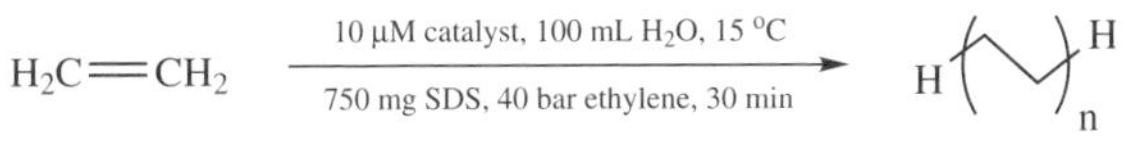

$$H_2C = CH_2 \quad \xrightarrow[\text{750 mg SDS, 40 bar ethylene, 30 min}]{10\ \mu M\ \text{catalyst, 100 mL H}_2\text{O, 15 }^{\circ}\text{C}} \quad \text{(3.35)}$$

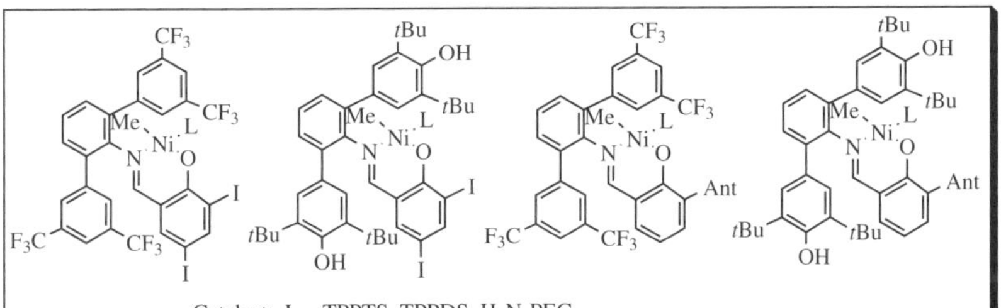

Catalyst: L = TPPTS, TPPDS, H$_2$N-PEG

3.6.2   Heck Reactions and Related Vinylation/Arylation

The reaction between aryl (or alkenyl) halides with alkenes in the presence of a catalytic amount of a palladium compound to give substitution of the halides by the alkenyl group is commonly referred to as the Heck reaction.[143] Both inter- and intramolecular Heck reactions of simple alkenes have been performed in aqueous media. Palladium-catalyzed reactions of aryl halides with acrylic acid or acrylonitrile gave the corresponding coupling products in high yields with a base (NaHCO$_3$ or K$_2$CO$_3$) in water. However, these reactions generally involved electron-deficient alkenes and will be discussed in detail in Chapter 10, on conjugated carbonyl compounds. For simple alkenes, Parsons investigated the viability of the aqueous Heck reactions of aromatic halides coupled with styrenes under superheated conditions.[144] The reaction proceeded to approximately the same degree at 400°C as at 260°C. Some 1,2-substituted alkanes can be used as alkene equivalents for the high-temperature Heck-type reaction in water (Eq. 3.36).[145] The Heck-type reaction can also use arenediazonium salts instead of aryl halides.[146]

$$(3.36)$$

The palladium-catalyzed Heck reaction of (S)-4-bromotryptophan with 1,1-dimethylallylalcohol in aqueous media was applied to the synthesis of optically active clavicipitic acid. By using Pd(AcO)$_2$ and water-soluble ligand TPPTS, the reaction could be carried out in alkaline aqueous media to give a high yield of the coupling product (up to 91%), but in organic solvent (dioxane or DMF) the reaction gave a complex mixture (Eq. 3.37). It is not necessary to protect the functional group in the substrate in the course of palladium-catalyzed reactions. (S)-4-Bromotryptophan was in turn prepared by biomimetic synthesis in two steps from 4-bromoindole with protected serine.[147]

$$(3.37)$$

The aqueous Heck coupling reaction was also used for the synthesis of unprotected branched-chain sugar. In the media of DMF-H_2O (5:1) and the use of Pd(dba)$_2$ and P(o-tol)$_3$ the Heck reaction proceeded smoothly to give the coupling product with high yield (up to 84%) (Eq. 3.38).[148]

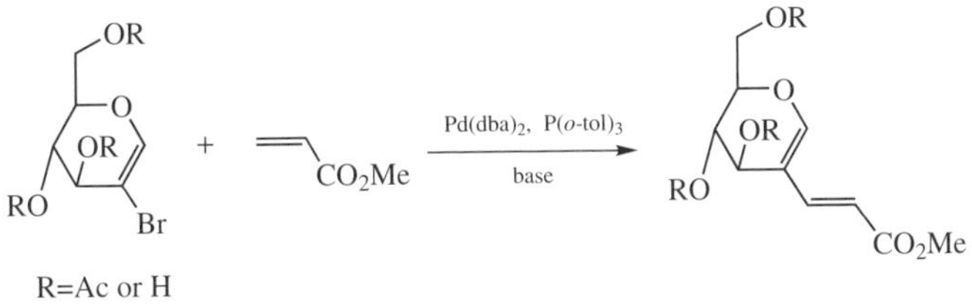

$$(3.38)$$

R=Ac or H

Water-soluble phosphine ligands containing m-guanidinium moieties were synthesized and applied to aqueous Heck coupling reactions.[149] High temperature appears beneficial for Heck-type coupling of simple alkenes in water.[150]

Recently, a Pd/Cu-catalyzed three-component coupling reaction of aryl halides, norbornadiene, and alkynols was reported to generate 2,3-disubstituted norbornenes in high yields in the presence of aqueous NaOH and a phase-transfer catalyst in toluene at 100°C (Eq. 3.39).[151]

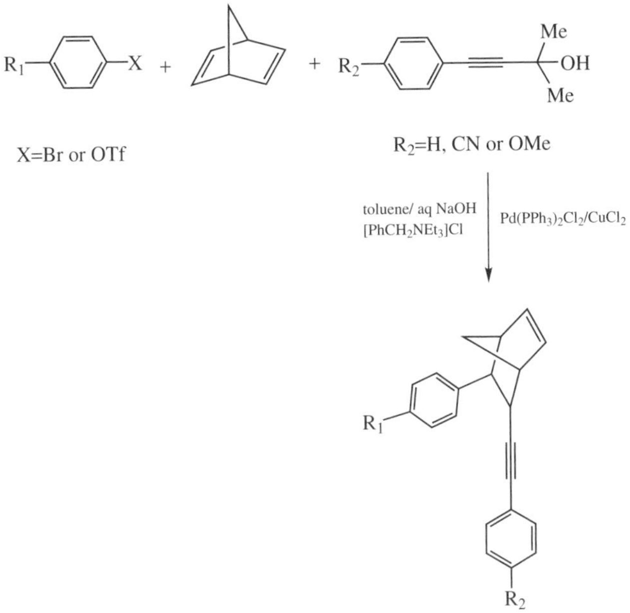

X=Br or OTf

R$_2$=H, CN or OMe

toluene/ aq NaOH
[PhCH$_2$NEt$_3$]Cl   Pd(PPh$_3$)$_2$Cl$_2$/CuCl$_2$

$$(3.39)$$

In addition to Heck reactions, other transition-metals have also been used for arylation and vinylaton of arylalkene. Lautens[152] as well as Genet[153] studied the coupling of phenylboronic acid with alkenes in the presence of a rhodium catalyst in aqueous media. It was found that the product of the addition was strongly affected by the nature of the aryl group. In the presence of a catalytic amount of [Rh(COD)Cl] together with three equivalents of Na_2CO_3 and a surfactant (SDS), the addition-hydrolysis product was formed when N-heteroarylalkenes were used, whereas the use of styrene derivatives generated addition-elimination products under the same reaction conditions (Eq. 3.40).

$$(3.40)$$

More recently, Chang reported a ruthenium-based Heck-type reaction in DME/H_2O (1:1) by using alumina-supported ruthenium catalysts.[154]

3.6.3 Hydrovinylation

The homo- and cross-addition of alkenes catalyzed by a transition-metal provided another economical way of forming C–C bonds.[155] These reactions are carried out by using nickel, palladium, or ruthenium phosphine complexes to yield vinylarenes and some can occur in aqueous media. By using carbohydrate-derived ligands, asymmetric hydrovinylations can be carried out in aqueous conditions.[156]

3.6.4 Reaction with Arenes

Under Lewis-acid-catalyzed conditions, electron-rich arenes can be added to alkenes to generate Friedel-Crafts reaction products. This subject will be discussed in detail in Chapter 7, on aromatic compounds. However, it is interesting to note that direct arylation of styrene with benzene in aqueous CF_3CO_2H containing H_2PtCl_6 yielded 30–5% trans-PhCH:CHR via the intermediate PhPt(H_2O)Cl_4.[157] Hydrophenylation of olefins can be catalyzed by an Ir(III) complex.[158]

3.6.5 Hydroformylation

Hydroformylation is a major industrial process that produces aldehydes and alcohols from olefins, carbon monoxide, and hydrogen (Eq. 3.41).[159] The reaction was discovered in 1938 by Roelen,[160] who detected the formation of aldehydes in the presence of a cobalt-based catalyst. A major improvement was made by joint efforts of Ruhrchemie and Rhone-Poulenc by using rhodium in aqueous media. Extensive research has been carried out related to this process.[161] The Rh/TPPTS catalyst was employed in the hydroformylation of N-allylacetamide in water, which proceeds at a much faster rate and in a much higher selectivity (>99%) than the Rh/PPh$_3$-catalyzed reaction in organic solvents. In water, at 90°C and 50 bar H$_2$/CO, turnover frequencies (TOF) are >10,700 h-1. The relations of regioselectivity with reaction conditions have been investigated in detail.[162] By using this method, the separation of catalyst and product is based on the use of transition-metal complexes with water-soluble phosphine ligands and water as an immiscible solvent for the hydroformylation. Initially, the water-soluble complex [HRh(CO)(Ph$_2$PPhSO$_3$Na)$_3$][163] was used. However, with this monosulfonated ligand, some leaching of rhodium into the organic phase was observed.[164] The highly water-soluble *tris*-sulfonated ligand, P(*m*-PhSO$_3$Na)$_3$,[165,166] was found to be highly recyclable. A variety of 1-alkenes were hydroformylated with this catalyst in high linear selectivity, generating the corresponding terminal aldehyde.[167] More effective catalysts involving the use of other sulfonated phosphine ligands have also been reported.[168] Examples include [Rh$_2$(μ-SR)$_2$(CO)$_2$(P(*m*-PhSO$_3$Na)$_3$)$_2$],[169] Ph$_2$PCH$_2$CH$_2$NMe$_3^+$,[170] *p*-carboxylatophenylphosphine,[171] and sulfoalkylated *tris*(2-pyridyl)-phosphine.[172]

$$RCH=CH_2 + CO + H_2 \xrightarrow{\text{``Co''}} RCH_2CH_2CHO + R(CH_3)CHCHO + \text{etc.} \qquad (3.41)$$

For long chain olefins, the hydroformylation generally proceeds slowly and with low selectivity in two-phase systems due to their poor solubility in water. Monflier et al. recently reported a conversion of up to 100% and a regioselectivity of up to 95% for the Rh-catalyzed hydroformylation of dec-1-ene in water, free of organic solvent, in the presence of partially methylated ß-cyclodextrins (Eq. 3.42).[173]

$$(3.42)$$

These interesting results are attributed to the formation of an alkene/cyclodextrin inclusion complex as well as the solubility of the chemically modified cyclodextrin in both phases. Prior to this study, hydroformylation in the presence of unmodified cyclodextrins had been studied by Jackson, but the results were rather disappointing.[174]

In another interesting area in the study of hydroformylation, Davis developed the concept of *supported aqueous phase* (SAP) catalysis.[175] A thin, aqueous film containing a water-soluble catalyst adheres to silica gel with a high surface area. The reaction occurs at the liquid–liquid interface. Through SAP catalysis, the hydroformylation of very hydrophobic alkenes, such as octene or dicyclopentadiene, is possible with the water-soluble catalyst $[HRh(CO)(tppts)_3]$.

Other metal complexes containing Pd, Ru, Co, or Pt have also been used.[176] The hydroformylation reaction can also be performed by using methyl formate instead of carbon monoxide and hydrogen.[177]

Hydroaminomethylation of alkenes occurred to give both *n*- and *iso*-aliphatic amines catalyzed by $[Rh(cod)Cl]_2$ and $[Ir(cod)Cl]_2$ with TPPTS in aqueous NH_3 with CO/H_2 in an autoclave. The ratio of *n*- and *iso*-primary amines ranged from 96:4 to 84:16.[178] The catalytic hydroaminomethylation of long-chain alkenes with dimethylamine can be catalyzed by a water-soluble rhodium-phosphine complex, $RhCl(CO)(TPPTS)_2$ [TPPTS: $P(m\text{-}C_6H_4SO_3Na)_3$], in an aqueous-organic two-phase system in the presence of the cationic surfactant cetyltrimethylammonium bromide (CTAB) (Eq. 3.43). The addition of the cationic surfactant CTAB accelerated the reaction due to the micelle effect.[179]

$$R \diagup\!\!\!\diagdown \xrightarrow[\substack{\text{catalyst} \\ H_2O}]{CO/H_2,\ R^1R^2NH} R \diagup\!\!\!\diagdown\!\!\!\diagup NR^1R^2 \;+\; R \diagdown\!\!\!\diagup NR^1R^2 \qquad (3.43)$$

3.6.6   Reaction with Alkynes

In aqueous media, the addition of unactivated alkynes to unactivated alkenes to form Alder-ene products has been realized by using a ruthenium catalyst (Eq. 3.44).[180] A polar medium (DMF:H_2O = 1:1) favors the reaction and benefits the selectivity. The reaction was proposed to proceed via a ruthenacycle intermediate.

$$C_2H_5O_2C\diagup\!\!\!\diagdown\!\!\!\equiv \;+\; \diagup\!\!\!\diagdown\!\!\!\diagup\!\!\!\diagdown \xrightarrow[\substack{\text{3:1 DMF-}H_2O \\ 90\%}]{CpRu(COD)Cl} C_2H_5O_2C\diagup\!\!\!\diagdown\!\!\!\diagup\!\!\!\diagdown\!\!\!\diagup\!\!\!\diagdown$$

$$(3.44)$$

Mascareñas developed a synthetic method to 1,5-oxygen-bridged medium-sized carbocycles through a sequential ruthenium-catalyzed alkyne-alkene coupling and a Lewis-acid-catalyzed Prins-type reaction (Eq. 3.45). The ruthenium-catalyzed reaction can be carried out in aqueous media (DMF/H$_2$O = 10:1).[181]

$$(3.45)$$

The addition of allyl alcohol to alkynes to form γ,δ-unsaturated ketones and aldehydes (Eq. 3.46) in aqueous media was developed by both Trost and Dixneuf.[182]

$$(3.46)$$

Other transition-metals have also been used. For example, Trost[183] reported that heating a 1:1 mixture of 1-octene and 1-octyne in DMF-water (3:1) at 100°C with a ruthenium complex for 2 h generated a 1:1 mixture of two products corresponding to the addition of the alkene to the acetylene (Eq. 3.47). The presence of a normally reactive enolate does not interfere with the reaction.

$$(3.47)$$

3.6.7 Carbonylation

Water-soluble dicationic palladium(II) complexes $[(R_2P(CH_2)_3PR_2)Pd-(NCMe)_2][BF_4]_2$ proved to be highly active in the carbon monoxide/ethene copolymerization under biphasic conditions (water-toluene). In the presence of an emulsifier and methanol as activator, the catalytic activity increased by a factor of about three. Also higher olefins could be successfully incorporated into the copolymerization with CO and the terpolymerization with ethene and CO.[184]

3.6.8 Cycloaddition Reactions of Alkenes

The thermo- and photocycloaddition of alkenes will be discussed in Chapter 12, on pericyclic reactions. On the other hand, transition-metals have effectively catalyzed some synthetically useful cycloaddition reactions in water. For example, Lubineau and co-worker reported a [4 + 3] cycloaddition by reacting α,α-dibromo ketones with furan or cyclopentadiene mediated by iron or copper, or α-chloro ketones in the presence of triethylamine (Eq. 3.48).[185]

$$Z = O \qquad 89:11$$
$$Z = CH_2 \qquad 67:33$$

(3.48)

3.7 OLEFIN METATHESIS

Olefin-metathesis is a useful tool for the formation of unsaturated C–C bonds in organic synthesis.[186] The most widely used catalysts for olefin metathesis include alkoxyl imido molybdenum complex (Schrock catalyst)[187] and benzylidene ruthenium complex (Grubbs catalyst).[188] The former is air- and moisture-sensitive and has some other drawbacks such as intolerance to many functional groups and impurities; the latter has increased tolerance to water and many reactions have been used in aqueous solution without any loss of catalytic efficiency.

The olefin-metathesis in aqueous media has been applied to the synthesis of various polymers.[189]

3.7.1 Ring-Opening Metathesis Polymerization (ROMP)

Living ring-opening metathesis polymerization was developed by Grubbs and co-workers in water.[190] The polymerization has been used in making a variety of materials such as dental products.[191] Novak and Grubbs reported the ring-opening metathesis polymerization (ROMP) of 7-oxanobornene derivatives initiated by $Ru(H_2O)_6(tos)_2$ in aqueous media (Eq. 3.49).[192]

$$(3.49)$$

Compared with the same reaction carried out in organic solvent, the initiation time was greatly decreased. After the polymerization, the aqueous catalyst solution not only was reused but also became more active in subsequent polymerizations. Living ROMP, by using some well-defined ruthenium carbene complexes in aqueous media in the presence of a cationic surfactant, have been used to prepare polymer latex.[193] Recent developments include the synthesis of new water-soluble ruthenium alkylidene catalysts and their application to olefin metathesis in water.[194] The addition of acid made the polymerization rate up to 10 times faster than without acid. Kiessling has extended the use of ruthenium alkylidenes catalyzed ROMP in aqueous media to give new, biologically active neoglycopolymers (Eq. 3.50).[195]

$$(3.50)$$

3.7.2 Ring-Closing Metathesis (RCM)

Whereas ROMP is an important method for making polymers, ring-closing metathesis (RCM) is an important method for construction of medium and macrocyclic compounds. Ring-closing metathesis (RCM) by using Grubb's catalyst has been used extensively in synthesis in aqueous conditions. For many biologically related substrates, the use of RCM in aqueous media can allow them to keep their important higher-order structures.[196] For example, RCM of σ,ω-dienes proceeded efficiently in aqueous media (Eq. 3.51).

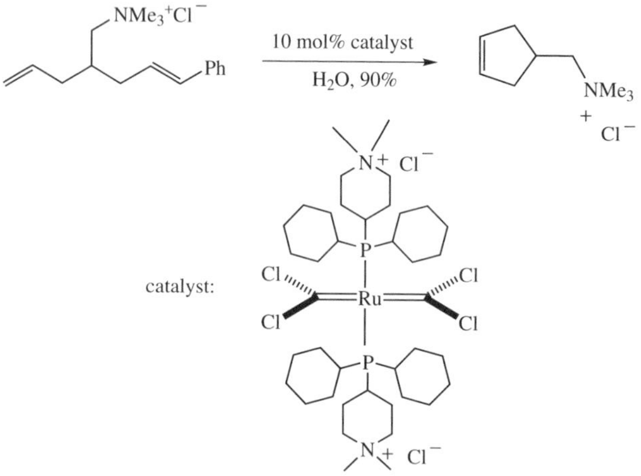

$$(3.51)$$

More recently, a new metathesis catalyst involving a ruthenium-alkylidene complex with a sterically bulky and electron-rich phosphine ligand has been synthesized and applied to RCM in aqueous media (Figure 3.5).[197] This catalyst has the benefit of being soluble in almost

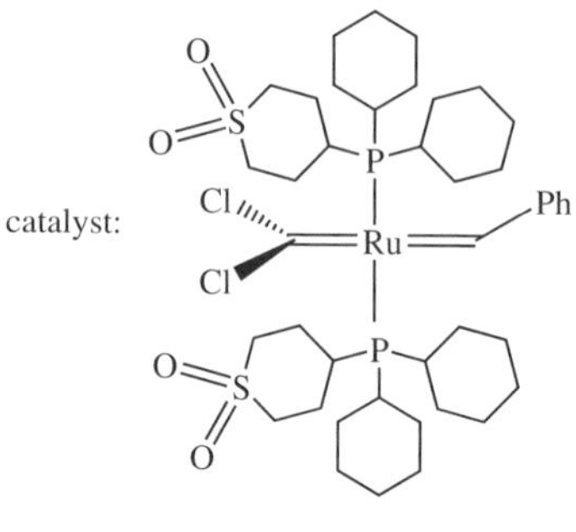

Figure 3.5

any solvent, (e.g., methanol, methanol-water, methylene chloride, benzene), while still behaving as an active catalyst for RCM.

A general synthesis of α-methylene-γ-lactones *cis-* or *trans-*fused to larger rings has been reported. The protocol originates with two unsaturated aldehydes of the same or different chain length. One of these is initially transformed by way of the Baylis-Hillman reaction into a functionalized allylic bromide. Merger of the two building blocks is subsequently accomplished in aqueous solution with powdered indium metal serving as the mediator. Once the lactone ring is grafted, the end products are generated by application of ring-closing metathesis (Eq. 3.52). The central issues surrounding this final step are the effects of the stereochemical disposition of the side chains, the consequences of ring strain, and the locus of the double bonds on cyclization efficiency.[198]

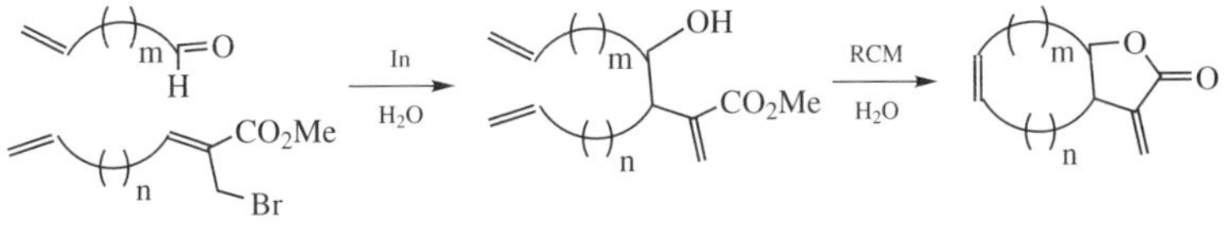

(3.52)

4-Deoxy-4,4-difluoro-glycopyrans have been synthesized for the first time via a direct sequence involving an indium-mediated difluoroallylation with 1-bromo-1,1-difluoropropene in water followed by ring-closing metathesis in methylene chloride (Scheme 3.10). Two protecting group strategies were explored, one to allow protection of the primary C-6 hydroxyl group throughout the sequence, while the second was intended to allow deprotection after RCM and before dihydroxylation. The benzyl

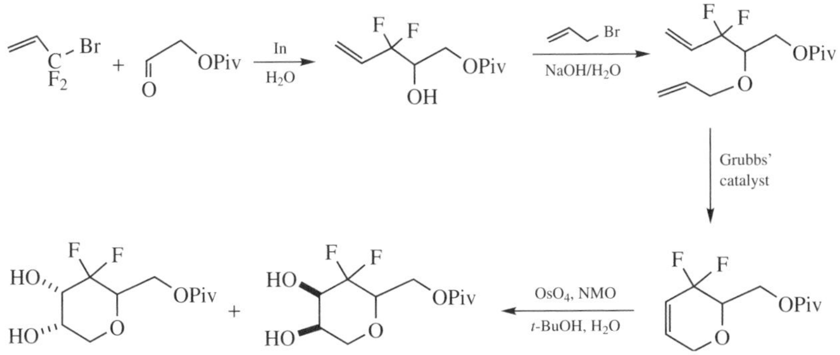

Scheme 3.10

ether could be used in the first role, and pivaloyl is effective in the second. Dihydroxylations were highly stereoselective and -controlled by the orientation of the glycosidic C–O bond.[199]

The synthesis and olefin metathesis activity in protic solvents of a phosphine-free ruthenium alkylidene bound to a hydrophilic solid support have been reported. This heterogeneous catalyst promotes relatively efficient ring-closing and cross-metathesis reactions in both methanol and water.[200] The catalyst-catalyzed cross-metathesis of allyl alcohol in D_2O gave 80% $HOCH_2CH=CHCH_2OH$.

3.8   REACTION OF ALLYLIC C–H BOND

3.8.1   Allylic Oxidation

In 2002, Muzart et al. reported an efficient allylic oxidation of olefins with organic peroxyesters, also known as the Kharasch-Sosnovsky reaction, in water. The reaction involves a combination of a hydrophilic N-donor ligand (Figure 3.6) with $Cu(MeCN)_4BF_4$ and using *tert*-butyl perbenzoate as the oxidant (Eq. 3.53). The water-soluble catalytic system is recyclable.[201] The oxidation of cyclohexene was repeated four times with the same aqueous phase without a decrease in the yield. The use of the water-soluble ligand also avoided the loss of the copper complex into the organic phase, obtaining an effective recycling of the catalyst.

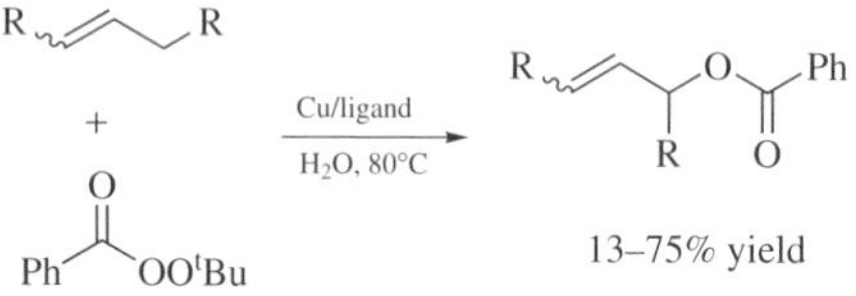

$$(3.53)$$

Subsequently, the enantioselective variant of this reaction[202] was carried out in a biphasic medium (water + diethylene glycol) by using a mixture of amino acids and copper complexes (Eq. 3.54). When the reaction was carried out under an argon atmosphere, the recycling of the catalyst was also possible. The enantiomeric excess decreased slightly

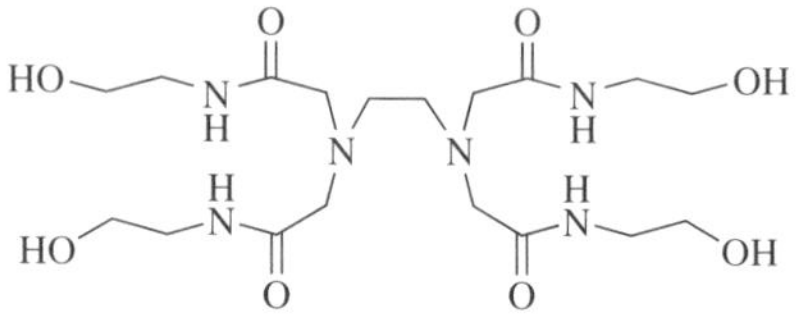

Figure 3.6

from run to run, while the yields were not affected. The use of diethylene glycol was necessary to carry out the reaction at room temperature; no reaction occurred in its absence.

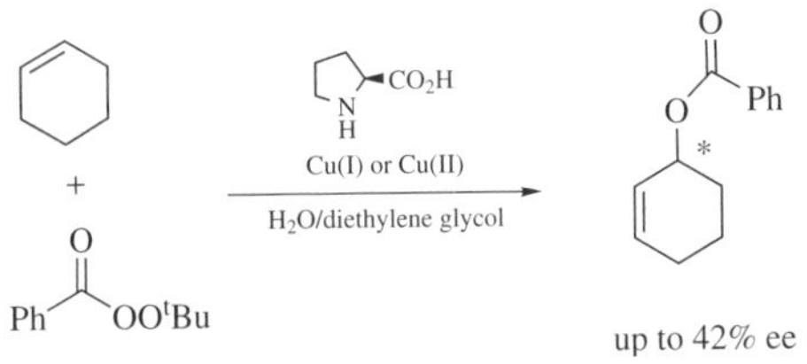

$$(3.54)$$

3.8.2 C–C Bond Formations

Reactions of the allylic position of alkenes with carbonyl or imine electrophiles are known as Prins reactions and have been discussed in previous sections (3.2.9). More examples of similar Alder-ene-type reactions (the Prins reaction) will be discussed in Chapter 8.

In Section 3.5 on alkene isomerization, it was mentioned that Li and co-workers reported a $RuCl_2(PPh_3)_3$-catalyzed shuffling of functional groups of allylic alcohols in water (Eq. 3.35).[140] Since the reaction proceeds through an enol intermediate, allyl alcohols can thus be considered as enol equivalents.[203] This has been developed into an aldol-type reaction by reacting allyl alcohols with aldehyde (Scheme 3.11).[204] The presence of $In(OAc)_3$ promoted the aldol reaction with α-vinylbenzyl alcohol and aldehyde.[205]

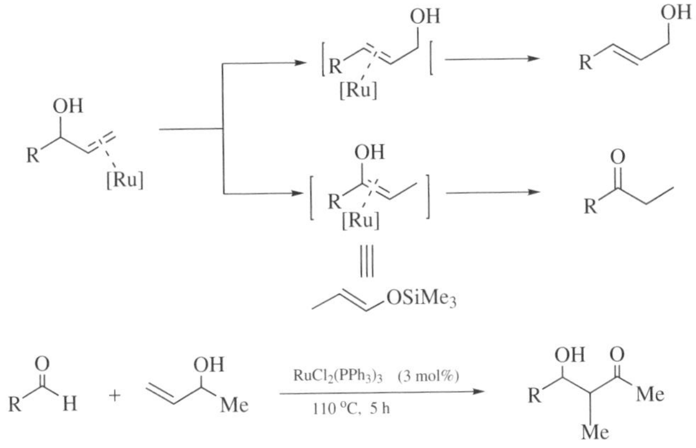

Scheme 3.11

REFERENCES

1. Quinn, P. J., Joó, F., Vigh, L., *Prog. Biophys. Mol. Biol.* **1989**, *53*, 71.
2. Spencer, M. S., Dowden, D. A. U.S. Patent 3 009 969 (1961); Kwiatek, J., Madok, I. L., Syeler, J. K., *J. Am. Chem. Soc.* **1962**, *84*, 304; Joo, F., Beck, M. T., *React. Kinet. Catal. Lett.* **1975**, *2*, 257; Tyurenkova, O. A., *Zhurnal Fizicheskoi Khimii* **1969**, *43*, 2088.
3. Dror, Y., Manassen, J., *J. Mol. Catal.* **1977**, *2*, 219.
4. Borowski, A. F., Cole-Hamilton, D. J., Wilkinson, G., *Nouv. J. Chim.* **1978**, *2*, 137.
5. Larpent, C., Dabard, R., Patin, H., *Tetrahedron Lett.* **1987**, *28*, 2507.
6. Larpent, C., Patin, H., *J. Organomet. Chem.* **1987**, *335*, C13.
7. Grosselin, J. M., Mercier, C., Allmang, G., Grass, F., *Organometallics* **1991**, *10*, 2126.
8. Smith, R. T., Ungar, R. K., Sanderson, L. J., Baird, M. C., *Organometallics*, **1983**, *2*, 1138.
9. Okano, T., Kaji, M., Isotani, S., Kiji, J., *Tetrahedron Lett.* **1992**, *33*, 5547; Kotzabasakis, V., Georgopoulou, E., Pitsikalis, M., Hadjichristidis, N., Papadogianakis, G., *J. Mol. Catal. A: Chem.* **2005**, *231*, 93.
10. Parmar, D. U., Bhatt, S. D., Bajaj, H. C., Jasra, R. V., *J. Mol. Catal. A: Chem.* **2003**, *202*, 9.
11. Ellis, D. J., Dyson, P. J., Parker, D. G., Welton, T., *J. Mol. Catal. A: Chem.* **1999**, *150*, 71.
12. Makihara, N., Ogo, S., Watanabe, Y., *Organometallics* **2001**, *20*, 497.
13. Kovacs, J., Todd, T. D., Reibenspies, J. H., Joo, F., Darensbourg, D. J., *Organometallics* **2000**, *19*, 3963.
14. Nakao, R., Rhee, H., Uozumi, Y., *Org. Lett.*, **2005**, *7*, 163.
15. Jennings, J. M., Bryson, T. A., Gibson, J. M., *Green Chemistry* **2000**, *2*, 87.
16. Niu, Y., Yeung, L. K., Crooks, R. M., *J. Am. Chem. Soc.* **2001**, *123*, 6840.
17. Ohde, M., Ohde, H., Wai, C. M., *Langmuir* **2005**, *21*, 1738.
18. Tour, J. M., Pendalwar, S. L., *Tetrahedron Lett.* **1990**, *31*, 4719; Tour, J. M., Cooper, J. P., Pendalwar, S. L., *J. Org. Chem.* **1990**, *55*, 3452.
19. Acharya, D., Das, M. N., *J. Org. Chem.* **1969**, *34*, 2828.
20. Worley, R., *School Science Review* **1986**, *68*, 74.
21. Shkaraputa, L. N., Kononov, A. V., Tishchenko, L. A., *Neftepererabotka Neftekhimiya (Kiev)* **1991**, *41*, 67.
22. De la Mare, P. B. D., Galandauer, S., *J. Chem. Soc.* **1958**, 36.
23. Groszkowski, S., Sienkiewicz, J., *Roczniki Chemii* **1971**, *45*, 1779.
24. Jereb, M., Zupan, M., Stavber, S., *Green Chemistry* **2005**, *7*, 100.

25. Burkhardt, G. N., Cocker, W., *Rec. Trav. Chim. Pays-Bas Belg.* **1931**, *50*, 837.

26. Sherrill, M. L., Mayer, K. E., Walter, G. F., *J. Am. Chem. Soc.* **1934**, *56*, 926; Michael, A., Weiner, N., *J. Org. Chem.* **1940**, *5*, 389.

27. Levy, J. B., Taft, R. W. Jr., Hammett, L. P., *J. Am. Chem. Soc.* **1953**, *75*, 1253.

28. Seitz, H., *Prax. Naturwiss. Chem.* **1984**, *33*, 40.

29. For a review, see: Kuhara, T., Kimura, M., Kawai, T., Xu, Z., Nakato, T., *Catalysis Today*, **1998**, *45*, 73; Okuhara, T., Kimura, M., Nakato, T., *Chem. Lett.* **1997**, 839.

30. Jensen, C. M., Trogler, W. C., *Science*, **1986**, *233*, 1069.

31. Shimizu, K., Takahashi, K., Ikushima, Y., *AIChE Annual Meeting, Conference Proceedings*, **2004**, 545E/1.

32. Jia, B., Yang, X., Huang, M.-Y., Jiang, Y.-Y., *React. Funct.Polym.* **2003**, *57*, 163.

33. Brown, H. C., Hammar, W. J., *J. Am. Chem. Soc.* **1967**, *89*, 1524.

34. Abreu, A. R., Costa, I., Rosa, C., Ferreira, L. M., Lourenco, A., Santos, P., *Tetrahedron* **2005**, *61*, 11986.

35. Uemura, S., Fukuzawa, S., Toshimitsu, A., *J. Organomet. Chem.* **1983**, *250*, 203.

36. Meunier, B., *Chem. Rev.* **1992**, *92*, 1411.

37. Bernadou, J., Fabiano, A. S., Robert, A., Meunier, B., *J. Am. Chem. Soc.* **1994**, *116*, 9375.

38. Pasc-Banu, A., Sugisaki, C., Gharsa, T., Marty, J.-D., Gascon, I., Kraemer, M., Pozzi, G., Desbat, B., Quici, S., Rico-Lattes, I., Mingotaud, C., *Chem. Eur. J.* **2005**, *11*, 6032.

39. Venturello, C., Alneri, E., Ricci, M., *J. Org. Chem.* **1983**, *48*, 3831.

40. Sato, K., Aoki, M., Ogawa, M., Hashimoto, T., Noyori, R., *J. Org. Chem.* **1996**, *61*, 8310.

41. Yao, H. R., Richardson, D. E., *J. Am. Chem. Soc.* **2000**, *122*, 3220.

42. Lane, B. S., Burgess, K., *J. Am. Chem. Soc.* **2001**, *123*, 2933.

43. Wei, M., Musie, G. T., Busch, D. H., Subramaniam, B., *J. Am. Chem. Soc.* **2002**, *124*, 2513; Busch, D. H., *Ind. Eng. Chem. Res.* **2003**, *42*, 6505.

44. Masuyama, A., Fukuoka, K., Katsuyama, N., Nojima, M., *Langmuir* **2004**, *20*, 82.

45. Tong, K.-H., Wong, K.-Y., Chan, T. H., *Tetrahedron* **2005**, *61*, 6009.

46. Neumann, R., Wang, T.-J., *Chem. Comm.* **1997**, 1915.

47. Fringuelli, F., Germani, R., Pizzo, F., Savelli, G., *Tetrahedron Lett.* **1989**, *30*, 1427.

48. Fringuelli, F., Germani, R., Pizzo, F., Santinelli, F., Savelli, G., *J. Org. Chem.* **1992**, *57*, 1198.

49. Berti, G., in *Topics in Stereochemistry*, Allinger, N. L., Eliel, E. L., eds., Vol 7, 93.

50. Kirshenbaum, K. S., Sharpless, K. B., *J. Org. Chem.* **1985**, *50*, 1979.

51. Torii, S., Uneyama, K., Tanaka, H., Yamanaka, T., Yasuda, T., Ono, M., Kohmoto, Y., *J. Org. Chem.* **1981**, *46*, 3312.

52. Masuyama, A., Yamaguchi, T., Abe, M., Nojima, M., *Tetrahedron Lett.* **2005**, *46*, 213.

53. (a) Curci, R., Fiorentino, M., Serio, M. R., *Chem. Comm.* **1984**, 155.
(b) Curci, R., D'Accolti, L., Fiorentino, M., Rosa, A., *Tetrahedron Lett.* **1995**, *36*, 5831.

54 Shi, Y., *Acc. Chem. Res.* **2004**, *37*, 488.

55 Yang, D., *Acc. Chem. Res.* **2004**, *37*, 497.

56. Gunstone, F. D., *Adv. Org. Chem.* **1960**, *1*, 103.

57. Hofmann, K. A., *Ber.* **1912**, *45*, 3329; Grieco, P. A., Ohfune, Y., Yokoyama, Y., Owens, W., *J. Am. Chem. Soc.* **1979**, *101*, 4749.

58. Milas, N. A., Sussman, S., *J. Am. Chem. Soc.* **1936**, *58*, 1302; Daniels, R., Fischer, J. L., *J. Org. Chem.* **1963**, *28*, 320.

59. Mugdan, M., Young, D. P., *J. Chem. Soc.* **1949**, 2988.

60. Sharpless, K. B., Akashi, K., *J. Am. Chem. Soc.* **1976**, *98*, 1986.

61. Van Rheenen, V., Kelly, R. C., Cha, D. Y., *Tetrahedron Lett.* **1976**, 1973.

62. Lee, D. G., *The Oxidation of Organic Compounds by Permanganate Ion and Hexavalent Chromium*. Open Court, La Salle, Illinois, 1980.

63. Mailhot, G., Pilichowski, J. F., Bolte, M., *New J. Chem.* **1995**, *19*, 91.

64. Minato, M., Yamamoto, K., Tsuji, J., *J. Org. Chem.* **1990**, *55*, 766.

65. Mugdan, M., Young, D. P., *J. Chem. Soc.* **1949**, 2988.

66. For reviews, see: Kolb, H. C., Van Nieuwenhze, M. S., Sharpless, K. B., *Chem. Rev.* **1994**, *94*, 2483.

67. Hanessian, S., Meffre, P., Girard, M., Beaudoin, S., Sanceau, J. Y., Bennani, Y., *J. Org. Chem.* **1993**, *58*, 1991; Corey, E. J., Jardine, P. D., Virgil, S., Yuen, P. W., Connell, R. D., *J. Am. Chem. Soc.* **1989**, *111*, 9243; Tomioka, K., Nakajima, M., Koga, K., *J. Am. Chem. Soc.* **1987**, *109*, 6213; Tokles, M., Snyder, J. K., *Tetrahedron Lett.* **1986**, *27*, 3951; Yamada, T., Narasaka, K. *Chem. Lett.* **1986**, 131.

68. Hentges, S. G., Sharpless, K. B., *J. Am. Chem. Soc.* **1980**, *102*, 4263; Oishi, T., Hirama, M., *Tetrahedron Lett.* **1992**, *33*, 639; Imada, Y., Saito, T., Kawakami, T., Murahashi, S. I., *Tetrahedron Lett.* **1992**, *33*, 5081.

69. Kwong, H. L., Sorato, C., Ogino, Y., Chen, H., Sharpless, K. B., *Tetrahedron Lett.* **1990**, *31*, 2999.

70. Gobel, T., Sharpless, K. B., *Angew. Chem., Int. Ed. Engl.* **1993**, *32*, 1329.

71. Crispino, G. A., Jeong, K. S., Kolb, H. C., Wang, Z. M., Xu, D., Sharpless, K. B., *J. Org. Chem.* **1993**, *58*, 844.

72. Kolb, H. C., Andersson, P. C., Sharpless, K. B., *J. Am. Chem. Soc.* **1994**, *116*, 1278.

73. Hale, K. J., Manaviazar, S., Peak, S. A., *Tetrahedron Lett.* **1994**, *35*, 425.

74. Abedel-Rahman, H., Adams, J. P., Boyes, A. L., Kelly, M. J., Mansfield, D. J., Procopiou, P. A., Roberts, S. M., Slee, D. H., Sidebottom, P. J., Sik, V., Watson, N. S., *J. Chem. Soc., Chem. Commun.* **1993**, 1841.

75. Huang, J., Corey, E. J., *Org. Lett.* **2003**, *5*, 3455.

76. Sharpless, K. B., Chong, A. O., Oshima, K., *J. Org. Chem.* **1976**, *41*, 177.

77. Bäckvall, J. E., *Tetrahedron Lett.* **1975**, *26*, 2225.

78. Hentges, S. G., Sharpless, K. B., *J. Am. Chem. Soc.* **1980**, *102*, 4263; Rubinstein, H., Svendsen, J. S., *Acta Chem. Scand.* **1994**, *48*, 439.

79. Li, G., Chang, H. T., Sharpless, K. B., *Angew. Chem., Int. Ed. Engl.* **1996**, *35*, 451.

80. Phillips, F. C., *Am. Chem. J.* **1894**, *16*, 255.

81. Anderson, J. S., *J. Chem. Soc.* **1934**, 971.

82. Smidt, J., Hafner, W., Jira, R., Sedlmeier, J., Sieber, R., Ruttinger, R., Kojer, H., *Angew. Chem.* **1959**, *71*, 176; **1962**, *74*, 93.

83. Henry, P. M., *Adv. Organomet. Chem.* **1975**, *13*, 363.

84. Bäckvall, J. E., Åkermark, B., Lijunggren, S. O., *J. Am. Chem. Soc.* **1979**, *101*, 2411.

85. Stille, J. K., Divakaruni, R., *J. Am. Chem. Soc.* **1978**, *100*, 1303.

86. Nagashima, H., Sakai, K., Tsuji, J., *Chem. Lett.* **1982**, 859.

87. Tsuji, J., Shimizu, I., Suzuki, H., Naito, Y., *J. Am. Chem. Soc.* **1979**, *101*, 5070.

88. Labinger, J. A., Herring, A. M., Bercaw, J. E., *J. Am. chem. Soc.* **1990**, *112*, 5628.

89. Luinstra, G. A., Labinger, J. A., Bercaw, J. E., *J. Am. chem. Soc.* **1993**, *115*, 3004.

90. For some general reviews, see: Bailey, P. S., *Chem. Rev.* **1958**, *58*, 925; Criegee, R., *Rec. Chem. Progr.* **1957**, *18*, 111.

91. Fremery, M. I., Fields, E. K., *J. Org. Chem.* **1963**, *28*, 2537.

92. Chan, T. H., Lee, M.-C., *J. Org.Chem.* **1995,** *60*, 4228.

93. Dowideit, P., von Sonntag, C., *Envir. Sci. Technol.* **1998**, *32*, 1112.

94. Ishizaki, K., Shinriki, N., Ikehata, A., Ueda, T., *Chem. Pharm. Bull.* **1981**, *29*, 868.

95. Matsui, M., Nakazumi, H., Kamiya, K., Yatome, C., Shibata, K., Muramatsu, H., *Chem. Lett.* **1989**, 723.

96. John, T. B., Flint, R. B., *J. Am. Chem. Soc.* **1931**, *53*, 1082; Kolonko, K. J., Shapiro, R. H., Barkley, R. M., Sievers, R. E., *J. Org. Chem.* **1979**, *44*, 3769.

97. Baeumer, U.-St., Schaefer, H. J., *J. Appl. Electrochem.* **2005**, *35*, 1283.

98. Poklukar, N., Mittelbach, M., *Monat. Chem.* **1991**, *122*, 719; Fremery, M. I., Fields, E. K., *J. Org. Chem.* **1963**, *28*, 2537.

99. Fields, E. K., *Adv. Chem. Ser.* **1965**, No. 51, 99.

100. DeLong, M., McCorkle, R., Fuentes, G., *Abstracts of Papers*, 221st ACS National Meeting, San Diego, CA, April 1–5, 2001, ORGN-502.

101. Cainelli, G., Contento, M., Manescalchi, F., Plessi, L., *Synthesis* **1989**, 47.

102. Coleman, J. E., Swern, D., *J. Am. Oil Chem. Soc.* **1958**, *35*, 675.

103. Lemieux, R. U., Von Rudloff, E., *Can. J. Chem.* **1955**, *33*, 1701.

104. Cornely, J., Ham, L. M. S., Meade, D. E., Dragojlovic, V., *Green Chemistry* **2003**, *5*, 34.

105. Xing, D., Guan, B., Cai, G., Fang, Z., Yang, L., Shi, Z., *Org. Lett.* **2006**, *8*, 693.

106. An, J. M., Yang, S. J., Yi, S.-Y., Jhon, G.-J., Nam, W., *Bull. Kor. Chem. Soc.* **1997**, *18*, 117.

107. For a review, see: Kazansky, V. B., *Catal. Today* **2002**, *73*, 127.

108. Prins, H. J., *Chem. Weekbl.* **1919**, *16*, 1072; for a review, see: Adams, D. R., Bhatnagar, S. P., *Synthesis*, **1977**, 661.

109. Stapp, P. R., *J. Org. Chem.* **1970**, *35*, 2419.

110. Keh, C. C. K., Namboodiri, V. V., Varma, R. S., Li, C.-J., *Tetrahedron Lett.* **2002**, *43*, 4993.

111. Keh, C. C. K., Li, C.-J., *Green Chemistry*, **2003**, *5*, 80.

112. Cho, Y. S., Kim, H. Y., Cha, J. H., Pae, A. N., Koh, H. Y., Choi, J. H., Chang, M. H., *Org. Lett.* **2002**, *4*, 2025.

113. Aubele, D. L., Lee, C. A., Floreancig, P. E., *Org. Lett.* **2003**, *5*, 4521.

114. Agami, C., Couty, F., Poursoulis, M., Vaissermann, J., *Tetrahedron* **1992**, *48*, 431.

115. Takebayashi, M., Shingaki, T., *Bull. Chem. Soc. Jap.* **1953**, *26* 137; Qiu, J., Charleux, B., Matyjaszewski, K., *Polimery*, **2001**, *46*, 663; Qiu, J., Charleux, B., Matyjaszewski, K., *Polimery*, **2001**, *46*, 575.

116. For examples, see: Schornick, G., Kistenmacher, A., Ritter, H., Jeromin, J., Noll, O., Born, M. (BASF A. G., Germany), Ger. Offen. (1997), 7 pp. Appl. DE 95–19533269 19950908; Abusleme, J. A., Guarda, P. A., De Pasquale, R. J. (Ausimont S.p.A., Italy), Eur. Pat. Appl. (1995), 10 pp. Appl. EP 94–116994 19941027; Takehisa, M., Senrui, S. (Japan Atomic Energy Research Institute), U.S. (1973), 4 pp. Appl. US 70–36682 19700512.

117. Opstal, T., Verpoort, F., *New J. Chem.* **2003**, *27*, 257.

118. Yorimitsu, H., Wakabayashi, K., Shinokubo, H., Oshima, K., *Tetrahedron Lett.* **1999**, *40*, 519; Yorimitsu, H., Wakabayashi, K., Shinokubo, H., Oshima, K., *Bull. Chem. Soc. Jpn* **2001**, *74*, 1963.

119. Sanchez, V., Greiner, J., *Tetrahedron Lett.* **1993**, *34*, 2931; Huang, B., Liu, J., *Chin. J. Chem.* **1990**, *4*, 358; Huang, W., Zhuang, J., *Chin. J. Chem.* **1991**, *9*, 270.

120. Broxterman, Q. B., Kaptein, B., Kamphuis, J., Schoemaker, H. E., *J. Org. Chem.* **1992**, *57*, 6286.

121. Yorimitsu, H., Nakamura, T., Shinokubo, H., Oshima, K., Omoto, K., Fujimoto, H., *J. Am. Chem. Soc.* **2000**, *122*, 11041.

122. Yoshifuji, M., Tagawa, J., Inamoto, N., *Tetrahedron Lett.* **1979**, *26*, 2415; Bethell, D., Newall, A. R., Stevens, G., Whittaker, D., *J. Chem. Soc. B: Phys. Org.* **1969**, *6*, 749.

123. Keilbaugh, S. A., Thornton, E. R., *Biochemistry* **1983**, *22*, 5063; Regitz, M., Rueter, J., *Chem. Ber.* **1969**, *102*, 3877; Zeller, K. P., *Angew. Chem.* **1977**, *89*, 827; Sanger, M., Borle, F., Heller, M., Sigrist, H., *Bioconj. Chem.* **1992**, *3*, 308.

124. For an example, see: Dehmlow, E. V., Lissel, M., *Tetrahedron Lett.* **1976**, *21*, 1783.

125. Amyes, T. L., Diver, S. T., Richard, J. P., Rivas, F. M., Toth, K., *J. Am. Chem. Soc.* **2004**, *126*, 4366.

126. Grabner, G., Richard, C., Koehler, G., *J. Am. Chem. Soc.* **1994**, *116*, 11470; Liu, M. T. H., Romashin, Y. N., Bonneau, R., *Int. J. Chem. Kinet.* **1994**, 26, 1179; Durand, A.-P., Brown, R. G., Worrall, D., Wilkinson, F., *J. Photochem. Photobio., A: Chem.* **1996**, *96*, 35; Bonnichon, F., Grabner, G., Guyot, G., Richard, C., *J. Chem. Soc. Perkin Trans. 2: Phys. Org. Chem.* **1999**, 1203; Chiang, Y., Kresge, A. J., Schepp, N. P., Xie, R.-Q., *J. Org. Chem.* **2000**, *65*, 1175; Othmen, K., Boule, P., Szczepanik, B., Rotkiewicz, K., Grabner, G., *J. Phys. Chem. A* **2000**, *104*, 9525.

127. Fedurco, M., Sartoretti, C. J., Augustynski, J., *Langmuir* **2001**, *17*, 2380.

128. Gonzalez, C., Restrepo-Cossio, A., Marquez, M., Wiberg, K. B., *J. Am. Chem. Soc.* **1996**, *118*, 5408.

129. Pliego, J. R. Jr., De Almeida, W. B., *J. Phys. Chem.* **1996**, *100*, 12410.

130. Bernasconi, C. F, Ruddat, V., *J. Am. Chem. Soc.* **2002**, *124*, 14968 and refs cited therein.

131. Ciardi, C., Reginato, G., Gonsalvi, L., de Rios, I., Romerosa, A., Peruzzini, M., *Organometallics* **2004**, *23*, 2020; Saoud, M., Romerosa, A., Peruzzini, M., *Organometallics* **2000**, *19*, 4005.

132. *Handbook of Metathesis*, Grubbs, R. H., ed. John Wiley & Sons, New York, 2003.

133. Chen, J., Li, D., Yu, Y., Jin, Z., Zhou, Q., Wei, G., *Organometallics* **1993**, *12*, 3885.

134. For a representative review, see: Gstöttmayr, C. W. K., Böhm, V. P. W., Herdtweck, E., Grosche, M., Herrmann, W. A., *Angew. Chem., Int. Ed.* **2002**, *41*, 1363.

135. Iwasa, S., Takezawa, F., Tuchiya, Y., Nishiyama, H., *Chem. Commun.* **2001**, 59; Wurz, R. P., Charette, A. B., *Org. Lett.* **2002**, *4*, 4531.

136. Dehmlow, E. V., *Tetrahedron* **1971**, *27*, 4071.

137. McQuillin, F. J., Parker, D. G., Stephenson, G. R., *Transition Metal Organometallics for Organic Synthesis.* Cambridge Press, 1991; Crabtree, R. H., *The Organometallic Chemistry of Transition Metals*, 2nd ed. John Wiley & Sons, New York, 1994.

138. McGrath, D. V., Grubbs, R. H., *Organometallics*, **1994**, *13*, 224.

139. See Ref. 203; see also: Cadierno, V., Garcia-Garrido, S. E., Gimeno, J., *Chem. Commun.* **2004**, 232.

140. Bricout, H., Mortreux, A., Monflier, E., *J. Organomet. Chem.* **1998**, *553*, 469.

141. For an example, see: Huttel, R., Bechter, M., *Angew. Chem.* **1959**, *71*, 456.

142. For reviews, see: (a) Mecking, S., Held, A., Bauers, F. M., *Angew. Chem., Int. Ed.* **2002**, *41*, 544. (b) Mulhaupt, R., *Macromol. Chem. Phys.* **2003**, *204*, 289. (c) Göttker-Schentmann, I., Korthals, B., Mecking, S., *J. Am. Chem. Soc.* **2006**, *128*, 7708.

143. For reviews, see: Heck, R. F., *Acc. Chem. Res.* **1979**, *12*, 146.

144. Reardon, P., Metts, S., Crittendon, C., Daugherity, P., Parsons, E. J., *Organometallics*, **1995**, *14*, 3810.

145. Diminnie, J., Metts, S., Parsons, E. J., *Organometallics* **1995**, *14*, 4023.

146. Kikukawa, K., Nagira, K., Wada, F., Matsuda, T., *Tetrahedron* **1981**, *37*, 31.

147. Yokoyama, Y., Hikawa, H., Mitsuhashi, M., Uyama, A., Murakami, Y., *Tetrahedron Lett.* **1999**, *40*, 7803.

148. Hayashi, M., Amano, K., Tsukada, K., Lamberth, C., *J. Chem. Soc., Perkin Trans. 1*, **1999**, 239.

149. Hessler, A., Stelzer, O., Dibowski, H., Worm, K., Schmidtchen, F. P., *J. Org. Chem.* **1997**, *62*, 2362.

150. Gron, L. U., Tinsley, A. S., *Tetrahedron Lett.* **1999**, *40*, 227.

151. Choi, C.-K., Tomita, I., Endo, T., *Chem. Lett.* **1999**, 1253.

152. Lautens, M., Roy, A., Fukuoka, K., Fagnou, K., Martin-Matute, B., *J. Am. Chem. Soc.* **2001**, *123*, 5358.

153. Amengual, R., Michelet, V., Genet, J.-P., *Tetrahedron Lett.* **2002**, *43*, 5905.

154. Park, S., Kim, M., Koo, D. H., Chang, S., *Adv. Synth. Cat.* **2004**, *346*, 1638.

155. For reviews on hydrovinylation of alkenes, see: RajanBabu, T. V., *Chem. Rev.* **2003**, *103*, 2845; Goossen, L. J., *Angew. Chem., Int. Ed.* **2002**, *41*, 3775.

156. RajanBabu, T. V., 225th ACS National Meeting, New Orleans, LA, United States, March 23–27, 2003.

157. Nizova, G. V., Shul'pin, G., *Izv. Akad. Nauk SSSR, Ser. Khim.* **1981**, 1436.

158. Matsumoto, T., Periana, R. A., Taube, D. J., Yoshida, H., *J. Mol. Cat. A; Chem.* **2002**, *180*, 1.

159. Cornils, B., *Hydroformylation, Oxo Synthesis, Roelen Reaction: New Synthesis with Carbon Monoxide*. Springer-Verlag, Berlin, Heidelberg, New York, 1980, p. 1.

160. Roelen, D., D. E. 84584, 1938; Ruhrchemie.

161. Kunz, E. G., *Chemtech* **1987**, 570; Herrmann, W. A., Kohlpaintner, C. W., *Angew. Chem. Int. Ed. Eng.* **1993**, *32*, 1524.

162. Verspui, G., Elbertse, G., Papadogianakis, G., Sheldon, R. A., *J. Orgnomet. Chem.* **2001**, *621*, 337.

163. Borowski, A. F., Cole-Hamilton, D. J., Wilkinson, G., *Nouv. J. Chim.* **1978**, *2*, 137.

164. Smith, R. T., Ungar, R. K., Baird, M. C., *Trans. Met. Chem.* **1982**, *7*, 288.

165. Bahrmann, H., Bach, H., *Phosphorus Sulfur* **1987**, *30*, 611.

166. Kuntz, E., U.S. Patent 4,248,802, Rhone-Poulenc Ind., 1981; *Chem. Abstr.* **1977**, *87*, 101944n.

167. Jenck, J., Fr. Patent 2,478,078 to Rhone-Poulenc Industries (03-12-1980); Kuntz, E., Fr. Patent 2,349,562 to Rhone-Poulenc Industries (04-29-1976).

168. Herrmann, W. A., Kohlpainter, C. W., Manetsberger, R. B., Bahrmann, H. (Hoechst AG), DE-B 4220,267A, **1992**.

169. Kalck, P., Escaffre, P., Serein-Spirau, F., Thorez, A., Besson, B., Colleuille, Y., Perron, R., *New. J. Chem.* **1988**, *12*, 687.

170. Smith, R. T., Ungar, R. K., Sanderson, L. J., Baird, M. C., *Organometallics* **1983**, *2*, 1138.

171. Russell, M. J. H., Murrer, B. A., U.S. Patent 4,399,312 to Johnson Matthey Company (08-27-1981); Russell, M. J. H., *Platinum Met. Rev.* **1988**, *32*, 179.

172. Fell, B., Pagadogianakis, G., *J. Mol. Catal.* **1991**, *66*, 143.

173. Monflier, E., Fremy, G., Castanet, Y., Mortreux, A., *Angew. Chem., Int. Ed. Engl.* **1995**, *34*, 2269.

174. Anderson, J. R., Campi, E. M., Jackson, W. R., *Catal. Lett.* **1991**, *9*, 55.

175. Arhancet, J. P., Davis, M. E., Merola, J. S., Hanson, B. E., *Nature* **1989**, *339*, 454; Haggin, J. *Chem. Eng. News*, **1992**, *70(17)*, 40; Herrmann, W. A., *Hoechst High Chem. Magazine* **1992**, (13), 14.

176. Khna, M. M. T., Halligudi, S. B., Abdi, S. H. R., *J. Mol. Catal.* **1988**, *48*, 313.

177. Jenner, G., *Tetrahedron Lett.* **1991**, *32*, 505.

178. Zimmermann, B., Herwig, J., Beller, M., *Angew. Chem., Int. Ed. Engl.* **1999**, *38*, 2372.

179. Wang, Y. Y., Luo, M. M., Li, Y. Z., Chen, H., Li, X. J., *Appl. Catal. A: Gen.* **2004**, *272*, 151.

180. Trost, B. M., Indolese, A. F., Müller, T. J. J., Treptow, B., *J. Am. Chem. Soc.* **1995**, *117*, 615.

181. López, F., Castedo, L., Mascareñas, J. L., *J. Am. Chem. Soc.* **2002**, *124*, 4218.

182. Trost, B. M., Martinez, J. A., Kulawiec, R. J., Indolese, A. F., *J. Am. Chem. Soc.* **1993**, *115*, 10402; Dérien, S., Jan, D., Dixneuf, P. H., *Tetrahedron* **1996**, *52*, 5511.

183. Trost, B. M., Indolese, A., *J. Am. Chem. Soc.* **1993**, *115*, 4361.

184. Lindner, E., Schmid, M., Wald, J., Queisser, J. A., Geprags, M., Wegner, P., Nachtigal, C., *J. Organometal. Chem.* **2000**, *602*, 173.

185. Lubineau, A., Bouchain, G., *Tetrahedron Lett.* **1997**, *38*, 8031.

186. Grubbs R. H., Pine S. H., in *Comprehensive Organic Synthesis*, Trost, B. M., Fleming, I., Paquette, L. A., eds. Pergamon, New York, 1991, Chapter 9.3, p. 1115; Schrock, P. R., in *The Strem Chemiker*, Vol XIV, 1992, p. 1; Ivin, K. J., Mol, J. C., *Olefin Metathesis and Metathesis Polymerization*. Academic. San Diego, 1997; Connon, S. J., Blechert, S., *Angew. Chem., Int. Ed.* **2003**, *42*, 1900.

187. Bazan, G. C., Oskam, J. H., Cho, H. N., Park, L. Y., Schrock, R. R., *J. Am. Chem. Soc.* **1991**, *113*, 6899.

188. Grubbs, R. H., Chang, S., *Tetrahedron*, **1998**, *54*, 4413.

189. Claverie, J. P., Soula, R., *Progr. Polym. Sci.*, **2003**, *28*, 619; Lynn, D. M., Mohr, B., Grubbs, R. H., Henling, L. M., Day, M. W., *J. Am. Chem. Soc.* **2000**, *122*, 6601; Grubbs, R. J., Lynn D. M., in *Aqueous-Phase Organometallic Catalysis*, Cornils, B., Herrmann, W. A. eds. Wiley-VCH, Weinheim, 1998, 466; Pariya, C., Jayaprakash, K. N., Sarkar, A., *Coord. Chem. Rev.* **1998**, *168*, 1; Mohr, B., Lynn, D. M., Grubbs, R. H., *Organometallics*, **1996**, *15*, 4317; France, M. B., Grubbs, R. H., McGrath, D. V., Paciello, R. A. *Macromolecules*, **1993**, *26*, 4742; Feast, W. J., Harrison, D. B., *J. Mol. Cat.* **1991**, *65*, 63.

190. Lynn, D. M., Mohr, B., Grubbs, R. H., *J. Am. Chem. Soc.* **1998**, *120*, 1627; Wagaman, M. W., Grubbs, R. H., *Macromolecules*, **1997**, *30*, 3978; Lynn, D. M., Kanaoka, S., Grubbs, R. H., *J. Am. Chem. Soc.* **1996**, *118*, 784; Nguyen, S. T., Johnson, L. K., Grubbs, R. H., Ziller, J. W., *J. Am. Chem. Soc.* **1992**, *114*, 3974; Lynn, D. M., Dias, E. L., Grubbs, R. H., Mohr, B., PCT Int. Appl. 1999 WO 9922865; Grubbs, R. H.,

Marsella, M. J., Maynard, H. D., PCT Int. Appl. 1998 WO 9830557; For a review on ROPM, see: Hafner, A., van der Schaaf, P. A., Muhlebach, A., Bernhard, P., Schaedeli, U., Karlen, T., Ludi, A., *Progr. Org. Coat.* **1997**, *32*, 89.

191. Bissinger, P., *Eur. Pat. Appl.* **1999**, 15 pp. EP 904767.

192. Novak, B., Grubbs, R. H., *J. Am. Chem. Soc.* **1988**, *110*, 7542.

193. Lynn, D. M., Kanaoka, S., Grubbs, R. H., *J. Am. Chem. Soc.* **1996**, *118*, 784.

194. Lynn, D. M., Mohr, B., Grubbs, R. H., *J. Am. Chem. Soc.* **1998**, *120*, 1627; Lynn, D. M., Mohr, B., Grubbs, R. H., Henling, L. M., Day, M. W., *J. Am. Chem. Soc.* **2000**, *122*, 6601.

195. Mortell, K. H., Weatherman, R. V., Kiessling, L. L., *J. Am. Chem. Soc.* **1996**, *118*, 2297; Kanai, M., Mortell, K. H., Kiessling, L. L., *J. Am. Chem. Soc.* **1997**, *119*, 9931; Manning, D. D., Hu, X., Beck, P., Kiessling, L. L., *J. Am. Chem. Soc.* **1997**, *119*, 3161; Manning, D. D., Strong, L. E., Hu, X., Beck, P., Kiessling, L. L., *Tetrahedron*, **1997**, *53*, 11937.

196. Kirkland, T. A., Lynn, D. M., Grubbs, R. H., *J. Org. Chem.* **1998**, *63*, 9904.

197. Rölle, T, Grubbs, R. H., *Chem. Commun.* **2002**, 1070.

198. Mendez-Andino, J., Paquette, L. A., *Adv. Synth. Catal.* **2002**, *344*, 303.

199. Audouard, C., Fawcett, J., Griffiths, G. A., Percy, J. M., Pintat, S., Smith, C. A., *Org. Biomol. Chem.*, **2004**, *2*, 528.

200. Connon, S. J., Blechert, S., *Bioorg. Med. Chem. Lett.* **2002**, *12*, 1873.

201. Le Bras, J., Muzart, J., *Tetrahedron Lett.* **2002**, *43*, 431.

202. Le Bras, J., Muzart, J., *Tetrahedron : Asymmetry* **2003**, *14*, 1911.

203. Li, C. -J., Wang, D., Chen, D. L., *J. Am. Chem. Soc.* **1995**, *117*, 12867.

204. Wang, M., Li, C. -J., *Tetrahedron Lett.* **2002**, *43*, 3589.

205. Wang, M., Yang, X.-F., Li, C.-J., *Eur. J. Org. Chem.* **2003**, 998.

CHAPTER 4

ALKYNES

Alkynes have three general reaction types in organic chemistry: the reaction of the terminal C–H bond, the reaction of the C–C triple bond, and reactions at the propargyl position and beyond. This chapter mostly describes the first two types of reactions.

4.1 REACTION OF TERMINAL ALKYNES

Because of the slightly acidic nature of the sp C–H bonds, the reaction of metal acetylides with various electrophiles is one of the most general strategies in organic transformations.[1] Traditionally, such reactions are carried out by using alkali metal acetylides which are air and water sensitive. On the other hand, there is much interest in developing transition-metal catalyzed terminal alkyne reactions involving soft and more stable C–M bonds as reaction intermediates, because many such reactions can tolerate water.

4.1.1 Alkyne Oxidative Dimerization

The dimerization of terminal alkynes, known as the Glaser coupling, the Eglinton coupling, and the Cadiot-Chodkiewicz coupling, is one

Comprehensive Organic Reactions in Aqueous Media, Second Edition, by Chao-Jun Li and Tak-Hang Chan
Copyright © 2007 John Wiley & Sons, Inc.

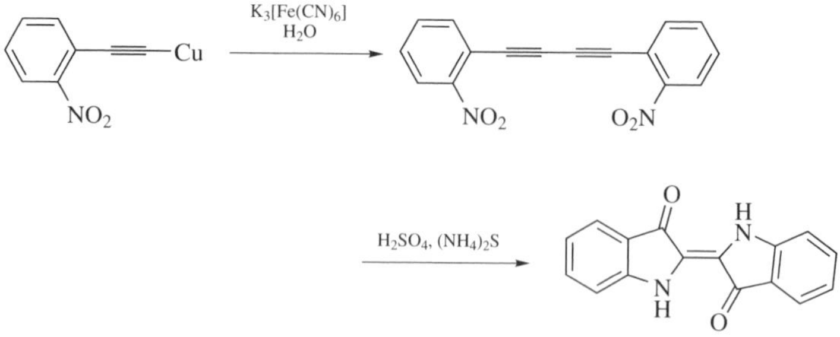

Scheme 4.1

of the oldest reactions in organic chemistry. The reaction was initially discovered in 1869 by Carl Glaser with the oxidative formation of diyne from copper(I) phenylacetylide and air.[2] Subsequently, extensive modifications have been made on the reaction. The Glaser coupling in water dated back at least to 1882, when Baeyer synthesized indigo using potassium ferricyanide as the oxidant (Scheme 4.1).[3]

The Glaser coupling has been used extensively in the synthesis of advanced organic materials.[4] For example, water-soluble conjugated rotaxanes and "naked" molecular dumbbells were synthesized from such couplings in water using hydrophobic interaction to direct rotaxane formation (Scheme 4.2).[5] Although such couplings are mainly catalyzed by copper, other transition-metal catalysts are also effective. Examples include a water-soluble palladium/TPPTS-catalyzed coupling, affording diynes in moderate yields (Eq. 4.1).[6] Cu(I)Cl-catalyzed Glaser coupling reaction of terminal alkynes without bases under near-critical water has also been reported recently.[7]

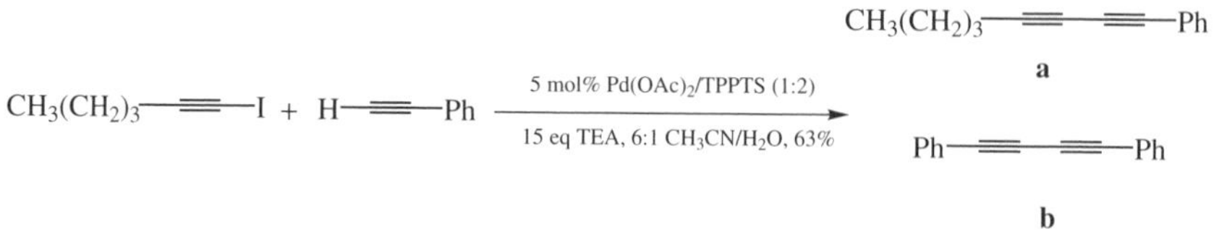

$$(4.1)$$

4.1.2   Alkyne-Alkyne Addition

The additive coupling of terminal alkynes catalyzed by copper is a classical reaction known as the Strauss coupling. Unfortunately, under the

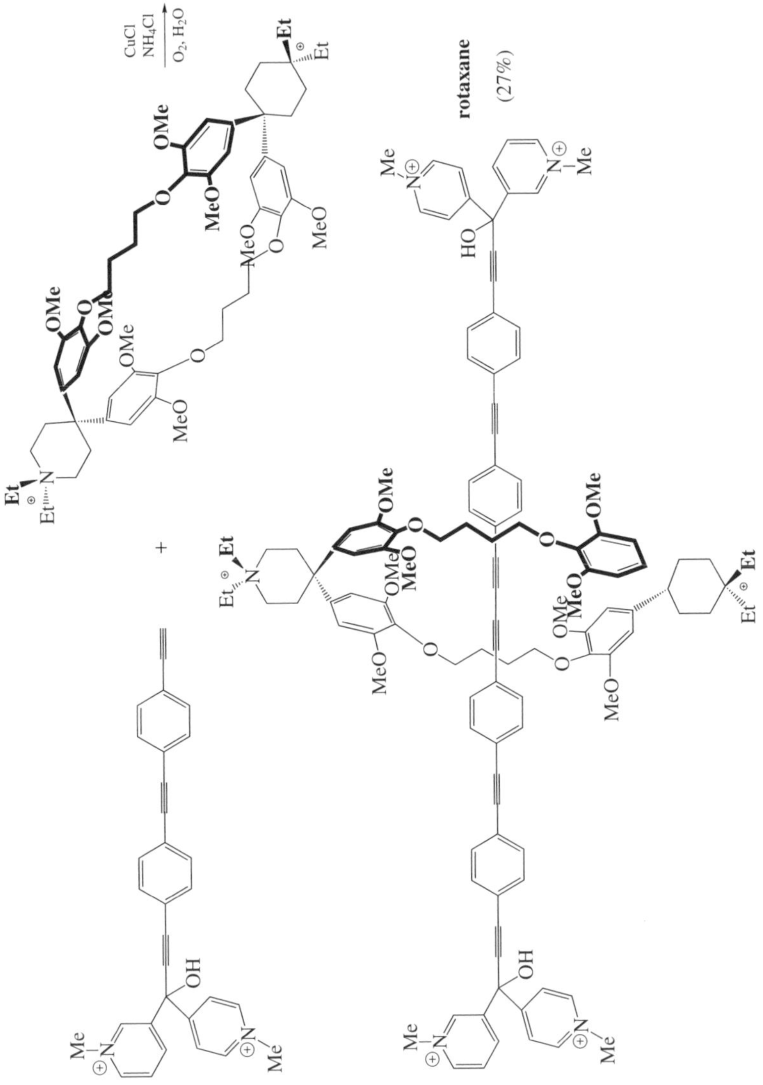

Scheme 4.2

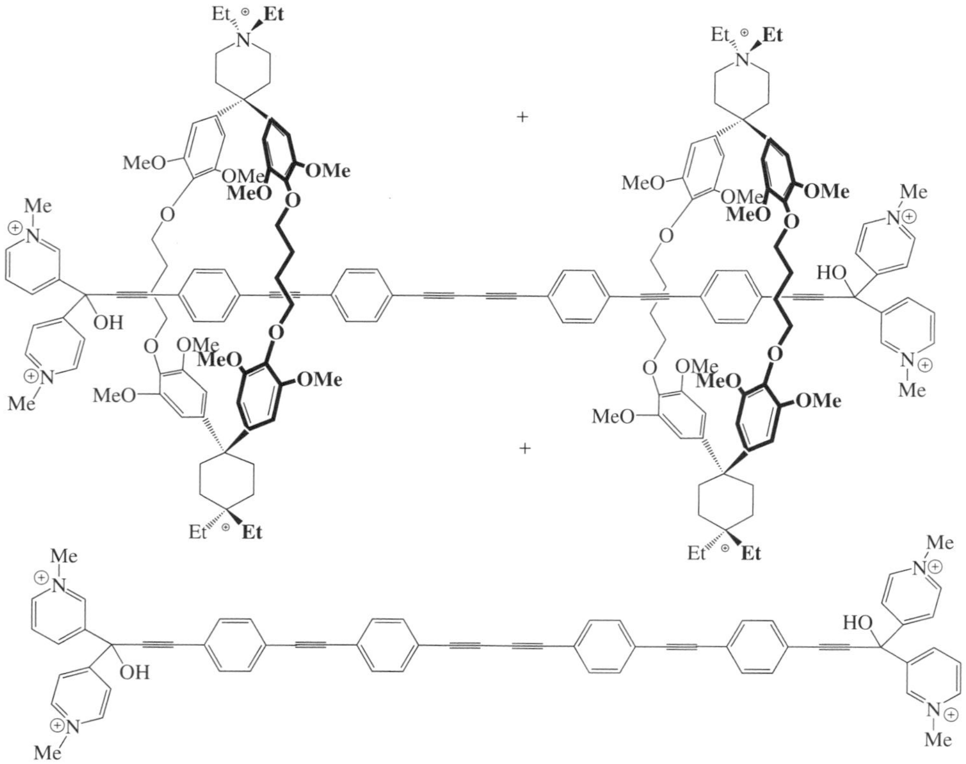

rotaxane

(30%)

naked dumbbell

(9%)

Scheme 4.2 *(continued)*

classical conditions, a mixture of regioisomers was obtained.[8] Recently, a rhodium-catalyzed homopolymerization and oligomerization of alkynes has been reported.[9] A major development to overcome this limitation was made by Trost using a catalytic amount of Pd(OAc)$_2$ and triphenylphosphine in dichloroethane. This resulted in a high yield of the homocoupling of terminal alkynes.[10] Recently, Li reported a simple and convenient copper/palladium-catalyzed addition of terminal alkynes to activated alkynes in water (Eq. 4.2).[11] The reaction gave the *cis* adduct regio- and stereoselectively. Importantly, the reaction proceeded better in water than in toluene.

$$R\text{====}H \;+\; R'\text{====}EWG \quad \xrightarrow[\text{water/60°C or r.t.}]{\substack{5\text{ mol\% CuBr/} \\ 2.5\text{ mol\% PdCl}_2(\text{Ph}_3\text{P})_2}} \quad$$

R = Ph; TMS; n-C$_4$H$_9$;

R' = Ph; Me
EWG = COMe;
CO$_2$Me

yield: 54–88%

Cl

(4.2)

It is interesting to note that a copper(II)-mediated coupling reaction of alkenyldialkyl- or trialkylboranes with alkynylcopper compounds, generated in situ, in the presence of various solvents and a small amount of water, gives (E)-1,3-enynes (or disubstituted alkynes) with various functional groups in reasonable yields (Eq. 4.3).[12]

$$Bu\text{====} \quad \xrightarrow[\text{THF }-15°C \text{ then } 0°C]{Bu_2BH} \quad \xrightarrow[\text{2. Cu(OAc)}_2, \text{ Cu(NO}_3)_2 \text{ 3H}_2\text{O, THF, 85\%}]{\substack{1.\; Bu\text{====}\;CuI, \text{ pyridine, DMA} \\ 0°C \text{ then } 20°C}}$$

BBu$_2$

(4.3)

4.1.3 Reaction of Alkynes with Organic Halides

The coupling of terminal alkynes with organic halides, known as the Castro-Stephens-Sonogashira reaction, has wide applications in synthesis. The most widely used method is the Sonogashira coupling, using a combination of palladium and copper as the catalyst.[13] Recently, the reaction was investigated extensively in aqueous media, in some cases without using copper as the co-catalyst. For example, by using a water-soluble palladium complex as the catalyst, unprotected nucleosides, nucleotides, and amino acids underwent coupling with terminal acetylenes in aqueous acetonitrile.[14] Compound T-505, part of a family of chain-terminating nucleotide reagents used in DNA sequencing and labeling, was synthesized by this route in 50% yield (Eq. 4.4).

5-iododideoxyuridine
5'-triphosphate

$$\text{(4.4)}$$

Genet[15] and Beletskaya[16] carried out more detailed studies of the aqueous reaction recently. A variety of aryl and vinyl iodides were coupled with terminal acetylenes in aqueous medium with either a water-soluble catalyst or non-water-soluble catalyst. The reaction can also be carried out without the use of a phosphine ligand in a water-alcohol emulsion in the presence of cetyltrimethylammonium bromide (CTAB) as an emulsifier.[17] Diynes were prepared similarly by reaction of alkynyl bromide with terminal alkynes.[18] Pd(OAc)$_2$/TPPTS has been used as an efficient catalyst in a sequential two-step reaction for the coupling of 2-iodoaniline or 2-iodophenol with terminal alkynes to give the corresponding indoles or benzofurans in good yield (Eq. 4.5).[19] This methodology can tolerate a number of different functional groups and does not require the use of a phase transfer catalyst or water-soluble phosphine ligands and can be performed directly in water.

$$\text{(4.5)}$$

Li et al. reported a highly efficient coupling of acetylene gas with aryl halides in a mixture of acetonitrile and water (Eq. 4.6).[20] The conditions are generally milder and the yields are better than previously reported results in organic solvents. A variety of aromatic halides are coupled to give the corresponding *bis*-arylacetylenes. Both a water-soluble palladium catalyst (generated in situ from Pd(OAc)$_2$/TPPTS) and water-insoluble catalyst (generated in situ from Pd(OAc)$_2$/PPh$_3$) can be used for the reaction.

$$(4.6)$$

The reaction can be carried out in water alone. The reaction proceeds equally well with or without CuI as a co-catalyst. By using the palladium-catalyzed coupling between aryl halides with acetylene gas, a variety of poly(arene ethynylene)s were prepared by Li et al.[21] from aryl diiodides. Copolymerization of 3,5-diiodobenzoic acid with acetylene gas in a basic aqueous medium gave a high-molecular-weight (~60 000) phenylethynylene polymer (Eq. 4.7).[22] The polymer has a high thermostability, is soluble in basic solutions, and could be switched from soluble to hydrogel states reversibly by changing the pH in water (Eq. 4.8).

$$(4.7)$$

hydrogel   soluble

$$(4.8)$$

Alkenyl and aryl iodonium salts have also been coupled with terminal alkynes. The reaction of (E)-[β-(trifluoromethanesulfonyloxy)-1-alkenyl](phenyl)iodonium trifluoromethanesulfonate with terminal alkynes in the presence of catalytic amounts of dichloro(triphenylphosphine)palladium(II) and CuI in aqueous medium proceeds stereospecifically to give the corresponding enynes in good yields (Eq. 4.9).[23] Ni(PPh$_3$)$_2$Cl$_2$/CuI was also reported as the catalyst for such couplings in water.[24]

$$
\underset{R^1}{\overset{TfO}{\diagdown}} \diagup \overset{H}{\underset{I^+Ph\ ^-OTf}{}} \quad + \quad R^2 \!-\!\!\!\equiv\!\!\!-\! H \quad \xrightarrow[\text{K}_2\text{CO}_3,\ \text{DMF/H}_2\text{O},\ \text{Et}_3\text{N}]{\text{PdCl}_2(\text{PPh}_3)_2,\ \text{CuI}} \quad \underset{R^1}{\overset{TfO}{\diagdown}} \diagup \overset{H}{\underset{R^2}{}}
$$

$$(4.9)$$

A potentially interesting development is the microwave-assisted transition-metal-free Sonogashira-type coupling reaction (Eq. 4.10). The reactions were performed in water without the use of copper(I) or a transition metal-phosphine complex. A variety of different aryl and hetero-aryl halides were reactive in water.[25a] The amount of palladium or copper present in the reaction system was determined to be less than 1 ppm by AAS-MS technique. However, in view of the recent reassessment of a similarly claimed transition-metal-free Suzuki-type coupling reaction, the possibility of a sub-ppm level of palladium contaminants found in commercially available sodium carbonate needs to be ruled out by a more sensitive analytical method.[25b]

$$
\underset{\text{MeOC}}{}\!\!\!\diagdown\!\!\!\diagup\!\!\!\diagdown\!\!\!\!-\!\text{Br} \quad + \quad \xrightarrow[\text{H}_2\text{O, MW, 15 min, 85\%}]{\text{TBAB, Na}_2\text{CO}_3,} \quad \underset{\text{MeOC}}{}
$$

$$(4.10)$$

A highly effective direct coupling of acid chlorides with terminal alkynes catalyzed by PdCl$_2$(PPh$_3$)$_2$/CuI together with a catalytic amount of sodium laurylsulfate as the surfactant and K$_2$CO$_3$ as the base provided ynones in high yields in water (Eq. 4.11). No reaction was observed when either Cu(I) or Pd(II) alone was used as the catalyst. The use of surfactant is also critical for the success of the reaction; without a surfactant/phase transfer reagent the yield dropped from 98% to 9%.[26]

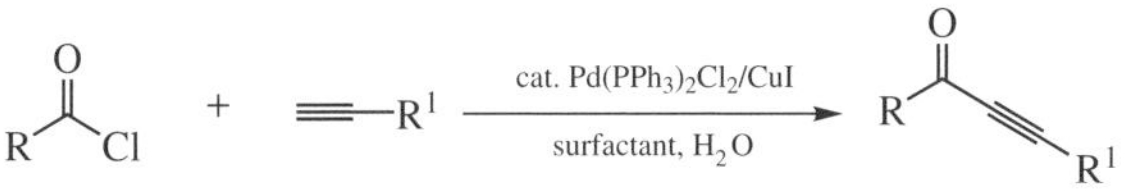

$$(4.11)$$

Various polymer-bound palladium catalysts have been used in the Sonogashira reaction in water. Bergbreiter et al. developed an effective polymer-bound Pd(0)-phosphine catalyst for the coupling of terminal alkynes with aryl halides in CH_3CN/H_2O under mild conditions. The polymer-bound Pd(0)-phosphine catalyst was made by ligand exchange of Pd(0)(dba)$_2$ with a phosphine ligand that is based on the water-soluble polymer poly(N-isopropyl)acrylamide (PNIPAM)[27] (Figure 4.1). Later, Xia et al. reported a polyethylene glycol supported palladium-catalyzed Sonogashira reaction.[28] However, recently, Quignard et al. have found that palladium catalysts immobilized in an aqueous film supported on mesoporous silica may leach.[29] After that, Uozumi et al. explored an efficient copper-free Sonogashira reaction in water catalyzed by an amphiphilic resin-supported palladium-phosphine complex (Figure 4.2), which was prepared by using polystyrene-poly(ethylene glycol) graft copolymer (1% DVB cross-linked) beads (PS-PEG).[30]

Aqueous medium is necessary for most biological processes where substrate anions usually are necessarily formed by the action of base. Schmidtchen et al. found that a palladium catalyst with a water-soluble guanidino phosphine **4.1** as a cationic ligand is more effective than $[P(m\text{-}C_6H_4SO_3Na)]_3$ (TPPTS) for the coupling reaction between anionic substrates in aqueous media (Eq. 4.12).[31] The method was used in the regioselective C–C coupling of free and unprotected peptides.[32] For example, the coupling of an unprotected Pro(p-I-Phe)-bradykinin

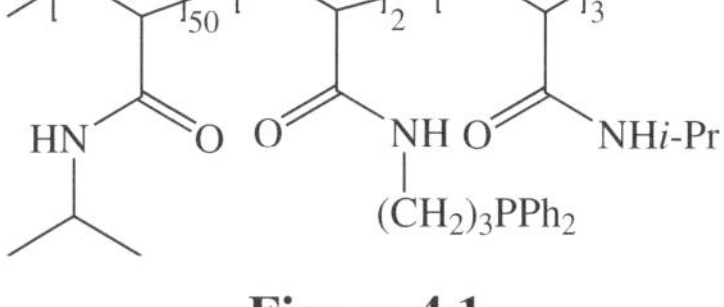

Figure 4.1

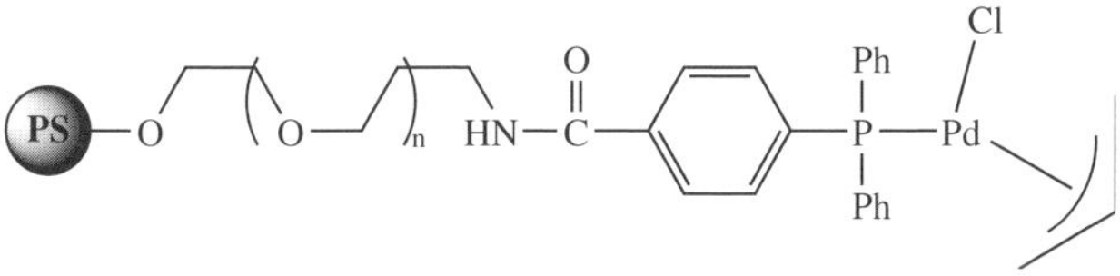

Figure 4.2

with propiolic acid in aqueous 3-[*tris*(hydroxymethyl)methylamino]-1-propanesulfonic acid (TAPS) buffer (pH 8.3) at 35°C gave 91% yield of the target product after 3 h (Eq. 4.13). The coupling of biotinylglutamoylpropargylamide, a water-soluble biotin derivative, with Pro(*p*-I-Phe)-bradykinin in aqueous TAPS buffer (pH 8.3) at 35°C gave 75% yield of target product after 4 h (Eq. 4.14). It has been shown that the specific protection for protein was not necessary.

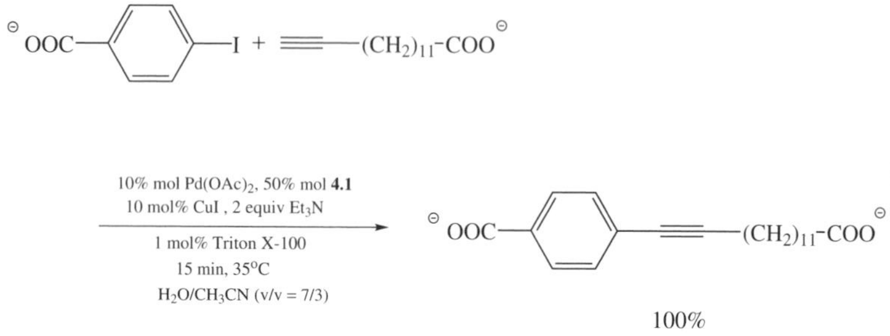

$$^{\ominus}OOC \overbracket{} I + \equiv\!\!\!-(CH_2)_{11}\text{-}COO^{\ominus}$$

(4.12)

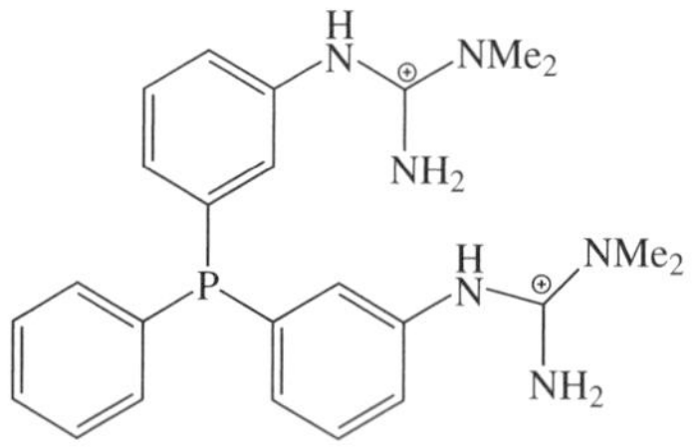

10% mol Pd(OAc)$_2$, 50% mol **4.1**
10 mol% CuI , 2 equiv Et$_3$N
———————————————→
1 mol% Triton X-100
15 min, 35°C
H$_2$O/CH$_3$CN (v/v = 7/3)

$$^{\ominus}OOC \overbracket{} \equiv\!\!\!-(CH_2)_{11}\text{-}COO^{\ominus}$$

100%

4.1

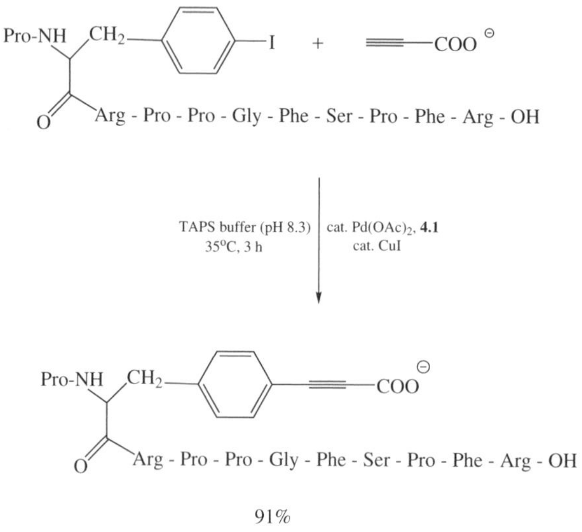

Pro-NH ⟍ CH$_2$— ⬡ —I   +   ≡—COO$^{\ominus}$
       |
       O⟍ Arg - Pro - Pro - Gly - Phe - Ser - Pro - Phe - Arg - OH

TAPS buffer (pH 8.3) │ cat. Pd(OAc)$_2$, **4.1**
35°C, 3 h   │ cat. CuI

(4.13)

Pro-NH ⟍ CH$_2$— ⬡ —≡—COO$^{\ominus}$
       |
       O⟍ Arg - Pro - Pro - Gly - Phe - Ser - Pro - Phe - Arg - OH

91%

$$(4.14)$$

Later, Ghadiri et al. reported a tricoupling of highly charged peptides $(-6 \sim +9)$ of considerable length (from 17 to 33 residues) with a trialkyne by using a Pd(0) catalyst under both acidic (pH 5.0) and basic (pH 7.5) conditions in water, giving protein-sized structures (12,000 mol wt) efficiently (Eq. 4.15).[33] These peptides, containing amines, carboxylates, guanidines, hydroxyls, and thiolesters, can react directly without any protection of the functional groups. However, the free thiols, thioethers, and bipyridyl moieties were shown not to be tolerated, perhaps because these functional groups play as competitive metal ligands for palladium as well as copper.

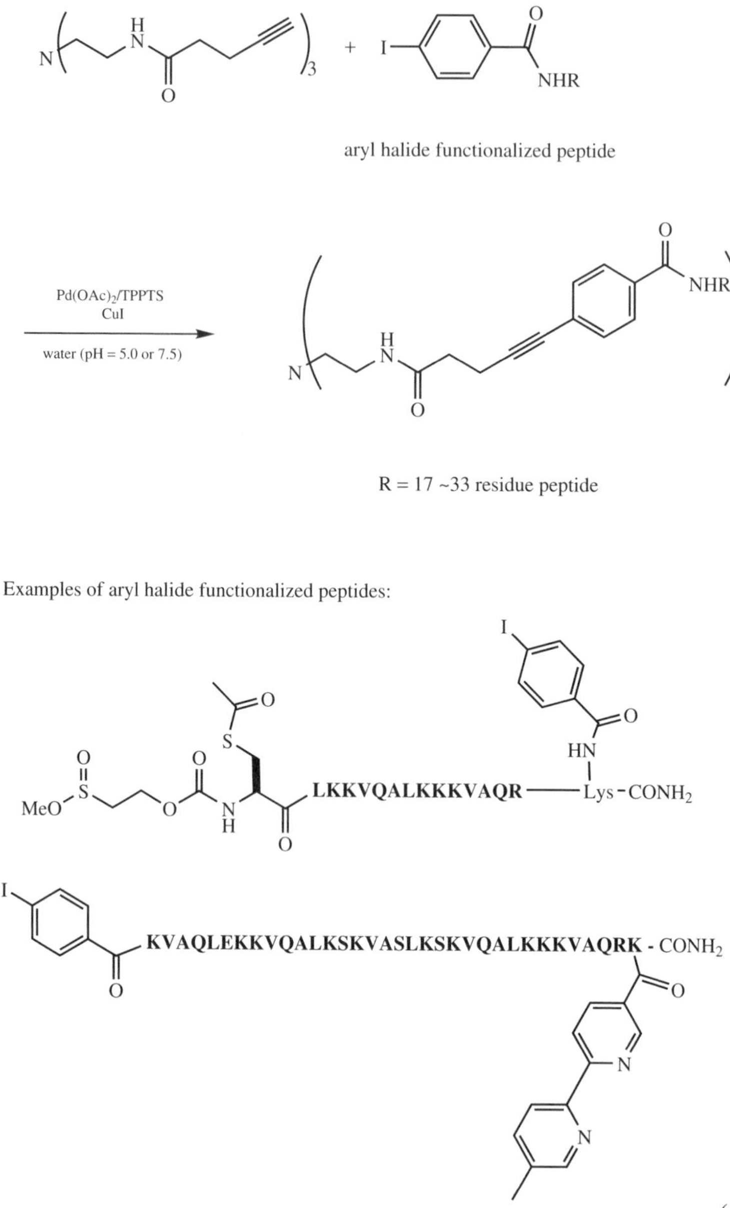

aryl halide functionalized peptide

R = 17 ~33 residue peptide

Examples of aryl halide functionalized peptides:

LKKVQALKKKVAQR——Lys–CONH₂

KVAQLEKKVQALKSKVASLKSKVQALKKKVAQRK - CONH₂

(4.15)

Palladium/charcoal also could serve as a catalyst for Sonogashira reactions of peptides in aqueous media. Recently, Granja et al. used palladium/carbon associated with 4-diphenylphosphinobenzoic acid (4-DPPBA) or triphenylphosphine ligand to catalyze such a reaction in aqueous DMF (Eq. 4.16).[34]

$$(4.16)$$

Kotschy et al. also reported a palladium/charcoal-catalyzed Sono-gashira reaction in aqueous media. In the presence of Pd/C, CuI, PPh$_3$, and i-Pr$_2$NH base, terminal alkynes smoothly reacted with aryl bromides or chlorides, such as 2-pyridyl chloride, 4-methylphenyl bromide, and so on, to give the expected alkyne products in dimethyl-acetamide (DMA)-H$_2$O solvent. Wang et al. reported an efficient cross-coupling of terminal alkynes with aromatic iodides or bromides in the presence of palladium/charcoal, potassium fluoride, cuprous iodide, and triphenylphosphine in aqueous media (THF/H$_2$O, v/v, 3/1) at 60°C.[35] The palladium powder is easily recovered and is effective for six consecutive runs with no significant loss of catalytic activity.

Recently, Pal et al. found that (S)-prolinol could facilitate the coupling reaction of terminal alkynes with 3-iodoflavone under palladium-copper catalysis in aqueous DMF to give 3-alkynyl substituted flavones of potential biological interest (Eq. 4.17). The coupling of iodobenzene with terminal alkynes at room temperature in water without any cosolvent was completed within 30 minutes, affording the desired product in good yield.[36]

$$(4.17)$$

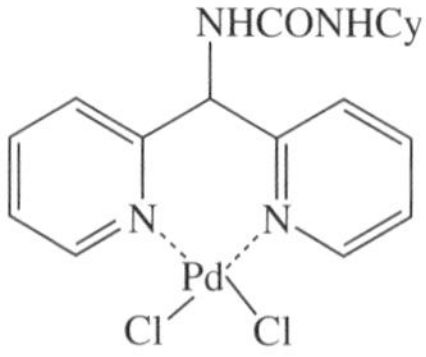

NHCONHCy

4.2

A palladium catalyst with a less electron-rich ligand, 2,2-dipyridyl-methylamine-based palladium complexes (**4.2**), is effective for coupling of aryl iodides or bromides with terminal alkynes in the presence of pyrrolidine and tetrabutylammonium acetate (TBAB) at 100°C in water.[37] However, the reactions were shown to be faster in NMP solvent than in water under the reaction conditions. Palladium-phosphinous acid (POPd) was also reported as an effective catalyst for the Sonogashira cross-coupling reaction of aryl alkynes with aryl iodides, bromides, or chlorides in water (Eq. 4.18).[38]

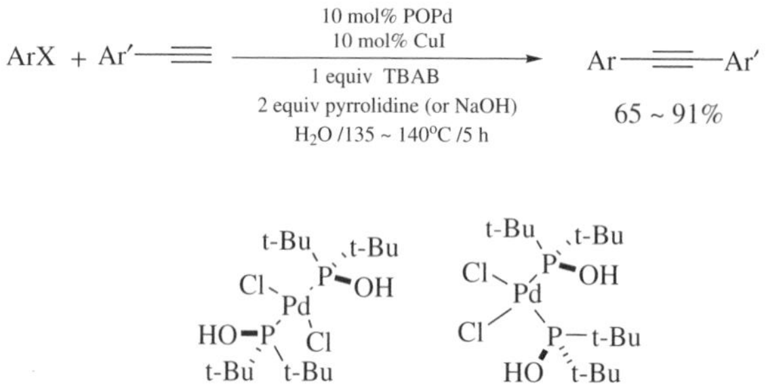

$$ArX + Ar'—\!\!\!\equiv\!\!\!— \xrightarrow[\substack{1 \text{ equiv TBAB} \\ 2 \text{ equiv pyrrolidine (or NaOH)} \\ H_2O /135 \sim 140°C /5 \text{ h}}]{\substack{10 \text{ mol\% POPd} \\ 10 \text{ mol\% CuI}}} \underset{65 \sim 91\%}{Ar—\!\!\!\equiv\!\!\!—Ar'}$$

(4.18)

POPd

Besides palladium catalysts, nickel was also found to be an effective catalyst for the Sonogashira reaction in aqueous media. Recently, Beletskaya et al. reported a $Ni(PPh_3)_2Cl_2$/CuI-catalyzed Sonogashira coupling reaction of terminal acetylenes with aryl iodides in aqueous dioxane in high yields (Eq. 4.19).[39]

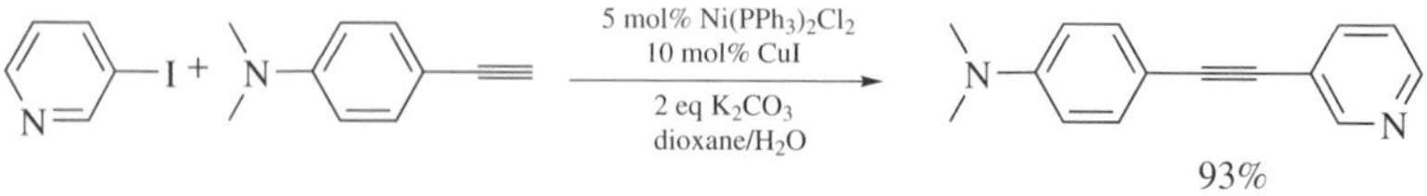

$$\xrightarrow[\substack{2 \text{ eq } K_2CO_3 \\ \text{dioxane/}H_2O}]{\substack{5 \text{ mol\% } Ni(PPh_3)_2Cl_2 \\ 10 \text{ mol\% CuI}}}$$

93%

(4.19)

For the development of the oxidative homocoupling reaction, in 1955 Chodkiewicz and Cadiot explored a Cu(I)-catalyzed hetero-coupling reaction of terminal alkynes with 1-bromoalkyne in the presence of an amine in aqueous medium, to produce the expected products with good yields (50 ~ 90%) (Eq. 4.20).[40]

$$
\begin{array}{c}
R\!\!-\!\!\!\equiv\!\!\!-X \\
+ \\
Br\!\!-\!\!\!\equiv\!\!\!-R'
\end{array}
\quad
\xrightarrow[\substack{H_2O/MeOH \\ 30 \sim 40°C}]{\substack{CuCl,\ NH_2OH\cdot HCl \\ EtNH_2}}
\quad
R\!\!-\!\!\!\equiv\!\!\!-\!\!\!\equiv\!\!\!-R'
\qquad (4.20)
$$

Later, Jones et al. applied this Cadiot-Chodkiewicz coupling reaction to the synthesis of triynoic acid, a fungal polyacetylene (Eq. 4.21).[41]

$$
\xrightarrow[\text{aq EtNH}_2]{\text{CuCl, NH}_2\text{OH}\cdot\text{HCl}}
\qquad (4.21)
$$

70%

Amatore et al. developed an aqueous cross-coupling reaction of terminal alkynes with 1-iodoalkynes using a water-soluble Pd(0) catalyst prepared in situ from $Pd(OAc)_2$ and sulfonated triphenylphosphine $P(C_6H_4 - m\text{-}SO_3Na)_3$ (TPPTS) without Cu(I) promoter, giving diynes with moderate yields (43–65%)(Eq. 4.22).[42]

$$
\begin{array}{c}
R\!\!-\!\!\!\equiv\!\!\!-I \\
+ \\
H\!\!-\!\!\!\equiv\!\!\!-R'
\end{array}
\quad
\xrightarrow[\substack{H_2O/MeCN\ (1/6) \\ 25 \sim 30°C}]{\substack{5\ mol\%\ Pd(OAc)_2 \\ 10\ mol\%\ TPPTS \\ 2.5\ eq\ Et_3N}}
\quad
R\!\!-\!\!\!\equiv\!\!\!-\!\!\!\equiv\!\!\!-R'
$$

$$(4.22)$$

$R = CH_3(CH_2)_3$,	$R' = C(CH_3)_2(OH)$	60%
$R = (CH_3)_3Si$,	$R' = CH(OH)(CH_2)_4CH_3$	43%
$R = (CH_3)_3Si$,	$R' = C(CH_3)_2(OH)$	57%
$R = CEt_2NH_2$,	$R' = CH(OH)(CH_2)_4CH_3$	65%

At the same time, Schmidtchen et al. compared cationic phosphine ligands containing the hydrophilic guanidinium (**4.3, 4.4**) and the anionic phosphine ligand TPPTS for this palladium-catalyzed coupling reaction. They found that the cationic ligands were effective for the coupling reaction but less efficient than TPPTS.[43]

guanidino phosphines

4.3   **4.4**

Recently, Marino et al. reported a Cadiot-Chodkiewicz cross-coupling reaction of bulky trialkylsilyl-protected alkynes with 1-bromoalkynes in aqueous amine to form a variety of unsymmetrical diynes in good yields (75 ~ 95%) (Eq. 4.23).[44]

$$
R\!\!\equiv\!\!H \;+\; Br\!\!\equiv\!\!R' \quad
\xrightarrow[\substack{aq\ n\text{-BuNH}_2 \\ \text{r.t.}}]{\substack{2\ mol\%\ CuCl \\ NH_2OH\cdot HCl}} \quad
R\!\!\equiv\!\!\equiv\!\!R'
$$

$$75 \sim 95\%$$

R = TES, TBS, TIPS

R' = C(CH$_3$)$_2$(OH), CH$_2$OH, CH$_2$NMe$_2$, CH$_2$CH$_2$CH$_3$, (CH$_3$)C=CH$_2$CH$_2$OH, 1-cyclohexenyl

TES: triethylsilyl, TBS: *tert*-butyldimethylsilyl, TIPS: triisopropylsilyl

$$(4.23)$$

Carbonylative Sonogashira coupling is an important way to construct conjugated ynones. Aqueous medium was found to be effective for the carbonylative Sonogashira coupling of terminal alkynes. Kang et al. have found that iodonium salts can readily undergo carbonylative coupling with terminal alkynes in the presence of Pd/Cu catalysts and one atmosphere pressure of carbon monoxide, in aqueous media at room temperature, affording expected α,β-acetylenic ketone products (Eq. 4.24).[45] In the cases of terminal arylalkynes, palladium or copper catalyst alone was also effective.

$$
R'PhI^+\ X^- \;+\; \equiv\!\!-R \quad
\xrightarrow[\substack{NaHCO_3 \\ \text{r.t., 1 atm CO} \\ DEM/H_2O\ (4:1)}]{\substack{Pd(OAc)_2\ (0.5\ mol\%) \\ CuI\ (10\ mol\%)}} \quad
R'\!-\!\overset{O}{\underset{\|}{C}}\!\!-\!\equiv\!\!-R
$$

$$76 \sim 89\%$$

$$(4.24)$$

R': Ph, 4-MeOC$_6$H$_4$, 2-thienyl, (*E*)-PhCH=CH
R: Ph, Bu
X: BF$_4$, OTs

Recently, Mori et al. reported a carbonylative coupling of phenylethyne with aryl iodide in the presence of PdCl$_2$(PPh$_3$)$_2$/(CuI), aqueous ammonia, and CO (1 atm) to give the corresponding conjugated alkynyl ketone in good yields at room temperature. A noncarbonylative coupling product **b** was not obtained (Eq. 4.25). Aqueous ammonia (a mixture of 2 mL water and 3 mL tetrahydrofuran) was found to be much more effective for such a coupling reaction than the use of an excess of tertiary amine as a solvent. For example, treatment of phenylethyne with 4-methoxy-1-iodobenzene in the presence of 1 mol% PdCl$_2$(PPh$_3$)$_2$ at room temperature under an ambient pressure of carbon monoxide in tertiary amine gave the carbonylative coupling product in only 11% yield. In contrast, when the reaction solvent was changed to 2 mL of aqueous ammonia (0.5 M) solution together with 3 mL THF, the carbonylative coupling product was obtained in 72% yield.[46]

without CuI:
R: aromatic
R′: H, OMe, Me, COMe, Cl
a: 50–81%
b: 0–7%

with CuI:
R: aliphtic
R′: Me, NH₂, COMe, Cl
a: 55–78%
b: 0–5%

$$(4.25)$$

4.1.4 Reaction of Alkynes with Carbonyl Compounds

Propargylic alcohols are well-known as versatile building blocks in organic synthesis. Their traditional synthesis through alkynylation of ketones and aldehydes using a stoichiometric amount of organometallic reagents such as Grignard reagents or alkyllithiums in an anhydrous organic solvent has recently been replaced with a catalytic system. Recently, Li et al. found an effective addition reaction via C–H activation where various aldehydes reacted with phenylacetylene catalyzed by a bimetallic Ru-In catalytic system to give Grignard-type nucleophilic addition products in water (Eq. 4.26).[47] It is believed that in the reaction, ruthenium catalyzes the overall reaction and indium activates the carbonyl function. Later, Li and co-workers found that silver complexes such as Cy_3PAgCl were more effective catalysts for the addition of terminal alkynes to aldehydes in water.[48] Interestingly, the phosphine ligand served as a remarkable chemo-switch for the reaction of aldehydes with alkynes in the presence of amines in water. In the presence of a phosphine ligand, silver catalyzed the exclusive aldehyde-alkyne coupling, whereas exclusive aldehyde-alkyne-amine (A^3) coupling occurred in the absence of phosphine (Scheme 4.3). Similarly, a highly efficient alkynylation-cyclization of terminal alkynes with *ortho*-alkynylaryl aldehydes leading to 1-alkynyl−1*H*-isochromenes was developed by Li and co-worker by using a gold-phosphine complex as catalyst in water (Eq. 4.27).[49] The reaction was dually promoted by

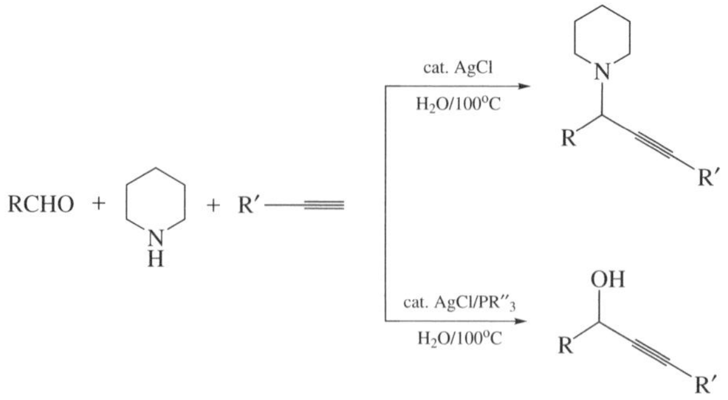

Scheme 4.3

an electron-donating phosphine ligand and water, as well as chelation-controlled.

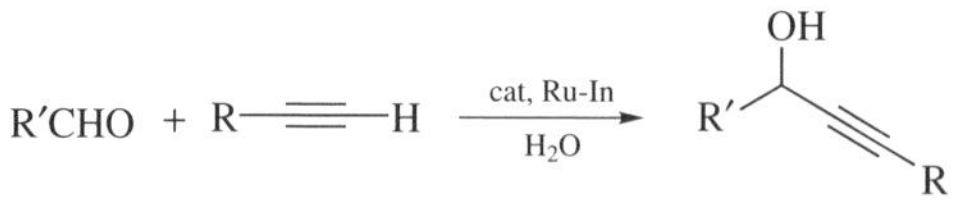

$$R'CHO \ + \ R\text{---}\!\!\equiv\!\!\text{---}H \xrightarrow[\text{H}_2\text{O}]{\text{cat, Ru-In}} \qquad (4.26)$$

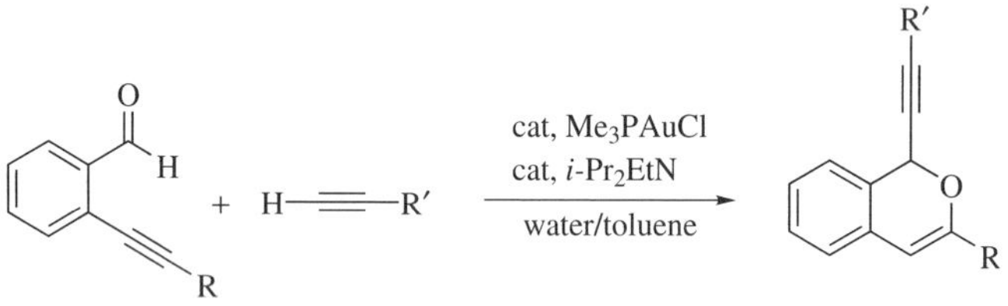

$$\qquad (4.27)$$

4.1.5   Reaction of Alkynes with C=N Compounds

4.1.5.1   *Addition to C=N*   Addition of terminal alkynes to imines is a very important reaction since it generates propargyl amines, which are good synthetic intermediates and show broad biological activities.[50] In 2002, Li et al. reported the first efficient catalytic addition of acetylene to various imines to generate propargyl amines via C–H activation by a Ru/Cu catalyst in water under mild conditions (Eq. 4.28).[51] However, this method was largely limited to imines generated from arylamines and aldehydes. A year later, they developed a gold-catalyzed addition reaction of terminal alkynes to imines generated from various amines and aldehydes in water (Eq. 4.29).[52] Recently, they found that Ag(I) salts, especially AgI, are also very effective for such a reaction, particularly for the reactions of imines generated from aliphatic aldehydes.[53]

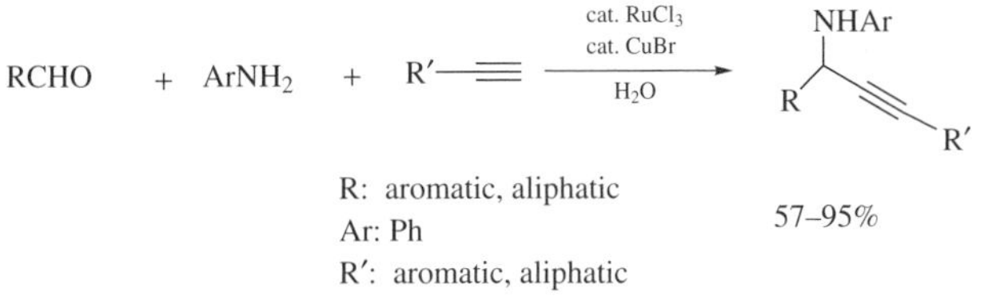
cat. RuCl3
cat. CuBr
H2O
RCHO
+ ArNH2
+ R′
NHAr
R
R′
R: aromatic, aliphatic
Ar: Ph
R′: aromatic, aliphatic
57–95%

(4.28)

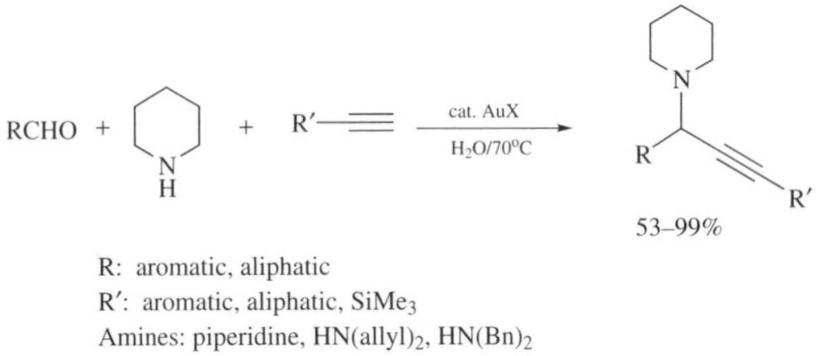
RCHO +
N
H
+ R′
cat. AuX
H2O/70°C
N
R
R′
53–99%
R: aromatic, aliphatic
R′: aromatic, aliphatic, SiMe3
Amines: piperidine, HN(allyl)2, HN(Bn)2

(4.29)

Li et al. subsequently extended this to the enantioselective addition of terminal alkynes to imines catalyzed by a chiral Cu(I)–*bis* (oxazolinyl)pyridine (pybox) complex in water. They observed that Cu(I) alone could provide the desired product in low conversions and postulated that the low catalytic activity with Cu(I) alone could be due to the strong and low reactivity of the C–Cu bond of copper acetylides. They proposed that the addition of a strongly coordinating and electron-rich ligand may weaken the strong C–Cu bond. Therefore, in order to get the enantioselective addition products, the combinations of Cu(I) catalysts with several chiral *bis*(oxazolinyl) ligands were examined in water. Finally, it was found that the use of **4.5**–Cu(OTf) complex provided the first enantioselective addition of terminal alkyne to imines and afforded the corresponding (+)-propargyl amine products in both high reactivity (48–86% yields) and enantioselectivity (78–91% ee in water and up to 99.6% ee in organic solvent) (Eq. 4.30).[54]

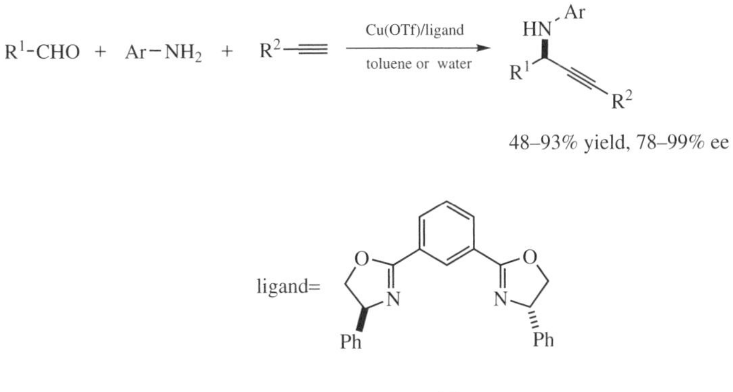
R1-CHO + Ar−NH2 + R2
Cu(OTf)/ligand
toluene or water
HN
Ar
R1
R2
48–93% yield, 78–99% ee
ligand=
O
N
Ph
O
N
Ph

(4.30)

4.5

Subsequently, various efforts were reported to improve the conditions of this reaction, including a microwave-promoted reaction. Because of its high dielectric constant, water as a solvent is particularly advantageous for microwave reactions.[55] Tu et al. reported an efficient three-component coupling of aldehyde, alkyne, and amine to generate propargylamines catalyzed by CuI alone under microwave irradiation in water. In addition, using (*S*)-proline methyl ester as a chiral amine source, highly pure chiral propargylamines were obtained (Eq. 4.31).[56]

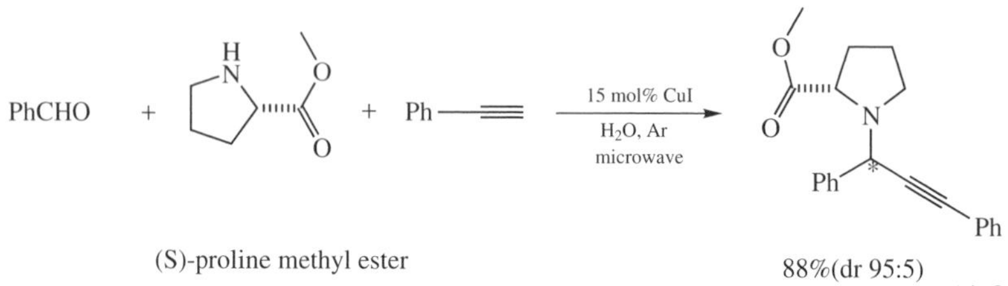

(*S*)-proline methyl ester

88%(dr 95:5)

(4.31)

Another example of the addition of terminal alkynes to C=N in water is the coupling of alkynes with in-situ-generated *N*-acylimines (Eq. 4.32) and *N*-acyliminium ions (Eq. 4.33). In 2002, Li et al. developed a coupling reaction of alkynes with *N*-acylimines and *N*-acyliminium ions mediated by Cu(I) in water to generate propargyl amide derivatives.[57] Either an activated imine derivative or imininum derivative was proposed as the intermediate, respectively.

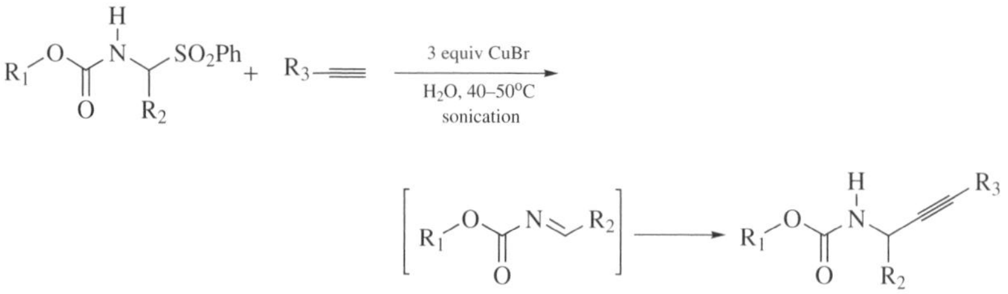

(4.32)

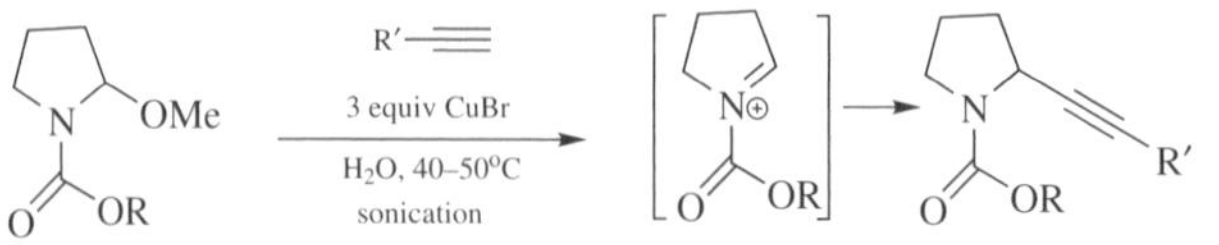

(4.33)

Recently, the three-component couplings of α-oxyaldehydes, alkynes, and amines in water were investigated by using gold, silver, and copper catalysts (Eq. 4.34).[58] Gold(I) was found to be the most effective catalyst in this reaction; affording propargylamines in good yields and moderate diastereoselectivities. On the other hand, silver catalysts show the best catalytic activities on noncoordinating α-alkyl-substituted aldehydes.

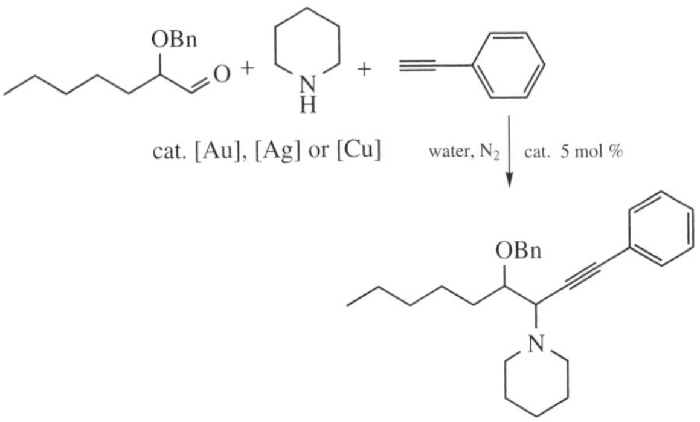

(4.34)

4.1.6 Conjugate Addition with Terminal Alkynes

The addition of terminal alkynes to carbon–carbon double bonds has not been explored until recently, possibly because C=C double bonds are not as good electrophiles as C=N or C=O. In 2003, Carreira et al. reported the first conjugate addition reaction of terminal alkynes to C=C catalyzed by copper in water. The reaction proceeded with derivatives of Meldrum's acid in water in the presence of $Cu(OAc)_2$ and sodium ascorbate (Eq. 4.35).[59] However, this method was limited to C=C double bonds with two electron withdrawing groups.

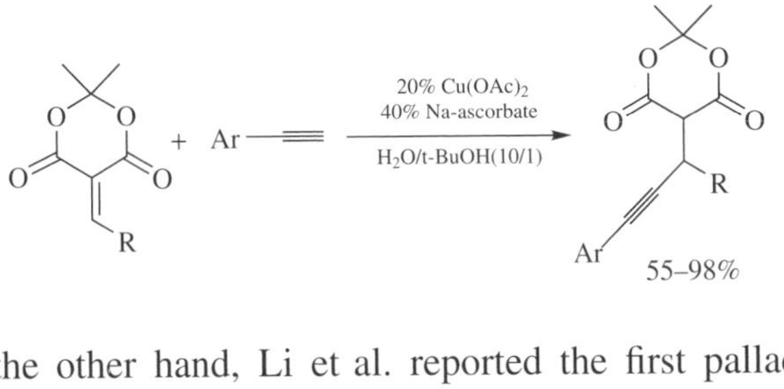

(4.35)

On the other hand, Li et al. reported the first palladium-catalyzed 1,4-addition of terminal alkynes to vinyl ketones in water with high yields (Eq. 4.36).[60] At nearly the same time, Chisholm et al. reported such a reaction catalyzed by $Rh(acac)(CO)_2$ (acac = acetylacetonoate) in the presence of *tris*(o-methoxyphenyl)phosphine in aqueous dioxane solution.[61] However, both methods are limited to unsubstituted vinyl ketones.

$$(4.36)$$

61–91%

4.2 ADDITIONS OF C≡C BONDS

4.2.1 Reduction

Like alkenes, various alkynes can be readily hydrogenated in aqueous or aqueous-organic biphasic media catalyzed by various transition metals. With certain catalysts, the reaction can stop at the alkene stage selectively. For example, diphenylacetylene and 1-phenyl-1-propyne were hydrogenated to 1,2-disubstituted alkenes in aqueous organic biphasic media using the water-soluble catalyst [{RuCl$_2$(mtppms)$_2$}$_2$] and an excess of the sulfonated phosphine ligand. The stereoselectivity of the reaction strongly depends on the pH of the catalyst-containing aqueous phase, and nearly complete Z selectivity was observed under acidic conditions.[62] Alkenes and alkynes can also be reduced with triethoxysilane and 5 mol% of palladium(II) acetate in a mixture of THF and water at ambient temperature to afford the corresponding hydrogenated products.[63] Water is essential for the process. The hydrogenation of non-activated internal alkynes with triethoxylsilane catalyzed by Pd(OAc)$_2$ gives the Z-alkenes without the formation of the corresponding alkanes. However, by the addition of catalytic amounts of methyl propynoate, internal alkynes can be reduced to the alkanes (Scheme 4.4). Hydrogenation of alkynes can also be carried out photochemically by using water and a titanium(II) complex.[64] In addition, hydrogenation of alkynes in supercritical H$_2$O or under aqueous photo-irradiation conditions has been explored as well.[65]

4.2.2 Addition of Water

The hydration of acetylenes, which usually gives aldehydes or ketones, has been widely used in industries for a long time. The hydration of acetylenes in aqueous media could date back to the 1930s, when it was reported that acetylene was converted into acetaldehyde in the presence

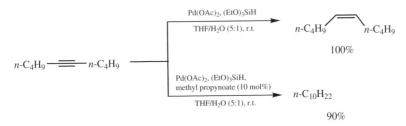

Scheme 4.4

of mercuric sulfate and sulfuric acid.[66] No further study was reported until recently. In 1990 Taqui Khan et al. reported the water-soluble complex K[Ru^{III}(EDTA-H)Cl]·2H$_2$O-catalyzed hydration of acetylene to give acetaldehyde cleanly (Eq. 4.37).[67]

$$\equiv \;+\; H_2O \;\xrightarrow[\text{H}_2\text{O, 80°C}]{\text{K[Ru}^{III}\text{(EDTA-H)Cl]•2H}_2\text{O}}\; \text{MeCHO} \qquad (4.37)$$

Recently, on the basis of the Markovnikov addition of water to alkynes, Trost et al. developed a three-component addition reaction of terminal alkynes, water, and methyl vinyl ketone, affording 1,5-diketones in DMF/water in the presence of ruthenium and indium catalysts (Eq. 4.38).

$$\text{R}\!-\!\!\!\equiv \;+\; H_2O \;+\; \text{(MVK)} \;\xrightarrow[\substack{\text{DMF/H}_2\text{O} \\ \text{100°C/ 4 h}}]{\substack{\text{CpRu(COD)Cl} \\ \text{In(OTf)}_3, \text{NH}_4\text{PF}_6}}\; \text{(1,5-diketone)} \qquad (4.38)$$

28–81%

Hydration of alkynes usually follows Markovnikov's rule to form the corresponding ketone compounds.[68] In contrast, it has been recently reported that the reaction can proceed regioselectively to form aldehydes in alcoholic aqueous media (a mixture of water and 2-propanol) by using ruthenium(II) complexes such as [RuCl$_2$(η^6-C$_6$H$_6$){PPh$_2$(C$_6$F$_5$)}][69] and [Ru(η^5-C$_9$H$_7$)Cl(PPh$_3$)$_2$] as catalysts (Eq. 4.39).[70]

$$\text{R}\!-\!\!\!\equiv \;+\; H_2O \;\xrightarrow[\text{2-propanol/H}_2\text{O}]{\text{[Ru(II)]}}\; \text{R} \overset{\text{O}}{\diagup}\!\!\diagdown\!\text{H} \qquad (4.39)$$

More recently, Atwood et al. developed a platinum complex of a water-soluble, bidentate phosphine ligand, *cis*-(TPPTS)$_2$PtCl$_2$ [TPPTS = tris(sodium *m*-benzenesulfonate)phosphine], as an effective hydration catalyst for the water-soluble alkynes 4-pentyn-1-ol and 3-pentyn-1-ol (Eq. 4.40). The reaction gave the same product 5-hydroxy-2-pentanone as the result of Markovnikov and anti-Markovnikov hydration respectively, presumably due to the participation of the hydroxyl function.[71]

$$
\begin{array}{c}
\text{or} \\
\xrightarrow[\text{r.t./H}_2\text{O}]{\text{cis-(TPPTS)}_2\text{PtCl}_2}
\end{array}
\qquad (4.40)
$$

Pt-catalyzed hydration of various aliphatic and aromatic alkynes under phase transfer conditions in $(CH_2Cl)_2/H_2O$ in the presence of Aliquat 336 led to either a Markovnikov product, mixtures of two ketones, or ketones with the carbonyl group positioned away from the bulky side.[72] In the absence of the phase transfer reagent, Aliquat 336, hardly any reaction took place. Recently, a hydrophobic, low-loading and alkylated polystyrene-supported sulfonic acid (LL–ALPS–SO$_3$H) has also been developed for the hydration of terminal alkynes in pure water, leading to ketones as the product.[73] Under microwave irradiation, the hydration of terminal arylalkynes was reported to proceed in superheated water (200°C) without any catalysts.[74]

4.2.3 Addition of Alcohols and Amines

Hydroalkoxylation of alkynes, or the addition of alcohol to alkynes, is a fundamental reaction in organic chemistry that allows the preparation of enol ethers and a variety of oxygen-containing heterocycles such as furan, pyran, and benzofuran derivatives.[75] Bergbreiter et al. found that a linear poly-(N-isopropylacrylamide) (PNIPAM) polymer exhibited inverse temperature solubility in water (i.e., soluble in cold water but insoluble in hot water). A recoverable homogeneous palladium catalyst was prepared based on the polymer.[76] The PNIPAM-bound Pd(0) catalyst was effective for the reaction of 2-iodophenol with phenylacetylene in aqueous THF media to give the target product 2-phenylbenzofuran (Eq. 4.41). This catalyst can be recovered simply by heating the water or by adding hexane (in aqueous THF). The catalyst could be reused up to 15 times with only a very modest loss of activity. Later, Uozumi et al. showed that an amphiphilic polystyrene-poly(ethylene glycol) (PS-PEG) resin-supported palladium-phosphine complex could also effect this reaction in water.[77]

$$\text{(structure: 2-iodophenol + phenylacetylene)} \xrightarrow[\substack{\text{THF/H}_2\text{O (v/v = 80/20)} \\ 50°\text{C/8 h}}]{\substack{0.5 \text{ mol% } \mathbf{L}_4\text{Pd(0)} \\ 0.5 \text{ mol% CuI/Et}_3\text{N}}} \text{(2-phenylbenzofuran)}$$

$$78\%$$

(4.41)

$$\text{PNIPAM}\text{—}\text{CH}_2\text{CH}_2\text{CH}_2\text{—PPh}_2 \equiv$$

structure L with repeating units (50, 3, 2): NH–CH(CH$_3$)$_2$ isopropyl amide, NH$_2$ amide, NH–(CH$_2$)$_3$–PPh$_2$

L

Pal et al. explored an efficient Pd/C-catalyzed reaction of 2-iodophenols with terminal alkynes in water without the use of any organic cosolvents in the presence of PPh$_3$, CuI, and prolinol to give the expected benzofuran products (Eqs. 4.42, 4.43).[78] The hydroxyl group was well tolerated during reactions. When used as a base, prolinol afforded better yields of products than triethylamine, a common organic base, possibly due to its better miscibility with water.

$$\text{R'—(2-iodophenol)} + \text{≡—R} \xrightarrow[\substack{3 \text{ equiv. (S)-prolinol} \\ \text{H}_2\text{O/80°C/3h}}]{\substack{10 \text{ mol% Pd/C} \\ 20 \text{ mol% CuI} \\ 40 \text{ mol% PPh}_3}} \text{R'—(benzofuran)—R}$$

$$68\text{~}88\%$$

(4.42)

R': H, NO$_2$
R: C(OH)(CH$_3$)$_2$, CH$_2$CH$_2$OH, CH(OH)C$_2$H$_5$, CH$_2$OH, CH(OH)CH$_3$, Ph

$$\text{O}_2\text{N—(diiodophenol)} + \text{≡—R} \xrightarrow[\substack{3 \text{ equiv. (S)-prolinol} \\ \text{H}_2\text{O/80°C/3h}}]{\substack{10 \text{ mol% Pd/C} \\ 20 \text{ mol% CuI} \\ 40 \text{ mol% PPh}_3}} \text{O}_2\text{N—(benzofuran)—R}$$

(4.43)

$$70\text{~}75\%$$

Hydroamination of alkynes offers a straightforward preparation of a variety of amines, enamines, and imines.[79] Numerous reports have appeared in the literature on this process. However, almost all these reactions have been carried out in organic solvents, which usually require the protection of functional groups or harsh conditions. Recently, Marinelli et al. have reported an Au(III)-catalyzed hydroamination of alkynes in aqueous media (EtOH/H_2O) in which the annulation of 2-alkynylanilines gave indole derivatives in good yields (Eq. 4.44).[80] However, in some cases, increasing the amount of water in the EtOH-water mixture resulted in a decreased product yield, perhaps due to the decreased solubility of the starting materials.

$$(4.44)$$

Vasudevan et al. have reported a microwave-promoted hydroamination of alkynes. Heating a mixture of 1-ethynyl-4-methoxybenzene and 4-bromoaniline in water at 200°C in a microwave reactor for 20 minutes without any catalyst gave an imine product in 87% yield (Eq. 4.45).[81]

$$(4.45)$$

4.2.4   Hydrosilation and Hydrometalations

Hydrosilation, hydrostannation, and hydrogermanation of alkynes are important processes to generate various reactive intermediates in organic synthesis.[82] Many examples[83] of hydrosilation of alkynes have appeared since the first report using Speier's catalyst.[84] However, these reactions are generally carried out in anhydrous conditions. Recently, Li et al. developed a highly regio- and stereoselective Pt(DVDS)-P catalyzed hydrosilation of terminal alkynes in water at room temperature. In most cases, 100% stereoselectivity (*trans*) was observed (Eq. 4.46).[85] It is worth noting that the presence of a hydroxyl group in the substrate does not need to be protected in the hydrosilation process.

$$R\!\!-\!\!\!\equiv\ +\ Et_3SiH\ \xrightarrow[\text{water/r.t./3 h}]{\text{Pt(DVDS)}-\mathbf{P}}\ \underset{trans}{R\diagup\!\!\!=\!\!\!\diagdown SiEt_3}\ +\ \underset{cis}{R\diagdown\!\!\!=\!\!\!\diagup SiEt_3}\ +\ \underset{\alpha-}{R\!\!-\!\!\!=\!\!\!<SiEt_3}$$

(4.46)

R= Ph, *p*–Me–C$_6$H$_4$, *n*–C$_5$H$_{11}$, *n*–C$_4$H$_9$, HOCH$_2$,
  HOCH$_2$CH$_2$, PhCH$_2$, 1-naphthalenyl,
  CH$_2$=CHCH$_2$OCH$_2$, PhCH$_2$OCH$_2$

90~98%
trans:*cis*:α– = (> 91):(<9):0

92%
trans:*cis*:α– = 35:65:0

R=SiMe$_3$

DVDS

P

Subsequently, cationic rhodium catalysts are also found to be effective for the regio- and stereoselective hydrosilation of alkynes in aqueous media. Recently, Oshima et al. reported a rhodium-catalyzed hydrosilylation of alkynes in an aqueous micellar system. A combination of [RhCl(nbd)]$_2$ and *bis*-(diphenylphosphino)propane (dppp) were shown to be effective for the (*E*)-selective hydrosilation in the presence of sodium dodecylsulfate (SDS), an anionic surfactant, in water.[86] An anionic surfactant is essential for this (*E*)-selective hydrosilation, possibly because anionic micelles are helpful for the formation of a cationic rhodium species via dissociation of the Rh–Cl bond. For example, Triton X-100, a neutral surfactant, gave nonstereoselective hydrosilation; whereas methyltrioctylammonium chloride, a cationic surfactant, resulted in none of the hydrosilation products. It was also found that the selectivity can be switched from *E* to *Z* in the presence of sodium iodide (Eq. 4.47).

$$R\!\!-\!\!\!\equiv\!\!\!-R'\ +\ Et_3SiH\ \xrightarrow[\text{SDS/water r.t.}]{\substack{0.5\ mol\%\\ [RhCl(nbd)]_2\\ dppp}}\ \underset{E}{}\ +\ \underset{Z}{}\ +\ \underset{\alpha}{}$$

(4.47)

R': H
R: Ph, CF$_3$–C$_6$H$_4$, *n*–C$_{10}$H$_{21}$, *t*-Bu, HOCH$_2$,
  HOCH$_2$CH$_2$, HO(CH$_2$)$_3$, *t*-BuMe$_2$SiO(CH$_2$)$_3$,
  BnO(CH$_2$)$_3$, EtOOC(CH$_2$)$_6$, HOOC(CH$_2$)$_6$, SiMe$_3$

E : 35 ~ 87%
Z :   < 2%
α– :   < 5%

R (= R'): Ph, Et

E : 53 ~ 66%
Z : 0
α– : 0

(In the presence of 1 equiv NaI):
R': H
R: *n*–C$_6$H$_{13}$, *n*–C$_{10}$H$_{21}$, *t*-BuMe$_2$SiO(CH$_2$)$_3$,
  BnO(CH$_2$)$_3$, EtOOC(CH$_2$)$_6$

E : 5 ~ 21%
Z : 50 ~ 70%
α– :   < 2%

Oshima et al. have reported a Pd(0)-catalyzed hydrogermanation of alkynes in water, providing mixtures of dienylgermanes and alkenyl-germanes efficiently, with the dienylgermanes as major products in most cases (Eq. 4.48).[87] It is worth noting that the reaction in water proceeded much faster than in organic solvents or under neat conditions. In addition, the catalyst loading could be as low as 0.0025 mol% in an aqueous system. Interestingly, the addition of surfactants such as sodium dodecyl sulfate, Triton X-100, or methyltrioctylammonium chloride decreased the reaction rates.

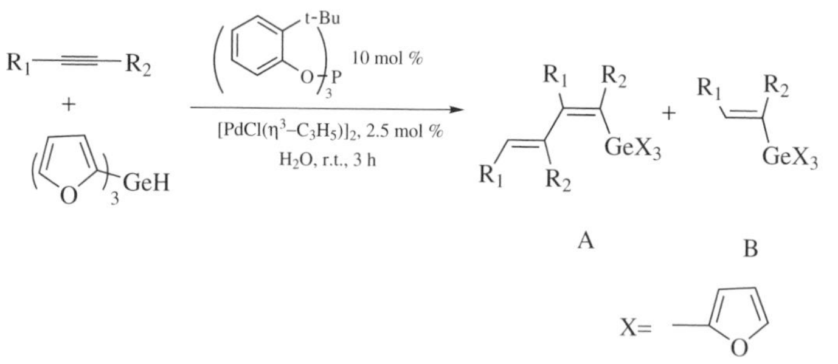

(4.48)

In addition, Wu and Li recently have developed an efficient rhodium-catalyzed cascade hydrostannation/conjugate addition of terminal alkynes and unsaturated carbonyl compounds in water stereoselectively (Scheme 4.5).[88]

4.2.5 Addition of Aryls

Arylation of alkynes via addition of arylboronic acids to alkynes represents an attractive strategy in organic synthesis. The first addition of arylboronic acids to alkynes in aqueous media catalyzed by rhodium was reported by Hayashi et al.[89] They found that rhodium catalysts associated with chelating bisphosphine ligands, such as 1,4-bis(diphenyl-phosphino)butane (dppb) and 1,1'-bis(diphenylphosphino)ferrocene

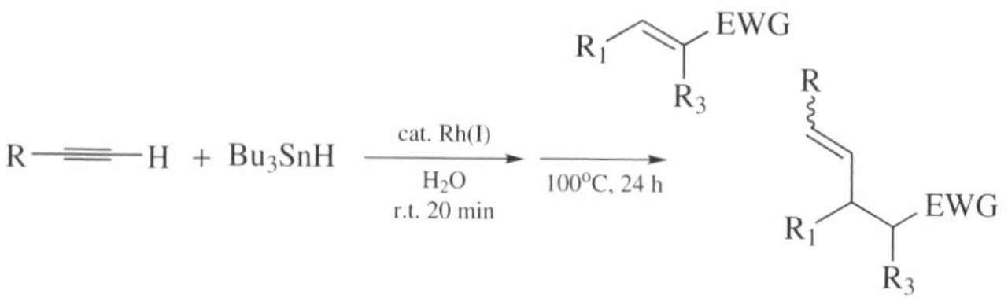

Scheme 4.5

(dppf), were highly effective for the arylation of alkynes with arylboronic acids in aqueous media (dioxane/water = 10/1), giving target products in high *syn*-selectivity. Triphenylcyclotriboroxane can also be used as the boronic acid equivalent (Eq. 4.49). The arylation of unsymmetrical alkynes such as 1-phenylpropyne, gave a mixture of *E* and *Z* isomers. However, unsymmetrical alkynes substituted with an ester or phosphonate group afforded desired products with high regio- and *syn*-selectivity. Surprisingly, isotope experiments showed that the hydrogen on the vinylic carbon did not come from water; instead it came from the phenyl group of boronic acid (Eq. 4.50), perhaps due to the process of 1,4-shift of rhodium from 2-aryl-1-alkenylrhodium to the 2-alkenylarylrhodium intermediate (Scheme 4.6).

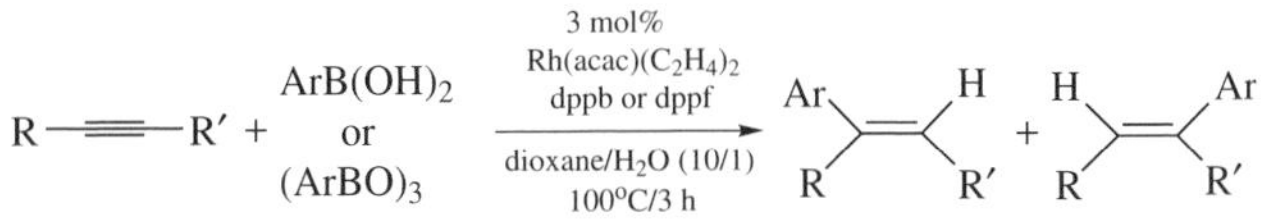

$$
R\!\!=\!\!=\!\!R' + \begin{array}{c} ArB(OH)_2 \\ \text{or} \\ (ArBO)_3 \end{array} \xrightarrow[\substack{\text{dioxane/H}_2\text{O (10/1)} \\ 100°\text{C/3 h}}]{\substack{3 \text{ mol\%} \\ Rh(acac)(C_2H_4)_2 \\ \text{dppb or dppf}}} \begin{array}{c} Ar \\ R \end{array}\!\!=\!\!\begin{array}{c} H \\ R' \end{array} + \begin{array}{c} H \\ R \end{array}\!\!=\!\!\begin{array}{c} Ar \\ R' \end{array} \quad (4.49)
$$

R = R′ = Et, Ph		90 ~ 97%
R = C₄H₉, Me₃Si, n-Hex R′ = COOMe, P(O)(OEt)₂	70 ~ 87%	0%
R = Me R′ = Ph	72%	24%

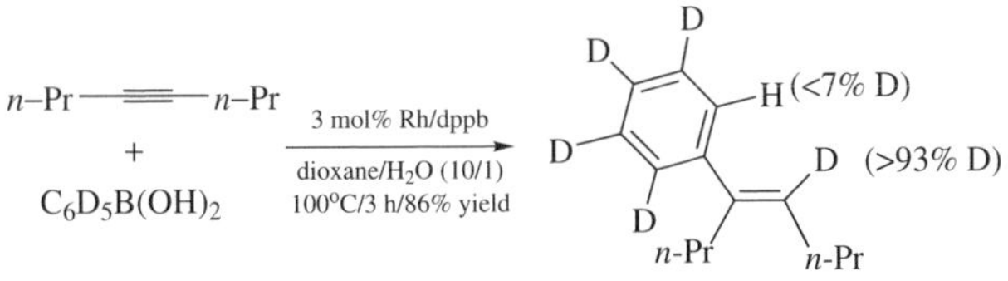

$$
\begin{array}{c} n\text{–Pr}\!\!=\!\!=\!\!n\text{–Pr} \\ + \\ C_6D_5B(OH)_2 \end{array} \xrightarrow[\substack{\text{dioxane/H}_2\text{O (10/1)} \\ 100°\text{C/3 h/86\% yield}}]{3 \text{ mol\% Rh/dppb}} \quad (4.50)
$$

H (<7% D)   D (>93% D) *n*-Pr   *n*-Pr

Lautens et al. explored a water-soluble pyridyl phosphine moiety–based ligand for such additions (Scheme 4.7).[90] [Rh(COD)Cl]₂

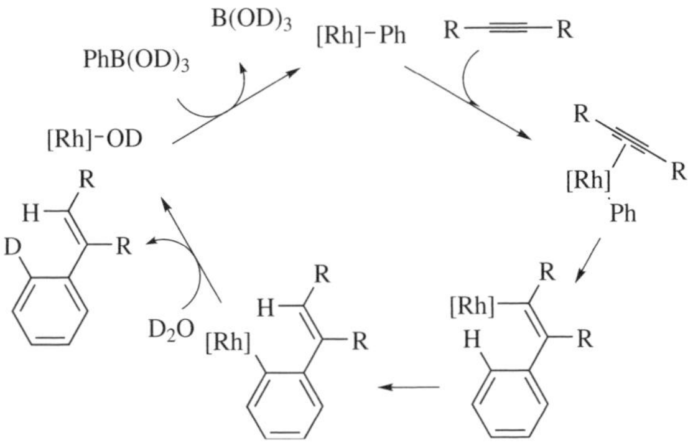

Scheme 4.6

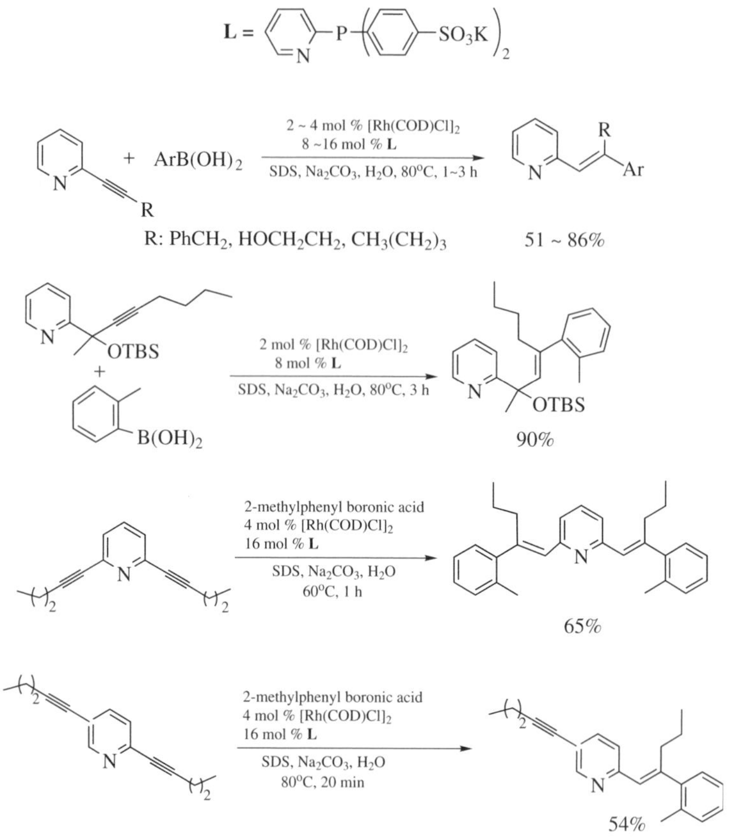

Scheme 4.7

together with this ligand was effective for the addition of arylboronic acids to 2-alkynyl pyridine compounds in the presence of sodium dodecyl sulfate (SDS) in water. The reaction was also effective for alkynes with a triple bond one carbon away from the pyridyl group. No reaction was observed for the alkynes substituted at positions other than the *ortho* position of pyridine, perhaps because this reaction involved a chelation-controlled process.

Biphasic systems were found to have a unique effect on the selectivity of the addition of arylboronic acids to alkynes. It was found that the use of [Rh(COD)OH]$_2$ associated with the water-soluble ligand, *m*-TPPTC, was highly effective for such a reaction in the biphasic water/toluene system (Eq. 4.51).[91] The reaction was completely stereo- and regioselective. In addition, the catalyst did not lose any activity after being reused four times.

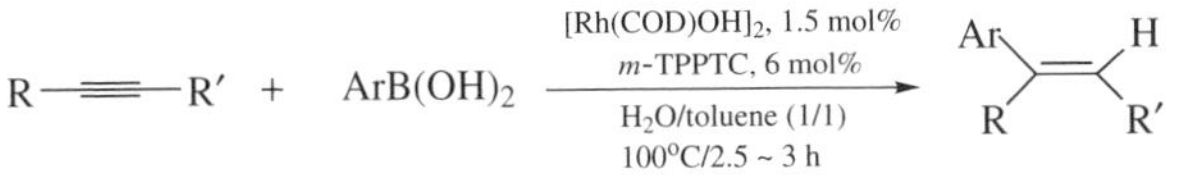

$$R{\equiv\!\!\!\equiv}R' + ArB(OH)_2 \xrightarrow[\substack{H_2O/toluene\ (1/1) \\ 100°C/2.5 \sim 3\ h}]{\substack{[Rh(COD)OH]_2,\ 1.5\ mol\% \\ m\text{-TPPTC},\ 6\ mol\%}} \quad \underset{R}{\overset{Ar}{\diagdown}}{=}\underset{R'}{\overset{H}{\diagup}}$$

(4.51)

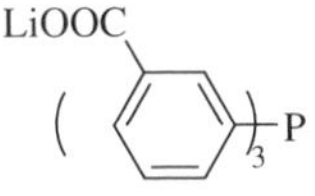

m-TPPTC

R = R′ = C₃H₇	80 ~ 99%
R = Me, R′ = Ph	95 ~ 96%
R = C₄H₉, R′ = 2-pyridyl, SiMe₃	94 ~ 98%

$R = R' = C_3H_7$ — 80 ~ 99%
$R = Me, R' = Ph$ — 95 ~ 96%
$R = C_4H_9, R' = 2\text{-pyridyl}, SiMe_3$ — 94 ~ 98%

4.2.6 Carbonylation of Internal Alkynes

Carbonylation of alkynes is a convenient method to synthesize various carbonyl compounds. Alper et al. found that carbonylation of terminal alkynes could be carried out in aqueous media in the presence of 1 atm CO by a cobalt catalyst, affording 2-butenolide products. This reaction can also be catalyzed by a cobalt complex and a ruthenium complex to give γ-keto acids (Scheme 4.8).[92]

Nickel-catalyzed carbonylation of α-haloalkynes with carbon monoxide under phase-transfer conditions gave either allenic monoacids or unsaturated diacids.[93] The carbonylation initially afforded monoacids, which reacted further to give diacids with high stereoselectivity (Eq. 4.52).

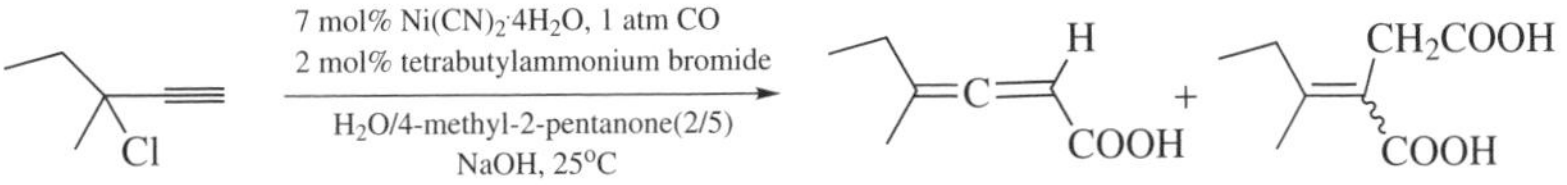

mono-acid   di-acid

1 h: 94% (overall yield) *mono-:di-* = 87/13 (*E/Z* = 5/95)
48 h: 96% (overall yield) *mono-:di-* = 0/100 (*E/Z* = 9/91)

(4.52)

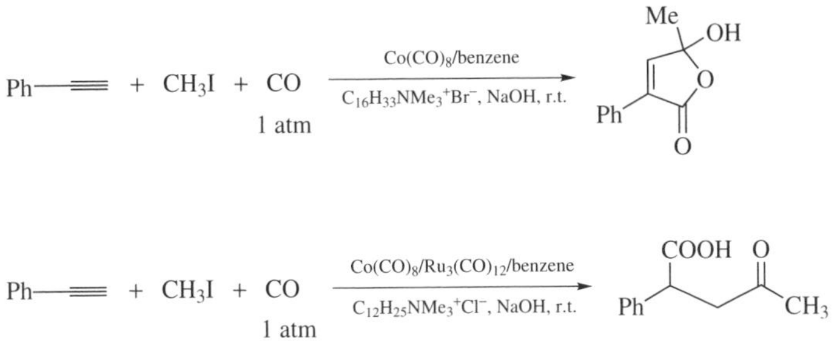

Scheme 4.8

Nickel-catalyzed carbonylation of α-ketoalkynes has also been reported by Arzoumanian et al. under phase-transfer conditions.[94] The carbonylation gave either furanone or unsaturated carboxylic acids depending on the substituents of substrates (Eq. 4.53). A similar reaction, nickel-catalyzed cyanation of α-ketoalkynes with KCN in water, was also reported to afford unsaturated hydroxylactams (Eq. 4.54).[95]

$$
R\text{---}\!\!\!\equiv\!\!\!\text{---}\overset{O}{\underset{}{\|}}R' \xrightarrow[\text{H}_2\text{O/toluene(1/1), NaOH}]{\substack{10 \text{ mol\% Ni(CN)}_2, 1 \text{ atm CO} \\ 3 \text{ mol\% tetrabutylammonium bromide}}} \underset{A}{\text{(furanone)}} + \underset{B}{\overset{R}{\underset{HOOC}{\big\rangle}}\!=\!CHCOR'} \quad (4.53)
$$

R = Ph,   R' = Me: 78% (A:B = 100/0)
R = Ph,   R' = CMe$_3$: 35% (A:B = 41/59)
R = C$_4$H$_9$, R' = Me: 53% (A:B = 0/100)
R = C$_4$H$_9$, R' = CMe$_3$: 67% (A:B = 0/100)

$$
R\text{---}\!\!\!\equiv\!\!\!\text{---}\overset{O}{\underset{}{\|}}R' \xrightarrow[\text{H}_2\text{O, NaOH (2.5N)}]{\substack{\text{Cat. Ni(CN)}_2, 1 \text{ atm CO} \\ \text{xs KCN}}} \text{(hydroxylactam)} \quad (4.54)
$$

R = n–C$_4$H$_9$, Ph
R' = Me, Et, n–C$_3$H$_7$, n–C$_4$H$_9$ 50 ~ 77%

Later, a nickel-catalyzed cascade conversion of propargyl halides and propargyl alcohol into a pyrone in water was reported. The reaction involved a carbonylation by CO and a cyanation by KCN (Eq. 4.55).[96] Recently, Gabriele et al. explored a facile synthesis of maleic acids by palladium catalyzed-oxidative carbonylation of terminal alkynes in aqueous DME (1,2-dimethoxyethane) (Eq. 4.56).[97]

$$
\equiv\!\!\!\diagup^{X} \xrightarrow[\substack{\text{NaOH, KCN} \\ \text{H}_2\text{O, r.t.}}]{\substack{\text{Cat. Ni(CN)}_2 \\ \text{CO (1 atm)}}} \text{(pyrone)} \quad (4.55)
$$

X = Br   86%
X = Cl   63%
X = OH   47%

$$
R\text{---}\!\!\!\equiv + 2CO + H_2O + 1/2\,O_2 \xrightarrow[\substack{\text{DME/H}_2\text{O (3/1)} \\ 80°\text{C/3 ~ 8 h}}]{\substack{0.05 \text{ mol\% PdI}_2 \\ 10 \text{ equiv KI}}} \underset{HOOC \quad COOH}{\overset{R}{\big\rangle}\!=\!} \quad (4.56)
$$

16 atm   4 atm air

R: Ph, 4-MeC$_4$H$_4$, 4-MeOC$_6$H$_4$, 4-ClC$_6$H$_4$, n-Bu, t-Bu   20 ~ 71%

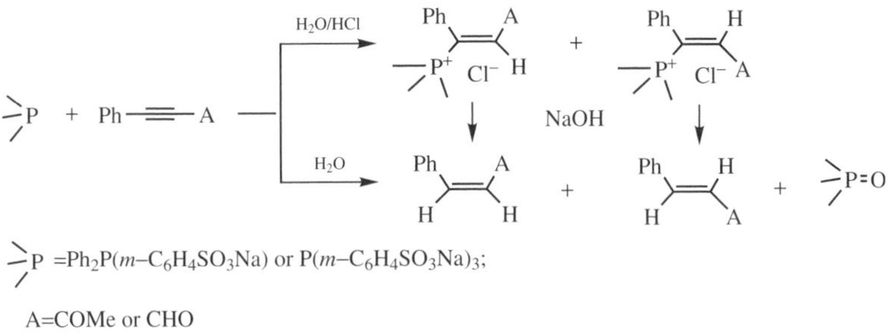

Scheme 4.9

4.2.7 Other Additions to Alkynes

Beside these reactions, there are other additions to alkynes carried out in water. For example, Lee et al. reported a synthesis of 1,2-diketones via the oxidation of alkynes by potassium permanganate in aqueous acetone.[98] Gorgues et al. found an efficient monohydrofluorination of electrophilic alkynes by the liquid biphasic $CsF-H_2O$ (8 equiv)–DMF (N, N-dimethylformamide) system, affording mainly *trans*-addition of HF products. However, in a homogeneous DMF-water system, no addition was observed.[99] Larpent et al. reported[100] that the reaction of water-soluble phosphines such as triphenylphosphine m-trisulfonate $[P(m\text{-}C_6H_4SO_3Na)_3 = TPPTS]$ and triphenylphosphine m-monosulfonate $[Ph_2P(m\text{-}C_6H_4SO_3Na) = TPPMS]$ with activated alkynes gave new vinylphosphonium salts or vinylphosphine oxides or alkenes depending on the pH of water solution (Scheme 4.9). Synthesis of γ,δ-unsaturated acetals and aldehydes via a ruthenium-catalyzed coupling of allyl alcohol with alkynes also could be carried out in water.[101] Indium catalyzed the addition of Ph_2Se_2 or Ph_2S_2 to alkynes in aqueous media.[102] It also has been found that the Pd(0)-catalyzed Kharasch reactions of alkynes could be enhanced by the use of a water solvent.[103] Carbon dioxide–promoted diiodination of alkynes also could be carried out in water.[104]

4.3 TRANSITION-METAL CATALYZED CYCLOADDITIONS

4.3.1 Pauson-Khand-Type Reactions

The Pauson-Khand reaction (PKR) is an efficient method to synthesize cyclopentenones.[105] The reaction is usually carried out in organic solvent. The first aqueous Pauson-Khand reaction was reported by

Sugihara et al. in 1997.[106] They utilized aqueous ammonium hydroxide as a reaction medium, which provided ammonia as a "hard ligand" to labilize the CO ligands and therefore enhance the rate of the PKR. The reaction of dicobalthexacarbonyl complexes of enynes and alkynes provided expected cyclopentenones via intramolecular and intermolecular modes respectively (Scheme 4.10).

Later, Chung et al. successfully developed an intramolecular Pauson-Khand reaction in water without any cosolvent by using aqueous colloidal cobalt nanoparticles as catalysts. The catalyst was prepared by reducing an aqueous solution of cobalt acetate containing sodium dodecyl sulfate (SDS) surfactant. The cobalt nanoparticle could be reused eight times without any loss of catalytic activity (Eq. 4.57).[107]

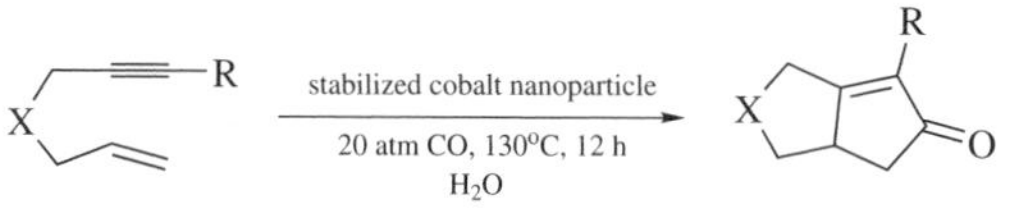

$$ (4.57) $$

R = Ph , X = C(COOEt)$_2$	90%
R = Ph , X = O	92%
R = Me, X = C(CO$_2$Et)$_2$	96%
R = Me, X = TsN	88%

Using cetyltrimethylammonium bromide (CTAB) as a surfactant, Krafft et al. have developed an efficient stoichiometric inter- and

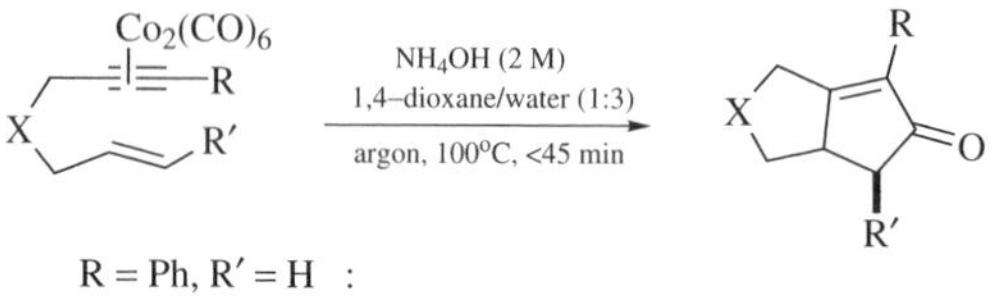

R = Ph, R′ = H :
    X = CH$_2$, CH$_2$CH$_2$, O, C(COOMe)$_2$ 78 ~ 90%

R = Ph, R′ = Me: X = CH$_2$ 92%
R = H , R′ = H : X = C(CO$_2$Me)$_2$ 67%

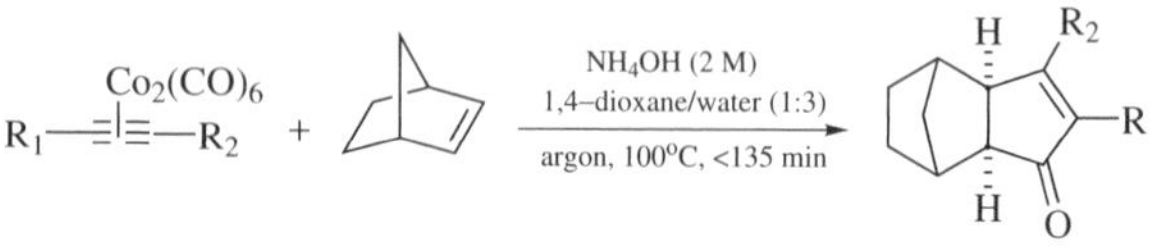

R$_1$ = Ph, R$_2$ = H	100%
R$_1$ = n-Bu, R$_2$ = H	92%
R$_1$ = n-Pr, R$_2$= n-Pr	42%

Scheme 4.10

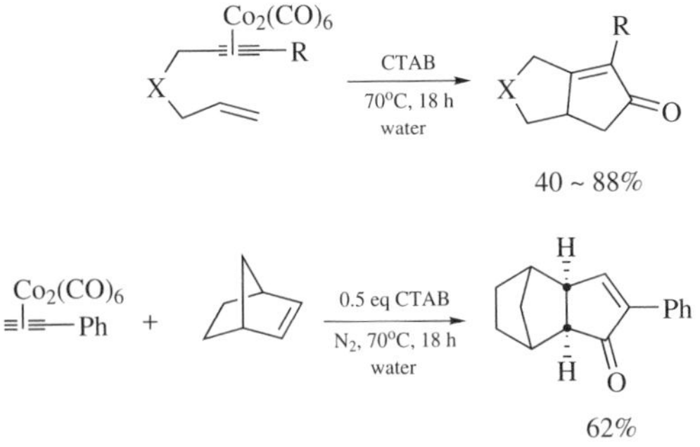

Scheme 4.11

intramolecular Pauson-Khand reaction in water without any cosolvent (Scheme 4.11).[108]

4.3.2 [2+2+2] Cyclotrimerization

Alkyne cyclotrimerization is an important synthetic methodology since aromatic rings are central to many pharmaceutical, biological, and polymer molecules. Numerous cyclotrimerization reactions in organic solvents have been reported using (cyclopentadienyl)cobalt dicarbonyl [CpCo(CO)$_2$].[109] Recently, alkyne cyclotrimerization catalyzed by CpCo was reported in supercritical water.[110] On the other hand, Sigman et al. reported alkyne cyclotrimerization in water at an elevated temperature by using a water-soluble cobalt catalyst, CpCo-η^4-cyclooctadiene (Eq. 4.58). Protection of functional groups such as amine, hydroxyl, ketone, ester, and carboxylic acid groups, was not required in this aqueous [2+2+2] cyclotrimerization chemistry.[111]

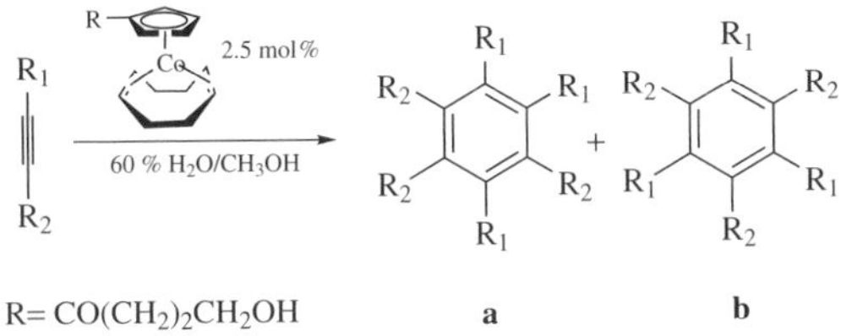

(4.58)

R= CO(CH$_2$)$_2$CH$_2$OH

R$_1$	R$_2$	yield (%)	a/b
H	COMe	44	0/44
H	COOMe	67	47/20
H	CH$_2$OH	85	62/23
H	CH$_2$NHCH$_3$	73	47/26
H	(CH$_2$)$_2$COOH	56	36/20

Yong et al. developed a cobalt-catalyzed [2+2+2] cyclotrimerization of terminal alkynes in good yields in aqueous media (80/20 mixture of water and ethanol) at room temperature. A cyclopentadienyl cobalt complex bearing a pendant phosphine ligand was used as a catalyst (Eq. 4.59). The cyclotrimerization of internal alkynes resulted in lower yields and required an elevated temperature, most likely due to steric interactions. For example, cyclotrimerization of 2,5-dimethyl-3-hexyne gave hexaisopropylbenzene in 51% yield and the reaction of diphenylethyne resulted in a 47% yield of hexaphenylbenzene.[112]

$$(4.59)$$

71–91% (major + minor)

R: Ph, *n*-Bu, *t*-Bu, HOCH$_2$, EtO$_2$C, Cl(CH$_2$)$_3$, *n*-decyl, *p*-tolyl

Rhodium also has been reported as a catalyst for [2+2+2] alkyne cycloaddition in water. Uozumi et al. explored the use of an amphiphilic resin-supported rhodium-phosphine complex as catalyst (Eq. 4.60). The immobilized rhodium catalyst was effective for the [2+2+2] cycloaddition of internal alkynes in water,[113] although the yields of products were not satisfactory.

$$(4.60)$$

R = COOMe   57%
R = Ph   29%

Rhodium catalysis in an aqueous-organic biphasic system was highly effective for intramolecular [2+2+2] cyclotrimerization. It has been shown that the use of a biphasic system could control the concentration of an organic hydrophobic substrate in the aqueous phase, thus increasing the reaction selectivity. The intramolecular cyclization for synthesis of medium- and large-sized ring systems usually requires a slow addition process or highly diluted reaction conditions to prevent intermolecular side reactions.[114] However, the process of slow addition is a time-consuming job and the high-dilution reaction conditions not only waste large quantities of solvent but also lower the reaction rate. Kinoshita et al. have successfully developed an intramolecular [2+2+2] cyclotrimerization in an aqueous-organic biphasic system catalyzed by a water-soluble rhodium catalyst that is prepared in situ from [RhCl(cod)]₂ and *tris*(sodium *m*-benzenesulfonate)phosphine (TPPTS) (Eq. 4.61).[115] Selective [2+2+2] cross-annulation between hydrophobic diynes such as 1,6-diyne and hydrophilic alkynes such as alkynols has also been achieved in good yields (Scheme 4.12). The low efficiency due to high-dilution conditions was possibly compensated for by the acceleration of such cycloaddition in water.[116] It was shown that medium and large rings were obtained in excellent yields with only trace amounts of polymeric compounds; no dimerization products were detected in the reaction mixture.

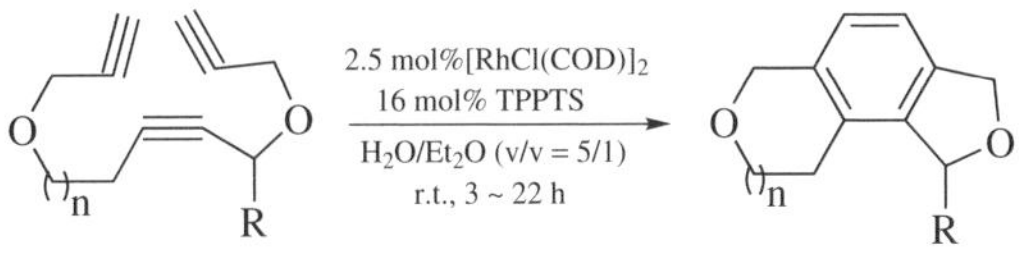

$$n = 0 \sim 5 \quad R = H \text{ or } CH_3 \qquad 79\text{–}93\%$$

(4.61)

Palladium has also been reported as an effective catalyst for [2+2+2] alkyne cyclotrimerization in water. Both aryl and alkylalkynes underwent

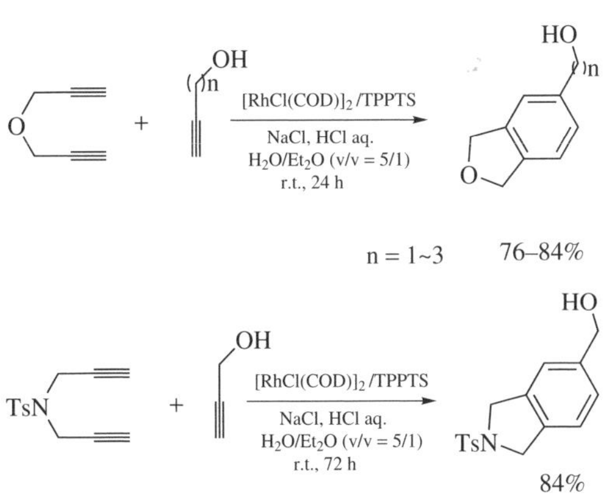

$$n = 1\text{~}3 \qquad 76\text{–}84\%$$

$$84\%$$

Scheme 4.12

cyclotrimerization to afford the corresponding products regioselectively in high yields in the presence of $PdCl_2$, $CuCl_2$ and CO_2 in water at room temperature.[117] For example, the cyclotrimerization of terminal alkynes gave symmetric benzenes and 1-phenylpropyne gave 1,3,5-triphenyl-2,4,6-trimethylbenzene. However, bulky substrates such as t-butylacetylene gave a dimerization product.

4.3.3 [2+2+2] Alkyne-Nitrile Cyclotrimerization

Alkyne-nitrile cyclotrimerization is a powerful synthetic methodology for the synthesis of complex heterocyclic aromatic molecules.[118] Recently, Fatland et al. developed an aqueous alkyne-nitrile cyclotrimerization of one nitrile with two alkynes for the synthesis of highly functionalized pyridines by a water-soluble cobalt(I) catalyst (Eq. 4.62). The reaction was chemospecific and several different functional groups such as unprotected alcohols, ketones, and amines were compatible with the reaction.[119] In addition, photocatalyzed [2+2+2] alkyne or alkyne-nitrile cyclotrimerization in water[120] and cyclotrimerization in supercritical H_2O[110,121] have been reported in recent years.

$$(4.62)$$

R: aliphatic, alkenyl, aromatic, amino   71–99%

4.3.4 [3+2] 1,3-Dipolar Cycloaddition

Triazoles show broad biological activities and have wide applications in organic synthesis.[122] One of the most important methods for the synthesis of triazoles is [3+2] 1,3-dipolar cycloaddition between an alkyne and an azide.[123] However, these cycloaddition reactions usually show low regioselectivities and result in 1,4- and 1,5-regioisomers.[124] Recently, Sharpless developed a Cu(I)-catalyzed reaction for the synthesis of 1,4-disubstituted 1,2,3-triazoles in excellent yields and near-perfect regioselectivity in aqueous media (a mixture of water and alcohol) or water without cosolvent (Eq. 4.63).[125] Various unprotected functional groups do not interfere with the reaction. The Cu(I) catalyst was generated in situ through the reduction of Cu(II) by sodium ascorbate. It

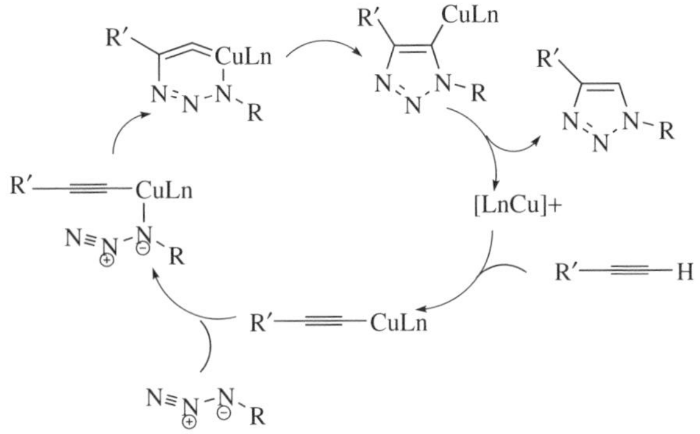

Scheme 4.13

was proposed that the reaction involved the process of a regioselective ligation of Cu(I) to azide (Scheme 4.13).

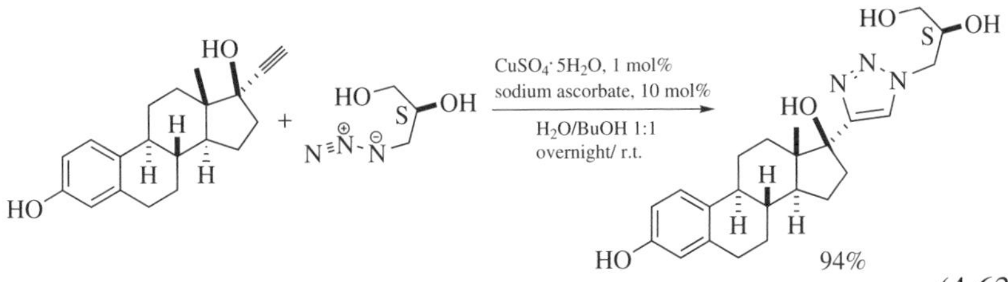

$$(4.63)$$

It is challenging to incorporate functional groups into biological molecules by organic synthesis to create a unique reactivity in large and complex targets since a large protein structure has numerous sensitive groups. However, Finn, Sharpless and co-workers applied the Cu(I)-catalyzed azide-alkyne [3+2] cycloaddition at room temperature to various biological molecules. Using a cowpea mosaic virus (CPMV) as the protein component, which is a readily available structurally rigid assembly of 60 identical copies of a two-protein asymmetric unit around a single-stranded RNA genome, a [3+2] cycloaddition occurred quantitatively toward the triazole product in the presence of a ligand of *tris*(triazolyl)amine, protected the virus from Cu-triazole-induced disassembly, and Cu(I) catalyst, which was generated in situ through the reduction of Cu(II)SO$_4$·5H$_2$O by the reductant of Cu wire or a water-soluble reducing agent of *tris*(carboxyethyl)phosphine (TCEP) (Eq. 4.64).[126]

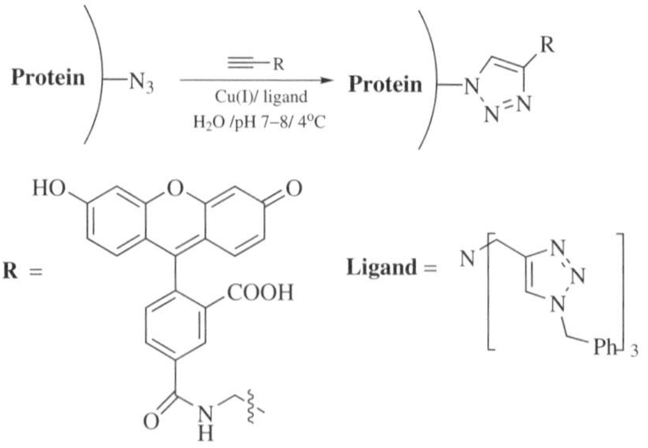

(4.64)

Water may enhance the regioselectivity of 1,3-dipolar cycloadditions. Wang et al. reported a regioselective 1,3-dipolar cycloaddition between terminal alkynes and azides to give triazoles without any catalysts in high yields in aqueous media.[127] The reaction of terminal arylalkynes afforded 1,4-disubstituted triazoles; whereas terminal aliphatic alkynes gave a mixture of regioisomers (Eq. 4.65). Water enhances the regioselectivity of the reaction. For example, phenylacetylene reacting with azidobenzene in water gave only the 1,4-disubstituted triazole product; whereas the same reaction in toluene under refluxing conditions afforded two regioisomers in approximately a 1:1 ratio.

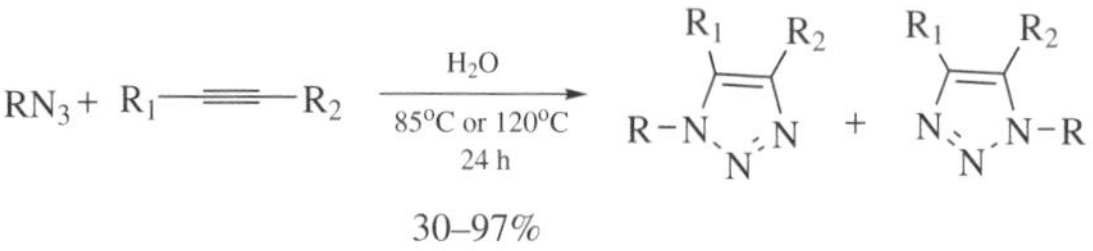

(4.65)

Recently, Li et al. have reported an efficient 1,3-dipolar cycloaddition of azides with electron-deficient alkynes without any catalysts at room temperature in water.[128] The reaction has been applied successfully to the coupling of an azido-DNA molecule with electron-deficient alkynes for the formation of [1,2,3]-triazole heterocycle (Eq. 4.66).

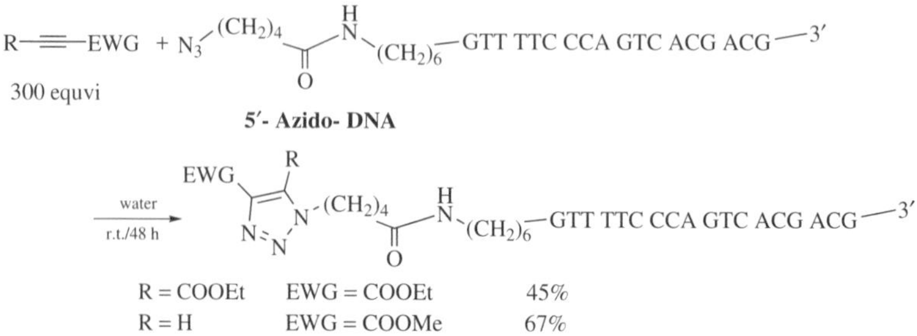

| R = COOEt | EWG = COOEt | 45% |
| R = H | EWG = COOMe | 67% |

(4.66)

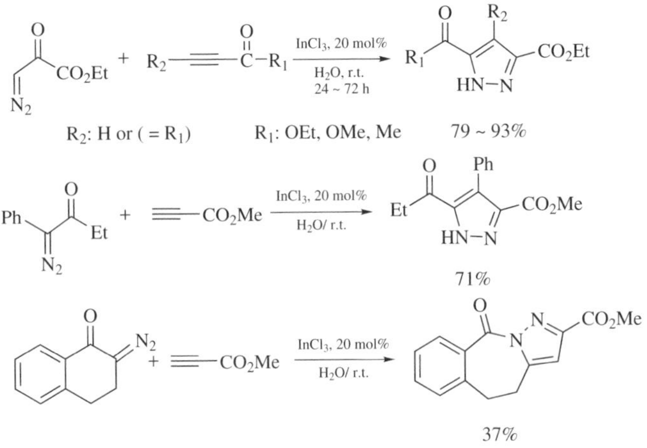

Scheme 4.14

Another type of [3+2] cycloaddition of alkynes with nitrogen-containing compounds is the 1,3-dipolar cycloaddition of diazo compounds with alkynes, which gives pyrazoles. Such structures show antitumor, anti-inflammatory, antimicrobial, and antipsychotic activities.[129] Recently, Li et al. reported the first intermolecular 1,3-dipolar cycloaddition of diazocarbonyl compounds with alkynes catalyzed by $InCl_3$ in water. The reaction proceeds by a domino 1,3-dipolar cycloaddition–hydrogen (alkyl or aryl) migration (Scheme 4.14).[130] It is worth mentioning that using water as solvent not only plays a crucial role in the reaction, but also enables the ready recycling of the $InCl_3$ catalyst. $InCl_3$ was reused two additional times for the reaction of ethyl diazoacetate and propiolate in water without loss of catalytic activity (yields were 87%, 89%, and 90% for the first, second, and third run respectively). In contrast, this reaction in organic solvents (CH_2Cl_2 or benzene, the common solvent for diazo chemistry) only gave a trace amount of the target product.

Another interesting [3+2] cycloaddition in aqueous media was recently reported by Murakami.[131] 2-Cyanophenylboronic acid reacted as a three-carbon component with alkynes or alkenes to afford substituted indenones or indanones (Eq. 4.67). The use of an alkynoate even produced benzotropone, a formal [3 + 2 + 2] adduct.

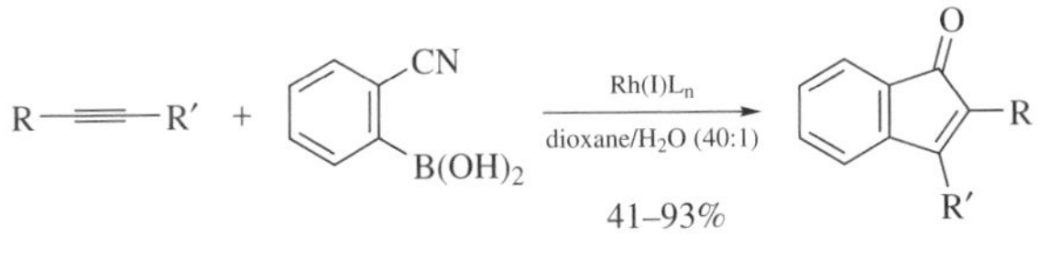

(4.67)

4.3.5 [4+2] Cycloaddition

Oshima et al. explored a cationic rhodium-catalyzed intramolecular [4+2] annulation of 1,3-dien-8-ynes in water in the presence of sodium dodecyl sulfate (SDS), an anionic surfactant.[132] When the substrate 1,3-dien-8-yne was a terminal alkyne, the reaction provided an intermolecular [2+2+2] product (Eq. 4.68). In water, a reactive cationic rhodium species was formed by the dissociation of the Rh–Cl bond in the presence of SDS. The SDS forms negatively charged micelles, which would concentrate the cationic rhodium species (Scheme 4.15).

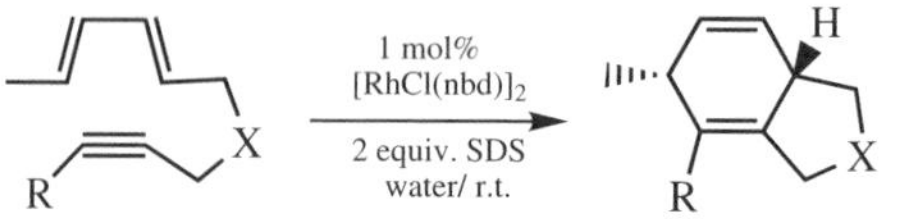

$$(4.68)$$

R = CH$_2$OH,	X = O,	93%
R = CH$_2$OBn,	X = O,	97%
R = SiMe$_3$,	X = O,	99%
R = CH$_3$,	X = NBn,	95%
R = Ph,	X = CH(COOEt)$_2$,	71%

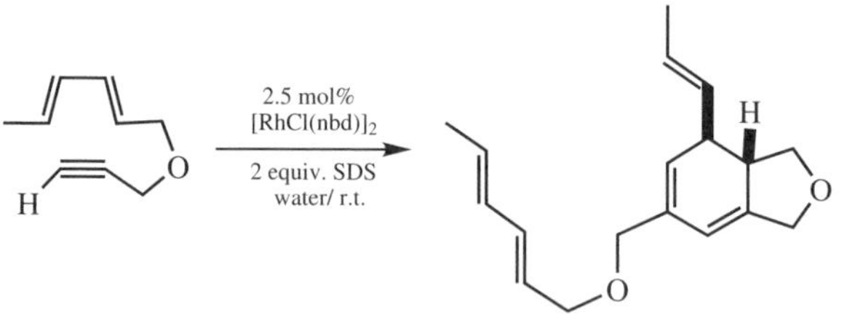

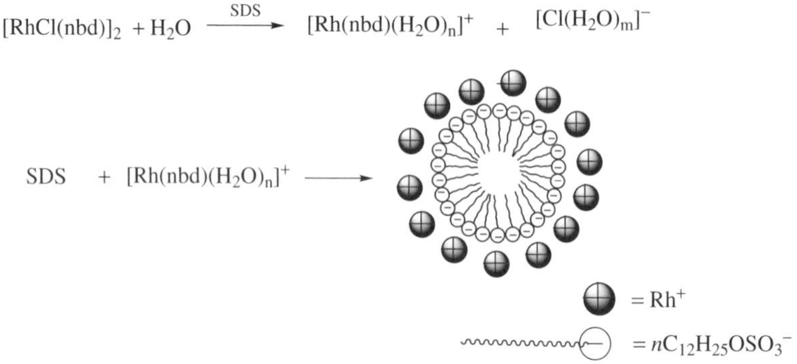

Scheme 4.15

4.3.6 [5+2] Cycloaddition

Wender et al. reported a [5+2] cycloaddition in water by using a water-soluble rhodium catalyst having a bidentate phosphine ligand to give a 7-membered ring product (Eq. 4.69). This water-soluble catalyst was reused eight times without any significant loss in catalytic activity.[133]

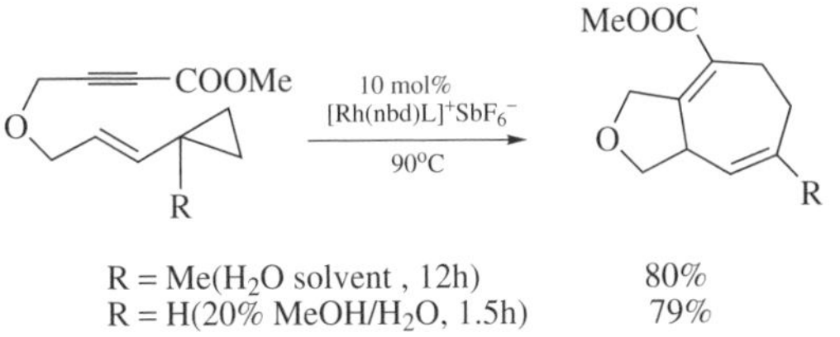

$$
\text{(4.69)}
$$

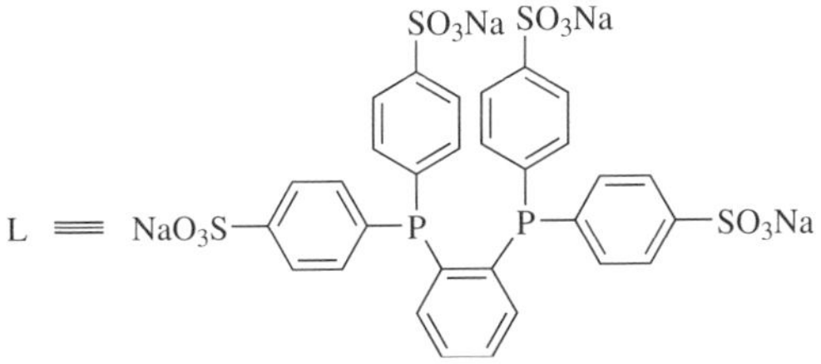

R = Me(H$_2$O solvent , 12h)   80%
R = H(20% MeOH/H$_2$O, 1.5h)   79%

$$
L \equiv
$$

4.3.7 Other Cycloadditions

It has been reported that nickel catalyzed the reactions of 6-amino-1,3-dimethyluracil with substituted alkynylketones in water to give substituted 2,4-dioxopyrido[2,3-d]pyrimidine derivatives in quantitative yields at room temperature (Eq. 4.70).[134] The products have potential pharmacological and biological activities. The reaction may have proceeded through an ionic process.

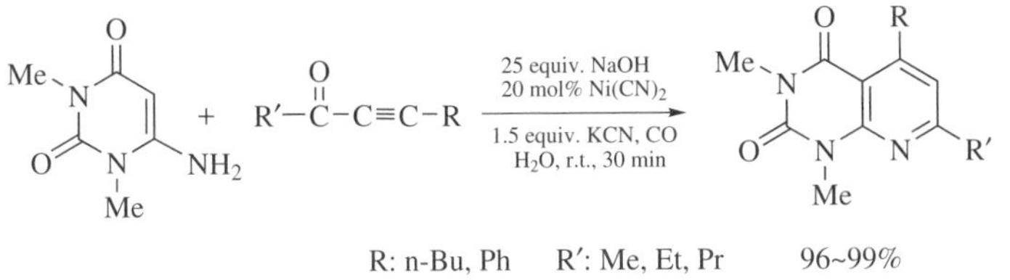

$$
\text{(4.70)}
$$

R: n-Bu, Ph   R′: Me, Et, Pr   96~99%

Poly-(acetylene)s are widely used in different fields, such as organic light-emitting diodes (OLEDS), solar cells, and lasers.[135] Synthesis of poly(acetylene)s in water is very attractive for many industrial applications since the reaction can form a stable aqueous dispersion that can be directly deposited on suitable supports. However, the synthesis of poly(acetylene) usually requires anhydrous conditions. Recently, Buchmeiser et al. explored the synthesis of poly(acetylene) in water catalyzed by a poly(2-oxazoline)-immobilized ruthenium catalyst.[136] The polymerization of diethyl dipropargyl malonate (DEDPM) and chiral 4-(ethoxycarbonyl)-4-(*1 S, 2 R, 5 S*)-(+)-menthoxycarbonyl-1,6-heptadiyne (ECMCH) gave quantitative yields of the expected polymers with molecular weight ranging from 10,000 to ~16,000 g mol^{-1} (Scheme 4.16). The immobilization of the Ru catalyst separated the product from the catalyst and provided Ru-free poly(acetylene)s.

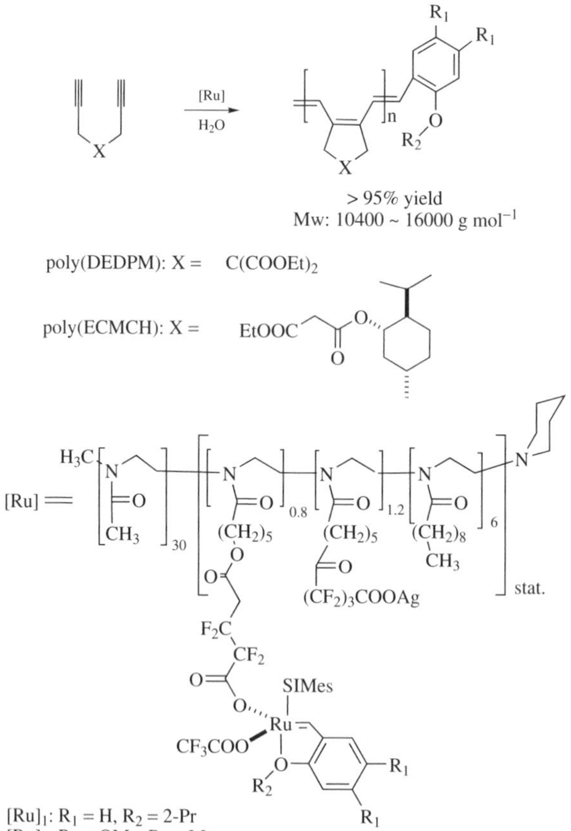

> 95% yield
Mw: 10400 ~ 16000 g mol^{-1}

poly(DEDPM): X = $C(COOEt)_2$

poly(ECMCH): X =

[Ru]$_1$: R$_1$ = H, R$_2$ = 2-Pr
[Ru]$_2$: R$_1$ = OMe, R$_2$ = Me
SIMes = 1,3-*bis*(2,4,6-trimethylphenyl)4,5-dihydroimidazolin-2-ylidene

Scheme 4.16

4.4 OTHER REACTIONS

In addition to the reactions discussed above, there are still other alkyne reactions carried out in aqueous media. Examples include the *Pseudomonas cepacia* lipase-catalyzed hydrolysis of propargylic acetate in an acetone-water solvent system,[137] the ruthenium-catalyzed cycloisomerization-oxidation of propargyl alcohols in DMF-water,[138] an intramolecular allylindination of terminal alkyne in THF-water,[139] and alkyne polymerization catalyzed by late-transition metals.[140]

REFERENCES

1. Brandsma, L., in *Preparative Acetylenic Chemistry*, Heus-Kloos, Y. A., Van Der Heiden, R., Verkruijsse, H. D., eds., 2nd ed. Elsevier, New York, 1988; Brandsma, L., Vasilevsky, S. F., Verkruijsse, H. D., *Application of Transition Metal Catalysts in Organic Synthesis*. Springer-Verlag; New York, 1999.

2. Glaser, C., *Ber. Dtsch. Chem. Ges.* **1869**, *2*, 422.

3. Baeyer, A., *Ber. Dtsch. Chem. Ges.* **1882**, *15*, 50.

4. For an example, see: Anderson, S., Anderson, H. L., *Angew. Chem., Int. Ed. Engl.* **1996**, *35*, 1956.

5. For reviews, see: Cadiot, P., Chodkiewicz, W., *Chemistry of Acetylenes* Viehe, H. G., ed. (1969), 597.

6. Amatore, C., Blart, E., Genet, J. P., Jutand, A., Lemaire-Audoire, S., Savignac, M., *J. Org. Chem.* **1995**, *60*, 6829.

7. Li, P-H., Yan, J-C., Wang, M., Wang, L., *Chin. J. Chem.* **2004**, *22*, 219.

8. Straus, F., *Annalen* **1905**, *342*, 190.

9. Baidossi, W., Goren, N., Blum, J., *J. Mol. Cat.* **1993**, *85*, 153.

10. Trost, B. M., Chan, C., Ruhter, G., *J. Am. Chem. Soc.* **1987**, *109*, 3486.

11. Chen, L., Li, C.-J., *Tetrahedron Lett.* **2004**, *45*, 2771.

12. Masuda, Y., Murata, M., Sato, K., Watanabe, S., *Chem. Commun.* **1998**, 807.

13. Sonogashira, K., in *Handbook of Organopalladium Chemistry for Organic Synthesis*, Negishi, E., ed. Wiley, Hoboken, 2002, p. 493.

14. Casalnuovo, A. L., Calabrese, J. C., *J. Am. Chem. Soc.* **1990**, *112*, 4324.

15. Genet, J. P., Blart, E., Savignac, M., *Synlett*, **1992**, 715.

16. Davydov, D. V., Beletskaya, I. P., *Russian Chem. Bull.* **1995**, *44*, 965; Bumagin, N. A., Bykov, V. V., Beletskaya, I. P., *Russ. J. Org. Chem.* **1995**, *31*, 348.

17. Davydov, D. V., Beletskaya, I. P., *Russ. Chem. Bull.* **1995**, *44*, 965.

18. Vlassa, M., Ciocan-Tarta, I., Margineanu, F., Oprean, I., *Tetrahedron* **1996**, *52*, 1337.

19. Pal, M., Subramanian, V., Yeleswarapu, K. R., *Tetrahedron Lett.* **2003**, *44*, 8221; Uozumi, Y., Kobayashi, Y., *Heterocycles* **2003**, *59*, 71.

20. Li, C.-J., Chen, D. L., Costello, C. W., *Org. Res. Proc. Devel.* **1997**, *1*, 315.

21. Li, C.-J., Slaven, W. T. IV, John, V. T., Banerjee, S., *Chem. Commun.* **1997**, 1569.

22. Li,C.-J., Slaven, W. T. IV, Chen, Y. P., John, V. T., Rachakonda, S. H., *Chem. Commun.* **1998**, 1351.

23. Pirguliyev, N. Sh., Brel, V. K., Zefirov, N. S., Stang, P. J., *Tetrahedron* **1999**, *55*, 12377; Sheremetev, A. B., Mantseva, E. V., *Tetrahedron Lett.* **2001**, *42*, 5759.

24. Beletskaya, I. P., Latyshev, G. V., Tsvetkov, A. V., Lukashev, N. V., *Tetrahedron Lett.* **2003**, *44*, 5011.

25. (a) Appukkuttan, P., Dehaen, W., Van der Eycken, E., *Eur. J. Org. Chem.* **2003**, 4713. (b) Arvela, R. K., Leadbeater, N. E., Sangi, M. S., Williams, V. A., Granados, P., Singer, R. D., *J. Org. Chem.* **2005**, *70*, 161.

26. Chen L., Li, C.-J., *Org. Lett.* **2004**, *6*, 3151.

27. Bergbreiter, D. E., Liu, Y-S., *Tetrahedron Lett.* **1997**, *38*, 7843.

28. Xia, M., Wang, Y. G., *Chinese Chemical Lett.*. **2001**, *12*, 941.

29. Quignard, F., Larbot, S., Goutodier, S., Choplin, A., *J. Chem. Soc., Dalton Trans.*, **2002**, 1147.

30. Uozumi, Y., Kobayashi, Y., *Heterocycles* **2003**, *59*, 71.

31. Dibowski, H., Schmidtchen, F. P., *Tetrahedron Lett.* **1998**, *39*, 525.

32. Dibowski, H., Schmidtchen, F. P., *Angew. Chem. Int. Ed.* **1998**, *37*, 476.

33. Bong, D. T., Ghadiri, M. R., *Org. Lett.* **2001**, *3*, 2509.

34. Lopez-Deber, M. P., Castedo, L., Granja, J. R., *Org. Lett.*, **2001**, *3*, 2823.

35. Wang, L., Li, P., *Syn. Comm.* **2003**, *33*, 3679.

36. Pal, M., Subramanian, V., Parasuraman, K., Yeleswarapu, K. R., *Tetrahedron* **2003**, *59*, 9563.

37. Nájera, C., Gil-Moltó, J., Karlström, S., Falvello, L. R., *Org. Lett.* **2003**, *5*, 1451.

38. Wolf, C., Lerebours, R., *Org. Biomol.Chem.* **2004**, *2*, 2161.

39. Beletskaya, I. P., Latyshev, G. V., Tsvetkov, A. V., Lukashev, N. V., *Tetrahedron Lett.* **2003**, *44*, 5011.

40. Chodkiewicz, W., *Ann. Chim. (Paris)* **1957**, *2*, 819; Chodkiewicz, W., Cadiot, P., *Corpt. Rend. Hebd. Seances Acad. Sci.* **1955**, *241*, 1055.

41. Gardner, J. N., Jones, E. R. H., Leeming, P. R., Stephenson, J. S., *J. Chem. Soc. 1960*, 691.

42. Amatore, C., Blart, E., Genet, J. P., Jutand, A., Lemaire-Audoire, S., Savignac, M., *J. Org. Chem.* **1995**, *60*, 6829.

43. Dibowski, H., Schmidtchen, F. P., *Tetrahedron* **1995**, *51*, 2325.

44. Marino, J. P., Nguyen, H. N., *J. Org. Chem.* **2002**, *67*, 6841.

45. Kang, S-K., Lim, K-H., Ho, P-S., Kim, W-Y., *Synthesis.* **1997**, 874.

46. Mohamed Ahmed, M. S., Mori, A., *Org. Lett.* **2003**, *5*, 3057; see also Liang, B., Huang, M., You, Z., Xiong, Z., Lu, K., Fathi, R., Chen, J., Yang, Z., *J. Org. Chem.* **2005**, *70*, 6097.

47. Wei, C., Li, C.-J., *Green Chem.* **2002**, *4*, 39; Li, C. J., *Acc. Chem. Res.* **2002,** *35*, 533.

48. Yao, X, Li, C.-J., *Org. Lett.* **2005**, *7*, 4395.

49. Yao, X, Li, C.-J., *Org. Lett.* **2006**, *8*, 1953.

50. Bloch, R. *Chem. Rev.*, **1998**, *98*, 1407; Aubrecht, K. B., Winemiller, M. D., Collum, D. B., *J. Am. Chem. Soc.*, **2000**, *122*, 11084.

51. Li, C.-J., Wei, C., *Chem. Comm.* **2002**, 268.

52. Wei, C., Li, C.-J., *J. Am. Chem. Soc.* **2003**, *125*, 9584.

53. Wei, C., Li, Z., Li, C.-J., *Org. Lett.* **2003**, *5*, 4473.

54. Wei, C., Mague, J. T., Li, C.-J., *Proc. Natl. Acad. Sci. (USA)* **2004**, *101*, 5749; Wei, C., Li, Z., Li, C.-J., *Synlett.* **2004**, 1472; Wei, C., Li, C.-J., *J. Am. Chem. Soc.* **2002**, *124*, 5638.

55. For recent reviews on microwave irradiation-assisted organic synthesis, see: Kuhnert, N., *Angew. Chem., Int. Ed.* **2002**, *41*, 1863; Lidstrom, P., Tierney, J., Wathey, B., Westman, J., *Tetrahedron* **2001**, *57*, 9225.

56. Shi, L., Tu, Y-Q., Wang, M., Zhang, F-M., Fan, C-A., *Org. Lett.* **2004**, *6*, 1001.

57. Zhang, J., Wei, C., Li, C.-J., *Tetrahedron Lett.* **2002**, *43*, 5731.

58. Huang, B., Yao, X., Li, C.-J., *Adv. Synth. Catal.* **2006**, *348*, 1528.

59. Knopfel, T. F., Carreira, E. M., *J. Am. Chem. Soc.* **2003**, *125*, 6054.

60. Chen, L., Li, C.-J., *Chem. Comm.* **2004**, 2362.

61. Lerum, R. V., Chisholm, J. D., *Tetrahedron Lett.* **2004**, *45*, 6591.

62. Horvath, H. H., Joo, F., *React. Kinet. Catal. Lett.* **2005**, *85*, 355.

63. Tour, J. M., Cooper, J. P., Pendalwar, S. L., *J. Org. Chem.* **1990**, *55*, 3452.

64. Demerseman, B., Dixneuf, P. H., *J. Chem. Soc. Chem. Commun.* **1981**, 665.

65. Jennings, J. M., Bryson, T. A., Gibson, J. M., *Green Chemistry* **2000**, *2*, 87; Cai, Z-S., Kuntz, R. R., *Langrnuir* **1988**, *4*, 830; Degani, Y., Willner, I., *J. Chem. Soc., Perkin Trans. 2*, **1986**, 37.

66. Frieman, R. H., Kennedy, E. R., Lucas, H. J., *J. Am. Chem. Soc.* **1937**, *59*, 722.

67. Taqui Khan, M. M., Halligudi, S. B., Shukla, S., *J. Mol. Catal.* **1990**, *58*, 299.

68. March, J., *Advanced Organic Chemistry*, 4th ed. Wiley, New York, 1992, p. 762; Damiano, J. P., Postel, M., *J. Organomet. Chem.* **1996**, *522*, 303.

69. Tokunaga, M., Wakatsuki, Y., *Angew. Chem., Int. Ed.* **1998**, *37*, 2867.

70. Alvarez, P., Bassetti, M., Gimeno, J., Mancini, G., *Tetrahedron Lett.* **2001**, *42*, 8467.

71. Lucey, D. W., Atwood, J. D., *Organometallics* **2002**, *21*, 2481.

72. Baidossi, W., Lahav, M., Blum, J., *J. Org. Chem.* **1997**, *62*, 669.

73. Iimura, S., Manabe, K., Kobayashi, S., *Org. Biomol. Chem.* **2003**, *1*, 2416.

74. Vasudevan, A., Verzal, M. K., *Synlett.* **2004**, 631.

75. For recent reviews, see: Beller, M., Seayad, J., Tillack, A., Jiao, H., *Angew. Chem. Int. Ed.* **2004**, *43*, 3368; Alonso, F., Beletskaya, I. P., Yus, M., *Chem. Rev.* **2004**, *104*, 3079.

76. Bergbreiter, D. E., Case, B. L., Liu, Y.-S., Caraway, J. W., *Macromolecules* **1998**, *31*, 6053.

77. Uozumi, Y., Kobayashi, Y., *Heterocycles* **2003**, *59*, 71.

78. Pal, M., Subramanian, V., Yeleswarapu, K. R., *Tetrahedron Lett.* **2003**, *44*, 8221.

79. (a) Beller, M., Seayad, J., Tillack, A., Jiao, H., *Angew. Chem. Int. Ed.* **2004**, *43*, 3368. (b) Alonso, F., Beletskaya, I. P., Yus, M., *Chem. Rev.* **2004**, *104*, 3079.

80. Arcadi, A., Bianchi, G., Marinelli, F., *Synthesis* **2004**, 610.

81. Vasudevan, A., Verzal, M. K., *Synlett.* **2004**, 631.

82. Denmark, S. E., Neuville, L., *Org. Lett.* **2000**, *2*, 3221; Hatanaka, Y., Hiyama, T., *Synlett*, **1991**, 845; McIntosh, M. C., Weinreb, S. M., *J. Org. Chem.* **1991**, *56*, 5010; Blumenkopf, T. A., Overman, L. E., *Chem. Rev.* **1986**, *86*, 857; Corey, E. J., Seibel, W. L., *Tetrahedron Lett.* **1986**, *27*, 905; Wolfsberger, W., *J. Prakt. Chem.* **1992**, *334*, 453.

83. For recent examples: Wang, F., Neckers, D. C., *J. Organomet. Chem.* **2003**, *665*, 1; Motoda, D., Shinokubo, H., Oshima, K., *Synlett*, **2002**, 1529; Kawanami, Y., Sonoda, Y., Mori, T., Yamamoto, K., *Org. Lett.* **2002**, *4*, 2825.

84. Speier, J. L., Webster, J. A., Barnes, G. H., *J. Am. Chem. Soc.* **1957**, *79*, 974.

85. Wu, W., Li, C.-J., *Chem. Comm.* **2003**, 1668.

86. Sato, A., Kinoshita, H., Shinokubo, H., Oshima, K., *Org. Lett.* **2004**, *6*, 2217.

87. Kinoshita, H., Nakamura, T., Kakiya, H., Shinokubo, H., Matsubara, S., Oshima, K., *Org. Lett.* **2001**, *3*, 2521.

88. Wu, W., Li, C.-J., *Lett. Org. Chem.* **2004**, *1*, 122.

89. Hayashi, T., Inoue, K., Taniguchi, N., Ogasawara, M., *J. Am. Chem. Soc.* **2001**, *123*, 9918.

90. Lautens, M., Yoshida, M., *Org. Lett.* **2002**, *4*, 123; Lautens, M., Yoshida, M., *J. Org. Chem.* **2003**, *68*, 762.

91. Genin, E., Michelet, V., Genet, J.-P., *Tetrahedron Lett.* **2004**, *45*, 4157.

92. Alper, H., Currie, J. K., des Abbayes, H., *J. Chem. Soc., Chem. Commun.* **1978**, 311; Alper, H., Petrignani, J.-F., *J. Chem. Soc., Chem. Commun.* **1983**, 1154.

93. Arzoumanian, H., Cochini, F., Nuel, D., Rosas, N., *Organometallics* **1993**, *12*, 1871; Arzoumanian, H., Cochini, F., Nuel, D., Petrignani, J. F., Rosas, N., *Organometallics* **1992**, *11*, 493.

94. Arzoumanian, H., Nuel, D., Jean, M., Cabrera, A., Garcia, J. L., Rosas, N., *Organometallics* **1995**, *14*, 5438.

95. Arzoumanian, H., Jean, M., Nuel, D., Cabrera, A., Guiterrez, J. L. G., Rosas, N., *Organometallics* **1997**, *16*, 2726.

96. Rosas, N., Salmón, M., Sharma, P., Alvarez, C., Ramirez, R., Garcia, J. L., Arzoumanian, H., *J. Chem. Soc., Perkin Trans. 1*, **2000**, 1493.

97. Gabriele, B., Veltri, L., Salerno, G., Costa, M., Chiusoli, G. P., *Eur. J. Org. Chem.* **2003**, 1722.

98. Srinivasan, N. S., Lee, D. G., *J. Org. Chem.* **1979**, *44*, 1574.

99. Gorgues, A., Stephan, D., Cousseau, J., *J. Chem. Soc., Chem. Commun.* **1989**, 1493.

100. Larpent, C., Meignan, G., *Tetrahedron Lett.* **1993**, *34*, 4331; Larpent, C., Meignan, G., Patin, H., *Tetrahedron* **1990**, *46*, 6381.

101. Derien, S., Jan, D., Dixneuf, P. H., *Tetrahedron* **1996**, *52*, 5511.

102. Galindo, A. C., Oliveira, J. M., Barboza, M. A. G., Goncalves, S. M. C., Henezes, P. H., *Phos. Sulf. Silic. Relat. Elem.* **2001**, *171–172*, 383.

103. Motoda, D., Kinoshita, H., Shinokubo, H., Oshima, K., *Adv. Synth. Catal.* **2002**, *344*, 261.

104. Li, J-H., Xie, Y-X., Yin, D-L., *Green Chemistry* **2002**, *4*, 505.

105. Pauson, P. L., in *Organometallics in Organic Synthesis,* de Meijere, A., tom Dieck, H., eds. Springer, Berlin, 1987, p. 233; Pauson, P. L., *Tetrahedron* **1985**, *41*, 5855; Khand, I. U., Knox, G. R., Pauson, P. L., Watts, W. E., Foreman, M. I., *J. Chem. Soc., Perkin Trans. 1*, **1973**, 977;

Pauson, P. L., Khand, I. U., Knox, G. R., Watts, W. E., *J. Chem. Soc., Chem. Commun.* **1971**, 36.

106. Sugihara, T., Yamada, M., Ban, H., Yamaguchi, M., Kaneko, C., *Angew. Chem., Int. Ed. Engl.* **1997**, *36*, 2801.

107. Son, S. U., Lee, S. I., Chung, Y. K., Kim, S-W., Hyeon, T., *Org. Lett.* **2002**, *4*, 277.

108. Krafft, M. E., Wright, J. A., Bonaga, L. V. R., *Tetrahedron Lett.* **2003**, *44*, 3417.

109. Vollhardt, K. P. C., Bergman, R. G., *J. Am. Chem. Soc.* **1974**, *96*, 4996; Vollhardt, K. P. C., *Angew. Chem., Int. Ed. Engl.* **1984**, *23*, 539. For excellent reviews on metal-mediated cycloaddition, see: Grotjahn, D. B., in *Comprehensive Organometallic Chemistry II,* Abel, E. W., Stone, F. G. A., Wilkinson, G., Hegedus, L. S., eds. Pergamon, Tarrytown, NY, 1995, Vol. 12, pp. 741–770; Schore, N. E., in *Comprehensive Organic Synthesis,* Trost, B. M., Fleming, I., Paquette, L. A., eds. Pergamon, Elmsford, NY, 1991, Vol. 5, pp. 1129–1162.

110. Jerome, K. S., Parsons, E. J., *Organometallics* **1993**, *12*, 2991.

111. Sigman, M. S., Fatland, A. W., Eaton, B. E., *J. Am. Chem. Soc.* **1998**, *120*, 5130.

112. Yong, L., Butenschöen, H., *Chem. Comm.* **2002**, 2852.

113. Uozumi, Y., Nakazono, M., *Adv. Synth. Catal.* **2002**, *344*, 274.

114. Dietrich, B., Viout, P., Lehn, J.-M., *Macrocyclic Chemistry*, Wiley-VCH: Weinheim, 1993.

115. Kinoshita, H., Shinokubo, H., Oshima, K., *J. Am. Chem. Soc.* **2003**, *125*, 7784.

116. For rate enhancement of the Diels-Alder reaction in aqueous media, see: Lubineau, A., Auge, J., in *Modern Solvent in Organic Synthesis*, Knochel, P., ed. Springer-Verlag, Berlin, 1999, p 1; *Organic Synthesis in Water*, Grieco, P. A., Ed., Blackie Academic & Professional, London, 1998; Breslow, R., *Acc. Chem. Res.* **1991**, *24*, 159; Grieco, P. A., *Aldrichimica Acta* **1991**, *24*, 59.

117. Li, J-H., Xie, Y-X., *Synth. Commun.* **2004**, *34*, 1737.

118. For examples, see: Grotjahn, D. B., Vollhardt, K. P. C., *Synthesis* **1993**, 579; Vollhardt, K. P. C., *Angew. Chem., Int. Ed. Engl.* **1984**, *23*, 539; Vollhardt, K. P. C., Bergman, R. G., *J. Am. Chem. Soc.* **1974**, *96*, 4996; Wakatsuki, Y., Yamazaki, H., *Tetrahedron Lett.* **1973**, *14*, 3383.

119. Fatland, A. W., Eaton, B. E., *Org. Lett.,* **2000**, *2*, 3131.

120. Heller, B., Sundermann, B., Buschmann, H., Drexler, H.-J., You, J., Holz-grabe, U., Heller, E., Oehme, G., *J. Org. Chem.* **2002**, *67*, 4414; Heller, B., *Nachr. Chem. Tech. Lab.* **1999**, *47*, 9; Heller, B., Heller, D., Wagler, P., Oehme, G., *J. Mol. Catal. A: Chem.* **1998**, *136*, 219; Heller, B.,

Heller, D., Oehme, G., *EPA Newsl.* **1998**, *62*, 37; Wagler, P., Heller, B., Ortner, J., Funken, K. H., Oehme, G., *Chemie-Ingenieur-Technik.* **1996**, *68*, 823; Heller, B., Oehme, G., *J. Chem. Soc., Chem. Commun.* **1995**, 179; Karabet, F., Heller, B., Kortus, K., Oehme, G., *Appl. Organomet. Chem.* **1995**, *9*, 651; Heller, B., Reihsig, J., Schulz, W., Oehme, G., *Appl. Organomet. Chem.* **1993**, *7*, 641.

121. Borwieck, H., Walter, O., Dinjus, E., Rebizant, J., *J. Organomet.Chem.* **1998**, *570*, 121.

122. Sikora, D., Gajda, T., *Tetrahedron* **1998**, *54*, 2243; Wamhoff, H., in *Comprehensive Heterocyclic Chemistry*, Katritzky, A. R., Rees, C. W., eds. Pergamon: Oxford, 1984, Vol. 5, pp. 669–732.

123. Finley, K. T., in *The Chemistry of Heterocyclic Compounds,* Montgomery, J. A., Ed. John Wiley & Sons, New York, 1980, *Vol 39.*

124. Howell, S. J., Spencer, N., Philp, D., *Tetrahedron*, **2001**, *57*, 4945; Gothelf, K. V., Jorgensen, K. A., *Chem. Rev.* **1998**, *98*, 863; Padwa A., in *Comprehensive Organic Synthesis*, Trost, B. M., Ed. Pergamon, Oxford, 1991, Vol. 4, pp. 1069–1109; Huisgen, R. in *1,3-Dipolar Cycloaddition Chemistry,* Padwa. A., Ed. Wiley, New York, 1984, pp. 1—176.

125. Rostovtsev, V. V., Green, L. G., Fokin, V. V., Sharpless, K. B., *Angew. Chem. Int. Ed.* **2002**, *41*, 2596.

126. Wang, Q., Chan, T. R., Hilgraf, R., Fokin, V. V., Sharpless, K. B., Finn, M. G., *J. Am. Chem. Soc.* **2003**, *125*, 3192.

127. Wang, Z-X., Qin, H-L., *Chem. Commun.* **2003**, 2450.

128. Li, Z., Seo, T. S., Ju, J., *Tetrahedron Lett.* **2004**, *45*, 3143.

129. Daidone, G., Maggio, B., Plescia, S., Raffa, D., Musiu, C., Milia, C., Perra, G., Marongiu, M. E., *Eur. J. Med. Chem.* **1998**, *33*, 375; Tsuji, K., Nakamura, K., Konishi, N., Tojo, T., Ochi, T., Senoh, H., Matsuo, M., *Chem. Pharm. Bull.* **1997**, *45*, 987; Nauduri, D., Reddy, G. B., *Chem. Pharm. Bull.* **1998**, *46*, 1254.

130. Jiang, N., Li, C.-J., *Chem. Comm.* **2004**, 394.

131. Miura, T., Murakami, M., *Org. Lett.* **2005**, *7*, 3339.

132. Motoda, D., Kinoshita, H., Shinokubo, H., Oshima, K., *Angew. Chem. Int. Ed.* **2004**, *43*, 1860.

133. Wender, P. A., Love, J. A., Williams, T. J., *Synlett.* **2003**, 1295.

134. Rosas, N., Sharma, P., Alvarez, C., Cabrera, A., Ramírez, R., Delgado, A., Arzoumanian, H., *J. Chem. Soc., Perkin Trans. 1*, **2001**, 2341.

135. MacDiarmid, A. G., *Angew. Chem. Int. Ed.* **2001**, *40*, 2581; Heeger, A. J., *Angew. Chem. Int. Ed.* **2001**, *40*, 2591; Shirakawa, H., *Angew. Chem. Int. Ed.* **2001**, *40*, 2575; *Handbook of Conducting Polymers*, Skotheim, T. A., Elsenbaumer, R. L., Reynolds, J. R., eds., 2nd ed. Dekker, New York, 1998.

136. Krause, J. O., Zarka, M. T., Anders, U., Weberskirch, R., Nuyken, O., Buchmeiser, M. R., *Angew. Chem. Int. Ed.* **2003**, *42*, 5965.

137. Kamezawa, M., Raku, T., Tachibana, H., Ohtani, T., Naoshima, Y., *Biosci. Biotechn. Biochem.* **1995**, *59*, 549.

138. Trost, B. M., Rhee, Y. H., *J. Am. Chem. Soc.* **1999**, *121*, 11680.

139. Salter, M. M., Sardo-Inffiri, S., *Synlett* **2002**, 2068.

140. Claverie, J. P., Soula, R., *Progr. Polym. Sci.* **2003**, *28*, 619.

CHAPTER 5

ALCOHOLS, PHENOLS, ETHERS, THIOLS, AND THIOETHERS

5.1 OXIDATION OF ALCOHOLS

Of all the reactions involving alcohols in water, the most extensively studied are the oxidation reactions. Such reactions can be carried out by using a variety of oxidizing reagents. Enzyme-catalyzed oxidation of alcohol is a natural process in water.[1] Early studies of alcohol oxidation in water have been on the use of stoichiometric high-valent chromium and manganese compounds. For chromium-based oxidation in aqueous media, strong acid can catalyze the oxidation.[2] In a binary solvent mixture of acetic acid and water, the rate of the oxidation for secondary alcohols decreases with the increase of the proportion of water.[3] Recently, the preparation of aldehydes and ketones by the oxidation of the corresponding primary and secondary alcohols with potassium dichromate in a neutral benzene-water system has been reported.[4] Another oxidizing reagent used for oxidation of alcohol in aqueous media is CAN (cerium ammonium nitrate). Hydroquinones, catechols, and their derivatives are oxidized to quinones by this system in aqueous acetonitrile.[5] While unactivated primary alcohols are not reactive under the Ce(IV) conditions, a variety of secondary and benzylic alcohols are

Comprehensive Organic Reactions in Aqueous Media, Second Edition, by Chao-Jun Li and Tak-Hang Chan
Copyright © 2007 John Wiley & Sons, Inc.

oxidized by CAN to carbonyl compounds. Tertiary alcohols are oxidatively cleaved to give a variety of products (Eq. 5.1).[6] The subject has been reviewed in detail by Molander.[7]

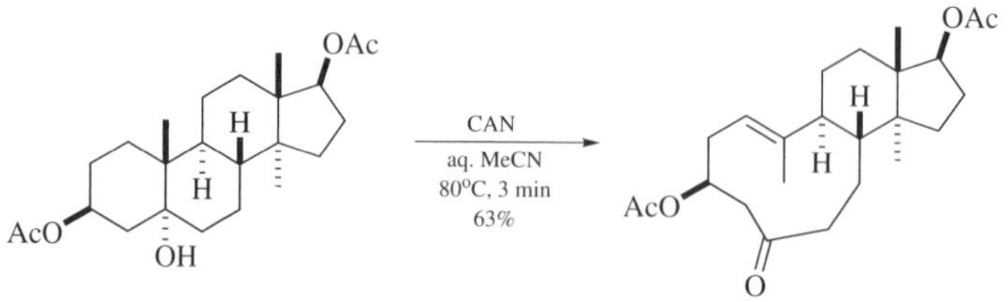

$$(5.1)$$

Besides metal-based oxidizing reagents, halogen-based reagents have also been used stoichiometrically to oxidize alcohols into aldehydes and ketones. For example, oxidation of secondary alcohols (e.g., cyclohexanol, 2-octanol, etc.), with NaOCl in aqueous acetic acid gave the corresponding ketones in >90% yields.[8] Primary alcohols were less reactive. Other halogen-based reagents for oxidizing alcohols in aqueous media include bromine,[9] iodine,[10] N-bromosuccinimide (NBS),[11] N-chloronicotinamide,[12] dimethyldioxirane,[13] hypohalites together with phase-transfer catalysts such as Bu_4NCl or Bu_4NHSO_4,[14] and N-bromohydantoins.[15] TEMPO (2,2,6,6-tetramethyl-1-piperidinyloxyl) can catalyze such oxidations.[16] Such catalytic processes have been widely used in carbohydrate chemistry.[17] A particularly effective type of oxidizing reagent for alcohol in water is the hypervalent iodine reagent, without the overoxidation of the initially formed carbonyl products.[18] Under the reaction conditions, the secondary hydroxyl group was chemoselectively oxidized to the corresponding ketone in moderate-to-good yields at room temperature, in the presence of primary hydroxyl group within the same molecule (Scheme 5.1).[19] A water-soluble derivative of o-iodoxybenzoic acid (IBX) as a selective oxidizing agent for allylic and benzylic alcohols in water was also reported.[20] Additionally, direct oxidative conversion of primary alcohols to nitriles by using molecular iodine in ammonia water was also reported.[21] Primary benzylic alcohols are selectively oxidized to the corresponding aromatic aldehydes by molecular oxygen at

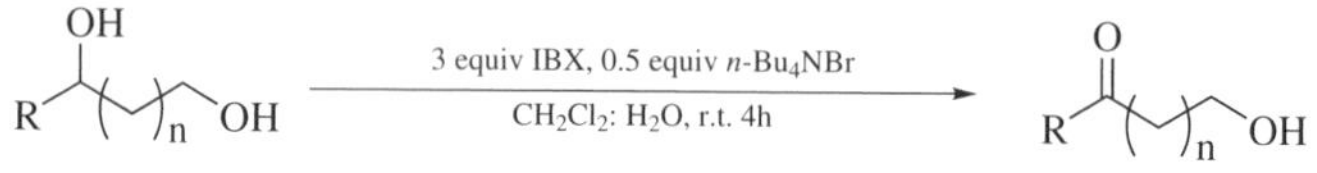

Scheme 5.1

atmospheric pressure, under Br_2-HNO_3 catalysis in a biphasic medium (1,2-dichloroethane-water, 5:1) at 60°C. Under paragonable experimental conditions, the aliphatic alcohols are oxidized to esters.[22] Ruthenium complexes can catalyze such oxidations effectively.[23]

Hydrogen peroxide can be used as a greener oxidizing reagent for alcohols in water with many transition-metal catalysts. *tert*-butyl hydroperoxide and oxone also can be used similarly.[24] These catalysts include iron tetrasulfophthalocyanine (FePcTs),[25] manganese porphyrin complexes,[26] $CuCl_2$-BQC,[27] tungsten-based polyoxometalate,[28] and zirconium alkoxide.[29] Secondary alcohols were catalytically oxidized with diacetoxyiodobenzene as oxidant in the presence of salen-Mn(III) complex to afford the corresponding ketones in up to 99% yield using CH_2Cl_2 or water as reaction media (Eq. 5.2).[30] A particularly effective catalyst for the oxidative decomposition of organic compounds with hydrogen peroxide in aqueous media is Colins's TAMP catalyst **5.1**.

$$(5.2)$$

5.1

Recently, great advancement has been made in the use of air and oxygen as the oxidant for the oxidation of alcohols in aqueous media. Both transition-metal catalysts and organocatalysts have been developed. Complexes of various transition-metals such as cobalt,[31] copper [Cu(I) and Cu(II)],[32] Fe(III),[33] Co/Mn/Br-system,[34] Ru(III and IV),[35] and $VOPO_4 \cdot 2H_2O$,[36] have been used to catalyze aerobic oxidations of alcohols. Cu(I) complex-based catalytic aerobic oxidations provide a model of copper(I)-containing oxidase in nature.[37] Palladium complexes such as water-soluble Pd-bathophenanthroline are selective catalysts for aerobic oxidation of a wide range of alcohols to aldehydes, ketones, and carboxylic acids in a biphasic

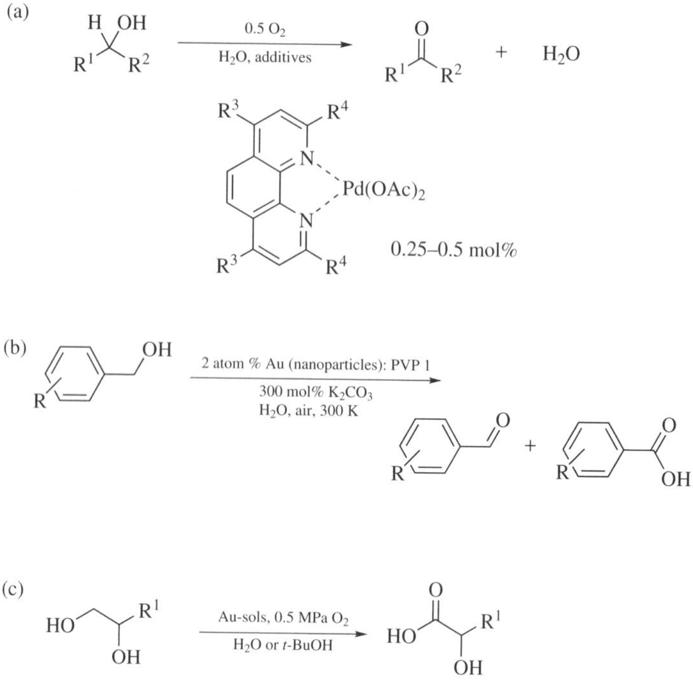

(a)

(b)

(c)

Scheme 5.2

water-alcohol system (Scheme 5.2, a).[38] Allylic and benzylic alcohols were oxidized to the corresponding aldehydes. Nonactivated 1-hexanol was oxidized to hexanoic acid, which can be stopped at the aldehyde stage by adding TEMPO. Recently, gold nanoparticles have been found to be highly effective for the oxidation of various alcohols in aqueous conditions. Gold nanoclusters ($\varphi = 1.3$ nm) stabilized by poly(N-vinyl-2-pyrrolidone) readily oxidize benzylic alcohols to the corresponding aldehydes and/or carboxylic acids under ambient temperature in water (Scheme 5.2, b).[39] Kinetic measurement revealed that smaller nanoparticles exhibit higher catalytic activity than larger homologues. An amphiphilic resin-dispersion of a nanopalladium catalyst was also highly effective for such oxidations in water.[40] The gold nanoparticles were found to be more effective than the palladium ones of comparable size (1.5 and 2.2 nm). The proposed mechanism for the oxidation involves a superoxo-like molecular oxygen species adsorbed on the surface of the small Au NCs, which abstracts a hydrogen atom from the alkoxide. Under 0.5 MPa O_2

pressure, 1,2-diols with varying chain lengths such as 1,2-propanediol or 1,2-octanediol are oxidized by gold nanocolloids stabilized with poly(vinyl alcohol) to the corresponding α-hydroxy carboxylic acids with total chemoselectivity (Scheme 5.2, c).[41] Pt-nanoparticles are also effective.[42] Other transition-metal catalyzed oxidations of alcohols include the Ir-catalyzed Oppenauer-type oxidation of secondary alcohols[43] and electrochemical oxidation of primary alcohols to aldehydes at the nickel hydroxide electrode.[44]

More recently, aerobic oxidations in water with organocatalysts have been developed that are beneficial to the pharmaceutical industry. A highly efficient catalytic system without transition-metals in water has been developed for aerobic oxidations of benzylic alcohols with 1 mol% TEMPO as a catalyst and with a catalytic amount of 1,3-dibromo-5,5-dimethylhydantoin and $NaNO_2$ as cocatalysts (Scheme 5.3, a).[45] I_2-KI-K_2CO_3-H_2O (Scheme 5.3, b)[46] and (PhIO$_2$)/Br$_2$/NaNO$_2$ (Scheme 5.3, c)[47] selectively oxidize alcohols to aldehydes and ketones under anaerobic conditions in water at 90°C with excellent yields. Alternatively, oxidation of simple alcohols can be performed in near-critical and supercritical water with air as the oxidant (Scheme 5.3, d).[48] The experimental data are consistent with a purely radical chain oxidation mechanism. Analysis of the mechanism identifies the primary oxidation pathway proceeding through acetaldehyde involving the initial abstraction of the hydroxyl hydrogen or a hydrogen atom from the secondary carbon. $CaCO_3$ can

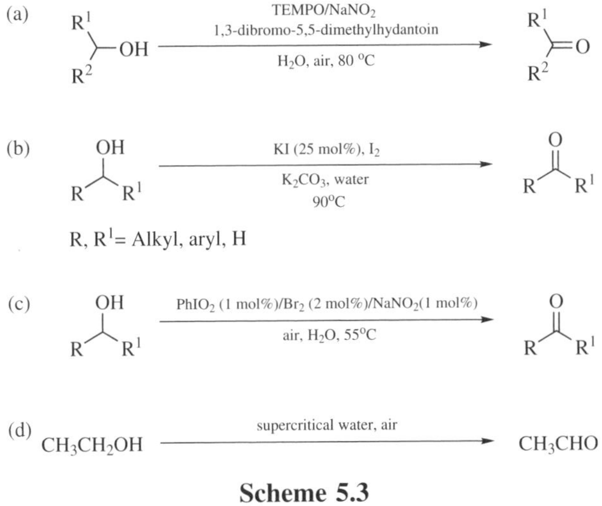

Scheme 5.3

catalyze such oxidations and formation of H_2O_2 was observed during the oxidation process.[49]

5.1.1   Diol Cleavage

The oxidative cleavage of *vic*-diols to give two carbonyl functions (Eq. 5.3) by periodates was first observed by Malaprade and has since been widely applied to the carbohydrate area.[50] Since both the reagent sodium periodate and the carbohydrate substrate are water soluble, the reaction is usually carried out in aqueous media.[51] The reaction has been applied to polysaccharides such as starch.[52] The periodate oxidations of sodium alginate in water as well as a dispersion in 1:1 ethanol-water mixture have been compared.[53] Because sodium alginate forms a highly viscous solution, the oxidation was observed to be more extensive in ethanol-water.

$$\text{HO}\quad\text{OH}\qquad\xrightarrow[\text{H}_2\text{O}]{\text{NaIO}_4}\qquad \text{>=O} \quad + \quad \text{O=<} \tag{5.3}$$

5.2   SUBSTITUTIONS/ELIMINATION

Classical types of reaction for alcohols in aqueous media are the acid-catalyzed nucleophilic substitution and elimination reactions. In the presence of a strong acid such as concentrated sulfuric acid, alcohols undergo nucleophilic substitution reactions to give ethers and elimination reaction to give alkenes. In supercritical water at 385°C and 34.5 MPa, ethanol is converted cleanly into ethylene and diethyl ether.[54] Recently, Kobayashi and co-worker reported Brönsted acid-surfactant-combined catalyst for ether, thioether, and dithioacetal formations via the dehydration of alcohols and thiols at 35–80°C in water in the presence of a surfactant, dodecylbenzenesulfonic acid (DBSA).[55] The reaction is milder and more selective compared with the classically strong acid-catalyzed reactions. For example, the reactions of benzylic alcohols in water using 10 mol% of DBSA as a catalyst were found to proceed smoothly to give the corresponding symmetrical ether in high yields (~90%). Under the same reaction conditions, etherification of the same substrate using TsOH instead of DBSA gave only a trace amount of the product. Unsymmetrical etherification of two alcohols, a primary alcohol and a benzylic alcohol, can be carried out selectively when two equivalents of the benzylic alcohol are used

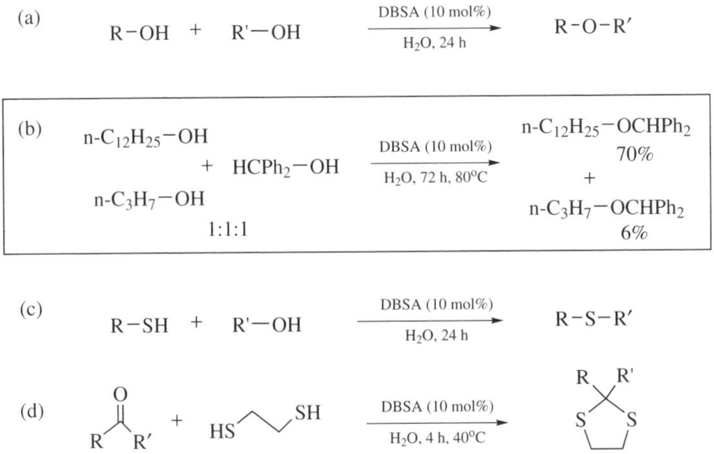

(a)

$$R-OH \; + \; R'-OH \xrightarrow[\text{H}_2\text{O, 24 h}]{\text{DBSA (10 mol\%)}} R-O-R'$$

(b)

$$\begin{aligned} &\text{n-C}_{12}\text{H}_{25}-\text{OH} \\ &\qquad + \; \text{HCPh}_2-\text{OH} \\ &\text{n-C}_3\text{H}_7-\text{OH} \\ &\qquad\qquad 1{:}1{:}1 \end{aligned} \xrightarrow[\text{H}_2\text{O, 72 h, 80°C}]{\text{DBSA (10 mol\%)}} \begin{aligned} &\text{n-C}_{12}\text{H}_{25}-\text{OCHPh}_2 \\ &\qquad 70\% \\ &\qquad + \\ &\text{n-C}_3\text{H}_7-\text{OCHPh}_2 \\ &\qquad 6\% \end{aligned}$$

(c)

$$R-SH \; + \; R'-OH \xrightarrow[\text{H}_2\text{O, 24 h}]{\text{DBSA (10 mol\%)}} R-S-R'$$

(d)

$$\underset{R \quad\; R'}{\overset{O}{R-\!\!\!\overset{\|}{C}\!\!\!-R'}} + \text{HS}\diagup\!\!\!\diagdown\text{SH} \xrightarrow[\text{H}_2\text{O, 4 h, 40°C}]{\text{DBSA (10 mol\%)}} \underset{S \quad\; S}{\overset{R \quad R'}{\diagdown\!\!/}}$$

Scheme 5.4

(Scheme 5.4, a). Selective etherification of two alcohols based on their difference in hydrophobicity can also be accomplished. When equal amounts of dodecanol and propanol were etherified with benzhydrol in the presence of DBSA in water, benzhydryl dodecyl ether was selectively obtained (Scheme 5.4, b). Thioetherification between a thiol and an alcohol can also be achieved (Scheme 5.4, c).

The DBSA-system is also applicable for the dithioacetalization of aldehdyes and ketones with 1,2-ethanedithiol to give the corresponding dithioacetals (Scheme 5.4, d). Increasing the reaction temperature decreases the yield of the products. Interestingly, increases in the concentration of the surfactant also decrease the yield of products formed, while shortening the alkyl chain of the surfactant abolishes its catalytic activity. Optical microscopy shows the formation of micelles, which are proposed to form hydrophobic environments and decrease the effective concentration of water and facilitate the dehydrative condensation reactions.

Alternatively, the S_N2 nucleophilic substitution reaction between alcohols (phenols) and organic halides under basic conditions is the classical Williamson ether synthesis. Recently, it was found that water-soluble calix[n]arenes (n = 4, 6, 8) containing trimethylammonium groups on the upper rim (e.g., calix[4]arene **5.2**) were inverse phase-transfer catalysts for alkylation of alcohols and phenols with alkyl halides in aqueous NaOH solution to give the corresponding alkylated products in good-to-high yields.[56]

5.2

Alkylations of phenoxide ions in water have recently been carefully studied by Breslow et al.[57] Alkylation can occur both at the phenoxide oxygen and on ortho and para positions of the ring when the phenoxide has at least one alkyl substituent (Eq. 5.4). Carbon alkylation occurs in water, but not in nonpolar organic solvents. This is attributed to the antihydrophobic effect of the organic solvents.

| 46% | 18% | 36% |

$$(5.4)$$

5.3   ADDITION OF ALCOHOLS, PHENOLS, AND THIOLS TO ALKENE AND ALKYNE BONDS

The addition of alcohols to alkenes and alkynes can occur via either nucleophilic or electrophilic methods depending on the nature of the alkene and alkyne molecules. The electrophilic addition of alcohols to alkenes and alkynes are often catalyzed by metal complexes such as Hg(II). In water, these reactions are often competing with hydration reactions. However, it was found that alkoxybutenes, $H_3CCH=CHCH_2$-OR [R = (un)substituted alkyl, alkenyl, cycloalkyl, cycloalkenyl, aryl] and $H_2C=CHCH(OR)CH_3$, can be prepared in high yield and selectivity with reduced dimer formation by the etherification of 1,3-butadiene with alcohols in the presence of water using a Brönsted acid catalyst. For example, 1,3-butadiene was converted into butyl ethers with 1-butanol in water in the presence of a cation-exchange resin in 62.3% combined yield versus 59.6% in a control etherification conducted without water.[58] The addition reaction of 2,2,2-trifluoroethanol to fluorinated olefins gave fluorinated ethers **A** and **B** (Eq. 5.5). The reaction that gave **A** was found to be facilitated by increasing the amount of the proton source in the reaction. Water as solvent was found to be an effective proton source for the reaction and it also allowed for a simple workup and easy scaling up.[59] When an organic solvent such as 1,4-dioxane was used, a substantial amount of the addition–elimination product **B** was formed as well.

$$CF_3CH_2-OH + F_2C=CXY \xrightarrow[H_2O]{KOH} CF_3CH_2OCF_2-CHXY + CF_3CH_2OCF=CXY$$

$$\begin{array}{ll} X,Y=F, F; & \qquad\qquad \mathbf{A} \qquad\qquad\qquad\qquad \mathbf{B} \\ =H, H; \\ =H, F; \\ =F, CF_3 \end{array}$$

$$(5.5)$$

5.4   ADDITION OF ALCOHOLS TO C=O BONDS: ESTERIFICATION AND ACETAL FORMATIONS

Esterification and acetal formations in aqueous conditions are among the most common reactions in nature because of the common ester bonds and acetal structures in various organic materials in nature.

Such esterifications and acetal formations are achieved through enzyme catalyses. However, such reactions are relatively rare in aqueous conditions chemically. This is because the reversed reactions, hydrolysis, are much more favorable entropically. Kobayashi and co-workers found that the same surfactant (DBSA) that can catalyze the ether formation in water (**5.2** above) can also catalyze the esterification and acetal formations reactions in water.[52] Thus, various alkanecarboxylic acids can be converted to the esters with alcohols under the DBSA-catalyzed conditions in water (Eq. 5.6). Carboxylic acid with a longer alkyl chain afforded the corresponding ester better than one with a shorter chain at equilibrium. Selective esterification between two carboxylic acids with different alkyl chain lengths is therefore possible.

$$R-CO_2H \quad + \quad R'-OH \quad \xrightarrow[\text{H}_2\text{O, 48 h, 40°C}]{\text{DBSA (10 mol\%)}} \quad R-CO_2-R' \quad\quad (5.6)$$

Smaller aldehydes form cyclic acetal-type oligomers readily in aqueous conditions.[60] Diols and polyols also form cyclic acetals with various aldehydes readily in water, which has been applied in the extraction of polyhydroxy compounds from dilute aqueous solutions.[61] I_2 in water was found to be an efficient catalyst for chemoselective protection of aliphatic and aromatic aldehydes with $HSCH_2CH_2OH$ to give 1,3-oxathiolane acetals under mild conditions (Eq. 5.7).[62]

$$RCHO + \underset{HO}{\overset{}{\diagup\diagdown}} SH \quad \xrightarrow{I_2/H_2O} \quad R-\overset{O}{\underset{S}{\diagup\diagdown}} \quad\quad (5.7)$$

5.5. REACTION OF ETHERS AND CYCLIC ETHERS

5.5.1 Ethers and Cyclic Ethers

In the presence of a strong oxidizing reagent such as chromic acid in aqueous H_2SO_4, diisopropyl ether can be oxidized to acetone without prior hydrolysis to isopropanol.[63] Without the oxidizing reagent, the hydrolysis product is obtained. Prenyl ethers, serving as the protecting group for hydroxyl functions, can be readily cleaved at room temperature with DDQ in dichloromethane-water (9:1) (Eq. 5.8).[64] The reaction conditions were compatible with the presence of other ethereal functionalities such as acetonides, allyl, benzyl, TBS, and TBDPS groups. Although the reaction time is longer, a catalytic amount of DDQ in the presence of three equivalents of $Mn(OAc)_3$ is a good alternative to the use of a stoichiometric amount of DDQ.

$$RO\diagup\diagdown\diagup< \quad \xrightarrow[\substack{CH_2Cl_2\text{-}H_2O\ (9:1) \\ \text{r.t. }0.75\text{--}9\,\text{h}}]{DDQ} \quad ROH \quad + \quad \diagup\diagdown CHO \qquad (5.8)$$

For cleaving TBDMS ethers, NBS in the presence of β-cyclodextrin in water is effective (Eq. 5.9).[65] On the other hand, SmI_2/H_2O/amine provides selective deallylation and bond cleavage of unsubstituted allyl ethers to give the corresponding alcohols in good-to-excellent yields. This method is useful in deprotection of alcohols and carbohydrates (Eq. 5.10).[66] Alternatively, under aqueous thermolysis conditions of subcritical and supercritical water, various aryl and benzyl ethers are cleaved to give the corresponding alcohols and phenols efficiently.[67] The rate of reaction for cleavage of diaryl ethers was inversely proportional to the concentration of various salts, which serves as a model reaction in coal liquefaction.

$$RCH_2-O-Si< \cdots + \underset{O}{\overset{O}{\bigcirc}}N-Br \xrightarrow[\text{r.t.}]{H_2O} \left[RCH_2-O-N\underset{O}{\overset{O}{\bigcirc}} \right] \longrightarrow RCHO \qquad (5.9)$$

R= aryl, alkyl

$$R\diagup O\diagup\diagdown \quad \xrightarrow{\text{i or ii}} \quad R-OH \qquad (5.10)$$

i, $SmI_2/H_2O/Et_3N$, r.t. 2 h, >99%
ii, $SmI_2/H_2O/i\text{-}PrNH_2$, r.t. 1 min, >99%

5.5.2   Reactions of Epoxides in Water and "Click Chemistry"

The cleavage of epoxides by water is a classical reaction. Such epoxide cleavage can be catalyzed by both acids and bases in aqueous media. In the presence of other nucleophiles, the corresponding nucleophilic ring-opening products are obtained with the nucleophiles being incorporated into the products.[68] Examples include azides, iodides, and thiols in the presence or absence of metal salts in aqueous media. The pH of the reaction medium controls the reactivity and regioselectivity of the process. At suitable pH values, even salts such as $AlCl_3$, $SnCl_4$ and $TiCl_4$ could be used as ring-opening catalysts.

In 2001, Sharpless and co-workers introduced the concept of "click" chemistry in a landmark paper.[69] In essence, *click chemistry* refers to a set of chemical reactions and processes that are easy to carry out, give high yields, require readily available starting materials and reagents, and have simple reaction conditions in no solvent or in a solvent that is benign (such as water) and simple product purification. Epoxides are readily available from epoxidation of alkenes and are considered as highly energetic "spring-loaded" molecules. Nucleophilic addition to epoxides, often with high regio- and stereoselectivity, is a key click reaction. An example of the regiocontrol of epoxide ring opening is the reaction of *trans*-cyclohexadiene diepoxide with ammonia in Scheme 5.5 to give 85% yield of the 2:3 adduct as a mixture of the three possible racemic diastereoisomers.

By combining several click reactions, click chemistry allows for the rapid synthesis of useful new compounds of high complexity and combinatorial libraries. The S_N2-type reaction of the azide ion with a variety of epoxides to give azido alcohols has been exploited extensively in click chemistry. First of all, azido alcohols can be converted into amino alcohols upon reduction.[70] On the other hand, aliphatic azides are quite stable toward a number of other standard organic synthesis conditions (orthogonality), but readily undergo 1,3-dipolar cycloaddition with alkynes. An example of the sequential reactions of

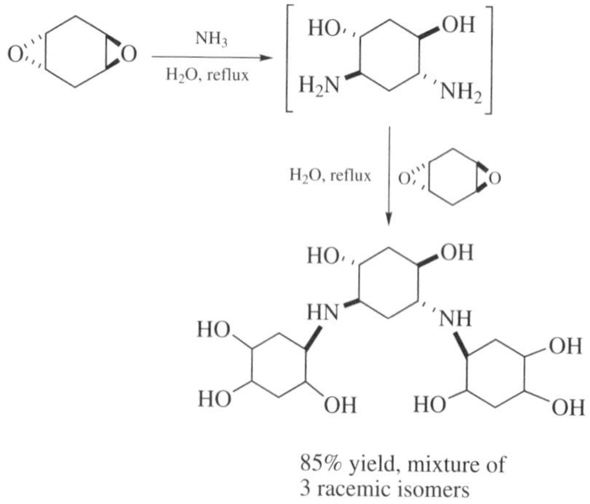

85% yield, mixture of
3 racemic isomers

Scheme 5.5

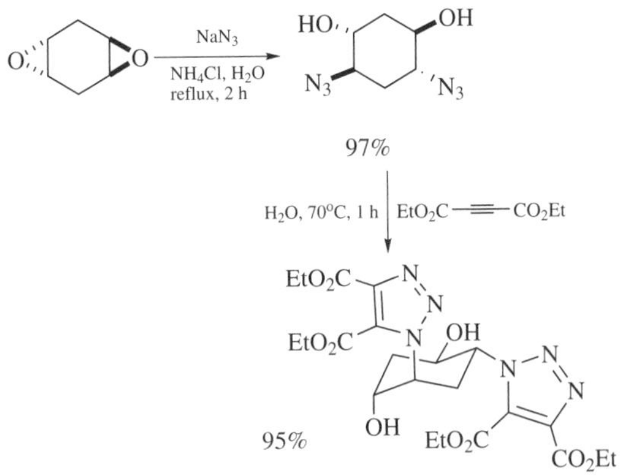

Scheme 5.6

epoxide with azide followed by dipolar cycloaddition, all carried out in water and giving highly functionalized complex adduct, is given in Scheme 5.6.

Li and co-workers developed a novel InCl-mediated coupling of epoxides with allyl bromide catalyzed by mesoporous silica–supported palladium to give homoallyl alcohols in THF.[71] The heterogeneous catalyst can be readily recovered and reused without significant loss of its catalytic activity. Subsequently, it was found that the coupling of various substituted epoxides with allyl-gallium reagent generated in situ in a mixture of Bu_4NBr-DMF-H_2O gave the corresponding homoallyl alcohols via an epoxide-aldehyde isomerization followed by allylation.[72] In the presence of a palladium catalyst, a similar allylation and propargylation of epoxides occurred in wet DMSO.[73]

5.6   REACTION OF SULFUR COMPOUNDS

5.6.1   Oxidation of Thiols and Thioethers

The oxidative couplings of thiols to disulfides and the reversed reductive cleavage reaction occur commonly in nature and have important biological importance because they assist the folding and unfolding of proteins in an aqueous environment. In subcritical water, molecular oxygen is an efficient oxidant for the oxidative coupling of thiols to disulfides in high yields (>90%) in the absence of catalysts (Eq. 5.11).[74] The reaction of an aliphatic thiol with aqueous H_2O_2

in the presence of a catalyst (ammonium paramolybdate) gives high yields of the corresponding aliphatic sulfonic acids.[75] When treated with bromine in water, EtSH and Et_2S_2 are oxidized to $EtSO_3H$ (Eq. 5.12).[76] Highly selective oxidation of sulfides to sulfoxides can be achieved with N-bromosuccinimide (NBS) catalyzed by β-cyclodextrin in water at room temperature in excellent yields without overoxidation to sulfone (Eq. 5.13).[77] The oxidation of organic sulfides in aqueous solutions by hydrogen peroxide was investigated via *ab initio* calculations, which shed new light on the mechanism of the reaction.[78] Further oxidation of sulfoxide can be obtained by using various oxidants such as aqueous hydrogen peroxide[79] and N-halosulfonamides (chloramine-B, chloramine-T, dichloramine-B, and dichloramine-T) conducted in 1:1 (vol/vol) water-methanol and water–acetic acid media in the presence of perchloric acid.[80] Peroxidase-catalyzed asymmetric oxidation of sulfide gives chiral sulfoxides in water.[81]

$$2RSH \xrightarrow[\text{subcritical water}]{O_2} RSSR \qquad (5.11)$$

$$RSH + Br_2 + H_2O \longrightarrow RSO_3H + HBr$$
$$RSSR + Br_2 + H_2O \longrightarrow RSO_3H + HBr \qquad (5.12)$$

$$R^1-S-R^2 \xrightarrow[\text{NBS/r.t.}]{\beta\text{-CD/H}_2\text{O}} R^1-\overset{\overset{\text{O}}{\|}}{S}-R^2 \qquad (5.13)$$

$$R^1=\text{aryl, naphthyl; } R^2=\text{alkyl}$$

5.6.2 Reduction of Disulfides

The reduction of disulfides in water can be performed with a variety of reducing reagents. The use of phosphines as reducing reagents for disulfides to give the corresponding thiols has been studied extensively. Water-soluble *tris*(2-carboxyethyl)phosphine (TCEP) (**5.3**) and dithiothreitol (DTT) (**5.4**) are widely used in the reduction of disulfide bonds in biochemical studies.[82] Selenols significantly catalyze the reduction by using dithiothreitol.[83] This reduction is an example of catalysis of thiolate-disulfide interchange. Second-order rate constants for the reduction of disulfides by Ph_3P in aqueous dioxane at 40°C were calculated via UV spectral data. A Brönsted plot for the reaction is linear (β = −0.98), indicating that rate differences do not arise from intramolecular catalytic effect, but merely reflect the simple polar effects of the

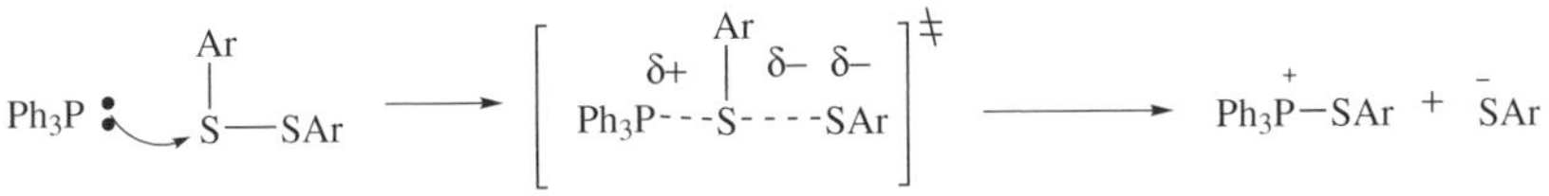

Scheme 5.7

substituent. A two-step mechanism was proposed, with nucleophilic cleavage of the S–S bond by Ph_3P being the rate-determining step (Scheme 5.7).[84] Metallic samarium-promoted reduction of sodium alkyl thiosulfates in the absence of an activating agent occurs to afford the corresponding disulfides with good yields in water at 90°C (Eq. 5.14).[85]

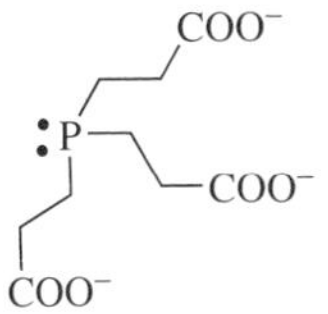

5.3 TCEP 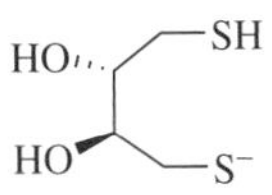   **5.4 DTT**

$$RSSO_3Na \xrightarrow[90°C]{Sm/H_2O} RSSR \qquad (5.14)$$

5.6.3 Native Chemical Ligation for Peptide Synthesis

Taking advantage of the earlier observations by Kemp,[86] a selective ligation strategy has been developed by Kent for the synthesis of proteins from unprotected peptide building blocks using the remarkable intramolecular acylating power of the thioester functionality.[87] In the first step, this native chemical ligation method involves the chemoselective *trans*-thioesterification of a peptide fragment **A**, which has a thioester end group, with another peptide fragment **B** which has an amino-terminal cysteine residue. The resulting thioester-linked intermediate undergoes spontaneous intramolecular reaction to form a native peptide bond at the ligation site (Scheme 5.8). Since the peptide fragments are unprotected, the whole process is carried out in aqueous solution. The approach has been applied to the synthesis of an 82-residue antibacterial glycoprotein.[88]

5.6.4 Other Reactions

Indium-mediated coupling of sulfonyl chlorides with alkyl bromides in water readily generates sulfones.[89] Sulfones can also be obtained via

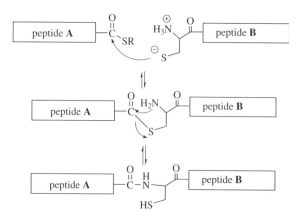

Scheme 5.8

the reaction of alkyl halides with $4\text{-MeC}_6\text{H}_4\text{SO}_2\text{Na}$ in DMF-water at room temperature under the influence of ultrasound (Eq. 5.15).[90] On the other hand, arylsulfinyl sulfones can be hydrolyzed readily in water in the presence of halide and acetate ions as catalysts (Eq. 5.16).[91]

$$RX \xrightarrow[\text{DMF-water mixt (1:2), r.t. ultrasound}]{p\text{-CH}_3\text{-C}_6\text{H}_4\text{SO}_2\text{Na}} p\text{-CH}_3\text{-C}_6\text{H}_4\text{SO}_2\text{-R} + NaX \qquad (5.15)$$

$$\underset{\overset{||}{O}\ \overset{||}{O}}{Ar-S-S-Ar} \xrightarrow[]{\text{Cl}^- \text{ or Br}^-,\ \text{AcOH:H}_2\text{O}=85/15} 2\ ArSO_2H \qquad (5.16)$$

REFERENCES

1. Groger, H., Hummel, W., Rollmann, C., Chamouleau, F., Husken, H., Werner, H., Wunderlich, C., Abokitse, K., Drauz, K., Buchholz, S., *Tetrahedron* **2004**, *60*, 633.

2. Bobtelsky, M., Radovensky-Cholatnikow, C., *Z. Anorg. Allgem. Chem.* **1931**, *199*, 241.

3. Venkatasubramanian, N., *J. Sci. Ind. Res., Section B: Phys. Sci.* **1961**, *20B*, 541.

4. Lou, J.-D., *J. Chem. Res., Synop.* **1997**, 206.

5. Ho, T.-L., *Synth. Commun.* **1979**, *9*, 237.

6. Balasubramanian, V., Robinson, C. H., *Tetrahedron Lett.* **1981**, *22*, 501.

7. Molander, G. A., *Chem. Rev.* **1992**, *92*, 29.

8. Blanzat, Y., Foulon, J. P., *Bull. Union Phys.* **1985**, *80*, 239; Ji, H.-B., Shi, D.-P., Shao, M., Li, Z., Wang, L.-F., *Tetrahedron Lett.* **2005**, *46*, 2517.

9. Swain, G. C., Wiles, R. A., Bader, R. F. W., *J. Am. Chem. Soc.* **1961**, *83*, 1945.

10. Gogoi, P., Sarmah, G. K., Konwar, D., *J. Org. Chem.* **2004**, *69*, 5153.

11. Krishnaveni, N. S., Surendra, K., Rao, K. R., *Adv. Synth. Catal.* **2004**, *346*, 346.

12. Ramkumar, B., *Oxid. Commun.* **2001**, *24*, 554.

13. Baumstark, A. L., Kovac, F., Vasquez, P. C., *Can. J. Chem.* **1999**, *77*, 308.

14. Lee, G. A., Freedman, H. H. U.S. (1976), 4 pp. US 3996259 19761207; Surendra, K., Krishnaveni, N., Srilakshmi, Rao, K. R., *Can. J. Chem.* **2004**, *82*, 1230, Xu, C. Z., Chen, L., Li, Z., Sun, W., Xia, C. G., *Chin. Chem. Lett.* **2004**, *15*, 1149.

15. Corral, R. A., Orazi, O. O., *Anal. Asoc. Quim. Argent.* **1967**, *55*, 205.

16. Tashino, Y., Togo, H., *Synlett* **2004**, 2010.

17. de Nooy, A. E. J., Besemer, A. C., van Bekkum, H., *Rec. Trav. Chim. Pays-Bas* **1994**, *113*, 165; Isogai, A., Kato, Y., *Cellulose (London)* **1998**, *5*, 153.

18. Tohma, H., Takizawa, S., Maegawa, T., Kita, Y., *Angew. Chem. Intern. Ed. Engl.* **2000**, *39*, 1306; Narender, M., Reddy, M. S., Kumar, V. P., Nageswar, Y. V. D., Rao, K. R., *Tetrahedron Lett.* **2005**, *46*, 1971; Tohma, H., Maegawa, T., Kita, Y., *Synlett* **2003**, 723; Liu, Z., Chen, Z.-C., Zheng, Q.-G., *Org. Lett.* **2003**, *5*, 3321; Kita, Y., Tohma, H., Maegawa, T., *Jpn. Kokai Tokkyo Koho* 2002, 11 pp. JP 2002020322 A2 20020123.

19. Kuhakarn, C., Kittigowittana, K., Pohmakotr, M., Reutrakul, V., *Tetrahedron* **2005**, *61*, 8995.

20. Vinod, T. K., Clinton, V., Nash, G. T., *Abstracts LAKES04-195*, 36th Great Lakes Regional Meeting of the American Chemical Society, Peoria, IL, USA, October 17–20, 2004.

21. Mori, N., Togo, H., *Synlett* **2005**, 1456.

22. Minisci, F., Porta, O., Recupero, F., Punta, C., Gambarotti, C., Pierini, M., Galimberti, L., *Synlett* **2004**, 2203.

23. For a review, see: Pagliaro, M., Campestrini, S., Ciriminna, R., *Chem. Soc. Rev.* **2005**, *34*, 837; Khan, M. M. T., Merchant, R. R., Chatterjee, D., Bhatt, K. N., *J. Mol. Catal.* **1991**, *67*, 309; Puttaswamy, Ramachandrappa, R., Gowda, N. M. M., *Synth. React. Inorg. Metal-Org. Chem.* **2002**, *32*, 1263; Genet, J. P., Pons, D., Juge, S., *Synth. Commun.* **1989**, *19*, 1721.

24. Boudreau, J., Doucette, M., Ajjou, A. N., *Tetrahedron Lett.* **2006**, *47*, 1695.

25. Hampton, Kenneth W., Ford, W. T., *J. Mol. Catal. A: Chem.* **1996**, *113*, 167.

26. Zheng, T.-C., Richardson, D. E., *Tetrahedron Lett.* **1995**, *36*, 833; Wietzerbin, K., Bernadou, J., Meunier, B., *Eur. J. Inorg. Chem.* **1999**, 1467; Wietzerbin, K., Meunier, B., Bernadou, J., *Chem. Commun.* **1997**, 2321.

27. Ferguson, G., Nait Ajjou, A., *Tetrahedron Lett.* **2003**, *44*, 9139.

28. Sloboda-Rozner, D., Witte, P., Alsters, P. L., Neumann, R., *Adv. Synth. Catal.* **2004**, *346*, 339; Sloboda-Rozner, D., Alsters, P. L., Neumann, R., *J. Am. Chem. Soc.* **2003**, *125*, 5280.

29. Krohn, K., *Synthesis* **1997**, 1115.

30. Sun, W., Wang, H., Xia, C., Li, J., Zhao, P., *Angew. Chem. Int. Ed.* **2003**, *42*, 1042.

31. Kervinen, K., Allmendinger, M., Repo, T., Rieger, B., Leskela, M., *Abstracts, 224th ACS National Meeting*, Boston, US, August 18–22, 2002, INOR-464.

32. Munakata, M., Nishibayashi, S., Sakamoto, H., *J. C. S., Chem. Commun.* **1980**, 219; Gamez, P., Arends, I. W. C. E., Reedijk, J., Sheldon, R. A., *Chem.Commun.* **2003**, 2414.

33. Xiong, Y., Xiao, D., Li, J., Xie, X., Yang, X., Wu, Y., *Zeit. Phys. Chem.* **2005**, *219*, 1121.

34. Nair, K., Sawant, D. P., Shanbhag, G. V., Halligudi, S. B., *Catal. Commun.* **2004**, *5*, 9.

35. Wolfson, A., Yehuda, C., Shokin, O., Tavor, D., *Lett. Org. Chem.* **2006**, *3*, 107; Ji, H. B., Qian, Y., *Chin. Chem. Lett.* **2003**, *14*, 615; Ji, H.; Mizugaki, T., Ebitani, K., Kaneda, K., *Tetrahedron Lett.* **2002**, *43*, 7179.

36. Carlini, C., Patrono, P., Raspolli Galletti, A. M., Sbrana, G., Zima, V., *Appl. Catal. A: General* **2005**, *289*, 197.

37. Gamez, P., Arends, I. W. C. E., Sheldon, R. A., Reedijk, J., *Adv. Synth. Catal.* **2004**, *346*, 805.

38. ten Brink, G.-j., Arends, I. W. C. E., Sheldon, R. A., *Science*, **2000**, *287(5458)*, 1636; Buffin, B. P., Clarkson, J. P., Belitz, N. L., Kundu, A., *J. Mol. Catal. A: Chem.* **2005**, *225*, 111; ten Brink, G.-j., Arends, I. W. C. E., Hoogenraad, M., Verspui, G., Sheldon, R. A., *Adv. Synth. Catal.* **2003**, *345*, 1341; ten Brink, G.-J., Arends, I. W. C. E., Hoogenraad, M., Verspui, G., Sheldon, R. A., *Adv. Synth. Catal.* **2003**, *345*, 497; ten Brink, G.-J., Arends, I. W. C. E., Sheldon, R. A., *Adv. Synth. Catal.* **2002**, *344*, 355.

39. Tsunoyama, H., Sakurai, H., Negishi, Y., Tsukuda, T., *J. Am. Chem. Soc.* **2005**, *127*, 9374.

40. Uozumi Y., Nakao, R., *Angew. Chem. Intern. Engl.* **2003**, *42*, 194; Biffis, A., Minati, L., *J. Catal.* **2005**, *236*, 405.

41. Mertens, P. G. N., Bulut, M., Gevers, L. E. M., Vankelecom, I. F. J., Jacobs, P. A., De Vos, D. E., *Catal. Lett.* **2005**, *102*, 57.

42. Ohtani, B., Takamiya, S., Hirai, Y., Sudoh, M., Nishimoto, S., Kagiya, T., *J. Chem. Soc. Perkin Tr. 2.* **1992**, 175.

43. Ajjou, A. N., Pinet, J.-L., *Can. J. Chem.* **2005**, *83*, 702.

44. Amjad, M., Pletcher, D., Smith, C., *J. Electrochem. Soc.* **1977**, *124*, 203; Schneider, R., Schaefer, H.-J., *Synthesis* **1989**, 742; Tanaka, H., Kubota, J., Miyahara, S., Kuroboshi, M., *Bull. Chem. Soc. Jpn.* **2005**, *78*, 1677; Kubota, J., Shimizu, Y., Mitsudo, K., Tanaka, H., *Tetrahedron Lett.* **2005**, *46*, 8975.

45. Liu, R., Dong, C., Liang, X., Wang, X., Hu, X., *J. Org. Chem.* **2005**, *70*, 729.

46. Gogoi, P., Konwar, D., *Org. Biomol. Chem.* **2005**, *3*, 3473.

47. Mu, R., Liu, Z., Yang, Z., Liu, Z., Wu, L., Liu, Z.-L., *Adv. Synth. Catal.* **2005**, *347*, 1333.

48. Rice, S. F., Croiset, E., *Ind. Eng. Chem. Res.* **2001**, *40*, 86.

49. Suppes, G. J., Roy, S., Ruckman, J., *AIChE J.* **2001**, *47*, 2102; Croiset, E., Rice, S. F., *Ind. Eng.Chem. Res.* **1998**, *37*, 1755.

50. Malaprade, L., *Bull. Soc. Chim.* **1928**, *43*, 683.

51. Sussich, F., Cesàro, A., *Carbohydr. Res.* **2000**, *329*, 87.

52. Veelaert, S., De Wit, D., Gotlieb, K. F., Verhe, R., *Carbohydr. Polym.* **1997**, *33*, 153.

53. Balakrishnan, B., Lesieur, S., Labarre, D., Jayakrishnan, A., *Carbohydr. Res.* **2005**, *340*, 1425.

54. Xu, X., De Almeida, C. P., Antal, M. J. Jr., *Ind. Eng. Chem. Res.* **1991**, *30*, 1478.

55. Manabe, K., Iimura, S., Sun, X.-M., Kobayashi, S., *J. Am. Chem. Soc.* **2002**, *124*, 11971.

56. Shimizu, S., Suzuki, T., Shirakawa, S., Sasaki, Y., Hirai, C., *Adv. Synth. Catal.* **2002**, *344*, 370.

57. Breslow, R., Groves, K., Mayer, M. U., *J. Am. Chem. Soc.* **2002**, *124*, 3622.

58. Kanand, J., Roeper, M. (BASF AG, Germany), Ger. Offen. (1998), 4 pp. DE 19637892 A1 19980319.

59. Murata, J., Tamura, M., Sekiya, A., *Green Chemistry* **2002**, *4*, 60.

60. Balashov, A. L., Krasnov, V. L., Danov, S. M., Chernov, A. Y., Sulimov, A. V., *J. Struct. Chem. (Transl. Zh. Strukt. Khim.)* **2001**, *42*, 398.

61. Tink, R. R., Neish, A. C., *Can. J. Technol.* **1951**, *29*, 243.

62. Bandgar, B. P., Bettigeri, S. V., *J. Chem. Res.* **2004**, 389.

63. Brownell, R., Leo, A., Chang, Y. W., Westheimer, F. H., *J. Am. Chem. Soc.* **1960**, *82*, 406.

64. Vatele, J.-M., *Synlett* **2002**, 507.

65. Reddy, M. S., Narender, M., Nageswar, Y. V. D., Rao, K. R., *Synthesis* **2005**, 714.

66. Dahlen, A., Sundgren, A., Lahmann, M., Oscarson, S., Hilmersson, G., *Org. Lett.* **2003**, *5*, 4085.

67. Siskin, M., Katritzky, A. R., Balasubramanian, M., *Fuel* **1993**, *72*, 1435; Siskin, M., Brons, G., Vaughn, S. N., Katritzky, A. R., Balasubramanian, M., *Energy & Fuels* **1990**, *4*, 488; Yokoyama, C., Nishi, K., Takahashi, S., *Sekiyu Gakkaishi* **1997**, *40*, 465.

68. For a review, see: Amantini, D., Fringuelli, F., Piermatti, O., Tortoioli, S., Vaccaro, L., *Arkivoc* **2002**, 293.

69. Kolb, H. C., Finn, M. G., Sharpless, K. B., *Angew. Chem. Int. Ed.* **2001**, *40*, 2004.

70. Vander Werf, C. A., Heisler, R.Y., McEwen, W. E., *J. Am. Chem. Soc.* **1954**, *76*, 1231.

71. Jiang, N., Hu, Q., Reid, C. S., Lu, Y., Li, C.-J., *Chem.Commun.* **2003**, 2318.

72. Gohain, M., Prajapati, D., *Chem. Lett.* **2005**, *34*, 90.

73. Roy, U. K., Roy, S., *Tetrahedron* **2006**, *62*, 678.

74. Ozen, R., Aydin, F., *Monatsh. Chem.* **2006**, *137*, 307.

75. Deschrijver, P., Ganhy, J. P., Blondeel, G. (Interox S. A., Fr.), Fr. Demande (1988), 8 pp. FR 2616786 A1 19881223.

76. Young, H. A., *J. Am. Chem. Soc.* **1937**, *59*, 811.

77. Surendra, K., Krishnaveni, N. S., Kumar, V. P., Sridhar, R., Rao, K. R., *Tetrahedron Lett.* **2005**, *46*, 4581.

78. Chu, J.-W., Trout, B. L., *J. Am. Chem. Soc.* **2004**, *126*, 900.

79. Tashlick, I. (Monsanto Chemical Co.), (1962), 2 pp. US 3069471, 19621218.

80. Gowda, B. T., Nambiar, P. V. V., D'Souza, J. D., *J. Ind. Chem. Soc.* **2001**, *78*, 403.

81. Dai, L., Klibanov, A. M., *Biotechn. Bioeng.* **2000**, *70*, 353.

82. Cline, D. J., Redding, S. E., Brohawn, S. G., Psathas, J. N., Schneider, J. P., Thorpe, C., *Biochemistry* **2004**, *43*, 15195.

83. Singh, R., Whitesides, G. M., *J. Org. Chem.* **1991**, *56*, 6931.

84. Overman, L. E., O'Connor, E. M., *J. Am. Chem. Soc.* **1976**, *98*, 771; Overman, L. E., Matzinger, D., O'Connor, E. M., Overman, J. D., *J. Am. Chem. Soc.* **1974**, *96*, 6081.

85. Wang, L., Li, P., Zhou, L., *Tetrahedron Lett.* **2002**, *43*, 8141.

86. Kemp, D. S., Carey, R. I., *J. Org. Chem.* **1993**, *58*, 2216.

87. Dawson, P. E., Muir, T. W., Clark-Lewis, I., Kent, S. B. H., *Science* **1994**, *266*, 776.

88. Shin, Y., Winans, K. A., Backes, B. J., Kent, S. B. H., Ellman, J. A., Bertozzi, C. R., *J. Am. Chem. Soc.* **1999**, *121*, 11684.

89. Wang, L., Zhang, Y., *J. Chem. Res. Synop.* **1998**, 588.

90. Biswas, G. K., Jash, S. S., Bhattacharyya, P., *Ind. J. Chem. Section B*: **1990**, *29B*, 491.

91. Kice, J. L., Guaraldi, G., *J. Org. Chem.* **1968**, *33*, 793.

CHAPTER 6

ORGANIC HALIDES

6.1 GENERAL

Organic halides play a fundamental role in organic chemistry. These compounds are important precursors for carbocations, carbanions, radicals, and carbenes and thus serve as an important platform for organic functional group transformations. Many classical reactions involve the reactions of organic halides. Examples of these reactions include the nucleophilic substitution reactions, elimination reactions, Grignard-type reactions, various transition-metal catalyzed coupling reactions, carbene-related cyclopropanations reactions, and radical cyclization reactions. All these reactions can be carried out in aqueous media.

6.2 REDUCTION

The reductive dehalogenation of organic halides in aqueous media is important in environmental remediation. For synthetic purposes, the C–X bonds in organic halides have been reduced to the corresponding C–H bonds in aqueous media by a variety of methods. For aliphatic organic halides, C–X bonds in CCl_4, $CHCl_3$ and 1-hexyl halides were transformed into C–H bonds by transfer hydrodehalogenation

Comprehensive Organic Reactions in Aqueous Media, Second Edition, by Chao-Jun Li and Tak-Hang Chan
Copyright © 2007 John Wiley & Sons, Inc.

catalyzed by water-soluble Ru(II) phosphine complexes. A turnover frequency of $1000 \ h^{-1}$ was achieved with aqueous HCO_2Na as the H donor and $RuCl_2(TPPMS)_2$ (TPPMS = $Ph_2P[m\text{-}C_6H_4SO_3Na]$) as the catalyst.[1] Kagan's reagent (SmI_2) is a mild and powerful reagent for the hydrogenolysis of organic halides. The reduction proceeds by one or more successive electron transfers.[2] Water-soluble and water-insoluble organic halides can be reduced in aqueous media in high yields using tetrabutyltin hydride at $90°C$.[3] For water-insoluble substrates such as 9-bromoanthracene, the presence of a detergent (CTAB) was necessary, otherwise the reaction did not go to completion even after an extended period. Halides and sulfonate esters are smoothly reduced under phase-transfer conditions in an aqueous-organic biphasic system to the corresponding alkanes with $NaBH_4$ in the presence of lipophilic quaternary ammonium and phosphonium salts (e.g., $n\text{-}C_{16}H_{33}PBu_3{}^+$ Br^-) as catalysts.[4] The reaction shows a high selectivity with respect to a number of functional groups (i.e., COOR, $CONR_2$, NO_2, cyano). The factors affecting the reactions, and the synthetic utility and limitations of the process have been discussed. A water soluble tin hydride **6.1** was recently prepared by Breslow.[5] The tin hydride has three hydrophilic methyoxyethoxypropyl groups. In the presence of light or AIBN, it reduces a variety of alkyl halides in water. A noteworthy example of the reduction is the reaction with a water-soluble sugar derivative. In water, the bromine was cleanly removed (Eq. 6.1).

6.1

(6.1)

Reduction of different organic halides, bromonucleosides among them, was successfully carried out in good yields, using $(TMS)_3SiH$ in a heterogeneous system with water as the solvent. The procedure, employing 2-mercaptoethanol as the catalyst and the hydrophobic diazo-compound ACCN as the initiator, illustrates that $(TMS)_3SiH$ can be the radical-based reducing agent of choice in an aqueous medium (Eq. 6.2).[6] $(TMS)_3SiH$ does not suffer from any significant reaction with water and can safely be used with additional benefit such as ease of purification and environmental compatibility. Indium metal in H_2O reduces α-halocarbonyl compounds and benzyl iodides to the corresponding dehalogenated products under sonication selectively, while simple alkyl and aryl iodides are inactive under these conditions (Eq. 6.3).[7] Similarly, bismuth metal was found to reduce various α-halocarbonyl compounds efficiently and chemoselectively in aqueous media to give the dehalogenated reduction products in excellent yields (Eq. 6.4).[8] Activation of the bismuth metal with ammonium hydrogen fluoride was necessary.

$$RX + (TMS)_3SiH \xrightarrow[\text{ACCN (cat.)/H}_2\text{O}]{\text{HOCH}_2\text{CH}_2\text{SH (cat.)}} RH + (TMS)_3SiX \qquad (6.2)$$

$$RCOCH_2X \xrightarrow{\text{In, H}_2\text{O, ultrasound}} RCOCH_3 \qquad (6.3)$$
$$X = I, Br$$

$$(6.4)$$

The hydrodehalogenation of aromatic halides, catalyzed by Pd/C in aqueous solution using molecular hydrogen as reducing agent, yields arenes in short reaction times at room temperature under normal pressure (Eq. 6.5).[9] The catalyst shows the highest activity in water. With 4-chlorohypnone in water, the C–Cl bond was more easily hydrogenated than the C=O bond; in isopropanol the reverse was true. Chemoselective reduction of halo nitroaromatic compounds by β-cyclodextrin-modified transition-metal catalysts in a biphasic (H_2O/organic) system can be used in the hydrogenation of halo nitroaromatic compounds (Eq. 6.6).[10] The Pt complex is the most active and selective, resulting in the chemoselective formation of halogenated anilines. The reductive dehalogenation of PhX (X = Cl, Br, I) was facilitated with $NiCl_2$-Zn-H_2O in $P(O)(NMe_2)_3$ with ultrasound irradiation.[11] The reductive coupling of p-chloroanisole takes place under mild conditions and employs water, zinc, and a catalytic amount of nickel chloride, triphenylphosphine, and iodide ion (Eq. 6.7).[12] The reaction rate was first order in nickel and aryl halide but independent of the water concentration. Hydrogenolysis of Ar-halogen bonds in a series of halogen-substituted benzoic acids, phenols, anilines, and N-heterocycles with excess $NaBH_4$ in the presence of catalytic $PdCl_2$ and NaOH in water gave a quantitative conversion and high selectivity.[13]

$$(6.5)$$

$$(6.6)$$

$$(6.7)$$

6.3 ELIMINATION REACTIONS

The elimination of hydrogen halide from organic halides under basic conditions generates alkenes and is a fundamental reaction in organic chemistry. It is sometimes carried out with a base in aqueous media.[14] In contrast, the corresponding Hofmann-type eliminations of quaternary ammonium hydroxides are frequently carried out in aqueous media,[15] which will be covered in Chapter 11.

The reductive elimination of vicinal dihalides has been accomplished by using many reagents, including the use of aqueous media.[16] An interesting method is the reductive elimination of vicinal dihalides by an electrochemical method using vitamin B_{12} in a water-in-oil microemulsion (Eq. 6.8).[17]

$$RCHBr\text{-}CHBrR' \xrightarrow[\text{water-in-oil}]{\text{electrode, vitamin } B_{12}} RCH\text{=}CHR' \qquad (6.8)$$

6.4 NUCLEOPHILIC SUBSTITUTIONS

6.4.1 Substituted by Heteroatoms

The polarized C–X bonds in aliphatic organic halides are electrophilic and can be attacked readily by various nucleophilies. These reactions generally proceed via the S_N2 mechanism. When the attacking nucleophile is anionic, such as alkoxide or halide ions, the rate of the nucleophilic substitution is generally reduced in water compared to organic solvents. This is due to the increased stabilization of the ionic species by water (hydration) compared with the transition state, which increases the activation energy of the reaction. Nucleophilic substitution through the formation of micelles in water has been reviewed.[18] On the other hand, the nucleophilic substitutions of organic halides by neutral nucleophiles are favored in water because the transition state in this case has more charge separation and is thus stabilized. Therefore, organic compounds are hydrolyzed readily in neutral or acid media.[19] The kinetics of the biphasic hydrolysis of 1-bromoadamantane and PhCHMeCl were studied in toluene-water and decane-water systems. Evidence has been obtained for the first examples of rate-limiting S_N1 reactions occurring at a hydrocarbon–water phase boundary.[20] Hydrolysis of alkyl halides in the presence of copper and its salts in aqueous DMSO (molar ratio of organic halide:$CuSO_4 \cdot 5H_2O$:DMSO:H_2O = 1:1:10–20:12-25) at 100–117°C gives alcohols in high yields. Aryl halides containing no activating substituents are inert under the reaction conditions.[21] Alkoxybenzenes are prepared by etherification of the corresponding phenols with linear or branched alkyl halides in the presence of phase-transfer catalysts in aqueous inorganic alkali solution.[22] A series of quaternary ammonium salts have been found effective as phase-transfer agents to promote reactions between phenols and alkyl halides in an aqueous solution of sodium hydroxide in the absence of organic solvent.[23]

The reaction of $MeC(=NH_2^+)SR$ X^- (R = Bu, n-octyl, allyl, $PhCH_2$, p-$MeC_6H_4CH_2$; X = Br, Cl) with R^1X (R^1 = Me_2CHCH_2, p-$MeC_6H_4CH_2$, Et, Bu, $PhCH_2$, EtO_2CCH_2, allyl, p-$ClC_6H_4CH_2$, p-$O_2NC_6H_4CH_2$, p-$MeOC_6H_4CH_2$) in a liquid–liquid biphasic system consisting of benzene, aqueous NaOH, and a catalytic amount of Bu_4NBr gave 77–100% yield of RSR^1 (Eq. 6.9).[24] An efficient procedure for the preparation of various sulfides was introduced through a simple reaction of disulfides with suitable alkyl or aryl halides that is promoted by commercial Zn powder in the presence of $AlCl_3$ in aqueous media at 65°C (Eq. 6.10).[25] The reaction of $(RS)_2CO$ (R = Me, Et, Bu, n-octyl, $PhCH_2$) with R^1X (R^1 = n-octyl, $PhCH_2$, 2-octyl, p-$O_2NC_6H_4$-, p-AcC_6H_4-, 2-benzothiazolyl; X = Cl, Br, I, tosylate, $MeSO_3$-) in the presence of Bu_4NBr and 30% aqueous KOH gave 93–100% yield of RSR^1 (Eq. 6.11).[26]

$$H_3C \underset{NH_2}{\overset{S}{\|}} + RX \xrightarrow[80-90\%]{CHCl_3/reflux,\ 1\ h} H_3C \underset{NH_2X^-}{\overset{S\diagdown R}{\|}} \xrightarrow[77-100\%]{R^1Y/\text{phase-transfer catalyst}\ \atop benzene/aq\ NaOH,\ r.t.\ 15min} R^1{\diagup}^S{\diagdown}R$$

$$(6.9)$$

$$RSSR + 2R'X \xrightarrow[DMF/H_2O]{Zn/AlCl_3} 2RSR' \qquad (6.10)$$

$$\underset{R-S}{\overset{R-S}{\diagdown}}C{=}O + 2\ R^1\text{-}X \xrightarrow[30\%\ aqueous\ KOH]{(n\text{-}C_4H_9)_4NBr} 2\ R^1\text{-}S\text{-}R \qquad (6.11)$$

In basic aqueous media, a kinetic study of the reaction between stannate(II) ions and alkyl halide shows that mono- and disubstituted organotin compounds are formed (Eq. 6.12a).[27] The monosubstituted organotin compound is obtained after a nucleophilic substitution catalyzed by a complexation between the tin(II) and the halide atom. The disubstituted compound results from an electrophilic substitution coupled with a redox reaction on a complex between the monosubstituted organotin compound and the stannate(II) ion. Stannate(IV) ions prevent the synthesis of the disubstituted compound by complexation. Similarly, when allyl bromide and tin were stirred in D_2O at 60°C, allyltin(II) bromide was formed first. This was followed by further reaction with another molecule of allyl bromide to give diallyltin(IV) dibromide (Eq. 6.12b).[28]

$$C_2H_5Br + Sn(OH)_3{}^- + OH^- \longrightarrow C_2H_5Sn(OH)_4{}^- + Br^- \qquad (6.12a)$$

$$\diagup\!\!\!\diagdown Br + Sn(0) \xrightarrow[60\ °C]{H_2O} \diagup\!\!\!\diagdown SnBr \xrightarrow{H_2O} \left(\diagup\!\!\!\diagdown\underset{Br}{SnBr_2}\right)_2 \qquad (6.12b)$$

Synthesis of 2-alkyl- or 2-aryl-substituted benzo[b]furans has been reported, involving a CuI·TMEDA complex which catalyzes the transformation of readily available ketone derivatives into the corresponding benzofurans in good-to-excellent yields in water (Eq. 6.13).[29]

$$R^1 \underset{Br}{\overset{O}{\diagdown}}\!\!\!\!\bigcirc\!\!\!\!{-}R^2 \xrightarrow[H_2O,\ 120°C]{CuI,\ TMEDA} R^1{-}\!\!\!\bigcirc\!\!\!\!\overset{O}{\diagup}\!\!\!\!{-}R^2 \qquad (6.13)$$

$$R^1=\text{Alkyl, Aryl} \qquad (74\text{-}99\%)$$

Benzyl-substituted quaternary ammonium salts (e.g., *N*-benzyl-*N*,*N*-dimethylanilinium chloride) are rapidly prepared in water or in a water-containing organic solvent (e.g., aqueous methanol) in high yield by reacting a tertiary amine (e.g., *N*,*N*-dimethylaniline) with a benzyl halide (e.g., benzyl chloride), optionally having one or more substituent groups inert to the reaction on the benzene ring and/or at the α-site (Eq. 6.14).[30] Aqueous *N*-heterocyclization of primary amines and hydrazines with dihalides has been reported by using microwave-assisted substitution reactions under aqueous conditions.[31] Alkyldibenzyl- and benzyldialkylphosphine oxides were prepared in one step by direct phosphorylation of a mixture of alkyl bromides (alkyl=Et, Pr) and benzyl chloride with red phosphorus under the conditions of phase-transfer catalysis (concentrated aqueous KOH solution–dioxane-benzyltriethylammonium chloride) (Eq. 6.15).[32]

$$(6.14)$$

R = Et, Pr

$$(6.15)$$

The nucleophilic displacement of the iodine moiety in 2-iodobenzoates mediated by triphenyltin hydride and di-n-butyltin dichloride in aqueous solution has been demonstrated (Eq. 6.16).[33] For example, 2-iodobenzoic acid reacts with a toluene solution of $Ph_3SnH/1,3-(NO_2)_2$ C_6H_4/aq. $NaHCO_3$ to give 89% yield of salicylic acid.

$$(6.16)$$

Diethyl arylphosphonates were synthesized by reactions of diethyl phosphonate with aryl iodides or bromides containing electron-donor or electron-acceptor substituents in the aromatic ring in aqueous MeCN or neat H_2O in the presence of Pd complexes with water-soluble ligands.[34] For example, $MeCN/H_2O$ (1 mL), PhBr (8.2 mmol), and $Ph_2P(C_6H_4SO_3Na$-*m*) (approximately 0.4 mmol) were successively added to $(EtO)_2P(O)H$ (10 mmol) and Et_3N (12 mmol); after the mixture had been stirred for 5 min, $Pd(OAc)_2$ (0.2 mmol) was added and the reaction mixture was heated at 70–80°C for 3 h to give 100% $PhP(O)(OEt)_2$ after workup.

The use of water as solvent interestingly influences the chemoselectivity in photochemical substitution reactions. For example, while the photochemical aromatic substitution of fluorine by the cyano group in ortho-fluoroanisole gives predominantly the hydoxylation product, the same reaction with para-fluoroanisole generates the cyanation product preferentially (Eq. 6.17).[35] The hydrogen bonding between water and the methoxyl group was attributed to the hydroxylation reaction in *ortho*-fluoroanisole. The effect of such a hydrogen bonding on the product distribution is much less in the latter case.

(6.17)

6.4.2 Carbon–Carbon Bond Formations

For carbon–carbon bond-formation purposes, S_N2 nucleophilic substitutions are frequently used. Simple S_N2 nucleophilic substitution reactions are generally slower in aqueous conditions than in aprotic organic solvents. This has been attributed to the solvation of nucleophiles in water. As previously mentioned in Section 5.2, Breslow and co-workers have found that cosolvents such as ethanol increase the solubility of hydrophobic molecules in water and provide interesting results for nucleophilic substitutions (Scheme 6.1). In alkylations of phenoxide ions by benzylic chlorides, S_N2 substitutions can occur both at the phenoxide oxygen and at the *ortho* and *para* positions of the ring. In fact, carbon alkylation occurs in water but not in nonpolar organic solvents and it is observed only when the phenoxide has at least one methyl substituent (*ortho*, *meta*, or *para*). The effects of phenol substituents and of cosolvents on the rates of the competing alkylation processes

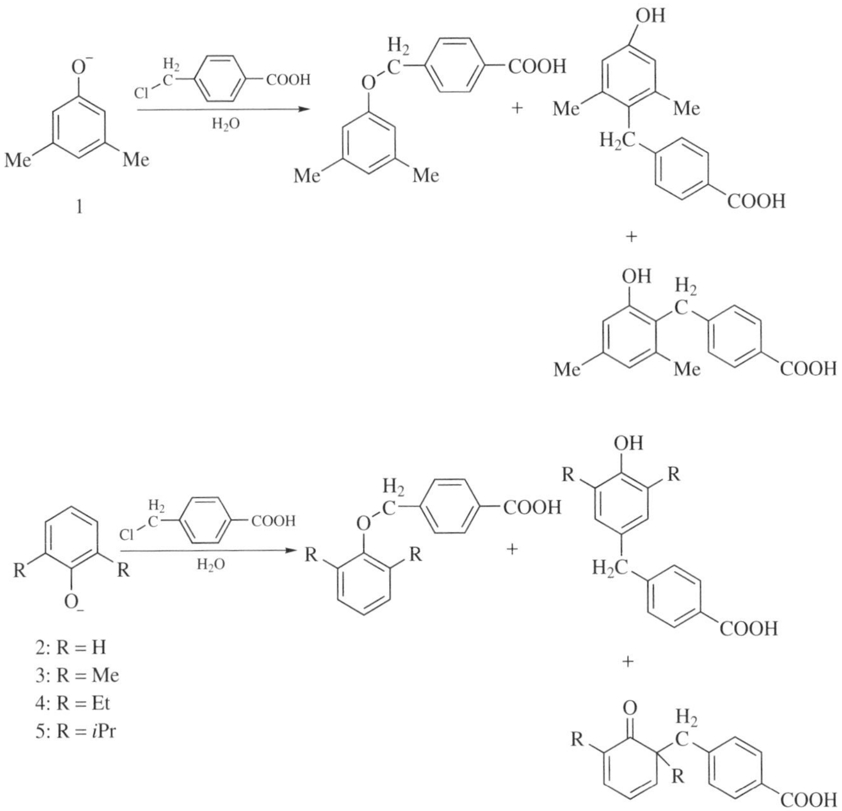

Scheme 6.1

TABLE 6.1   Product Distributions for the Reactions of Alkyl-Substituted Phenoxides with *p*-Carboxybenzyl Chloride in Water at 25°C

	Oxgen%	Ortho%	Para%
1	46	36	18
2	100		64
3	36		
4	30	25	45
5	75	9	16

indicate that in water the carbon alkylation involves a transition state with hydrophobic packing of the benzyl group onto the phenol ring (Table 6.1).[36]

Methylation of relatively acidic (pKa < 13) carbon nucleophiles occurs at neutral pH in aqueous media when substituted methylsulfonium and -selenonium salt are used as electrophiles (e.g., **6.2**).[37]

6.2

A simple one-pot preparation of *cis*-cyclopropanes from γ,δ-ketoalkenes using intramolecular alkylation under aqueous conditions was reported. Sequential treatment of γ,δ-keto alkenes with aqueous NBS in DMSO and then with solid KOH provides *cis*-cyclopropanes in good overall yields with a diastereoselective excess >99% (Eq. 6.18).[38]

(6.18)

Quantum mechanical/molecular mechanical study on the Favorskii rearrangement in aqueous media has been carried out.[39] The results obtained by QM/MM methods show that, of the two accepted mechanisms for Favorskii rearrangement, the semibenzilic acid mechanism (a) is favored over the cyclopropanone mechanism (b) for the α-chlorocyclobutanone system (Scheme 6.2). However, the study of the ring-size effects reveals that the cyclopropanone mechanism is the energetically preferred reactive channel for the α-chlorocyclohexanone ring, probably due to the straining effects on bicycle cyclopropanone, an intermediate that does not appear on the semibenzilic acid pathway. These results provide new information on the key factors responsible for the behavior of reactant systems embedded in aqueous media.

The water-soluble calix[n]arenes **6.3** (n = 4, 6 and 8) containing trimethylammonium groups act as efficient inverse phase-transfer catalysts in the nucleophilic substitution reaction of alkyl and arylalkyl halides with nucleophiles in water (Eq. 6.19).[40] In the presence of various surfactants (cationic, zwitterionic and anionic), the reactions of different halides and ketones show that the amount of ketone alkylation is much higher and that the reactions are faster in the presence than in the absence of surfactant aggregates.[41] The hydrolysis of the halide is minimized in the presence of cationic or zwitterionic surfactants.

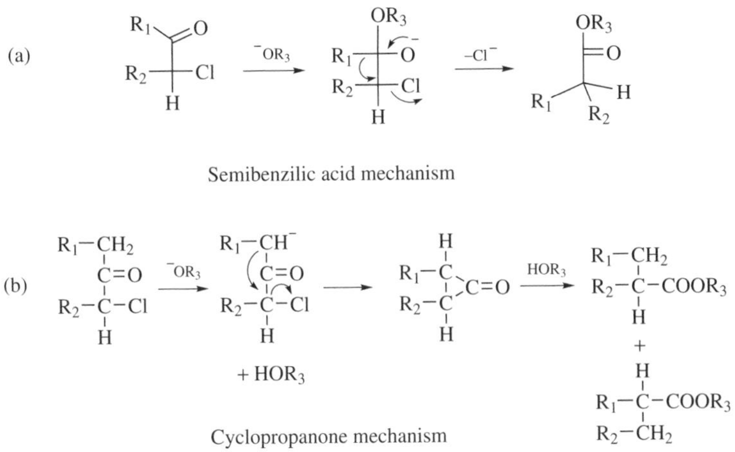

(a)

Semibenzilic acid mechanism

(b)

$+ HOR_3$

Cyclopropanone mechanism

Scheme 6.2

Nucleophilic aromatic photosubstitution reactions in aqueous solutions and in micellar media has been investigated extensively.[42]

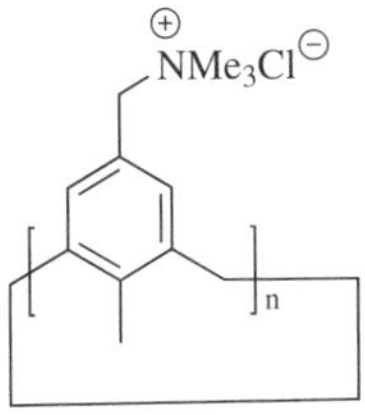

6.3

$$R-X + NaY \text{ (or KY)} \xrightarrow[\text{60 or 100°C}]{\text{6.3, H}_2\text{O}} R-Y + NaX \text{ (or KX)} \qquad (6.19)$$

$R = C_8H_{17}, PhCH_2CH_2,$ $X = Cl, Br, I$
   Bn, 2-naphthylmethyl $Y = CN, SCN, I$

6.5 REDUCTIVE COUPLING

6.5.1 Wurtz-Type Coupling

Homocoupling of alkyl halides in aqueous media can be mediated by manganese/cupric chloride to give the dimerization products in good yield. Cross-coupling can also be controlled to give the desired product.[43] Wurtz-type coupling of allyl halides was (in low yields) the normal outcome in refluxing alcohol.[44] A reductive coupling reaction of benzyl, allyl, and alkyl halides in an aqueous medium was promoted by zinc. Organic halides undergo reductive dimerization (Wurtz-type coupling) promoted by zinc at room temperature in an aqueous medium.[45] The reaction yields are strongly enhanced by copper catalysis. This coupling procedure provides an efficient and simple method for the homo-coupling of benzylic and allylic bromides and primary alkyl iodides. An allylgallium reagent is found to be effective for radical allylation of α-iodo or α-bromo carbonyl compounds. Treatment of benzyl bromoacetate with allylgallium, prepared from allylmagnesium chloride and gallium trichloride, in the presence of triethylborane in THF provided benzyl 4-pentenoate in good yield. The addition of water as a cosolvent improved the yields of allylated products (Eq. 6.20). [46]

$$
X \overset{\displaystyle O}{\underset{R^1}{\bigwedge}} Y \quad \xrightarrow[\text{Et}_3\text{B/O}_2,\ \text{THF/H}_2\text{O}]{R^2 \diagdown\diagup \text{GaL}_n} \quad \diagup\diagdown \overset{R^2}{\underset{R^1}{\bigwedge}} \overset{O}{\bigwedge} Y \qquad (6.20)
$$

$$X = \text{Br, I}$$
$$Y = NR^3R^4, OR^5$$

The Wurtz-type reductive coupling reaction of primary alkyl iodides or allyl halides and haloorganotins in cosolvent/$H_2O(NH_4Cl)$ and Zn media provides a route to mixed alkyl and allylstannanes. For example, mixed tetra-alkylstannanes R_3SnR' (R = Et, n-Pr, n-Bu and R' = Me, Et, n-Pr, n-Bu, n-Pent) and $R_2SnR'_2$ (R = n-Bu and R' = Me, Et, n-Pr, n-Bu) can be easily prepared in a one-pot synthesis via coupling reaction of alkyl iodides R'I with R_3SnX (X = Cl, I) and R_2SnCl_2 compounds in cosolvent/$H_2O(NH_4Cl)$ medium mediated by zinc dust. Coupling also occurs with $(Bu_3Sn)_2O$. Secondary alkyl iodides do not couple under the same reaction conditions (Eq. 6.21).[47]

$$
R_3SnCl \ + \ R'I \ \xrightarrow[\text{THF/H}_2\text{O(NH}_4\text{Cl saturated), reflux}]{\text{Zn dust,}} \ R_3SnR' \qquad (6.21)
$$

The reductive coupling of allyl halides to 1,5-hexadiene at glassy C electrodes was catalyzed by *tris*(2, 2'-bipyridyl)cobalt(II) and *tris*(4, 4'-dimethyl-2, 2'-bipyridyl)cobalt(II) in aqueous solutions of 0.1 M sodium dodecylsulfate (SDS) or 0.1 M cetyltrimethylammonium bromide (CTAB).[48] An organocobalt(I) intermediate was observed by its separate voltammetric reduction peak in each system studied. This intermediate undergoes an internal redox reaction to form 1,5-hexadiene and Co(II). The small micellar enhancements of reaction rates found for *tris*(2, 2′-bipyridyl)cobalt(II) in 0.1 M CTAB can be attributed to reactant compartmentalization in the micelles. Ph_2CHCl added to a refluxing solution of $CrCl_2$ in aqueous dioxane gave 100% Wurtz-type coupling product. Similar coupling occurs with other benzyl halides (Eq. 6.22).[49]

$$\text{(6.22)}$$

6.5.2 Ullmann-Type Coupling and Related Reactions

It is worth noting that the Ullmann-Goldberg condensation of aryl halides with phenols and anilines worked efficiently in the presence of copper in water.[50] For example, the coupling of 2-chlorobenzoic acid with 4-chlorophenol (K_2CO_3/pyridine/copper powder) gave 2-(4-chlorophenoxy)carboxylic acid (Eq. 6.23).[51] The Cu(I)-catalyzed transformation of 2-bromobenzoic acid into salicylic acid has also been studied in aqueous media (Eq. 6.24).[52]

$$\text{(6.23)}$$

$$\text{(6.24)}$$

The homocoupling of aryl halide to diaryl compounds, known as Ullmann coupling, is a synthetically useful reaction and has wide applications in material research. Such couplings have been studied in aqueous conditions. In 1970, arylsulfinic acids were coupled with Pd(II) in aqueous solvents to biaryls (Eq. 6.25).[53] However, the reaction required the use of a stoichiometric amount of palladium. In the presence of hydrogen gas, aryl halides homocoupled to give biaryl compounds in moderate yields (30–50%) in an aqueous/organic microemulsion (Eq. 6.26).[54]

$$2ArSO_2Na + Na_2PdCl_4 \xrightarrow[\text{1-36\%}]{H_2O} Ar–Ar + 2SO_2 + Pd + 4NaCl \qquad \text{(6.25)}$$

$$2ArX \xrightarrow[\text{H}_2\text{O-BuOH}]{\text{H}_2, \text{ cat. PdCl}_2, \text{ K}_2\text{CO}_3} Ar\text{-}Ar \qquad (6.26)$$

$$X = I, Br; Ar = Ph, p\text{-MePh}, p\text{-MePPh}, p\text{-ClPh}, m\text{-CF}_3\text{Ph}$$

In 1999, Venkatraman and Li reported a facile coupling of aryl halides using zinc in air and aqueous acetone at room temperature by using Pd/C as catalyst (Eq. 6.27).[55] Various aryl iodide and aryl bromide coupled effectively under these conditions. Subsequently, they found that the addition of surfactant such as crown ether in water alone provided better isolated yields of the product.[56] Sasson and co-workers further developed this reaction by using polyethyleneglycol (PEG) as an additive and performing the reaction under a higher temperature. In this case, aryl chlorides also work effectively.[57] Reductive homocoupling of chlorobenzenes to biphenyls affords high yields (93–95%) in the presence of catalytic PEG-400 and 0.4 mol% of a recyclable, heterogeneous trimetallic catalyst (4% Pd, 1% Pt, and 5% Bi on carbon). The competing reduction process is minimized.[58] They believe that dihydrogen is generated in situ. In addition to Pd/C, Rh/C is also effective as the catalyst.[59] Carbon dioxide was found to promote the palladium-catalyzed zinc-mediated reductive Ullmann coupling of aryl halides. In the presence of carbon dioxide, Pd/C, and zinc, various aromatic halides including less reactive aromatic chlorides were coupled to give the corresponding homocoupling products in good yields.[60]

$$2 ArX \xrightarrow[\substack{\text{Zn, H}_2\text{O/18-crown-6} \\ \text{air atmosphere, r.t.}}]{\text{Pd/C cat.}} Ar\text{-}Ar \qquad (6.27)$$

6.6 CARBONYLATION OF ORGANIC HALIDES

6.6.1 Carbonylation of Alkyl Halides

Transition-metal catalyzed carbonylation of 1-perfluoroalkyl-substituted 2-iodoalkanes has been carried out in aqueous media to give carboxylic acids with a perfluoroalkyl substituent at the β position (Eq. 6.28).[61]

$$RfCH_2CHR'I + CO + H_2O \xrightarrow[\text{base 42–89\%}]{\text{Pd, Co, or Rh cat.}} RfCH_2CHR'CO_2H \qquad (6.28)$$

$$Rf=\text{pefluoroalkyl group}$$

6.6.2   Carbonylation of Allylic and Benzylic Halides

The transition-metal catalyzed carbonylation of allylic and benzylic compounds offers a useful method for the synthesis of β,γ-unsaturated acids.[62] The requirement of a high carbon monoxide pressure and the low yield of the products limited the usefulness of the method in organic synthesis.[63] In 1977, it was found that the carbonylation of benzyl bromide and chloride could be carried out by stirring aqueous sodium hydroxide and an organic solvent using a phase transfer agent and a cobalt catalyst (Eq. 6.29).[64,65] Under high pressure and temperature, even benzylic mercaptans reacted similarly to give esters.[66] In the presence of a nickel catalyst, similar carbonylations of allyl bromide and chloride in aqueous NaOH can be carried out at atmospheric pressure.[67] The base concentration significantly influenced the yield and the product distribution. More recently, it was found that the palladium-catalyzed carbonylation of allyl chloride proceeded smoothly in a two-phase aqueous NaOH/benzene medium under atmospheric pressure at room temperature.[68] Catalysts with or without phosphorus ligands gave similar results and the presence of hydroxide was essential. The reaction seemed to occur at the liquid-liquid interface because no phase-transfer agent was used. However, the addition of surfactants such as n-$C_7H_{15}SO_3Na$ or n-$C_7H_{15}CO_2Na$ did accelerate the reactions.[69]

$$PhCH_2Cl + CO \xrightarrow[\text{PTC}]{\text{[Co(CO)}_4\text{]}^-,\text{ aq NaOH}} PhCH_2CO_2Na \qquad (6.29)$$

By using a water-soluble palladium catalyst, 5-hydroxymethylfurfural is selectively carbonylated to the corresponding acid at 70°C, together with reduced product (Eq. 6.30).[70]

$$HOH_2C \overbrace{}^{O} CHO \xrightarrow[\text{Pd/tppts}]{\text{CO, H}^+/\text{H}_2\text{O}} HO_2CH_2C \overbrace{}^{O} CHO \qquad (6.30)$$

6.6.3   Carbonylation of Aryl Halides

The palladium-catalyzed carbonylation of aryl halides in the presence of various nucleophiles is a convenient method for synthesizing various aromatic carbonyl compounds (e.g., acids, esters, amides, thioesters, aldehydes, and ketones). Aromatic acids bearing different aromatic fragments and having various substituents on the benzene ring have been prepared from aryl iodides at room temperature under 1 atm

CO in a mixed solvent of H_2O/DMF (1/1 or 1/2, v/v), and even in water alone, depending on the solubility of the substrate (Eq. 6.31).[71] The palladium(II) complexes $Pd(OAc)_2$, K_2PdCl_4, $PdCl_2(PPh_3)_2$, and $Pd(NH_3)_4Cl_2$ are used as the precursors of the catalyst, using either K_2CO_3 or NaOAc as the base. Iodoxyarenes can be carbonylated in water alone due to their solubility in the solvent.[72] Recent work has been done on the use of water-soluble catalysts.[73] Under the appropriate pressure and temperature conditions, aryl mercaptans (thiophenols) can also be carbonylated in aqueous media with cobalt carbonyl as the catalyst.[74]

$$\text{Ar-I} \xrightarrow[-I^-]{\text{CO, OH}^-, \text{[Pd]}} \text{ArCO}_2^- \xrightarrow{H^+} \text{ArCO}_2\text{H} \qquad (6.31)$$

$$\text{Ar} = \text{XC}_6\text{H}_4, \text{ naph, heteroaryl} \qquad (30-100\%)$$

6.7 TRANSITION-METAL CATALYZED COUPLING REACTIONS

6.7.1 The Heck Coupling

The reaction between aryl (or alkenyl) halides and alkenes in the presence of a catalytic amount of a palladium compound to give substitution of the halides by the alkenyl group is commonly referred to as the Heck reaction.[75] Both inter- and intramolecular Heck reactions have been performed in aqueous media. Palladium-catalyzed reactions of aryl halides with acrylic acid or acrylonitrile give the corresponding coupling products in high yields with a base in water (Eq. 6.32).[76] The reaction provides a new and simple method for the synthesis of substituted cinnamic acids and cinnamonitriles. Recently, such reactions have been carried out by using a water-soluble phosphine ligand.[77] Iodobenzoic acid can be used directly to couple with acrylic acid. Diaryliodonium salts react similarly.[78] Jeffery studied the reaction under phase-transfer conditions.[79] It was found that the presence of water is the determinant for the efficiency of quaternary ammonium salt in the palladium-catalyzed vinylation of organic halides using an alkaline metal carbonate as the base. The phase-transfer procedure can be performed without the organic cosolvent.

$$\text{Ar-X} + \underset{}{\overset{E}{=\!\!/}} \xrightarrow[\substack{\text{NaHCO}_3/\text{K}_2\text{CO}_3/80-100^\circ\text{C} \\ 87-97\%}]{\text{Pd(OAc)}_2 \, (1 \text{ mol}\%), \text{H}_2\text{O}} \underset{\text{Ar}}{\overset{E}{=\!\!/}} \qquad (6.32)$$

In aqueous DMF, the reaction can be applied to the formation of C–C bonds in a solid-phase synthesis with resin-bound iodobenzoates (Eq. 6.33).[80] The reaction proceeds smoothly and leads to moderate to high yield of product under mild conditions. The optimal conditions involve the use of 9:1 mixture of DMF-water. Parsons investigated the viability of the aqueous Heck reactions under superheated conditions.[81] A series of aromatic halides were coupled with styrenes under these conditions. The reaction proceeded to approximately the same degree at 400°C as at 260°C. Some 1,2-substituted alkanes can be used as alkene equivalents for the high-temperature Heck-type reaction in water.[82]

$$(6.33)$$

6.7.2 The Suzuki Coupling

The cross-coupling reaction of alkenyl and aryl halides with organoborane derivatives in the presence of a palladium catalyst and a base, known as the Suzuki reaction, has often been carried out in an organic/aqueous mixed solvent (Eq. 6.34).[83] The reaction was initially carried out in a mixture of benzene and aqueous Na_2CO_3.[84] However, the reaction proceeds more rapidly in a homogeneous medium (e.g., aqueous DME). This condition works satisfactorily in most aryl-aryl couplings.[85] Thus, dienes are conveniently prepared from the corresponding alkenylborane and vinyl bromide in refluxing THF in the presence of $Pd(PPh_3)_4$ and an aqueous NaOH solution.[86] The use of aqueous TlOH, instead of NaOH or KOH, significantly increased the rate of the coupling. In Kishi's palytoxin synthesis, the cross-coupling between the alkenylboronic acid and the iodoalkene was accomplished stereoselectively at room temperature.[87] The TlOH-modified coupling has also been used effectively in Roush's synthesis of kijanimicin,[88] Nicolaou's synthesis of (12R)-HETE,[89] and Evans's synthesis of rutamycin B.[90] Multigram-scale synthesis of a biphenyl carboxylic acid derivative has been achieved using a Pd/C-mediated Suzuki-coupling approach.[91] The Suzuki reaction in aqueous medium has also been used in other applications such as the synthesis of monosubstituted arylferrocenes,[92] insulated molecular wires (conjugated polyrotaxanes),[93] cyclodextrin

[2]rotaxanes,[94] unprotected halonucleosides,[95] 6-substituted N-Boc 3,4-dihydro-2H-pyridines,[96] heteroarylbenzoic acids,[97] alkenylpurines,[98] 6-alkyl-N-alkoxycarbonyl-3,4-dihydro-2H-pyridines,[99] coumarinic derivatives,[100] and biaryl colchicinoids.[101] The reaction has been used in synthesizing a combinatorial library of biaryls.[102]

$$RX + R'—B< \xrightarrow{[Pd]} R\text{-}R' \tag{6.34}$$

R = 1-alkenyl or aryl; R' = aryl or alkyl
X = Br or I

Casalnuovo and Calabrese reported that by using the water-soluble palladium(0) catalyst $Pd(PPh_2(m\text{-}C_6H_4SO_3M))_3$ (M = Na^+, K^+), various aryl bromides and iodides reacted with aryl and vinyl boronic acids, terminal alkynes, and dialkyl phosphites to give the cross-coupling products in high yields in water.[103] This reaction can tolerate a broad range of functional groups, including those present in unprotected nucleotides and aminoacids. The cross-coupling of boronic acids or esters with alkenyl iodides was conducted similarly, generating functionalized dienes.[104] Polyfunctional biaryls are prepared by a modified Suzuki cross-coupling reaction between arylboronic acids or sodium tetraphenylborate and aryl halides in aqueous solvents or neat water using a phosphine-free palladium catalyst, in the presence of bases with high catalytic efficiency (250,000 catalytic cycles). All four Ph groups of Ph_4BNa participate in the reaction.[105] The poly(ethylene glycol) ester of bromo-, iodo-, and triflate-*para*-substituted benzoates are smoothly cross-coupled with aryl boronic acids (Suzuki reaction) under "ligandless" palladium acetate catalysis in water. The reaction proceeds without organic cosolvent under conventional thermal conditions (70°C, 2 h) and under microwave irradiation (75 W, 2–4 min). Whereas conventional thermal conditions induced ester cleavage (up to 45%), this side reaction is suppressed when microwave conditions are employed.[106] The Suzuki coupling can be done without any organic cosolvent.[107]

Recently, iodobenzoates anchored onto an ionic liquid support (**6.4**) were coupled to various aryl boronic acids (**6.5**) in aqueous media using $Pd(OAc)_2$ as the catalyst at 80°C to give the coupled product **6.6** (Scheme 6.3). Compounds **6.6** were purified simply by washing the reaction mixture with ether, which removed the unreacted starting materials and the side product **6.7** without the need of chromatography. Compounds **6.6** were then cleaved from the ionic liquid support

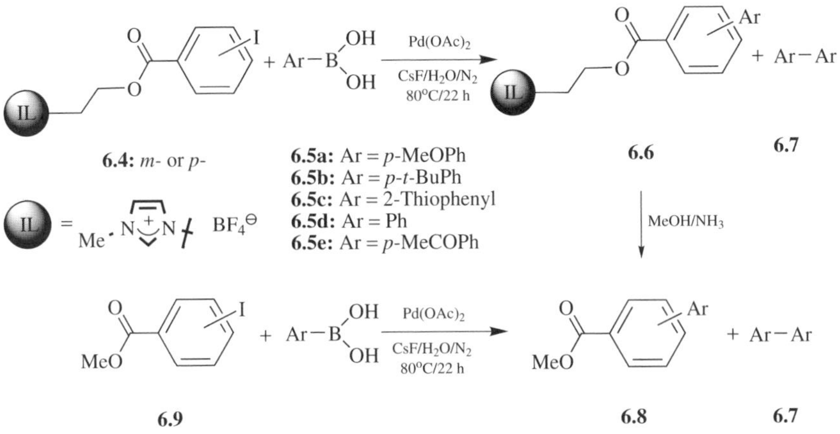
Pd(OAc)2
CsF/H2O/N2
80°C/22 h
+ Ar−Ar
6.4: m- or p-
6.5a: Ar = p-MeOPh
6.5b: Ar = p-t-BuPh
6.5c: Ar = 2-Thiophenyl
6.5d: Ar = Ph
6.5e: Ar = p-MeCOPh
6.6
6.7
MeOH/NH3
6.9
6.8
6.7

Scheme 6.3

by methanolic ammonia to give the pure methyl ester **6.8** in good yields. The ionic-liquid-supported reactions were compared with the conventional unsupported reactions starting from **6.9** to give **6.8** under identical conditions.[108] In the unsupported synthesis, purification of **6.8** from unreacted starting materials and side product **6.7** by chromatography was necessary, and the overall yield was found to be inferior to the ionic-liquid-supported synthesis.

Pd nanoparticles stabilized by poly(N-vinyl-2-pyrrolidone) (PVP) are efficient catalysts for the Suzuki reactions in an aqueous medium.[109] The initial rate depends linearly on the concentration of Pd catalyst, suggesting that the catalytic reaction occurs on the surface of the Pd nanoparticles. The role of capping materials on catalytic activity and stability of transition-metal nanoparticles used in catalysis in solution was studied.[110] The stability of the Pd nanoparticles was measured by the tendency of nanoparticles to give Pd black powder after the catalytic reaction. The Suzuki reactions between arylboronic acids and bromoarenes, catalyzed by Pd nanoparticles, result in byproducts due to the homocoupling of bromoarenes. The two properties were found to be anticorrelated (i.e., the most stable is the least catalytically active). The Pd/CD clusters were found to exhibit an excellent catalytic activity toward the Suzuki-Miyaura cross-coupling reactions in water reactions between iodophenols and phenylboronic acid.[111] Reverse-phase glass beads have been employed in Suzuki reactions to provide, in aqueous media, a route to diverse polar substrates in good yield with low levels of palladium leaching.[112] Pd(II)-exchanged

NaY zeolite showed high activity in the Suzuki cross-coupling reactions of aryl bromides and iodides without added ligands. The DMF:water ratio and the type and amount of base were found to be crucial for the efficiency of the reaction. The catalyst is reusable after regeneration. The addition of phosphine additives such as dimethyldiphenylphosphine and 3-methylbenzothiazolium iodide brought the reaction to a halt.[113] Cyclodextrins or calixarenes possessing extended hydrophobic host cavities and surface-active properties were found to be very efficient as mass-transfer promoters for the palladium-mediated Suzuki cross-coupling reaction of 1-iodo-4-phenylbenzene and phenylboronic acid in an aqueous medium. The cross-coupling rates were up to 92 times higher than those obtained without addition of any promoter compound.[114]

Sterically demanding, water-soluble alkylphosphines **6.10** and **6.11** as ligands have been found to have a high activity for the Suzuki coupling of aryl bromides in aqueous solvents (Eq. 6.35).[115] Turnover numbers up to 734,000 mmol/mmol Pd have been achieved under such conditions. Glucosamine-based phosphines were found to be efficient ligands for Suzuki cross-coupling reactions in water.[116]

$$\text{(6.35)}$$

Recently, Suzuki-type reactions in air and water have also been studied, first by Li and co-workers.[117] They found that the Suzuki reaction proceeded smoothly in water under an atmosphere of air with either Pd(OAc)$_2$ or Pd/C as catalyst (Eq. 6.36). Interestingly, the presence of phosphine ligands prevented the reaction. Subsequently, Suzuki-type reactions in air and water have been investigated under a variety of systems. These include the use of oxime-derived palladacycles[118] and tuned catalysts (TunaCat).[119] A preformed oxime-carbapalladacycle complex covalently anchored onto mercaptopropyl-modified silica is highly active (>99%) for the Suzuki reaction of p-chloroacetophenone and phenylboronic acid in water; no leaching occurs and the same catalyst sample can be reused eight times without decreased activity.[120]

By utilizing a solid support–based tetradentate N-heterocyclic carbene-palladium catalyst, cross couplings of aryl bromides with phenylboronic acid were achieved in neat water under air.[121] A high ratio of substrate to catalyst was also realized.

$$(6.36)$$

The use of microwave heating is a convenient way to facilitate the Suzuki-type reactions in water.[122] It is possible to prepare biaryls in good yield very rapidly (5–10 min) on small (1 mmol) and larger (10–20 mmol) scales from aryl halides and phenylboronic acid using water as solvent and palladium acetate as catalyst. Recently, Leadbeater[123] reported that the Suzuki-reaction can be carried out in aqueous sodium carbonate solutions at high temperature without using any transition-metal catalyst (Eq. 6.37).[124] At 150°C in a sealed tube, optimum yields of the product were obtained when the ratio of aryl bromide to boronic acid was 1:1.3. The required excess amount of boronic acid was attributed to some protodeboronation of the boronic acid, which produced benzene. Various aryl bromides bearing both electron-donating and electron-withdrawing groups proceeded readily. Aryl bromides provided better yields than aryl iodides or aryl chlorides; the latter showed no reactivity under the reaction conditions. Sterically demanding aryl bromides were also coupled in good yields. The reaction also proved to be regiospecific with respect to both the aryl bromide and the boronic acid. The reaction of 4-bromoacetophenone with 4-methylphenylboronic acid gave the desired coupling product in excellent yield. Microwave heating for 5 min provided comparable yields to conventional heating for 5 h with 4-bromoacetophenone. With unactivated and deactivated aryl bromides, conventional heating is not sufficient even after 16 h. To show that the reaction was truly metal-free, new glassware, apparatus, and reagents were used. No palladium was detected down to 0.1 ppm Pd by analysis of the crude reaction mixture. Induced coupled plasma atomic absorption (ICP-AA) spectroscopy analysis did not show other possible catalytic metals above the detection limit of the apparatus (0.5–1 ppm). Later reassessment by Leadbeater and others led to the conclusion that it was the extremely low contamination of palladium at the level of 50 ppb in commercially available sodium carbonate that was responsible for catalyzing the Suzuki reaction. The reaction was therefore not truly metal-free.[125]

The detection of the palladium contaminant was carried out by ICP-MS instead of ICP-AA.

$$(6.37)$$

Z=Me, OMe, Cl, NH$_2$, NO$_2$, CHO, COMe, CO$_2$H, CO$_2$R, etc.

6.7.3 The Stille Coupling

The coupling between alkenyl and aryl halides with organostannanes in the presence of a palladium catalyst is referred to as the Stille reaction.[126] Although it was known that in the palladium-catalyzed coupling reaction of organostannanes with vinyl epoxides, the addition of water to the organic reaction medium increased yields and affected the regio- and stereochemistry,[127] the Stille-coupling reaction in aqueous media was investigated only recently. Davis[128] reported a coupling in aqueous ethanol (Eq. 6.38). The reaction gave a high yield of the coupled product, which hydrolyzed in situ.

$$(6.38)$$

Collum[129] reported that while the Stille coupling can proceed without using a phosphine ligand, the addition of a water-soluble ligand improved the yield of the reaction. Water-soluble aryl and vinyl halides were coupled with alkyl-, aryl-, and vinyltrichlorostannane derivatives in this way (Eq. 6.39).

$$(6.39)$$

no ligand   81%

with PhP(m-SO$_3$Na)$_2$ 95%

Arenediazonium chlorides and hydrogen sulfates react with tetramethyltin in aqueous acetonitrile in the presence of a catalytic amount of palladium acetate to give high yields of substituted toluenes.[130] One-pot hydrostannylation/Stille couplings with catalytic amounts of tin were performed by a syringe-pump addition of 1.5 equiv of various Stille electrophiles to a 37°C ethereal mixture of alkyne, aqueous Na_2CO_3, polymethylhydrosiloxane, Pd_2dba_3, tri-2-furylphosphine, $PdCl_2(PPh_3)_2$, and 0.06 equiv of Me_3SnCl over 15 h to give the cross-coupled products in 75–91% yield (Eq. 6.40).[131] An efficient Stille cross-coupling reaction using a variety of aryl halides in neat water has been developed.[132] Employing a palladium–phosphinous acid catalyst, $[(t\text{-}Bu)_2P(OH)]_2PdCl_2$, allows the formation of biaryls from aryl chlorides and bromides in good-to-high yields. Li and co-workers reported a Stille-type coupling in water under an atmosphere of air. [133]

$$
\begin{array}{c}
\text{R}\!\!-\!\!\!\equiv\!\!\!-\!\!\text{H} \\
+ \\
\text{R}'\!-\!\text{X}
\end{array}
\quad
\xrightarrow[\substack{\text{1 mol\% Pd}_2\text{dba}_3,\ \text{4 mol\% (2-furyl)}_3\text{P} \\ \text{aq Na}_2\text{CO}_3,\ \text{PMHS, Et}_2\text{O, 37°C, 15 h}}]{\substack{\text{6 mol\% Me}_3\text{SnCl} \\ \text{1 mol\% PdCl}_2(\text{PPh}_3)_2}}
\quad
\text{R}\diagup\!\!\diagdown\!\!\diagup\text{R}'
\qquad (6.40)
$$

6.7.4 Other Transition-Metal Catalyzed Couplings

There are many other transition-metal catalyzed coupling reactions that are based on organic halides in aqueous media. One example is the coupling of terminal alkyne with aryl halides, the Sonogashira coupling, which has been discussed in detail in the chapter on alkynes (Chapter 4). An example is the condensation of 2-propynyl or allyl halides with simple acetylenes in the presence of copper salts.

Another example is the use of arylsilicon reagents for aryl-aryl coupling. Biaryls were obtained in good yields by reacting diphenyldifluorosilane or diphenyldiethoxysilane with aryl halides in aqueous DMF at 120°C in the presence of KF and a catalytic amount of $PdCl_2$ (Eq. 6.41).[134] Phosphine ligands are not required for the reaction. Li and co-workers reported a highly efficient palladium-catalyzed coupling of aryl halides with arylhalosilanes in open air and in water in the presence of base or fluoride. Both $Pd(OAc)_2$ and Pd/C are effective as catalysts. [135]

$$
\text{ArBr} + \text{Ph}_2\text{SiF}_2 \quad \xrightarrow[\substack{\text{DMF-H}_2\text{O, 120°C} \\ \text{KF}}]{\text{2 mol\% PdCl}_2} \quad \text{Ar-Ph} + [\text{PhSiF}_3] \qquad (6.41)
$$

Examples involving the use of organomercury reagents as nucleophiles in aqueous media are also known. Bergstrom studied the synthesis of C-5-substituted pyrimidine nucleosides in aqueous media via a mercurated intermediate using Li_2PdCl_4 as a catalyst (Eq. 6.42).[136]

Mertes investigated the coupling of the 5-mercuriuridines with styrenes in aqueous media resulting in alkylation of the uracil nucleotides.[137] Carbon alkylation of C-5 on the uracil ring in the ribo- and deoxyribonucleosides and nucleotides was obtained in high yields by this method. A similar reaction was used by Langer et al. in the synthesis of 5-(3-amino)allyluridine and deoxyuridine-5′-triphosphates (AA-UTP and AA-dUTP) (**6.12**).[138]

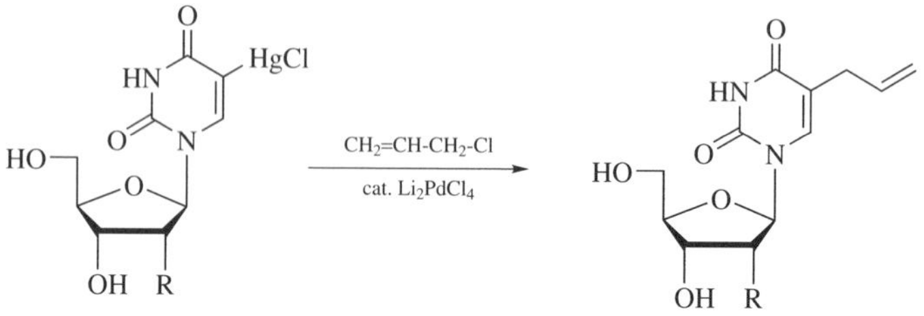

(6.42)

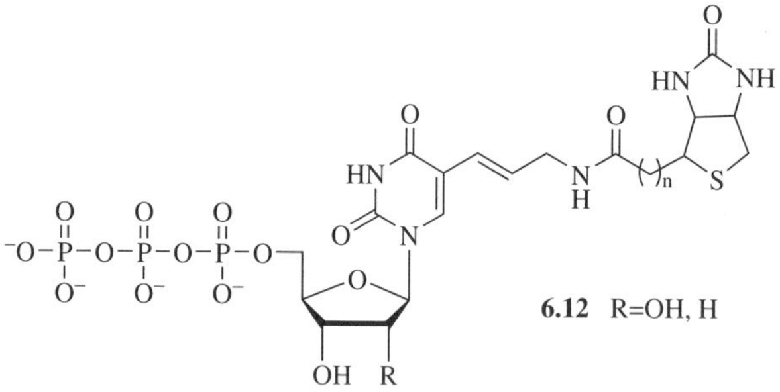

6.12 R=OH, H

REFERENCES

1. Benyei., A. C., Lehel, S., Joo, F., *J. Mol. Catal. A: Chem.* **1997**, *116*, 349.
2. For a review, see: Krief, A., Laval, A.-M., *Acros Organics Acta* **1996**, *2*, 17.
3. Maitra, U., Sarma, K. D., *Tetrahedron Lett.* **1994**, *35*, 7861.
4. Rolla, F., *J. Org. Chem.* **1981**, *46*, 3909.
5. Light, J., Breslow, R., *Tetrahedron Lett.* **1990**, *31*, 2957.
6. Postigo, A., Ferreri, C., Navacchia, M. L., Chatgilialoglu, C., *Synlett* **2005**, 2854.
7. Ranu, B. C., Dutta, P., Sarkar, A., *J. Chem. Soc. Perkin Trans. 1*, **1999**, 1139.
8. Lee, Y. J., Chan, T. H., *Can. J. Chem.* **2004**, *82*, 71.

9. Xia, C., Xu, J., Wu, W., Liang, X., *Catal. Commun.* **2004**, *5*, 383.

10. Reetz, M. T., Frombgen, C., *Synthesis* **1999**, 1555.

11. Yamashita, J., Inoue, Y., Kondo, T., Hashimoto, H., *Bull. Chem. Soc. Jpn* **1985**, *58*, 2709.

12. Colon, I., *J. Org. Chem.* **1982**, *47*, 2622.

13. Davydov, D. V., Beletskaya, I. P., *Metalloorgan. Khim.* **1993**, *6*, 111.

14. Smith, M. B., March, J., *March's Advanced Organic Chemistry*, 5th ed., John Wiley & Sons, New York, 2001.

15. Keeffe, J. R., Jencks, W. P., *J. Am. Chem. Soc.* **1983**, *105*, 265.

16. Baciocchi, E., Lillocci, C., *J. Chem. Soc. Perkin Trans. 2*, **1975**, 802.

17. Owlia, A., Wang, Z., Rusling, J. F., *J. Am. Chem. Soc.* **1989**, *111*, 5091.

18. Ruasse, M.-F., Blagoeva, I. B., Ciri, R., Garcia-Rio, R. C. L., Leis, J. R., Marques, A., Mejuto, J., Monnier, E., *Pure Appl. Chem.* **1997**, *69*, 1923.

19. For an early review, see: Olivier, S. C. J., *Chem. Weekbl.* **1929**, *26*, 518.

20. Ohtani, N., Besse, J. J., Regen, S. L., *Bull. Chem. Soc. Jpn* **1981**, *54*, 607.

21. Menchikov, L. G., Vorogushin, A. V., Korneva, O. S., Nefedov, O.M., *Mendel. Commun.* **1995**, 223.

22. Watanabe, T., Suzuki, A., Monma, Y. (Konika Chemical Corporation, Japan), Jpn. Kokai Tokkyo Koho (2004), 7 pp. JP 2004292323, A2: 20041021.

23. Vanden, E., Jean, J., Mailleux, I., *Synth. Commun.* **2001**, *31*, 1.

24. Takido, T., Itabashi, K., *Synthesis* **1987**, 817.

25. Lakouraj, M. M., Movassagh, B., Fadaei, Z., *Synth. Commun.* **2002**, *32*, 1237.

26. Degani, I., Fochi, R., Regondi, V., *Synthesis* **1983**, 630.

27. Devaud, M., Madec, M. C., *J. Organomet. Chem.* **1975**, *93*, 85.

28. Chan, T.-H., Yang, Y., Li, C.-J., *J. Org. Chem.* **1999**, *64*, 4452.

29. Carril, M., SanMartin, R., Tellitu, I., Dominguez, E., *Org. Lett.* **2006**, *8*, 1467.

30. Nakano, S. (Nippon Paint Co. Japan), Eur. Pat. Appl. (1997), 5 pp. EP 791575; A1: 19970827.

31. Ju, Y., Varma, R. S., *J. Org. Chem.* **2006**, *71*, 135.

32. Gusarova, N. K., Shaikhudinova, S. I., Ivanova, N. I., Reutskaya, A. M., Albanov, A. I., Trofimov, B. A., *Russ. J.Gen. Chem.* **2001**, *71*, 718.

33. Sarma, K. D., Maitra, U., *Ind. J. Chem. Sect. B* **2001**, *40B*, 1148.

34. Beletskaya, I. P., Kabachnik, M. M., Solntseva, M. D., *Russ. J. Org. Chem.* **1999**, *35*, 71.

35. Liu, J. H., Weiss, R. G., *J. Org. Chem.* **1985**, *50*, 3655.

36. Breslow, R., Groves, K., Mayer, M. U., *J. Am. Chem. Soc.* **2002**, *124*, 3622.

37. Winkler, J. D., Finck-Estes, M., *Tetrahedron Lett.* **1989**, *30*, 7293.

38. Dechoux, L., Ebel, M., Jung, L., Stambach, J. F., *Tetrahedron Lett.* **1993**, *34*, 7405.

39. Castillo, R., Andres, J., Moliner, V., *J. Phys. Chem. B* **2001**, *105*, 2453.

40. Shimizu, S., Kito, K., Sasaki, Y., Hirai, C., *Chem. Commun.* **1997**, 1629.

41. Cerichelli, G., Cerritelli, S., Chiarini, M., De Maria, P., Fontana, A., *Chem. Eur. J.* **2002**, *8*, 5204.

42. Tedesco, A. C., Nogueira, L. C., Bonilha, J. B. S., Alonso, E. O., Quina, F. H., *Quimica Nova* **1993**, *16*, 275; Marquet, J., Jiang, Z., Gallardo, I., Batle, A., Cayon, E., *Tetrahedron Lett.* **1993**, *34*, 2801.

43. Ma, J., Chan, T.-H., *Tetrahedron Lett.* **1998**, *39*, 2499.

44. Nosek, J., *Collect. Czech. Chem. Commun.* **1964**, *29*, 597.

45. De Sa, A. C. P. F., Pontes, G. M. A., Dos Anjos, J. A. L., Santana, S. R., Bieber, L. W., Malvestiti, I., *J. Braz. Chem. Soc.* **2003**, *14*, 429.

46. Usugi, S., Yorimitsu, H., Oshima, K. *Tetrahedron Lett.* **2001**, *42*, 4535.

47. Marton, D., Tari, M., *J. Organomet. Chem.* **2000**, *612*, 78; Furlani, D., Marton, D., Tagliavini, G., Zordan, M., *J. Organomet. Chem.* **1988**, *341*, 345.

48. Kamau, G. N., Rusling, J. F., *J. Electroanal. Chem. Interfac. Electrochem.* **1988**, 240, 217.

49. Slaugh, L. H., Raley, J. H., *Tetrahedron* **1964**, *20*, 1005; Kochi, J. K., Buchanan, D., *J. Am. Chem. Soc.* **1965**, *87*, 853.

50. Pellon, R. F., Carrasco, R., Rodes, L., *Synth. Commun.* **1993**, *23*, 1447; Pellon, R. F., Mamposo, T., Carrasco, R., Rodes, L., *Synth. Commun.* **1996**, *26*, 3877.

51. Pellon, R. F., Carrasco, R., Millian, V., Rodes, L., *Synth. Commun.* **1995**, *25*, 1077.

52. Saphier, M., Masarwa, A., Cohen, H., Meyerstein, D., *Eur. J. Inorg. Chem.* **2002**, 1226.

53. Garves, K., *J. Org. Chem.* **1970**, *35*, 3273.

54. Davydov, D. V., Beletskaya, I. P., *Russ. Chem. Bull.* **1995**, *44*, 1139.

55. Venkatraman, S., Li, C.-J., *Org. Lett.* **1999**, *1*, 1133.

56. Venkatraman, S., Li, C.-J., *Tetrahedron Lett.* **2000**, *41*, 4831; Venkatraman, S., Huang, T., Li, C.-J., *Adv. Synth. Cat.* **2002**, *344*, 399; Mukhopadhyay, S., Yaghmur, A., Baidossi, M., Kundu, B., Sasson, Y., *Org. Proc. Res. Devel.* **2003**, *7*, 641.

57. Mukhopadhyay, S., Rothenberg, G., Gitis, D., Sasson, Y., *Org. Lett.* **2000**, *2*, 211.

58. Mukhopadhyay, S., Rothenberg, G., Sasson, Y., *Adv. Synth. Cat.* **2001**, *343*, 274.

59. Mukhopadhyay, S., Joshi, A. V., Peleg, L., Sasson, Y., *Org. Proc. Res. Devel.* **2003**, *7*, 44.

60. Li, J.-H., Xie, Y.-X., Yin, D.-L., *J. Org. Chem.* **2003**, *68*, 9867.

61. Urata, H., Kosukegawa, O., Ishii, Y., Yugari, H., Fuchikami, T., *Tetrahedron Lett.* **1989**, *30*, 4403.

62. Heck, R.F., *Palladium Reagents in Organic Syntheses,* Academic press, London, 1985.

63. Kiji, J., Okano, T., Nishiumi, W., Konishi, H., *Chem. Lett.* **1988**, 957. Refs. are cited therein.

64. Alper, H., Abbayes, H. D., *J. Organomet. Chem.* **1977**, *134*, C11.

65. Cassar, L., Foa, M., *J. Organomet. Chem.* **1977,** *134*, C15.

66. Alper, H., *J. Organomet. Chem.* **1986**, *300*, 1. Refs. are cited therein.

67. Joo, F., Alper, H., *Organometallics* **1985**, *4*, 1775.

68. Bumagin, N. A., Nikitin, K. V., Beletskaya, I. P., *J. Organomet. Chem.* **1988**, *358*, 563.

69. Okano, T., Hayashi, T., Kiji, J., *Bull. Chem. Soc. Jpn.* **1994**, *67*, 2339.

70. Papadogianakis, G., Maat, L., Sheldon, R. A., *J. Chem. Soc., Chem. Commun.* **1994**, 2659.

71. Bumagin, N. A., Nikitin, K. V., Beletskaya, I. P., *J. Organomet. Chem.* **1988**, *358*, 563.

72. Grushin, V. V., Alper, H. H., *J. Org. Chem.* **1993**, *58*, 4798.

73. Monteil, F., Kalck, P., *J. Organomet. Chem.* **1994**, *482*, 45.

74. Shim, S. C., Antebi, S., Alper, H., *J. Org. Chem.* **1985**, *50*, 147.

75. For reviews, see: Heck, R. F., *Acc. Chem. Res.* **1979**, *12*, 146.

76. Bumagin, N. A., Andryuchova, N. P., Beletskaya, I. P., *Izv. Akad. Nauk SSSr*, **1988**, *6*, 1449; Bumagin, N. A., More, P. G., Beletskaya, I. P., *J. Organomet. Chem.* **1989**, *371*, 397; Bumagin, N. A., Bykov, V. V., Sukhomlinova, L. I., Tolstaya, T. P., Beletskaya, I. P., *J. Organomet. Chem.* **1995**, *486*, 259.

77. Genet, J. P., Blart, E., Savignac, M., *Synlett.* **1992**, 715.

78. Bumagin, N. A., Sukhomlinova, L. I., Banchikov, A. N., Tolstaya, T. P., Beletskaya, I. P., *Bull. Russ. Acad. Sci.* **1992**, *41*, 2130.

79. Jeffery, T., *Tetrahedron Lett.* **1994**, *35*, 3051.

80. Hiroshige, M., Hauske, J. R., Zhou, P., *Tetrahedron Lett.* **1995**, *36*, 4567.

81. Reardon, P., Metts, S., Crittendon, C., Daugherity, P., Parsons, E. J., *Organometallics* **1995**, *14*, 3810.

82. Diminnie, J., Metts, S., Parsons, E. J., *Organometallics* **1995**, *14*, 4023.

83. Miyaura, N., Ishiyama, T., Sasaki, H., Ishikawa, M., Sato, M., Suzuki, A., *J. Am. Chem. Soc.* **1989**, *111*, 314. For a recent review, see: Miyaura, N., Suzuki, A., *Chem. Rev.* **1995**, *95*, 2457.

84. Miyaura, N., Yanagi, T., Suzuki, A., *Synth. Commun.* **1981**, *11*, 513.

85. Katz, H. E., *J. Org. Chem.* **1987**, *52*, 3932.

86. Colberg, J. C., Rane, A., Vaquer, J., Soderquist, J. A., *J. Am. Chem. Soc.* **1993**, *115*, 6065.

87. Armstrong, R. W., Beau, J. M., Cheon, S. H., Christ, W. J., Fujioka, H., Ham, W. H., Hawkins, L. D., Jin, H., Kang, S. H., Kishi, Y., Martinelli, M. J., McWhorter, W. W., Mizuno, M., Nakata, M., Stutz, A. E., Talamas, F. X., Taniguchi, M., Tino, J. A., Ueda, K., Uenishi, J. I., White, J. B., Yonaga, M., *J. Am. Chem. Soc.* **1989**, *111*, 7525.

88. Roush, W. R., Brown, B. B., *J. Org. Chem.* **1993**, *58*, 2162.

89. Nicolaou, K. C., Ramphal, J. Y., Palazon, J. M., Spanevello, R. A., *Angew. Chem., Int. Ed. Engl.* **1989**, *28*, 587.

90. Evans, D. A., Ng, H. P., Rieger, D. L., *J. Am. Chem. Soc.* **1993**, *115*, 11446.

91. Ennis, D. S., McManus, J., Wood-Kaczmar, W., Richardson, J., Smith, G. E., Carstairs, A., *Org. Proc. Res. Develop.* **1999**, *3*, 248.

92. Imrie, C., Loubser, C., Engelbrecht, P., McCleland, C. W., *J. Chem. Soc., Perkin 1*, **1999**, 2513.

93. Taylor, P. N., O'Connell, M. J., McNeill, L. A., Hall, M. J., Aplin, R. T., Anderson, H. L., *Angew. Chem., Int. Ed.* **2000**, *39*, 3456.

94. Stanier, C. A., O'Connell, M. J., Anderson, H. L., Clegg, W., *Chem. Commun.* **2001**, 493.

95. Western, E. C., Daft, J. R., Johnson, E. M. II, Gannett, P. M., Shaughnessy, K. H., *J. Org. Chem.* **2003**, *68*, 6767.

96. Occhiato, E. G., Trabocchi, A., Guarna, A., *Org. Lett.* **2000**, *2*, 1241.

97. Gong, Y., Pauls, H. W., *Synlett* **2000**, 829.

98. Havelkova, M., Dvorak, D., Hocek, M., *Synthesis* **2001**, 1704.

99. Occhiato, E. G., Trabocchi, A., Guarna, A., *J. Org. Chem.* **2001**, *66*, 2459.

100. Hesse, S., Kirsch, G., *Tetrahedron Lett.* **2002**, *43*, 1213; Zhu, Q., Wu, J., Fathi, R., Yang, Z., *Org. Lett.* **2002**, *4*, 3333.

101. Morin Deveau, A., Macdonald, T. L., *Tetrahedron Lett.* **2004**, *45*, 803.

102. Uozumi, Y., Nakai, Y., *Org. Lett.* **2002**, *4*, 2997.

103. Casalnuovo, A. L., Calabrese, J. C., *J. Am. Chem. Soc.* **1990**, *112*, 4324.

104. Genet, J. P., Linquist, A., Blard, E., Mouries, V., Savignac, M., Vaultier, M., *Tetrahedron Lett.* **1995**, *36*, 1443.

105. Bumagin, N. A., Bykov, V. V., *Tetrahedron* **1997**, *53*, 14437.

106. Blettner, C. G., Koenig, W. A., Stenzel, W., Schotten, T., *J. Org. Chem.* **1999**, *64*, 3885.

107. Hesse, S., Kirsch, G., *Synthesis* **2001**, 755.

108. Miao, W., Chan, T.-H., *Org. Lett.* **2003**, *5*, 5003.

109. Li, Y., Hong, X. M., Collard, D. M., El-Sayed, M. A., *Org. Lett.* **2000**, *2*, 2385.

110. Li, Y., El-Sayed, M. A., *J. Phys. Chem. B* **2001**, *105*, 8938.

111. Sakurai, H., Hirao, T., Negishi, Y., Tsunakawa, H., Tsukuda, T., *Trans. Mater. Res. Soc. Jap.* **2002**, *27*, 185.

112. Lawson Daku, K. M., Newton, R. F., Pearce, S. P., Vile, J., Williams, J. M. J., *Tetrahedron Lett.* **2003**, *44*, 5095.

113. Bulut, H., Artok, L., Yilmazu, S., *Tetrahedron Lett.* **2002**, *44*, 289.

114. Hapiot, F., Lyskawa, J., Bricout, H., Tilloy, S., Monflier, E., *Adv. Synth. Catal.* **2004**, *346*, 83.

115. Shaughnessy, K. H., Booth, R. S., *Org. Lett.* **2001**, *3*, 2757; Moore, L. R., Shaughnessy, K. H. *Org. Lett.* **2004**, *6*, 225.

116. Parisot, S., Kolodziuk, R., Goux-Henry, C., Iourtchenko, A., Sinou, D., *Tetrahedron Lett.* **2002**, *43*, 7397.

117. Venkatraman, S., Li, C.-J., *Org. Lett.* **1999**, *1*, 1133; Venkatraman, S., Huang, T. S., Li, C.-J., *Adv. Syn. Catal.* **2002**, *344*, 399. See also Molander, G. A., Biolatto, B., *Org. Lett.* **2002**, *4*, 1867; Sakurai, H., Tsukuda, T., Hirao, T., *J. Org. Chem.* **2002**, *67*, 2721; Arcadi, A., Cerichelli, G., Chiarini, M., Correa, M., Zorzan, D., *Eur. J. Org. Chem.* **2003**, 4080; Friesen, R. W., Trimble, L. A., *Can. J. Chem.* **2004**, *82*, 206; Bedford, R. B., Blake, M. E., Butts, C. P., Holder, D., *Chem. Commun.* **2003**, 466.

118. Botella, L., Najera, C., *Angew. Chem., Int. Ed.* **2002**, *41*, 179.

119. Colacot, T. J., Gore, E. S., Kuber, A., *Organometallics* **2002**, *21*, 3301.

120. Baleizao, C., Corma, A., Garcia, H., Leyva, A., *Chem. Commun.* **2003**, 606.

121. Zhao, Y., Zhou, Y., Ma, D., Liu, J., Li, L., Zhang, T. Y., Zhang, H., *Org. Biomol. Chem.* **2003**, *1*, 1643; Byun, J.-W., Lee, Y.-S., *Tetrahedron Lett.* **2004**, *45*, 1837.

122. Leadbeater, N. E., Marco, M., *Org. Lett.* **2002**, *4*, 2973; Leadbeater, N. E., Marco, M., *J. Org. Chem.* **2003**, *68*, 888; Bai, L., Wang, J.-X., Zhang, Y., *Green Chemistry* **2003**, *5*, 615.

123. Leadbeater, N. E., Marco, M., *Angew. Chem. Int. Ed.* **2003**, *42*, 1407; Leadbeater, N. E., Marco, M., *J. Org. Chem.* **2003**, *68*, 5660.

124. For a review, see: Li, C.-J., *Angew. Chem., Int. Ed.* **2003**, *42*, 4856.

125. Arvela, R. K., Leadbeater, N. E., Sangi, M. S., Williams, V. A., Granados, P., Singer, R. D., *J. Org. Chem.* **2005**, *70*, 161.

126. For reviews, see: Stille, J. K., *Angew. Chem. Int. Ed. Engl.* **1986**, *25*, 508.

127. Tueting, D. R., Echavarren, A. M., Stille, J. K., *Tetrahedron* **1989**, *45*, 979.

128. Zhang, H. C., Daves, G. D. Jr., *Organometallics* **1993**, *12*, 1499.

129. Rai, R., Aubrecht, K. B., Collum, D. B., *Tetrahedron Lett.* **1995**, *36*, 3111.

130. Bumagin, N. A., Sukhomlinova, L. I., Tolstaya, T. P., Beletskaya, I. P., *Russ. J. Org. Chem.* **1994**, *30*, 1605.

131. Maleczka, R.E. Jr., Gallagher, W. P., Terstiege, I., *J. Am. Chem. Soc.* **2000**, *122*, 384; Gallagher, W. P., Terstiege, I., Maleczka, R. E. Jr., *J. Am. Chem. Soc.* **2001**, *123*, 3194.

132. Wolf, C., Lerebours, R., *J. Org. Chem.* **2003**, *68*, 7551.

133. Venkatraman, S., Li, C.-J., *Org. Lett.* **1999**, *1*, 1133; Venkatraman, S., Huang, T. S., Li, C.-J., *Adv. Syn. Catal.* **2002**, *344*, 399.

134. Roshchin, A. I., Bumagin, N. A., Beletskaya, I. P., *Doklady Chem.* **1994**, *334*, 47.

135. Huang, T. S., Li, C.-J., *Tetrahedron Lett.* **2002**, *43*, 403.

136. Bergstrom, D. E., Ruth, J. L., *J. Am. Chem. Soc.* **1976**, *98*, 1587; Huang, T. S., Li, C.-J., *J. Org. Chem.* **1978**, *43*, 2870.

137. Bigge, C. F., Kalaritis, P., Deck, J. R., Mertes, M. P., *J. Am. Chem. Soc.* **1980**, *102*, 2033; Bigge, C. F., Kalaritis, P., Mertes, M. P., *Tetrahedron Lett.* **1979**, 1653.

138. Langer, P. R., Waldrop, A. A., Ward, D. C., *Proc. Natl. Acad. Sci. USA* **1981**, *78*, 6633.

CHAPTER 7

AROMATIC COMPOUNDS

7.1 GENERAL

The association of aromatic surfaces with one another in water plays a significant role in the folding and complexation behavior of biopolymers such as proteins and DNA. Such associations are also important in molecular recognition studies for host–guest chemistry, self-assembly research in materials, and controlling regio- and stereoselectivity in organic reactions. Aromatic stacking is the common term describing such associations. Thermodynamic measurements in a self-complementary DNA duplex (5′-dXCGCGCG)$_2$, in which X is an unpaired natural or nonnatural deoxynucleoside, were used to study the forces that stabilize aqueous aromatic stacking in the context of DNA. It was shown that hydrophobic effects are found to be larger than other effects in stabilizing the stacking (such as electrostatic effects, dispersion forces).[1] In natural DNA, the dominant attractive forces between parallel aromatic groups (e.g., in nucleic acids) involve interactions between partial charges on atoms in the adjacent rings rather than dispersion forces or the classical hydrophobic effect.[2] Using conformational searches, molecular dynamics simulations, potential of mean force (PMF), and free energy perturbation (FEP) calculations combined with NMR studies of model compounds, it was suggested that

Comprehensive Organic Reactions in Aqueous Media, Second Edition, by Chao-Jun Li and Tak-Hang Chan
Copyright © 2007 John Wiley & Sons, Inc.

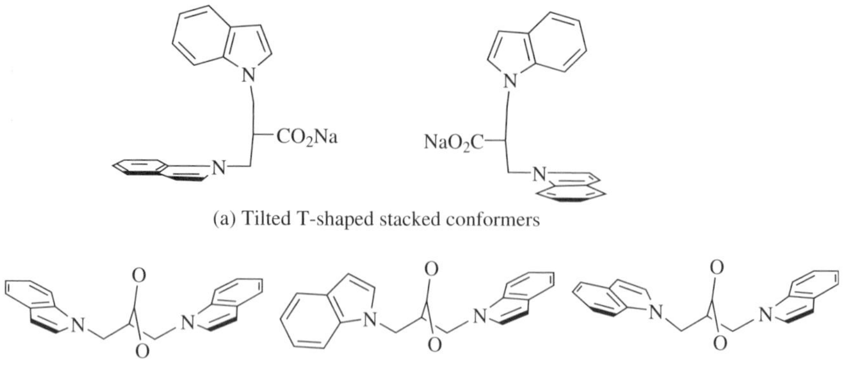

(a) Tilted T-shaped stacked conformers

(b) Nonstacked conformers

Figure 7.1 The most thermodynamically stable conformers of sodium (2,2)-*bis*(indol-1-yl-methyl)acetate in water at room temperature based on calculations.

the driving force of aromatic stacking interactions in water was the hydrophobic effect. The most thermodynamically stable conformers of sodium (2,2)-*bis*(indol-1-yl-methyl)acetate in water at room temperature are the tilted T-shaped stacked conformers (a) and nonstacked conformers (b), which are in a rapid equilibrium with the tilted T-shaped stacked conformation that is more populated (Figure 7.1).[3]

7.2 SUBSTITUTION REACTIONS

7.2.1 Electrophilic Substitutions

The electrophilic substitution of aromatic compounds is a classical type of organic reaction. In such reactions, the hydrogen atom(s) on the aromatic ring are replaced by other atoms (groups) through halogenation, nitration, sulfonation, and Friedel-Crafts type reactions. Increase in electron-density on the aromatic ring facilitates the reaction. Many such electrophilic aromatic substitutions can be carried out in aqueous media. However, the reactivity of such reactions is generally lower than in anhydrous conditions with a few exceptions. For example, alkylaromatics can be halogenated by hypochlorous and hypobromous acid, chlorine monoxide, hypobromous acidium ion, chlorine, bromine, and bromine chloride in water at 20°C in the pH range 2.5–7.5 and at ionic strength 0.1 M.[4] Facile bromination can be carried out in concentrated solution of zinc bromide (60–80% wt/wt ZnBr$_2$ in water)

(Eq. 7.1).[5] In this solution, all water molecules are bound to the metal center. It was proposed that the concentrated aqueous solution behaves similarly to anhydrous media and enhances the reactivity of molecular bromine by polarizing the Br–Br bond. In concentrated aqueous sulfuric acid solution, cinnamic acids and other styrene derivatives were nitrated with HNO_3 (Eq. 7.2).[6] On the other hand, electron-rich phenols and anilines[7] undergo electrophilic substitution more readily in water. Phenols are chlorinated by $CuCl_2$ in aqueous solution (Eq. 7.3).[8] The reaction is markedly accelerated by aluminum, chromium, and vanadyl chlorides as well as by HCl. Aromatic amines are similarly chlorinated preferentially at the p-position by $CuCl_2$. Reaction of iodine with phenol in an aqueous solution generated 2,4,6-triiodophenol (Eq. 7.4).[9] 4-Nitrosophenol, which is useful as an intermediate in the manufacturing of pharmaceuticals and dyes, is prepared in high yield and selectively by the reaction of phenol with sodium hydroxide and sodium nitrite in water.[10] Treatment of phenol, anisole, aniline, and N,N-dimethylaniline in deuterium oxide and sodium deuteroxide solution at 400°C and 250 bar showed ortho/para deuteration, which is characteristic of electrophilic substitution.[11]

$$Ar\text{-}H \xrightarrow[\text{concentrated aq ZnBr}_2]{Br_2} Ar\text{-}Br \qquad (7.1)$$

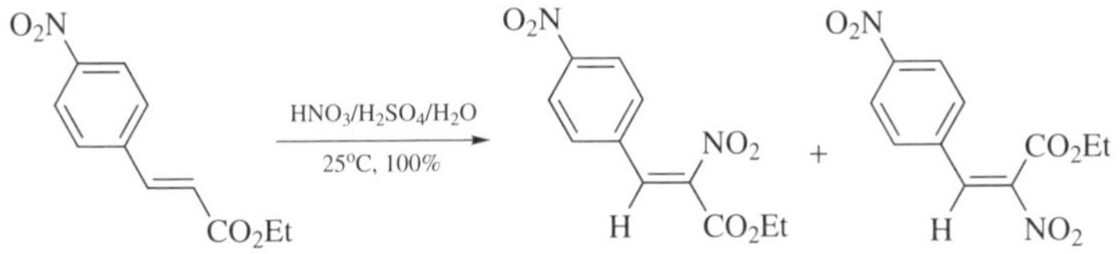

(7.2)

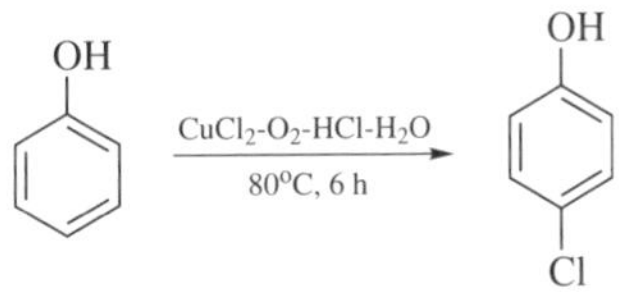

(7.3)

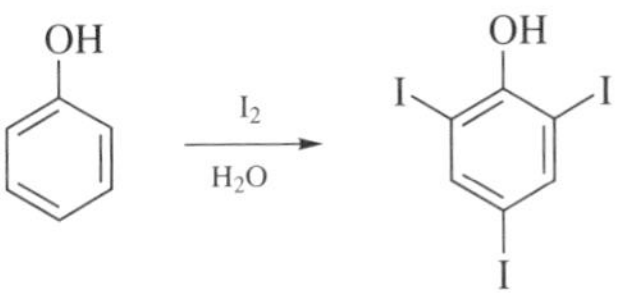

(7.4)

7.2.2 Friedel-Crafts C–C Bond Formations

This electrophilic substitution reaction is the most common reaction mode for aromatic compounds. Carbon–carbon bond formation via the electrophilic substitution of aromatic hydrogens can proceed under aqueous conditions. The most well-known example is the Friedel-Crafts-type reaction. Various indole derivatives reacted with an equimolar amount of 3% aqueous CH_2O and 33% aqueous Me_2NH at 70–75°C for 10 min in 96% ethanol to give Mannich-type products.[12] Reactions of furan, sylvan (2-methylfuran), and furfuryl alcohol with aqueous formaldehyde in two- and three-phase systems in the presence of cation-exchange resins in their H^+-form or soluble acids gave hydroxymethylation products.[13] Depending on the reactivity of the substrates, both acidity and lipophilicity of the catalysts were found to play roles in hydroxymethylation. Enzyme-catalyzed electrophilic aromatic substitution has been reported in prenyl-transfer reactions (Eq. 7.5).[14] The substitution of heteroaromatic compounds by superelectrophilic 4,6-dinitrobenzofuroxan (DNBF) proceeded in H_2O-Me_2SO mixtures (Eq. 7.6).[15] Synthesis of 2(3H)-benzofuranones from glyoxal and phenols was successful with 40% aqueous glyoxal and 38% HCl in AcOH and phenols or naphthols at 106°C for 16 h.[16]

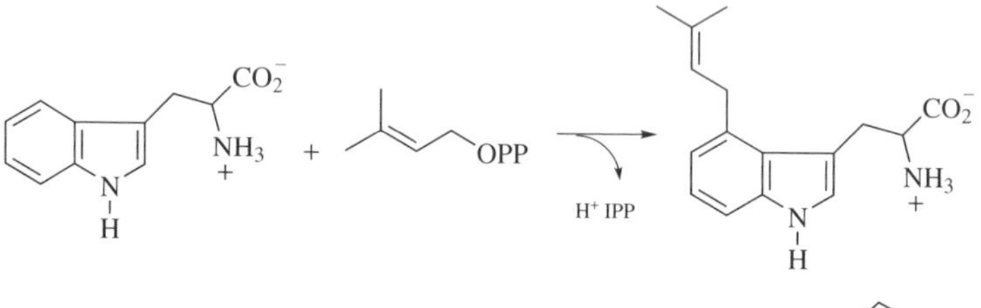

$$(7.5)$$

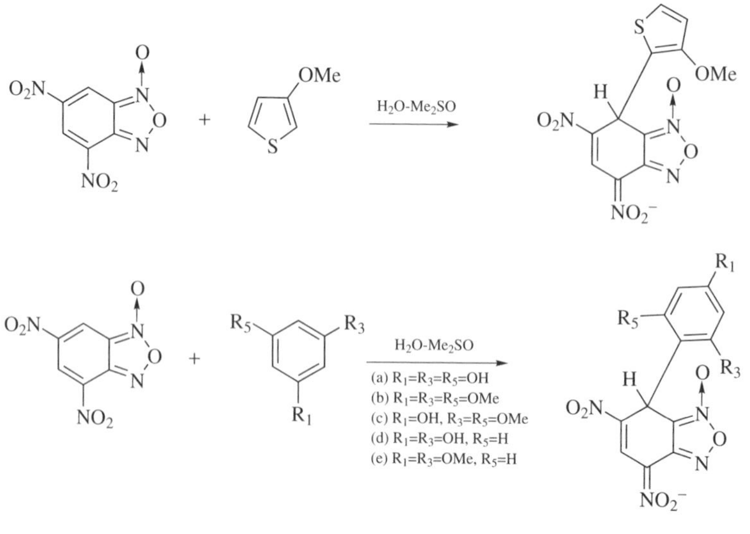

(a) $R_1=R_3=R_5=OH$
(b) $R_1=R_3=R_5=OMe$
(c) $R_1=OH, R_3=R_5=OMe$
(d) $R_1=R_3=OH, R_5=H$
(e) $R_1=R_3=OMe, R_5=H$

$$(7.6)$$

Instead of Brönsted acids, lanthanide triflates can be used to catalyze the reaction of indole with benzaldehyde (Eq. 7.7). The use of an ethanol/water system was found to be the best in terms of both yield and product isolation. The use of organic solvent such as chloroform resulted in oxidized byproducts.[17]

$$ (7.7) $$

The reaction of indole and N-methylpyrrole via Friedel-Crafts reactions with $OCHCO_2Et$ in various aqueous solutions generated substituted indoles and pyrroles without using any metal catalyst (Eq. 7.8).[18]

$$ (7.8) $$

62–84%

Bismuth *tris*-trifluoromethanesulfonate, $Bi(OTf)_3$, and $BiCl_3$ were found to be effective catalysts for the Friedel-Crafts acylation of both activated and deactivated benzene derivatives such as fluorobenzene.[19] Ga(III) triflate is also effective for Friedel-Crafts alkylation and acylation in alcohols and can tolerate water.[20] This catalyst is water-stable and its catalytic activity is much higher than that of other metallic tri-flates, M(OTf)$_3$, previously reported (M = Al, Ga, Ln or Sc). Scandium *tris*(dodecylsulfate), a Lewis-acid-surfactant-combined catalyst, can be used for conjugate addition of indoles to electron-deficient olefins in water (Eq. 7.9).[21] The 1,4-conjugate addition of indoles to nitroalkenes was efficiently carried out in aqueous media using a catalytic amount of indium tribromide (5 mol%) (Eq. 7.10).[22] The indium tribromide was recycled consecutively several times with the same efficacy. Rare-earth metal triflates such as Sc(OTf)$_3$, Yb(OTf)$_3$, and Sm(OTf)$_3$ work as highly effective catalysts for the chloromethylation of aromatic hydro-carbons with hydrochloric acid and trioxane.[23] They are active enough in aqueous solution at a concentration of less than 1–5% of the sub-strate under heterogeneous conditions of organic and aqueous phases. The triflate stays in the aqueous phase after the catalysis and the organic products are easily separated from the catalyst. The catalyst in the aque-ous solution could be recycled and used for further reactions without significant loss of activity. The catalysis occurred via the formation of a chloromethylated triflate complex, and electrophilic addition to an aro-matic hydrocarbon. The In(OTf)$_3$-catalyzed Friedel-Crafts reaction of aromatic compounds with methyl trifluoropyruvate in water generated various α–hydroxyl esters (Eq. 7.11).[24]

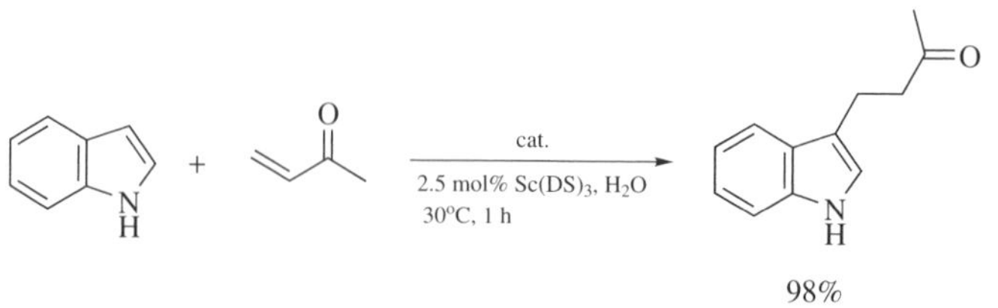

(7.9)

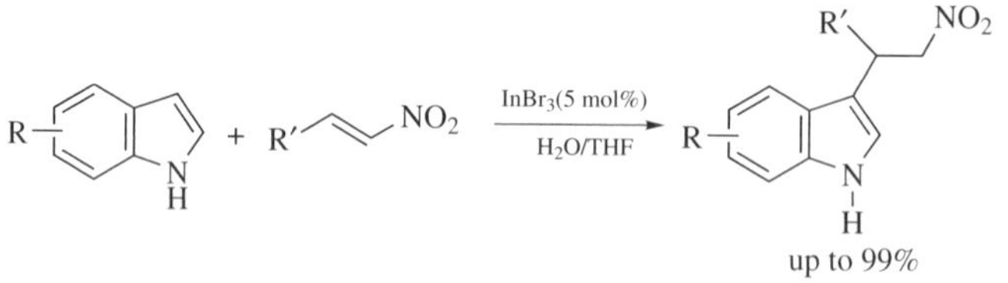

98%

(7.10)

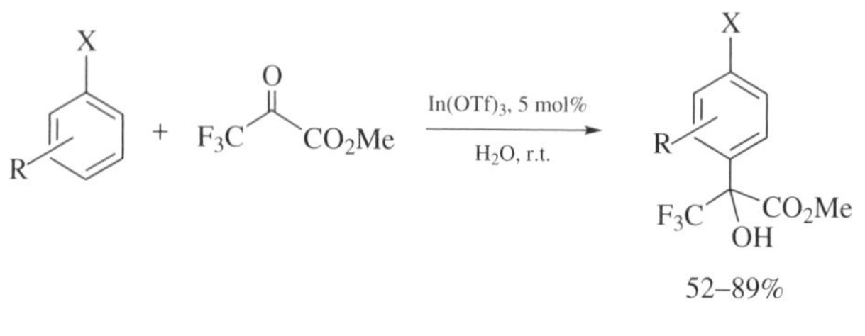

up to 99%

(7.11)

52–89%

An aqueous Friedel-Crafts reaction has also been used in polymer synthesis. The acid-catalyzed polymerization of benzylic alcohol and fluoride functionality in monomeric and polymeric fluorenes was investigated in both organic and aqueous reaction media.[25] Polymeric products are consistent with the generation of benzylic cations that participate in electrophilic aromatic substitution reactions. Similar reactions occurred in a water-insoluble Kraft pine lignin by treatment with aqueous acid. A Bisphenol A-type epoxy resin is readily emulsified in aqueous medium with an ethylene oxide adduct to a Friedel-Crafts reaction product of styrene and 4-(4-cumyl)phenol as emulsifier.[26] Electrophilic substitution reaction of indoles with various aldehydes and ketones proceeded smoothly in water using the hexamethylenetetramine-bromine complex to afford the corresponding *bis*(indolyl)methanes in excellent yields.[27] InF$_3$-catalyzed electrophilic substitution reactions of indoles with aldehydes and ketones are carried out in water.[28] Enzymatic Friedel-Crafts-type electrophilic substitution reactions have been reported.[29]

Near-critical water has been used as a medium for various C–C bond formation reactions including Friedel-Crafts alkylation and acylation (Eq. 7.12).[30] In these reactions, near-critical water solubilizes the organics and acts as a source of both hydronium and hydroxide ions, thereby replacing the normally required hazardous solvents and catalysts that require subsequent neutralization and disposal.

(7.12)

Pseudo-C$_3$-symmetrical trisoxazoline copper(II) complexes prove to be excellent catalysts in the Friedel-Crafts alkylation of indoles with alkylidene malonates (Eq. 7.13). Water tolerance of chiral catalyst trisoxazoline/Cu(OTf)$_2$ was examined, and it was found that the addition of up to 200 equivalents of water relative to the catalyst in *iso*-butyl alcohol had almost no effect on enantioselectivity, but it did slow down the reaction.[31]

$$(7.13)$$

Substituted 1,4-benzoquinone derivatives exist widely in nature and exhibit various important biological activities. An $In(OTf)_3$-catalyzed conjugate addition of aromatic compounds to 1,4-benzoquinones followed by in situ dehydrogenation in water gave aryl-substituted benzoquinone compounds (Eq. 7.14).[32] A highly efficient direct coupling of indole compounds with 1,4-benzoquinones was developed "on water" in the absence of any catalyst, organic cosolvent, or additives (Eq. 7.15).[33] The on-water condition provided the best yields of the corresponding products and was the only system to produce *bis*-coupling products (Eq. 7.16).

$$(7.14)$$

$$(7.15)$$

$$(7.16)$$

A Friedel-Crafts-type reaction of phenols under basic conditions is also possible. Aqueous alkaline phenol-aldehyde condensation is the reaction for generating phenol-formaldehyde resin.[34] The condensation of phenol with glyoxylic acid in alkaline solution by using aqueous glyoxylic acid generates 4-hydroxyphenylacetic acid. The use of tetraalkylammonium hydroxide instead of sodium hydroxide increases the para-selectivity of the condensation.[35] Base-catalyzed formation of benzo[b]furano[60]- and -[70]fullerenes occurred via the reaction of $C_{60}Cl_6$ with phenol in the presence of aqueous KOH and under nitrogen.[36]

Breslow and co-workers have found that cosolvents such as ethanol increase the solubility of hydrophobic molecules in water and provide interesting results for nucleophilic substitutions of phenoxide ions by benzylic chlorides: carbon alkylation occurs in water but not in nonpolar organic solvents, and it is observed only when the phenoxide has at least one methyl substituent (ortho, meta, or para). This has been discussed in Chapter 6 (Section 6.4.2).

7.2.3 Other Substitution Reactions

Alkyl radicals generated efficiently from allylsulfones in 80% aqueous formic acid induced a cyclization reaction on aromatic and heteroaromatic compounds to provide polycyclic aromatic and heteroaromatic derivatives (Eq. 7.17).[37]

$$(7.17)$$

7.3 OXIDATION REACTIONS

7.3.1 Simple Oxidation

The oxidation of the aromatic ring in aqueous environment is important in biology (e.g., the metabolic studies of pharmaceutical agents) and in environmental[38] research. Environmentally, benzene and PhOH in aqueous H_2O_2 solution at pH 5.2–6.7 underwent photooxidative decomposition by UV-irradiation via the generation of OH radicals.[39] The formation of mutagens was observed from polycyclic aromatic hydrocarbons and mono-substituted benzenes by UV irradiation in water containing NO_2^- by using a bacterial mutation assay.[40] No mutagenicity was observed for products obtained by photolysis in the absence of NO_2^- and NO_3^-. Synthetically, enzyme-catalyzed dihydroxylation of various aromatic compounds gave optically active cyclohexa-3,5-diene-1,2-diols.[41] These diols are used in the enantioselective synthesis of various natural products such as pancratistatin and 7-deoxypancratistatin, promising antitumor agents.[42]

Catalytic Ce(IV) reagent combined with an additional oxidant such as sodium bromate oxidizes hydroquinones, catechols, and their derivatives to quinones in aqueous acetonitrile (e.g., Eq. 7.18).[43]

$$(7.18)$$

On the other hand, the oxidation of the alkyl substituent in alkyl aromatic compounds can be carried out by various methods efficiently. For example, CAN has been used to oxidize substituted toluene to aryl aldehydes. Selective oxidation at one methyl group can be achieved (Eq. 7.19).[44] The reaction is usually carried out in aqueous acetic acid.

Other examples of oxidation systems for the alkyl groups include nitrate-polyoxophosphomolybdates[45] and air together with Co/Mn/Br[-].[46] Recently, supercritical and subcritical water has been shown to be a highly efficient and selective oxidation system for such oxidations. Under subcritical conditions, alkyl aromatic compounds were oxidized to aldehydes, ketones, and acids by molecular oxygen mediated by transition-metal catalysts.[47] It is possible to stop at the aldehyde stage with proper control of the reaction conditions. The reactions were found to be very sensitive to the nature of the catalysts with $MnBr_2$ and $CoBr_2$ providing the cleanest reactions and highest yields. Other metal salts led to severe charring or coupling reactions. In supercritical water (scH_2O) using manganese(II) bromide as catalyst, methylaromatic compounds are aerobically oxidized to give the corresponding carboxylic acids in a continuous mode over a sustained period of time in good yield.[48] Benzylic bromination of various 4-substituted toluenes (Me, tert-Bu, CO_2Et, Ac) was effectively conducted with NBS in pure water and with a 40 W incandescent lightbulb as an initiator of the radical chain process (Eq. 7.20).[49]

$$(7.19)$$

$$(7.20)$$

7.3.2 Oxidative Coupling

The oxidative polymerization of phenols and anilines by enzymatic and chemical methods is an important method for synthesizing polyphenols[50] and polyanilines[51] in material research. Such polymerizations are often carried out in aqueous conditions.

The combination of $Pd(OAc)_2$/molybdophosphoric acid ($H_3PMo_{12}O_{40}$)/O_2/AcOH-H_2O (2:1) has been found to oxidize benzene to give biphenyl by the oxidative dimerization with 100% selectivity and 19% yield under the conditions of 130°C, 10 atm, and 4 h (Eq. 7.21).[52] The use of $PdHPMo_{12}O_{40}$ itself as a catalyst was found to be effective and the reaction gave 95% of selectivity of biphenyl with a lower yield. Poly(oxy-1,4-phenylene) was obtained by electrooxidative polymerization of *p*-bromophenol in aqueous NaOH solution, the yield increasing when the aqueous NaOH was replaced with aqueous KOH or the reaction was performed at higher temperature; in contrast, *p*-chlorophenol dimerized to give 2,7-dichlorodibenzo[1,4]-dioxin.[53]

$$
\text{C}_6\text{H}_6 \xrightarrow[\text{AcOH (H}_2\text{O), 130\,°C, 4 h}]{\text{Pd(OAc)}_2,\ \text{co-catalyst, O}_2}
$$

Ph–Ph + Ph–OH + Ph–OAc   (7.21)

7.4 REDUCTIONS

Catalytic hydrogenation of aromatic rings in aqueous media has applications in the biofuel industry, the pulp and paper industry, and organic synthesis. Catalytic aromatic hydrogenation of 2-methoxy-4-propylphenol and a milled wood lignin was achieved using various Ru systems.[54] A number of transition-metal nanoparticles are shown to be highly efficient for such hydrogenations. For example, aqueous suspensions of iridium nanoparticles, produced from IrCl$_3$ and *N*,*N*-dimethyl-*N*-cetyl-*N*-(2-hydroxyethyl)ammonium chloride surfactant, show efficient activity in catalytic hydrogenation of various aromatic derivatives in biphasic media under mild conditions (Eq. 7.22).[55] An aqueous suspension of rhodium(0) nanoparticles (2–2.5 nm) generated from RhCl$_3$•3H$_2$O and *N*-alkyl-*N*-(2-hydroxyethyl)ammonium salts catalyzes the hydrogenation of arene derivatives under biphasic conditions at room temperature and 1 atm H$_2$ pressure (Eq. 7.23).[56] The nanoparticles can be reused without significant loss of activity. The ion pair generated from RhCl$_3$ and Aliquat 336 in an aqueous-organic two-phase medium catalyzed the hydrogenation and deuteration of arenes at 30°C/0.9 atm under apparently homogeneous conditions (Eq. 7.24).[57] Water proved to be essential for the hydrogenation. Raney Ni-Al alloy in dilute aqueous alkaline solution[58] and colloid-catalyst in aqueous/supercritical fluid biphasic media are also used for such hydrogenations.[59]

$$(7.22)$$

$$(7.23)$$

$$(7.24)$$

REFERENCES

1. Guckian, K. M., Schweitzer, B. A., Ren, R. X.-F., Sheils, C. J., Tahmassebi, D. C., Kool, E. T., *J. Am. Chem. Soc.* **2000**, *122*, 2213.

2. Newcomb, L. F., Gellman, S. H., *J. Am. Chem. Soc.* **1994**, *116*, 4993.

3. Pang, Y.-P., Miller, J. L., Kollman, P. A., *J. Am. Chem. Soc.* **1999**, *121*, 1717.

4. Voudrias, E. A., Reinhard, M., *Environ. Sci. Techn.* **1988**, *22*, 1056.

5. Epstein, D. M., Meyerstein, D., *Inorg. Chem. Commun.* **2001**, *4*, 705.

6. Moodie, R. B., Schofield, K., Taylor, P. G., Baillie, P. J., *J. C. S. Perkin Trans. 2*, **1981**, 842.

7. Crampton, M. R., Rabbitt, L. C., Terrier, F., *Can. J. Chem.* **1999**, *77*, 639.

8. Crocker, H. P., Walser, R., *J. Chem. Soc. C.* **1970**, 1982.

9. Rosenthaler, L., Capuano, L., *Pharm. Act. Helv.* **1946**, *21*, 225.

10. Ghica, M., Harles, L. S., Angheluta, F., Rom. 1994, 3 pp. RO: 108558; B1: 19940630.

11. Hibbs, M. R., Yao, Ji., Evilia, R. F., *High Temp. Mater. Sci.* **1996**, *36*, 9.

12. Thesing, J., Binger, P., *Chem. Ber.* **1957**, *90*, 1419.

13. Iovel, I., Goldberg, Y., Shymanska, M., *J. Mol. Cat.* **1989**, *57*, 91.

14. Gebler, J. C., Woodside, A. B., Poulter, C. D., *J. Am. Chem. Soc.* **1992**, *114*, 7354.

15. Kizilian, E., Terrier, F., Chatrousse, A.-P., Gzouli, K., Halle, J. C., *J. Chem. Soc. Perkin 2* **1997**, *12*, 2667; Terrier, F., Pouet, M.-J., Halle, J.-C., Kizilian, E., Buncel, E., *J. Phys. Org. Chem.* **1998**, *11*, 707; Crampton, M. R., Rabbitt, L. C., Terrier, F., *Can. J. Chem.* **1999**, *77*, 639.

16. Layer, R. W., *J. Hetero. Chem.* **1975**, *12*, 1067.

17. Chen, D., Yu, L., Wang, P. G., *Tetrahedron Lett.* **1996**, *37*, 4467.

18. Zhuang, W., Jorgensen, K. A., *Chem. Commun.* **2002**, 1336.

19. Desmurs, J. R., Labrouillere, M., Le Roux, C., Gaspard, H., Laporterie, A., Dubac, J., *Tetrahedron Lett.* **1997**, *38*, 8871; Repichet, S., Le Roux, C., Roques, N., Dubac, J., *Tetrahedron Lett.* **2003**, *44*, 2037.

20. Surya Prakash, G. K., Yan, P., Toeroek, B., Bucsi, I., Tanaka, M., Olah, G. A., *Catal. Lett.* **2003**, *85*, 1.

21. Manabe, K., Aoyama, N., Kobayashi, S., *Adv. Synth. Cat.* **2001**, *343*, 174.

22. Bandini, M., Melchiorre, P., Melloni, A., Umani-Ronchi, A., *Synthesis* **2002**, 1110.

23. Kishida, T., Yamauchi, T., Kubota, Y., Sugi, Y., *Green Chem.* **2004**, *6*, 57.

24. Ding, R., Zhang, H. B., Chen, Y. J., Liu, L., Wang, D., Li, C.-J., *Synlett.* **2004**, 555.

25. Hagaman, E. W., Lee, S. K., *Energy & Fuels* **1999**, *13*, 1006.

26. Hibino, K., Kimura, Y., *Macromol. Mat. Eng.* **2001**, *286*, 325.

27. Teimouri, M., Mivehchi, H., *Synth. Commun.* **2005**, *35*, 1835.

28. Bandgar, B. P., Shaikh, K. A., *J. Chem. Res.* **2004**, 34.

29. For a review, see: Retey, J., *Naturwissenschaften* **1996**, *83*, 439.

30. Eckert, C. A., Liotta, C. L., Brown, J. S., *Chem. Ind.* **2000**, 94.

31. Zhou, J., Ye, M.-C., Huang, Z.-Z., Tang, Y., *J. Org. Chem.* **2004**, *69*, 1309.

32. Zhang, H.-B., Liu, L., Chen, Y.-J., Wang, D., Li, C.-J., *Adv. Synth. Catal.* **2006**, *348*, 229.

33. Zhang, H.-B., Liu, L., Chen, Y.-J., Wang, D., Li, C.-J., *Eur. J. Org. Chem.* **2006**, 869.

34. Lambuth, A. L. (Monsanto Co.), U.S. (1967), 6 pp. US 3342776 19670919.

35. Wuthrick, M.-F., Maliverney, C., *Ind. Chem. Lib.* **1996**, *8*, 343.

36. Darwish, A. D., Avent, A. G., Kroto, H. W., Taylor, R., Walton, D. R. M., *J. Chem. Soc. Perkin Trans. 2*, **1999**, 1983.

37. Wang, S.-F., Chuang, C.-P., Lee, W.-H., *Tetrahedron* **1999**, *55*, 6109.

38. Beltran, F. J., Ovejero, G., Encinar, J. M., Rivas, J., *Ind. Eng. Chem. Res.* **1995**, *34*, 1596.

39. Mansour, M., *Bull. Environ. Contam. Toxicol.* **1985**, *34*, 89.

40. Suzuki, J., Hagino, T., Ueki, T., Nishi, Y., Suzuki, S., *Bull. Environ. Contam.Toxicol*. **1983**, *31*, 79.

41. Hudlicky, T., Gonzalez, D., Gibson, D. T., *Aldrichimica Acta* **1999**, *32*, 35.

42. Hudlicky, T., Tian, X., Koenigsberger, K., Maurya, R., Rouden, J., Fan, B., *J. Am. Chem. Soc*. **1996**, *118*, 10752.

43. Ho, T. L., *Synth. Commun*. **1979**, *9*, 237.

44. Trahanovsky, W. S., Young, L. B., *J. Org. Chem*. **1966**, *31*, 2033.

45. Khenkin, A. M., Neumann, R., *J. Am. Chem. Soc*. **2004**, *126*, 6356.

46. Nair, K., Sawant, D. P., Shanbhag, G. V., Halligudi, S. B., *Catal. Commun*. **2004**, *5*, 9.

47. Holliday, R. L., Jong, B. Y. M., Kolis, J. W., *J. Supercrit. Flu*. **1998**, *12*, 255; Kayan, B., Oezen, R., Gizir, A. M., Kus, N. S., *Org. Prep. Proc. Int*. **2005**, *37*, 83.

48. Garcia-Verdugo, E., Venardou, E., Thomas, W. B., Whiston, K., Partenheimer, W., Hamley, P. A., Poliakoff, M., *Adv. Synth. Catal*. **2004**, *346*, 307.

49. Podgorsek, A., Stavber, S., Zupan, M., Iskra, J., *Tetrahedron Lett*. **2006**, *47*, 1097.

50. Kim, Y.-J., Uyama, H., Kobayashi, S., *Macromol*. **2003**, *36*, 5058.

51. Xing, S., Chu, Y., Sui, X., Wu, Z., *J. Mater. Sci*. **2005**, *40*, 215.

52. Okamoto, M., Watanabe, M., Yamaji, T., *J. Organomet. Chem*. **2002**, *664*, 59; Burton, H. A., Kozhevnikov, I. V., *J. Mol. Catal. A: Chem*. **2002**, *185*, 285; Jintoku, T., Taniguchi, H., Fujiwara, Y., *Chem. Lett*. **1987**, 1865.

53. Taj, S., Ahmed, M. F., Sankarapapavinasam, S., *J. Chem. Res., Synop*. **1993**, 232.

54. Wong, T. Y. H., Pratt, R., Leong, C. G., James, B. R., Hu, T. Q., *Chem. Ind*. **2001**, *82*, 255.

55. Mevellec, V., Roucoux, A., Ramirez, E., Philippot, K., Chaudret, B., *Adv. Synth.Catal*. **2004**, *346*, 72.

56. Schulz, J., Roucoux, A., Patin, H., *Chem. Eur. J*. **2000**, *6*, 618.

57. Blum, J., Amer, I., Vollhardt, K. P. C., Schwarz, H., Hoehne, G., *J. Org. Chem*. **1987**, *52*, 2804.

58. Tsukinoki, T., Kanda, T., Liu, G.-B., Tsuzuki, H., Tashiro, M., *Tetrahedron Lett*. **2000**, *41*, 5865.

59. Jason B. R., Jessop, P. G., James, B. R., *Chem. Commun*. **2000**, 941.

CHAPTER 8

ALDEHDYE AND KETONES

8.1 REDUCTION

8.1.1 Hydrogenation

Hydrogenation of aldehydes and ketones can be catalyzed by a number of catalysts in an aqueous medium. The use of formic acid as a hydrogen source in the presence of a water-soluble rhodium complex, $ClRh(Ph_2Pm\text{-}C_6H_4SO_3Na)_3$, reduces both aldehydes and alkenes.[1] Ruthenium-based complexes seem to be more widely used for such hydrogenations than others. Joo and Benyei have shown that by using $RuCl_2(TPPMS)_2$ and sodium formate as a hydrogen donor, a variety of aromatic and α,β-unsaturated aldehydes were transformed into the corresponding saturated alcohols in aqueous solution.[2] Keto acids were hydrogenated to give hydroxyl acids with $[RuH(OAc)(TPPMS)_3]$ and $[RuHCl(TPPMS)_3]$ (Eq. 8.1).[3]

$$ (8.1) $$

By using the more water-soluble ligand, TPPTS, Grosselin et al. converted several unsaturated aldehydes into the corresponding unsaturated

Comprehensive Organic Reactions in Aqueous Media, Second Edition, by Chao-Jun Li and Tak-Hang Chan
Copyright © 2007 John Wiley & Sons, Inc.

alcohols with a 99% chemoselectivity on the carbonyl group (Eq. 8.2).[4] Basset et al. found that the addition of NaI, which would assist the rapid formation of a metal-carbon bond, enhanced the reactivity.[5] Water-soluble cis-RuCl$_2$(PTA)$_4$ (PTA = 1,3,5-triaza-7-phosphaadamantane) is also an effective catalyst for such hydrogenations using a biphasic aqueous/organic medium with sodium formate as the source of hydrogen.[6] Both aromatic and aliphatic aldehydes were reduced to the corresponding alcohols. Rhodium and iridium complexes are also effective.[7]

$$\text{[HRuCl(TPPTS)}_3] \quad \xrightarrow{\text{H}_2}$$

(8.2)

99%

The use of chiral ruthenium catalysts can hydrogenate ketones asymmetrically in water. The introduction of surfactants into a water-soluble Ru(II)-catalyzed asymmetric transfer hydrogenation of ketones led to an increase of the catalytic activity and reusability compared to the catalytic systems without surfactants.[8] Water-soluble chiral ruthenium complexes with a β-cyclodextrin unit can catalyze the reduction of aliphatic ketones with high enantiomeric excess and in good-to-excellent yields in the presence of sodium formate (Eq. 8.3).[9] The high level of enantioselectivity observed was attributed to the preorganization of the substrates in the hydrophobic cavity of β-cyclodextrin.

$$[\{\text{RuCl}_2(\text{C}_6\text{H}_5)\}_2]$$

catalyst:substrate:HCO$_2$Na
1 : 10 : 100

(8.3)

93%   47% ee

8.1.2 Other Reductions

Reduction of carbonyl compounds can be carried out in an aqueous medium by various reducing reagents. Among these reagents, sodium borohydride is the most frequently used. The reduction of carbonyl compounds by sodium borohydride can also use phase-transfer catalysts (Eq. 8.4),[10] inverse phase-transfer catalysts,[11] or polyvinylpyridines[12] in a two-phase medium in the presence of surfactants. Potassium boro-hydride can also be used as reductant in $[bmim]PF_6/H_2O$ biphasic media.[13]

$$(8.4)$$

Other reagents for the reduction in aqueous medium include a cadmium chloride-magnesium-THF-water system (Eq. 8.5),[14] samarium iodide in aqueous THF (Eq. 8.6),[15] sodium dithionite in aqueous DMF,[16] titanocene(IV)/Zn,[17] $FeS-NH_4Cl-CH_3OH-H_2O$,[18] $FeCl_3$-DMF-H_2O system,[19] molybdocene monohydride,[20] cobalt(II) chloride hexahydrate-zinc,[21] $SbCl_3$-(Al- and Zn),[22] sodium sulfide in the presence of polyethylene glycol,[23] and metallic zinc together with nickel chloride (Eq. 8.7).[24] Diaryl ketones were reduced selectively to the corresponding benzhydrols in good yields by using aluminum powder together with sodium hydroxide in MeOH:H_2O, "on the contrary," dialkyl ketones, α-tetralone, aryl alkyl ketones, and cycloalkanones, which are mostly unaffected (Eq. 8.8).[25] No reaction was observed in the absence of water in this case. Reduction of a mixture of a ketone and an ester by sodium in water generates ketols.[26] Aldehydes are reduced chemoselectively by using tributyltin hydride in methanol, aqueous organic solvents, or water to provide the corresponding alcohols in high yields (Eq. 8.9).[27]

$$(8.5)$$

$$(8.6)$$

$$(8.7)$$

$$(8.8)$$

$$\text{RCHO} \xrightarrow[\text{CH}_3\text{OH, H}_2\text{O-CH}_3\text{OH, H}_2\text{O-THF, or H}_2\text{O}]{\text{Bu}_3\text{SnH}} \text{R} \diagup \text{OH} \qquad (8.9)$$

$$9\text{--}100\%$$

Sodium hydrogen telluride, (NaTeH), prepared in situ from the reaction of tellurium powder with an aqueous ethanol solution of sodium borohydride, is an effective reducing reagent for many functionalities, such as azide, sulfoxide, disulfide, activated C=C bonds, nitroxide, and so forth. Water is a convenient solvent for these transformations.[28] A variety of functional groups including aldehydes, ketones, olefins, nitroxides, and azides are also reduced by sodium hypophosphite buffer solution.[29]

Sodium formate serves as a reducing reagent for aldehyde in subcritical water at 310–350°C and high pressures (Eq. 8.10).[31] The reduction of aldehydes in aqueous media can also be achieved by using an electrochemical method. The voltammetry of benzaldehyde in an acidic methanol/water mixture is affected strongly by the cathode material.[30]

$$\text{C}_5\text{H}_{11} \diagdown \!\!\! \overset{\text{O}}{\underset{\text{H}}{\diagup\!\!\!\diagdown}} \xrightarrow[\substack{\text{H}_2\text{O} \\ 340\text{--}350°\text{C} \\ 1200 \text{ psi}}]{\text{HCO}_2\text{Na}} \text{C}_5\text{H}_{11} \diagdown\!\!\diagup\!\!\diagdown \text{OH} \qquad (8.10)$$

Highly chemoselective reduction of aldehyde was catalyzed by a polymer-bound Rh_6 cluster complex under water–gas shift reaction conditions (Eq. 8.11).[32] The reaction system is triphasic, which makes the workup procedure simple. Hydrophobically directed selective reduction of ketones was discovered by using three hydrophobic borohydrides carrying phenyl, pentafluorophenyl, and β-naphthyl groups in water or in methanol (Eq. 8.12).[33] With these ketones, there was the preferential reduction in methanol of competing acetyl groups, but preferential reduction of the aryl ketones in water with up to a 40-fold selectivity reversal in the most extreme case. Lithium borohydride showed no such change in selectivity and favored acetyl reduction in both solvents. Salt and cosolvent effects showed that hydrophobic packing is involved in these reductions.

$$\text{(structure: benzaldehyde CHO)} \quad \text{vs} \quad \text{(structure: acetophenone)} \xrightarrow[\text{polymer-bound Rh}_6 \text{ cluster}]{\text{CO/H}_2\text{O, 10 atm, 80°C, benzene}} \text{(structure: benzyl alcohol CH}_2\text{OH)} \qquad (8.11)$$

$$95\%$$

$$\underset{Ph}{\overset{O}{\|}}\!\!\!-\!NMe_3^{\oplus} \;+\; \overset{O}{\|}\!\!\!-\!NMe_3^{\oplus} \;\xrightarrow{LiRBH_3}\; \underset{Ph}{\overset{HO\;H}{}}\!\!-\!NMe_3^{\oplus} \;+\; \overset{HO\;H}{}\!\!-\!NMe_3^{\oplus}$$

$$\mathbf{A} \qquad\qquad \mathbf{B}$$

R	D_2O	$LiCl/D_2O$	CD_3OD
	A : B	**A : B**	**A : B**
H	44:56	47:53	31:69
Ph	56:44	64:36	35:65
C_6H_5	74:26	84:16	40:60

$$(8.12)$$

For the asymmetric reduction of ketones in aqueous conditions, various enzymes are often used and can provide very high enantioselectivities.[34] Enantioselective reduction of aldehydes by using sodium borohydride together with chiral promoters is also possible in aqueous conditions. Asymmetric reduction of ketones with sodium borohydride in the presence of lecithin in aqueous solution generated the corresponding alcohols in low enantioselectivity (Eq. 8.13).[35] The enantioselective cathodic reduction of some prochiral ketones can be carried out at a mercury pool cathode in N,N-dimethylformamide (DMF)-water (90:10) using tetrabutylammonium tetrafluoroborate (TBA•BF_4) as a supporting electrolyte in the presence of $(1R, 2S)$-(-)-N,N-dimethylephedrinium tetrafluoroborate (DET) (Eq. 8.14).[36] β-Cyclodextrin catalyzes the asymmetric reduction of α-azido aryl ketones into the corresponding alcohols using sodium borohydride in water. The azido group appeared to be important for the reduction. The reduction of 2-azido-1-phenylethanone in the presence of β-cyclodextrin gave $(1S)$-2-azido-1-phenylethanol in 92% yield with 52% ee (Eq. 8.15).[37] Amphiphilic chiral dendrimers **8.1** derived from poly-amidoamine (PAMAM) and D-gluconolactone (generations 1–4) (Scheme 8.1) also induce chirality in such reductions. When the reduction of acetophenone was carried out in water by $NaBH_4$ in the presence of various generations of the chiral dendrimers, the fourth-generation dendrimer [G(4)G] gave the best results with the product 1-phenylethanol obtained in 92% yield with 98% (S) ee (Eq. 8.16). Other dendrimeric supports gave poor results. In contrast, when the same reduction was carried out in THF, the third generation [G(3)G] of these amphiphilic dendrimers shows the best results with 92% yield and good asymmetric inductions (99% ee). The dendrimers could be recovered by filtration and recycled (up to 10 times) without losing enantioselectivity.[38] On the other hand, the reduction gave very low ee in water with the formation of micelles or vesicles with sugar-headed surfactants.[39]

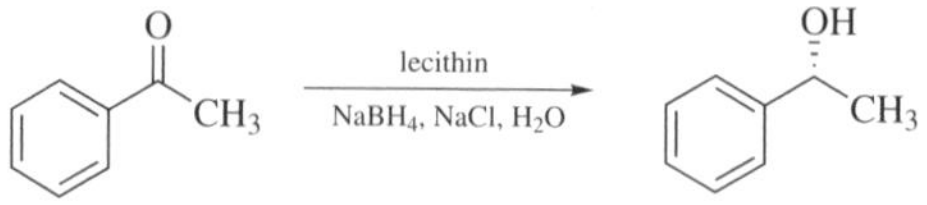
lecithin
NaBH4, NaCl, H2O

(8.13)

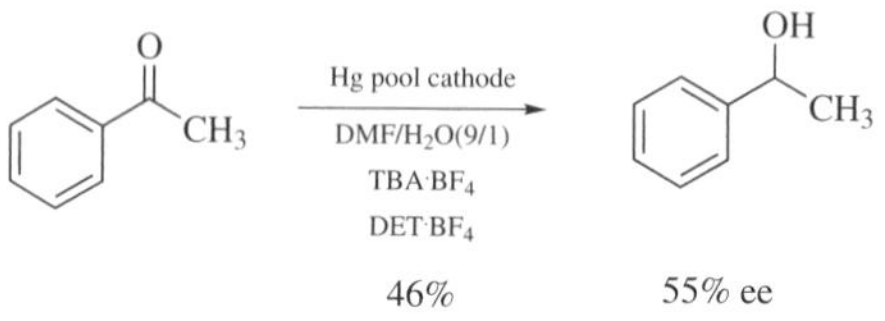
Hg pool cathode
DMF/H2O(9/1)
TBA·BF4
DET·BF4
46%
55% ee

(8.14)

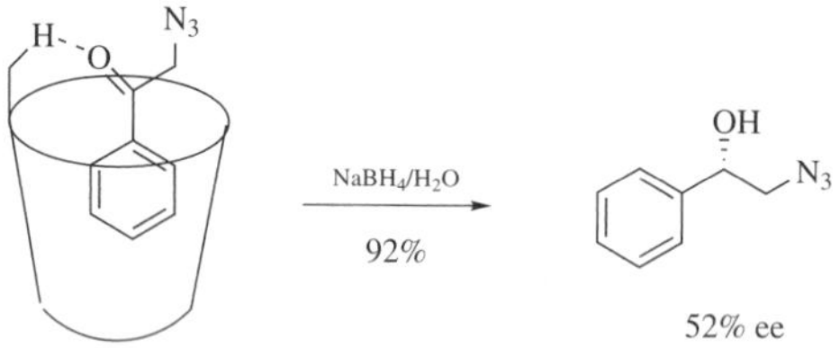
NaBH4/H2O
92%
52% ee

(8.15)

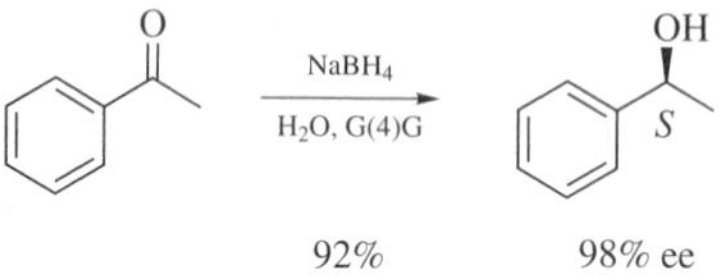
NaBH4
H2O, G(4)G
92%
98% ee

(8.16)

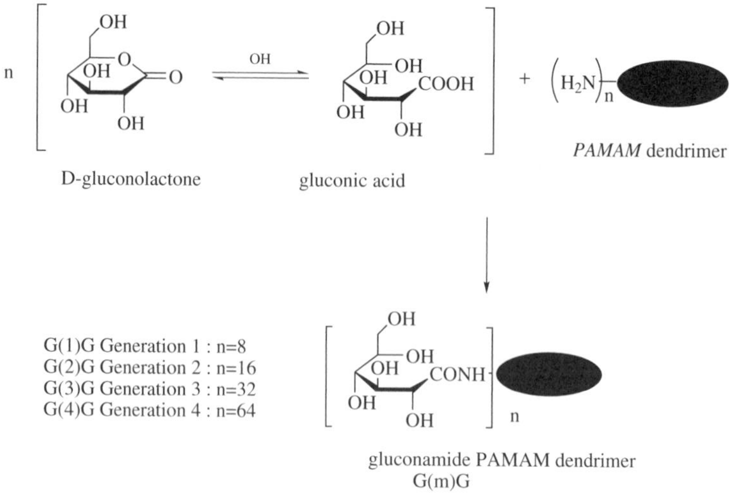
D-gluconolactone
gluconic acid
PAMAM dendrimer
G(1)G Generation 1 : n=8
G(2)G Generation 2 : n=16
G(3)G Generation 3 : n=32
G(4)G Generation 4 : n=64
gluconamide PAMAM dendrimer
G(m)G

8.1

Scheme 8.1

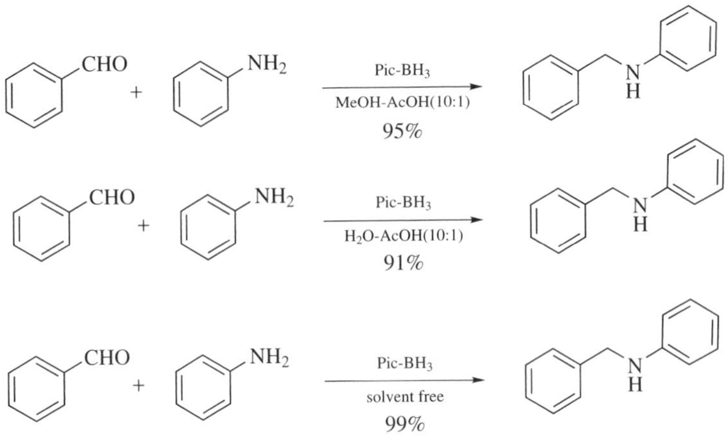

Scheme 8.2

Besides direct reduction, a one-pot reductive amination of aldehydes and ketones with α-picoline-borane in methanol, in water, and in neat conditions gives the corresponding amine products (Scheme 8.2).[40] The synthesis of primary amines can be performed via the reductive amination of the corresponding carbonyl compounds with aqueous ammonia with soluble Rh-catalyst (Eq. 8.17).[41] Up to an 86% yield and a 97% selectivity for benzylamines were obtained for the reaction of various benzaldehydes. The use of a bimetallic catalyst based on Rh/Ir is preferable for aliphatic aldehydes.

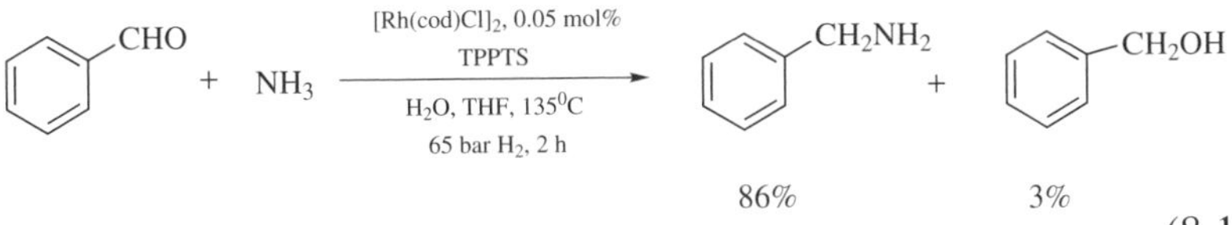

$$(8.17)$$

The reduction of aryl ketones by Ni-Al alloy in water under reflux proceeded to give the methylene compounds within 2 h in 89.0–99.8% relative yields (Eq. 8.18).[42]

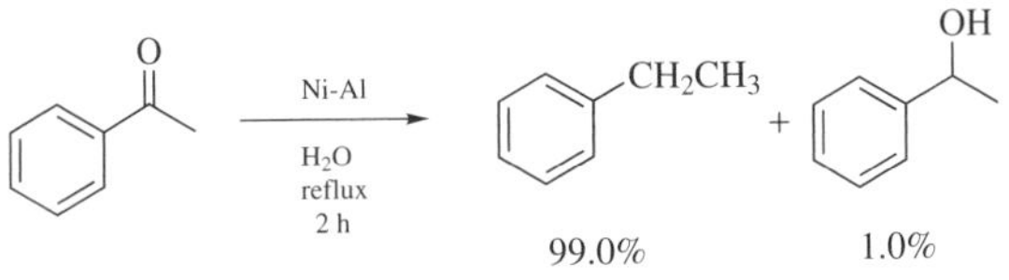

$$(8.18)$$

8.2 OXIDATION

Oxidation of both aldehydes and ketones has been carried out in aqueous conditions. For the oxidation of aldehydes to give carboxylic acids, classical methods involve strong oxidizing reagents such as the Jones reagent (based on high valent chromium) and $KMnO_4$. Quinolinium dichromate in sulfuric acid oxidizes heterocyclic aldehydes (2-furaldehyde, 2-pyrrolecarbaldehyde, 2-thiophenecarbaldehyde) to the corresponding acids in a 50% (vol/vol) acetic acid-water medium (Eq. 8.19).[43] Recently, milder conditions have been developed by using catalytic methods. The use of $[(n\text{-}C_8H_{17})_3NMe]_3^+$ $[PO_4[W(O)(O_2)_2]_4]^{3-}$ in combination with H_2O_2 in an aqueous/organic biphase system oxidizes aldehydes to carboxylic acids (Eq. 8.20).[44] Gold on carbon oxidizes aldehydes to carboxylic acids in a water solution under mild conditions without loss of activity on recycling (Eq. 8.21).[45] Although the reaction could also be carried out in the absence of a solvent, for solid aldehydes the use of water as solvent is quite convenient. Oxidation of aromatic aldehydes to aromatic carboxylic acids by the bacterium *Burkholderia cepacia*, TM1 isolated from humus, has been achieved in distilled water in high yields.[46] Thus, vanillic acid from vanillin and syringic acid from syringaldehyde were obtained in 94% and 72% yield respectively (Scheme 8.3).

$$(8.19)$$

$$(8.20)$$

$$(8.21)$$

For the oxidation of ketones, Baeyer-Villiger oxidation of cyclic ketones with monopersuccinic acid in water gives lactones in good results (Eq. 8.22).[47] Peroxy species generated from borax in 30% hydrogen peroxide is effective for the Baeyer-Villiger oxidation of

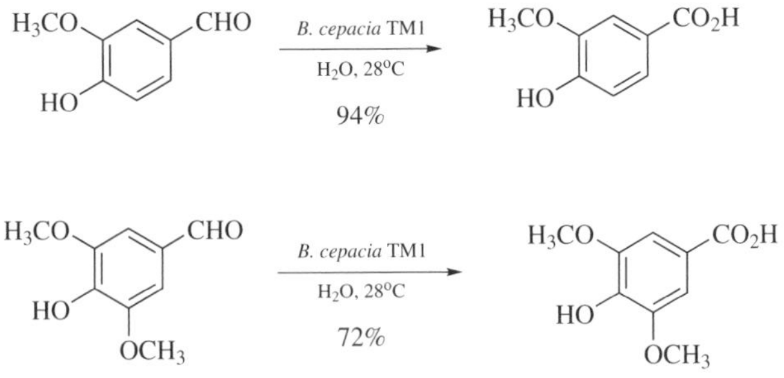

Scheme 8.3

several ketones insoluble in water at $80°C$, to give 100% conversion in 2–4 h (Eq. 8.23).[48] A one-pot synthesis of phenols from aromatic aldehydes can be achieved by the Baeyer-Villiger oxidation with H_2O_2 using water-tolerant Lewis acids such as Sn-Beta zeolite as catalysts (Eq. 8.24).[49]

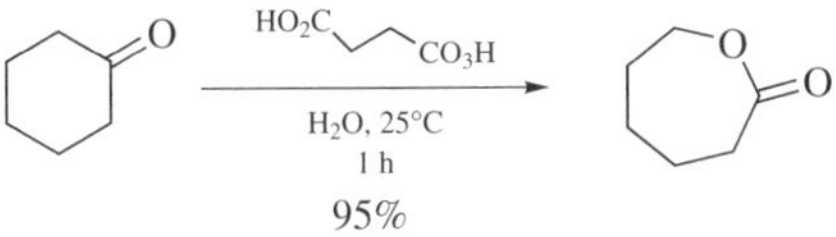

(8.22)

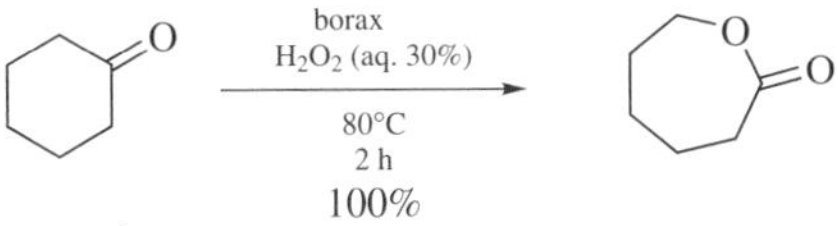

(8.23)

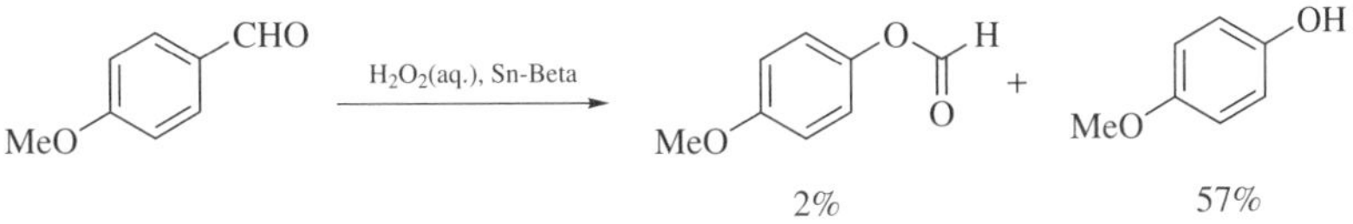

(8.24)

Another useful oxidative reaction in aqueous medium is the cleavage of cyclic ketones by hydrogen peroxide in the presence of Fe(II) salts (Eq. 8.25). The reaction proceeds through an α-hydroxy hydroperoxide, leading to a variety of products.[50] The presence of Fe(II) salts decomposes the intermediate, generating a radical. In the presence of halide ions, the radical leads to synthetically useful halocarboxylic acids.[51]

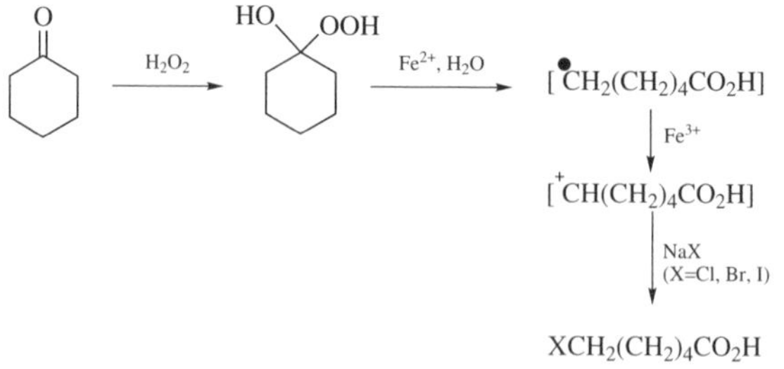

$$(8.25)$$

8.3 NUCLEOPHILIC ADDITION: C–C BOND FORMATION

The nucleophilic addition of organometallic reagents to carbonyl compounds is among the most important methods for forming carbon–carbon bonds. However, a major requisite in these reactions is the strict exclusion of moisture. On the other hand, some classes of organometallics remain viable in the presence of water. For example, the preparation of arylmercuric chlorides in aqueous media has been known since 1905.[52] And in the 1960s, tribenzylstannyl halide was produced in large scale in water.[53] Wurtz-type reductive coupling of allyl halides proceeded in aqueous alcohol.[54] These reports indicated the possibility of carrying out these kinds of reactions in water under special circumstances. Indeed, within the past two decades, extensive research has been carried out on developing organometallic-type nucleophilic additions of aldehydes and ketones in aqueous media. Although a large amount of the research has been on allylation reactions, the reaction has been extended to all type of substrates.

8.3.1 Allylation

Among all the nucleophilic addition reactions of carbonyl compounds, allylation reaction has been the most successful, partly due to the relatively high reactivity of allyl halides. Various metals have been found to be effective in mediating such a reaction (Scheme 8.4). Among them, indium has emerged as the most popular metal for such a reaction.

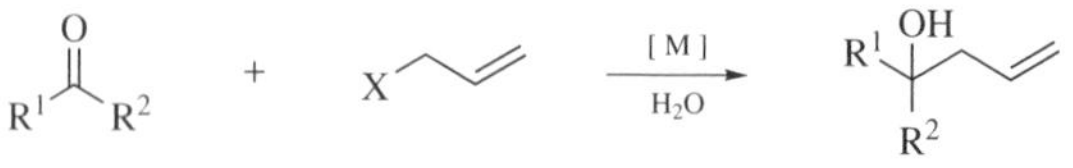

M: Zn, Sn, In, B, Si, Ga, Mg, Co, Mn, Bi, etc.

Scheme 8.4

8.3.1.1 Mediated by Zinc.   In 1977,[55] Wolinsky et al. reported that slow addition of allyl bromide to a stirred slurry of "activated" zinc dust and an aldehyde or a ketone in 95% ethanol at 78°C gave allylation products with yields comparable to those obtained in aprotic solvents (Eq. 8.26). Then, in 1985, Luche et al. found that allylation of aldehydes and ketones can be effected in aqueous media using zinc as the metal and THF as a cosolvent under magnetic stirring or soni-cation conditions (Eq. 8.27).[56,57] The replacement of water by aqueous saturated ammonium chloride solution enhanced the efficiency. In this case, comparable results were obtained either with or without the use of sonication. In the same year, Benezra et al. reported[58] that ethyl (2-bromomethyl)acrylate can couple with carbonyl compounds, mediated by metallic zinc, in a mixture of saturated aqueous NH_4Cl-THF under reflux to give α-methylene-γ-butyrolactones (Eq. 8.28). The same reac-tion in THF alone gives only a low yield (15%) of the product within the same time range and under the same conditions. Although it is much less effective, (2-bromomethyl)acrylic acid could also be used directly upon neutralization with triethylamine. Later, Wilson carried out a more detailed study of zinc-mediated reactions in water[59] through a modification involving the use of a solid organic support instead of the cosolvent THF. The solid organic supports included reverse-phase C-18 silica gel, biobeads S-X8, which is a spherical porous styrene divinylbenzene copolymer with 8% cross-links, and GC column pack-ing OV-101 on Chromosorb, and so on. The reactions proceed at about the same rate as reactions with THF as a cosolvent. Both allyl bro-mide and allyl chloride can be used. Kunz and Reißig reported[60] the zinc-mediated allylation of methyl γ-oxocarboxylates in a mixture of saturated aqueous ammonium chloride and THF. The reaction provides a convenient synthesis of 5-allyl-substituted γ-lactone.

(8.26)

66%

(8.27)

A: Zn, sat.aq. NH_4Cl / THF = 5:1, r.t., stirring;
B: Sn, H_2O / THF = 5:1, sonication.

$$(8.28)$$

Chan and Li reported that conjugated 1,3-butadienes were produced in moderate yields when carbonyl compounds reacted with 1,3-dichloropropene and zinc in water (Eq. 8.29).[61] The use of 3-iodo-1-chloropropene instead of 1,3-dichloropropene greatly improved the yields. When the reactions were interrupted after their initial allylations, subsequent base treatment of the intermediate compounds produced vinyloxiranes in high yields. Similarly, reactions of carbonyl compounds with 3-iodo-2-chloromethyl-1-propene followed by base treatment produced 2-methylenetetrahydrofurans (Eq. 8.30).[62] Thus, the 3-iodo-2-chloromethyl-1-propene served as a novel trimethylene-methane equivalent.[63]

$$(8.29)$$

tran:cis > 98:1

$$(8.30)$$

Oda et al. reported that under reflux conditions, the zinc-promoted reaction of 2,3-dichloro-1-propene with aldehydes and ketones in a two-phase system of water and toluene containing a small amount of acetic acid gave 2-chloroallylation products (Eq. 8.31).[64] No conversion occurred when tin was used as the promoter. The absence of water completely shuts down the reaction. Interestingly, the action of 2,3-dichloropropene plus zinc powder in aqueous ethanol gives the dechlorination product, allene.[65]

$$(8.31)$$

Reisse used "activated" zinc for aqueous Barbier-type reactions.[66] Submicromic zinc powder produced by pulsed sono-electroreduction is about three times more effective than the commercial variety. The stereochemical course of the allylation and propargylation of several aldehydes with crotyl and propenyl halides using zinc powder as the condensing agent in cosolvent/water(salt) media has been extensively studied.[67] The Zn-mediated reactions of cinnamyl chlorides with aldehydes and ketones in THF-NH$_4$Cl (aq) give α- and γ-addition products as well as phenylpropenes and dicinnamyls, indicating the presence of radical intermediates in the reaction.[68] Enolizable 1,3-dicarbonyl compounds can be allylated by zinc.[69]

An efficient route for the synthesis of the Phe-Phe hydroxyethylene dipeptide isostere precursors utilized for the design of potential inhibitors of renin and HIV-protease was developed. The key step is the zinc-mediated stereoselective allylation of N-protected α-amino aldehydes in aqueous solution (Eq. 8.32).[70] NaBF$_4$/M (M = Zn or Sn) showed facilitating allylation of a variety of carbonyl compounds in water, and α-and γ-addition products of crotylations could be alternatively obtained under the control of this novel mediator (Eq. 8.33).[71]

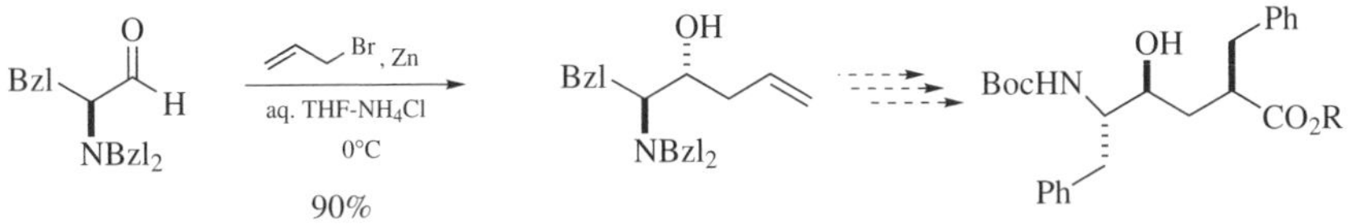

(8.32)

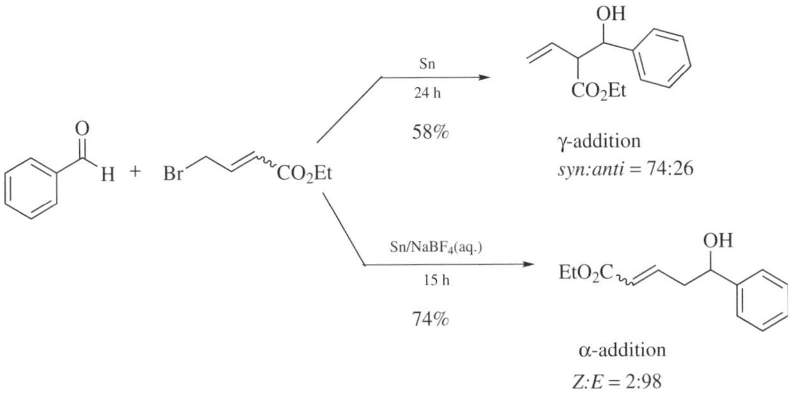

(8.33)

The aqueous Barbier-Grignard-type reaction has also been used in the synthesis of natural products. Chan and Li used the zinc mediated allylation as a key step in a total synthesis of (+)-muscarine (Scheme 8.5).[72] The strategy was based on the observation that the diastereoselectivity of the allylation reaction in water can be reversed through the protection of the α-hydroxyl group.

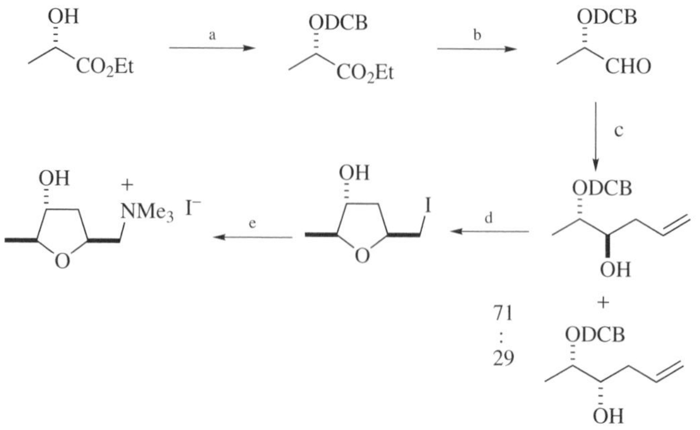

Scheme 8.5 (a) DCBBr/Ag$_2$O/Et$_2$O/reflux/6 h (90%); (b) DIBAL-H/Et$_2$O/ −78°C/2 h; (c) CH$_2$ = CHCH$_2$Br/Zn/H$_2$O/NH$_4$CI/3 h (85%, two steps); (d) I$_2$/CH$_3$CN/0°C/3 h (85%); (e) NMe$_3$/EtOH/80°C/4 h (60%).

Diastereoselective allylation under aqueous Barbier conditions of α-amino aldehydes with the chiral building block (Ss)-3-chloro-2-(p-tolylsulfinyl)-1-propene to give enantiomerically pure sulfinyl amino alcohols in good yields and with high diastereoselectivity was reported (Eq. 8.34).[73]

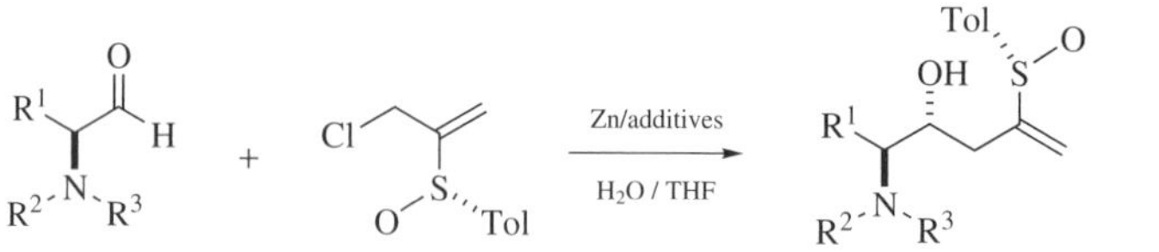

8.3.1.2 Mediated by Tin. In 1983, Nokami et al. observed an acceleration of the reaction rate during the allylation of carbonyl compounds with diallyltin dibromide in ether through the addition of water to the reaction mixture.[74] In one case, by the use of a 1:1 mixture of ether/water as solvent, benzaldehyde was allylated in 75% yield in 1.5 h, while the same reaction gave only less than 50% yield in a variety of other organic solvents such as ether, benzene, or ethyl acetate, even after a reaction time of 10 h. The reaction was equally successful with a combination of allyl bromide, tin metal, and a catalytic amount of hydrobromic acid. In the latter case, the addition of metallic aluminum powder or foil to the reaction mixture dramatically improved the yield of the product. The use of allyl chloride for such a reaction, however, was not successful. The reaction can also proceed intramolecularly. By use of the combination of tin, aluminum, and hydrobromic acid in an aqueous medium, ketones having allylic halide functionality were cyclized to form five-and six-membered rings.[75] Similar reactions occurred with aldehydes.[76] The intramolecular allylation of carbonyl compounds promoted by metallic tin proceeded in a stereocontrolled manner to give cyclic products with high diastereoselectivity (Eq. 8.35).

$$(8.35)$$

Later, Torii et al. found that the tin-aluminum-mediated allylation can be carried out with the less expensive allyl chloride, instead of allyl bromide, when a mixture of alcohol-water–acetic acid was used as the solvent.[77] When combined with stoichiometric amounts of aluminum powder, both stoichiometric and catalytic amounts of tin are effective. As reported by Wu et al., higher temperatures can be used instead of aluminum powder.[78] Under such a reaction condition, allyl quinones were obtained from 1,4-quinones, followed by oxidation with ferric chloride. Allylation reactions in water/organic solvent mixtures were also carried out electrochemically, with the advantage that the allyltin reagent could be recycled.[79]

Otera et al. extended the tin-mediated allylation to 2-substituted allyl bromides.[80] When 2-bromo and 2-acetoxy-3-bromo-1-propene were used, the allylation with tin produced the corresponding functionalized coupling products (Eqs. 8.36 and 8.37). In the case of 2,3-dibromopropene, the reaction occurred exclusively through allylation in the presence of the vinyl bromo group. The presence of other electrophiles such as a nitrile (–CN) or an ester (–COOR) did not interfere with the reaction.

$$(8.36)$$

$$(8.37)$$

The nature of the organotin intermediates has been studied. It was found that when allyl bromide and tin reacted in aqueous media, allyltin(II) bromide was first formed and then was followed by the formation of diallyltin(IV) dibromide (See also Section 6.4.1, Eq. 6.12b). Either of the two organotin intermediates can react with carbonyl compounds to give the corresponding homoallylic alcohols. However, the tin(II) species was found to be more reactive than the tin(IV) species (Eq. 8.38).[81]

$$\text{(8.38)}$$

Luche found that tin-mediated allylations can also be performed through ultrasonic radiation, instead of using aluminum powder and hydrobromic acid to promote the reaction.[82,83] The use of a saturated aqueous NH_4Cl/THF solution, instead of water/THF, dramatically increased the yield. When a mixture of aldehyde and ketone was subjected to the reaction, highly selective allylation of the aldehyde was achieved.

The allylation of carbonyl compounds in aqueous media with $SnCl_2$ can also employ allylic alcohols (Eq. 8.39)[84] or carboxylates[85] in the presence of a palladium catalyst. The diastereoselectivity of the reactions with substituted crotyl alcohols was solvent dependent. Improved diastereoselectivity was obtained when a mixture of water and THF or DMSO was used, instead of the organic solvent alone.

	syn	anti
THF / H_2O (10 equiv)	16	84
DMSO / H_2O (3 equiv)	86	14

$$\text{(8.39)}$$

Allylations, allenylations, and propargylations of carbonyl compounds in aqueous media can also be carried out with preformed organic tin reagent, rather than the use of metals.[86,87,88] For example, the allylation reaction of a wide variety of carbonyl compounds with tetraallyltin was successfully carried out in aqueous media by using scandium trifluoromethanesulfonate (scandium triflate) as a catalyst (Eq. 8.40).[89] A phase-transfer catalyst (PTC) was found to help the allylation mediated by tin at room temperature without any other assistance.[90]

Recently, nanometer tin-mediated allylation of aldehydes or ketones in distilled or tap water gave rise to homoallyl alcohols in high yield without any other assistance such as heat or supersonic or acidic media (Eq. 8.41).[91] Allylation of β-keto aldehydes and functionalized imines by diallyltin dibromide was carried out to generate skipped and conjugated dienes.[92] Aldehydes are allylated with $CH_2=CHCH_2SnBu_3$ using Sn catalysts in acidic aqueous media. Exclusive aldehyde selectivity was observed for competitive reactions of aldehydes and ketones in the presence of 5 mol% of $(CH_2=CHCH_2)_4Sn$ or $SnCl_4$ in a mixture of aqueous HCl and THF (Eq. 8.42).[93]

$$\text{Ph}\diagup\diagup\text{CHO} + (\text{allyl})_4\text{Sn} \xrightarrow[\substack{H_2O/EtOH\ (1:9) \\ 96\%}]{Sc(OTf)_3\ 5\ mol\%} \text{Ph}\diagup\diagup\diagdown\diagup\diagup \overset{OH}{} \qquad (8.40)$$

$$(8.41)$$

$$\diagup\diagup\diagdown\text{SnBu}_3 + \text{PhCHO} + \text{PhCOCH}_3 \xrightarrow[\text{aq. HCl/THF, 20°C}]{\left(\diagdown\diagup\diagdown\right)_4\text{Sn}\quad 5\ mol\%}$$

97%

$$\underset{Ph}{\overset{OH}{|}}\diagup\diagdown\diagup\diagup + \underset{Ph}{\overset{Me}{\underset{}{\overset{OH}{|}}}}\diagup\diagdown\diagup\diagup$$

>99: 1

$$(8.42)$$

Methyltin trichloride and In(III) chloride promote the addition of aldehydes to cyclic allylic stannanes providing good yields of the corresponding homoallylic alcohols.[94] Bu_4NBr/PbI_2 acts as an effective catalyst for the allylation of aldehydes with allylic tin reagents in water.[95] A high *syn*-selectivity was achieved in water without any aprotic solvents in the reaction of the aromatic aldehydes with crotyltri-*n*-butyltin irrespective of their E/Z geometry. A Lewis-acid-surfactant-combined catalyst (LASC) has been developed and applied to Lewis-acid-catalyzed allylation by allyltin reagents in water.[96] Polymer-supported scandium catalyst $PhSO_3Sc(O_3SCF_3)_2$, attached to divinylbenzene-cross-linked polystyrene by a pentylphenylpentyl spacer, was prepared. The catalyst is active in water and several C–C bond-forming reactions proceeded smoothly and in high yield using this catalyst (Eq. 8.43).[97]

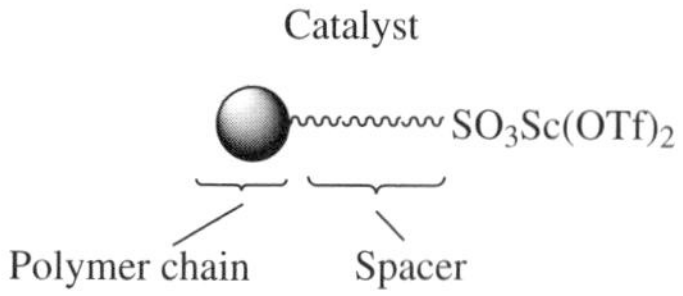

$$(8.43)$$

1-Bromobut-2-ene on a dichloromethane-water biphasic system at $25°C$ causes α-regioselective addition to aldehydes with $SnBr_2$ to produce 1-substituted pent-3-en-1-ols and also causes γ-regioselective addition to aldehydes with $SnBr_2$-Bu_4NBr to produce 1-substituted 2-methylbut-3-en-1-ols (Eq. 8.44).[98]

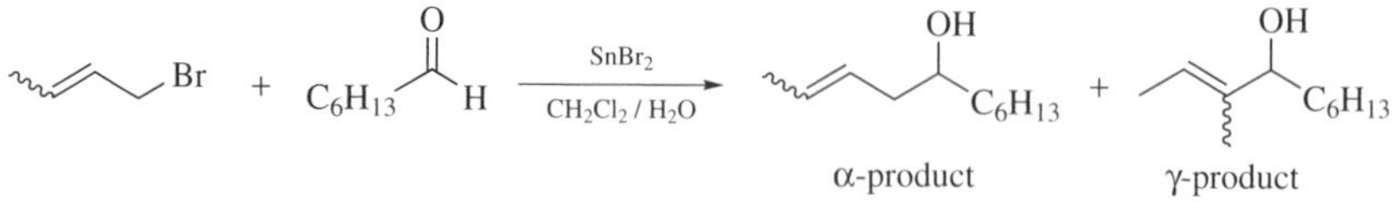

	α-product	γ-product
With 1 equiv Bu₄NBr	9	91
Without Bu₄NBr	91	9

$$(8.44)$$

The allylation of aldehydes can be carried out using stannous chloride and catalytic cupric chloride or copper in aqueous media.[99] In-situ probing provides indirect (NMR, CV) and direct (MS) evidence for the copper(I)-catalyzed formation of an allyltrihalostannane intermediate in very high concentration in water (Scheme 8.6). Hydrophilic palladium complex also efficiently catalyzes the allylation of carbonyl compounds with allyl chlorides or allyl alcohols with $SnCl_2$ under aqueous-organic

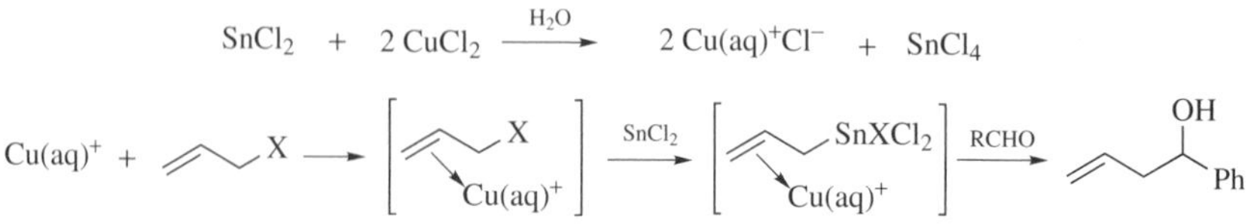

Scheme 8.6

biphase conditions that allow easy separation of the product and recovery of the organic solvent from the reaction mixture.[100] A combination of $TiCl_3$ and $SnCl_2$ is also effective for the allylation.[101]

A solid-phase version of the palladium-catalyzed carbonyl allylation of aldehydes by allylic alcohol has been described. Thus, allylation of resin-bound aldehyde (P = Merrifield resin) with allylic alcohols (e.g., $MeCH=CHCH_2OH$) in the presence of $SnCl_2$ afforded the homoallylic alcohols under different solvent conditions, in DMSO and aqueous DMSO respectively (Eq. 8.45).[102]

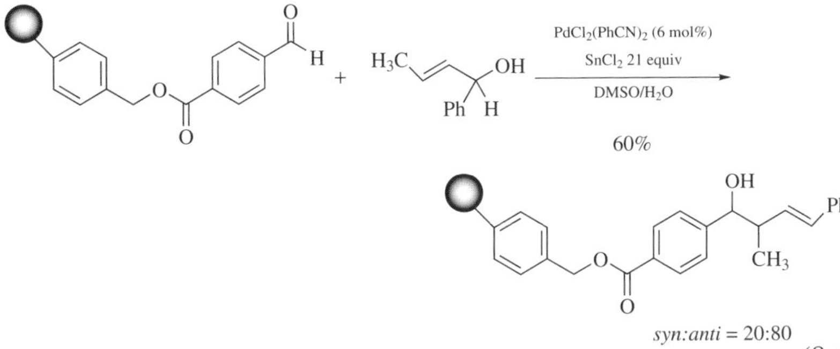

$$syn:anti = 20:80$$

(8.45)

Bis-homoallylic alcohols were prepared in good yields by allylation of dialdehydes or their acetals with allyl bromide, tin(II) chloride, and potassium iodide in water or water/THF (Eq. 8.46).[103] Under ultrasonication, it was found that $SnCl_2$ could efficiently mediate the aqueous Barbier reactions between carbonyl compounds and allyl bromide to give the corresponding homoallylic alcohols in high yields without using any Lewis-acid catalyst.[104]

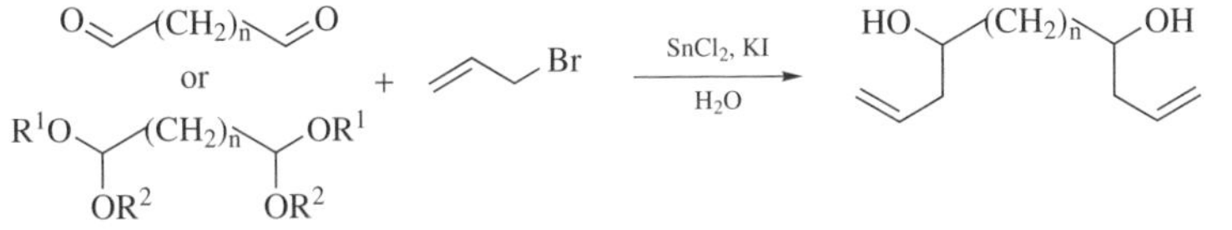

(8.46)

Highly regio- and stereoselective allylation of aldehydes by allenes proceeds smoothly in aqueous/organic media in the presence of $PdCl_2(PPh_3)_2$, HCl, and $SnCl_2$. For example, the reaction of 1,1'-dimethylallene and $SnCl_2$ with PhCHO under the above conditions, gave the corresponding carbonyl allylation product in 95% isolated yield (Eq. 8.47). The reaction likely occurs via hydrostannylation of allenes and allylation of aldehydes by the in-situ-generated allyltrichlorotins to afford the final products.[105]

$$R^1 \atop R^2 \!\!=\!\!\bullet\!\!= \;+\; {O \atop H\!\diagdown\!\!R^3} \xrightarrow[\text{HCl/SnCl}_2,\text{ DMF, r.t.}]{\text{PdCl}_2(\text{PPh}_3)_2} \quad {OH \atop \diagup\!\!\diagdown\!\!\diagdown\!\!R^3 \atop R^1 \; R^2} \qquad (8.47)$$

In 1991, Whitesides et al. reported the first application of aqueous medium Barbier-Grignard reaction to carbohydrate synthesis through the use of tin in an aqueous/organic solvent mixture (Eq. 8.48).[106] These adducts were converted to higher carbon aldoses by ozonolysis of the deprotected polyols followed by suitable derivatization. The reaction showed a higher diastereoselectivity when there was a hydroxyl group present at C-2. However, no reaction was observed under the reaction conditions when there was an N-acetyl group present at the C-2 position.

$$\begin{matrix} CHO \\ | \\ (CHOH)n \\ | \\ CH_2OH \end{matrix} \;+\; \diagup\!\!\diagdown\!\!\diagup\!\!Br \xrightarrow[\text{ultrasound}]{\substack{\text{Sn} \\ \text{H}_2\text{O/KOEt}}} \begin{matrix} CH\!=\!CH_2 \\ | \\ CH_2 \\ | \\ CH\!-\!OH \\ | \\ CH\!-\!OH \\ | \\ (CHOH)_{n\text{-}1} \\ | \\ CH_2OH \end{matrix} \;+\; \begin{matrix} CH\!=\!CH_2 \\ | \\ CH_2 \\ | \\ HO\!-\!CH \\ | \\ CH\!-\!OH \\ | \\ (CHOH)_{n\text{-}1} \\ | \\ CH_2OH \end{matrix} \qquad (8.48)$$

8.3.1.3 Mediated by Indium.

The transferring of electrons from metals to organic substrates plays an important role in many metal-mediated reactions. The ionization potential of an element is directly related to its ability to give off electrons. Thus, by examining the first ionization potentials of different elements,[107] it was found that indium has the lowest first ionization potential compared to the other metal elements near it in the periodic table. In fact, the ionization potential of indium is on the same level with the most active alkali metals and was much lower than that of zinc or tin, or even magnesium (Table 8.1). On the other hand, unlike those more reactive alkali metals, indium metal is not sensitive to boiling water or alkali and does not form oxides readily in air. Such special properties of indium indicate that it is perhaps a promising metal for aqueous Barbier-Grignard-type reactions.

TABLE 8.1 First Ionization Potential of Some Metals

Metal	Indium	Magnesium	Zinc	Tin	Lithium	Sodium
First ionization potential (ev)	5.79	7.65	9.39	7.43	5.39	5.12

Cited from *CRC Handbook of Chemistry and Physics*, 75th ed., CRC Press, 1994.

In 1991, Li and Chan reported the use of indium to mediate Barbier-Grignard-type reactions in water (Eq. 8.49).[108] When the allylation was mediated by indium in water, the reaction went smoothly at room temperature without any promoter, whereas the use of zinc and tin usually requires acid catalysis, heat, or sonication. The mildness of the reaction conditions makes it possible to use the indium method to allylate a methyl ketone in the presence of an acid-sensitive acetal functional group (Eq. 8.50). Furthermore, the coupling of ethyl 2-(bromomethyl)acrylate with carbonyl compounds proceeds equally well under the same reaction conditions, giving ready access to various hydroxyl acids including, for example, sialic acids.

$$\underset{R^1 \quad R^2}{\overset{O}{\|}} + X\diagdown\diagup\diagdown \quad \xrightarrow[X = I, Br, Cl]{In/H_2O} \quad \underset{R^2}{\overset{OH}{R^1\diagdown\diagup\diagdown}} \qquad (8.49)$$

$$\underset{CH_3O}{\overset{CH_3O \quad O}{\diagdown}} + \diagup\diagdown\diagup Br \quad \xrightarrow[H_2O]{In} \quad \underset{CH_3O}{\overset{CH_3O \quad OH}{\diagdown}} \qquad (8.50)$$

M	Yield %
Zn	0 and destruction of starting materials
Sb	10 (under sonication)
In	70%

Later, Araki et al. found that the allylation of aldehydes and ketones can be carried out by using catalytic amounts of indium(III) chloride in combination with aluminum or zinc metal.[109] This reaction was typically performed in a THF-water (5:2) mixture at room temperature, although the conversion was much slower compared to the same reaction mediated by use of a stoichiometric amount of indium and it required days to complete. When the reaction was carried out in anhydrous THF alone, the yield dropped considerably and side-reactions such as reduction to alcohol increased. The combinations of Al-InCl$_3$ or Zn-InCl$_3$ gave comparable results.

Whitesides et al. examined the effect of substituents on the allyllic moiety of the indium-mediated reactions in water and found that the use of indium at room temperature gave results comparable to those of tin-mediated reactions carried out at reflux.[110] Replacement of the aqueous phase with 0.1 N HCl further increased the rate of the reaction. The transformation can also be carried out with preformed allylindium chloride.

The combination of 2-halomethyl-3-halo-1-propene with carbonyl compounds mediated by indium in water generated *bis*-allylation products (Eq. 8.51).[111] The *bis*-allylation of 1,3-dibromo-propene with carbonyl compounds mediated by indium in water gave predominately 1,1-*bis*-allylation product.[112]

$$
\underset{R^1 \quad R^2}{\overset{O}{\|}} \;+\; X\diagdown\diagup\diagup\diagdown X \;\; \xrightarrow{\text{In/H}_2\text{O}} \;\; \underset{R^2}{\overset{OH}{R^1}}\diagdown\diagup\diagdown\overset{OH}{\diagup}\underset{R^2}{R^1} \qquad (8.51)
$$

The indium-mediated allylation carried out with allylstannanes in combination with indium chloride in aqueous medium was reported by Marshall et al.[113] Allylindium was proposed as the reaction intermediate. Various aldehydes can be alkylated very efficiently with 3-bromo-2-chloro-1-propene mediated by indium in water at room temperature. Subsequent treatment of the compound with ozone in methanol followed by workup with sodium sulfite provided the desired hydroxyl ester in high yield.[114]

Because of its superior reactivity, the indium-mediated reaction in water has found wider applications in natural product synthesis. Chan and Li reported[115] an efficient synthesis of (+)-3-deoxy-D-*glycero*-D-*galacto*-nonulosonic acid (KDN) (Scheme 8.7), using the indium-mediated allylation reaction in water. A similar synthesis of 3-deoxy-D-manno-octulonate (KDO) led primarily to the undesired diastereomer. However, through the disruption of the newly generated stereogenic center,[116] they completed a formal synthesis[117] of KDO. In contrast to the tin-mediated reactions, the indium-mediated reaction also occurred on a substrate with an *N*-acetyl group present at C-2. Whitesides et al. reported the synthesis of *N*-acetyl-neuraminic acid[118] as well as other sialic acid derivatives based on this strategy. The use of indium is essential for the carbon–carbon bond formation step in these sialic acid syntheses. KDO was synthesized via indium-mediated allylation of 2,3:4,5-di-O-isopropylidene-D-arabinose.[119] In this case, the desired product became the major product due to the protection of the α-hydroxyl group. Subsequently, Chan et al. further shortened the already concise sialic acids' synthesis to two steps through the indium-mediated reaction of α-(bromomethyl)acrylic acid with sugars. Both KDN and *N*-acetyl-neuraminic acid have been synthesized in such a way.[112] The indium-mediated allylation reaction was applied by Schmid et al. to de-oxy sugars[120] and to the elongation of the carbon chain of carbohydrates in forming higher analogs.[121] The carboxylic acid functionality on allyl halides is compatible with the indium-mediated reactions (Scheme 8.8).[122,123] Thus, when the 2-(bromomethyl)acrylic acid,

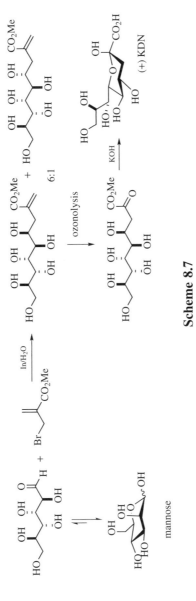

Scheme 8.7

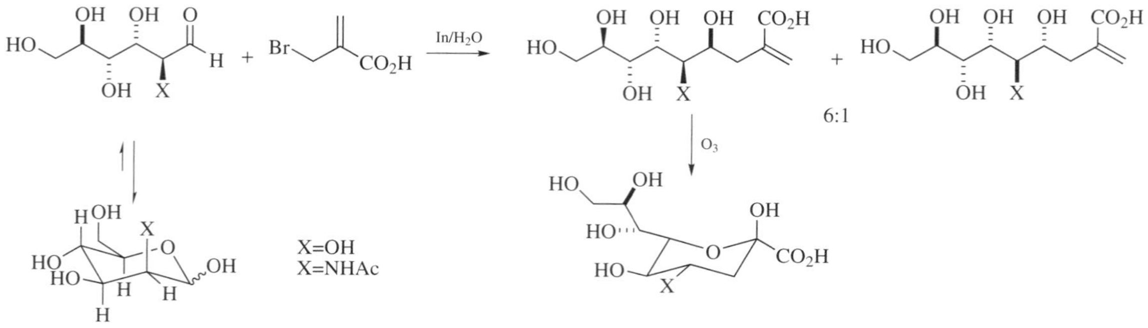

Scheme 8.8

instead of the ester, was treated directly with carbonyl compounds and indium in water, the corresponding γ-hydroxyl-α-methylenecarboxylic acids were generated in good yields.

Phosphonic acid analogues **8.2** and **8.3** of both KDN and N-acetylneuraminic acid have been synthesized using the indium-mediated coupling of the lower carbohydrates with dimethyl 3-bromopropenyl-2-phosphonate (**8.4**) in water (Scheme 8.9).[124]

Loh et al. reported the reaction of the glucose-derived aldehyde with allyl bromide mediated by indium.[125] The reaction gives a non-chelation product as the major diastereomer. The use of an organic cosolvent increases the diastereoselectivity. The addition of ytterbium trifluoromethanesulfonate [Yb(OTf)$_3$] enhances both the reactivity and the diastereoselectivity. Enolizable 1,3-dicarbonyl compounds undergo efficient carbonyl allylation reactions in an aqueous medium (Eq. 8.52).[126] A variety of 1,3-dicarbonyl compounds have been alkylated successfully using allyl bromide or allyl chloride in conjunction with either tin or indium. The reaction can be used readily for the synthesis of cyclopentane derivatives (Eq. 8.53).[127] The allylation reaction in water also could be used to prepare α,α-difluorohomoallylic alcohols from *gem*-difluoro allyl halides.[128] Application of aqueous Barbier-type reaction in a carbocycle ring enlargement methodology was developed by Li and co-workers (Eq. 8.54).[129] By using the indium-mediated Barbier-type reaction in water, 5-, 6-, 7-, 8-, and 12-membered rings are enlarged by two carbon atoms into 7-, 8-, 9-, 10-, and 14-membered ring derivatives respectively. The use of water as a solvent was found to be critical for the success of the reaction. Similar ring expansion in organic solvents was not successful. The ring expansion has also been applied to the synthesis of a heterocyclic medium ring.[130] One carbon-ring expansion was reported similarly.[131]

$$(8.52)$$

$$(8.53)$$

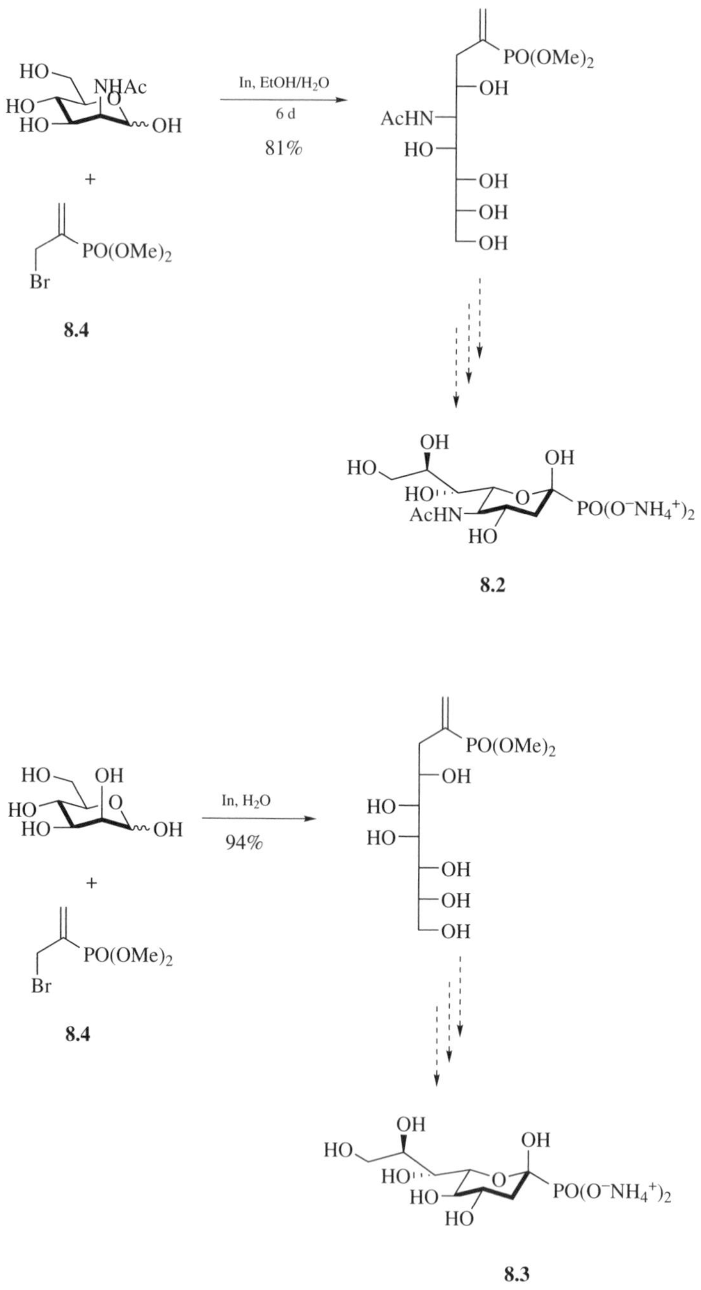

Scheme 8.9

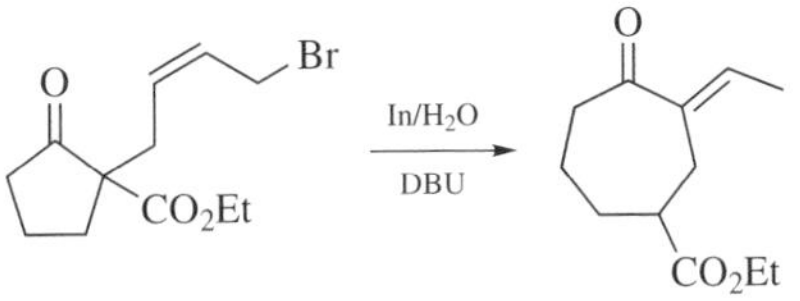

$$(8.54)$$

Indium-mediated allylation of an unreactive halide with an aldehyde[132] was used to synthesize an advanced intermediate in the synthesis of antillatoxin,[133] a marine cyanobacteria (*Lyngbya majuscula*) that is one of the most ichthyotoxic compounds isolated from a marine plant to date. In the presence of a lanthanide triflate, the indium-mediated allylation of Z-2-bromocrotyl chloride and aldehyde in saturated NH_4Cl under sonication yielded the desired advanced intermediate as a 1:1 mixture of diastereomers in 70% yield. Loh et al.[134] then changed the halide compound to methyl (Z)-2-(bromomethyl)-2-butenoate and coupled it with aldehyde under the same conditions to yield the desired homoallylic alcohol in 80% yield with a high 93:7 *syn:anti* selectivity (Eq. 8.55).

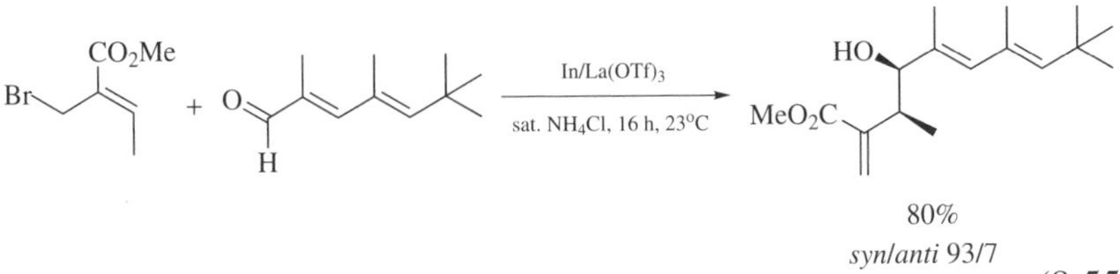

80%
syn/anti 93/7

$$(8.55)$$

The indium-mediated allylation of trifluoroacetaldehyde hydrate (R = H) or trifluoroacetaldehyde ethyl hemiacetal (R = Et) with an allyl bromide in water yielded α-trifluoromethylated alcohols (Eq. 8.56).[135] Lanthanide triflate-promoted indium-mediated allylation of aminoaldehyde in aqueous media generated β-aminoalcohols stereoselectively.[136] Indium-mediated intramolecular carbocyclization in aqueous media generated fused α-methylene-γ-butyrolactones (Eq. 8.57).[137] Forsythe and co-workers applied the indium-mediated allylation in the synthesis of an advanced intermediate for azaspiracids (Eq. 8.58).[138] Other potentially reactive functionalities such as azide, enone, and ketone did not compete with aldehyde for the reaction with the in situ-generated organo-indium intermediate.

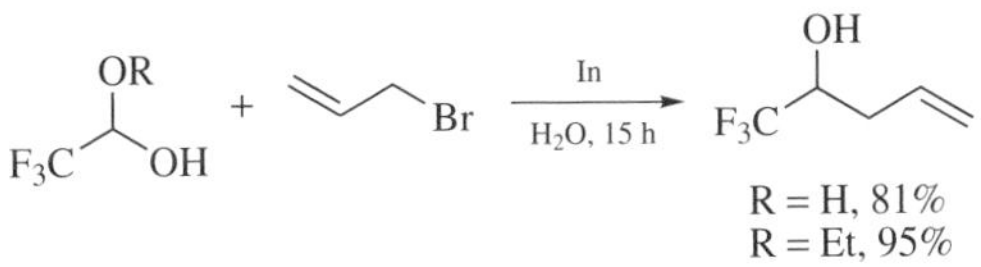

$$(8.56)$$

R = H, 81%
R = Et, 95%

$$(8.57)$$

$$(8.58)$$

8.3.1.4 Regio- and Stereoselectivity.

For the allylation of carbonyl compounds mediated by indium and other compounds in aqueous media, usually the carbon–carbon bond forms at the more substituted carbon of the allyl halide, irrespective of the position of halogen in the starting material. However, the carbon–carbon bond forms at the less-substituted carbon when the γ-substituents of allyl halides are large enough (e.g., trimethylsilyl or *tert*-butyl) as shown by Chan et al.[139] (Scheme 8.10). The following conclusions can be drawn:

1. In general, the reaction gives the regioisomer where the substituent is alpha to the carbon–carbon bond to be formed.
2. The regioselectivity is governed by the steric size of the substituent but not by the degree of substitution. When the substituent is sterically bulky (e.g., *t*-butyl or silyl), the preferred regioisomer formed has the substituent away (at the γ-position) from the carbon–carbon bond being created.
3. Regioselectivity is not governed by the conjugation of the double bond with the substituent.
4. Regioselectivity is independent of the geometry of the double bond and either *E*- or *Z*-cinnamyl bromide coupled with *i*-butanal to give the same regioisomer.
5. Regioselectivity is independent of the initial location of the substituent on the double bond.

The stereochemistry of the reaction in an aqueous medium is somewhat analogous to that in an organic medium. In terms of

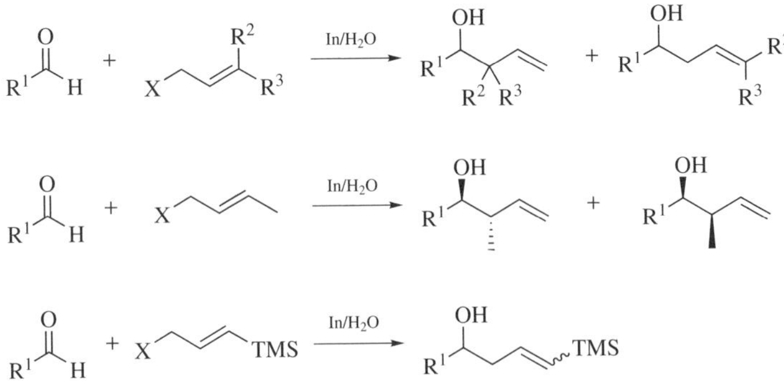

Scheme 8.10

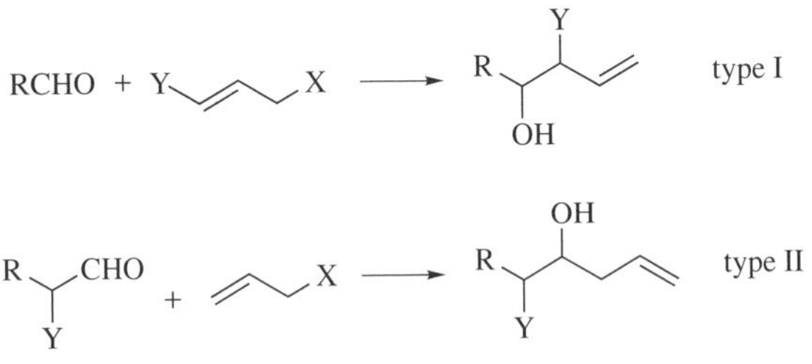

Scheme 8.11

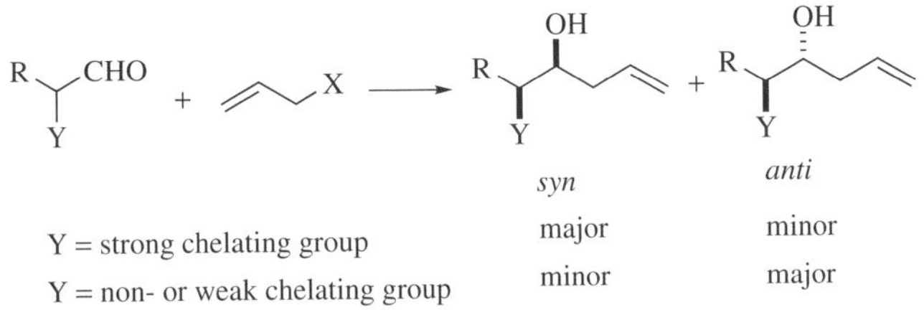

Y = strong chelating group

Y = non- or weak chelating group

Scheme 8.12

the diastereoselectivity, two types of situations prevail (type I and type II) (Scheme 8.11). Within the type I situation, the reaction can favor either *syn* or *anti*-diastereoselectivity, depending on the properties of the α-substituents (Scheme 8.12). The presence of a strong α-chelating group such as a hydroxyl leads to the *syn* product, whereas a non-α-chelating group such as a methyl produces the *anti* product. However, when a weak α-chelating group (e.g., alkoxyl) is present, allylation in an organic solvent usually favors a chelation-controlled product. The presence of an α-alkoxyl group will generate the non-chelating product through the aqueous reactions.[140] Thus, it is possible to reverse the diastereoselectivity of an allylation simply by use of a free hydroxyl group or by protecting it as an alkoxyl. In the type I situation, the stereogenic center can be far away from the carbonyl group. Such an example can be found in Waldmann's studies of the diastereoselectivity of allylations using proline benzyl ester as a chiral auxiliary to produce α-hydroxyl amides. The diastereoselectivity was around 4 ~ 5:1 (Scheme 8.13).[141] Separation of the diastereomers followed by reaction with methyl lithium produced the enantiomerically pure alcohol.

The type I situation usually gives an antidiastereoselectivity that is independent of the stereochemistry of the double bond in the allyl bromide moiety. The diastereoselectivity (*anti/syn* ratio) is governed by the steric size of the substituent on the aldehydes. *Anti/syn* ratio increases as the size of the aldehyde R group or the substituent on the allyl bromide increases (Scheme 8.14).

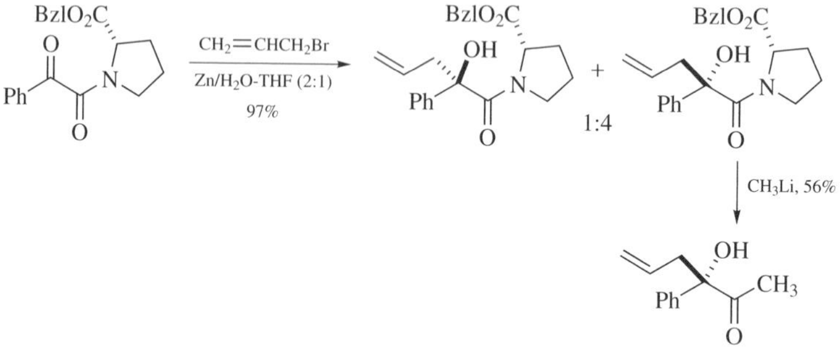

Scheme 8.13

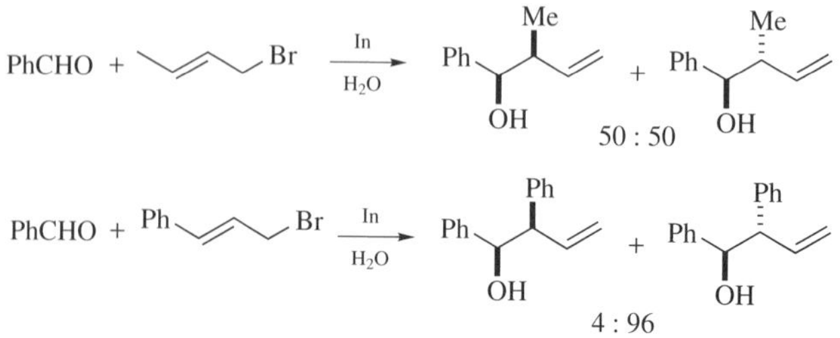

Scheme 8.14

Normally, the addition of C-nucleophiles to chiral α-alkoxyaldehydes in organic solvents is opposite to Cram's rule (Scheme 8.15). The *anti*-Cram selectivity has been rationalized on the basis of chelation control.[142] The same *anti* preference was observed in the reactions of α-alkoxyaldehydes with allyl bromide/indium in water.[143] However, for the allylation of α-hydroxyaldehydes with allyl bromide/indium, the *syn* isomer is the major product. The *syn* selectivity can be as high as 10:1 (*syn:anti*) in the reaction of arabinose. It is argued that in this case, the allylindium intermediate coordinates with both the hydroxy and the carbonyl function leading to the *syn* adduct.

The same coordination is used to account for the observed *anti* preference in the allylation of β-hydroxybutanal with allyl bromide/indium in water (Scheme 8.16). The intermediate leads to the *anti* product. In support of the intramolecular chelation model, it is found that if the hydroxy group is converted to the corresponding benzyl or *t*-butyldimethylsilyl ether, the reaction is not stereoselective at all and gives nearly equal amounts of *syn* and *anti* products.

It is possible to combine both the type I and the type II situations in the coupling of a chiral aldehyde with a substituted allylic halide. Such is the case in the coupling of unprotected aldoses (e.g., glyceraldehyde)

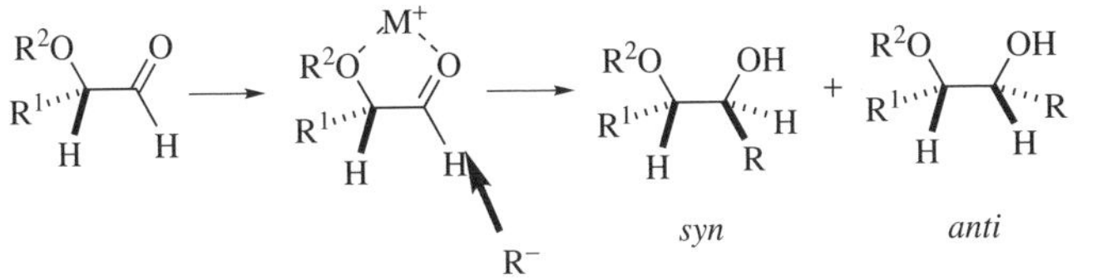

Scheme 8.15

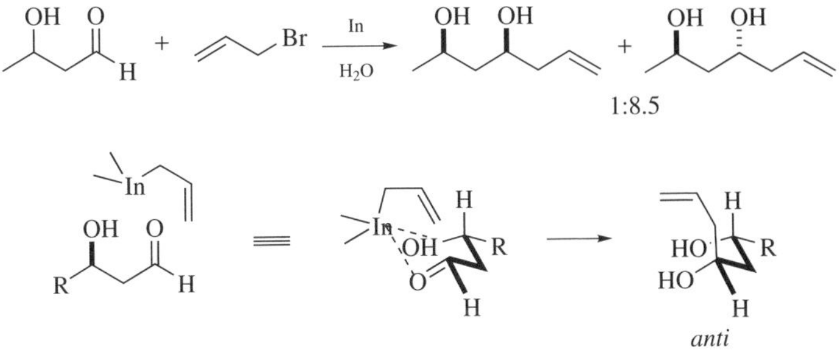

1:8.5

Scheme 8.16

with cinnamyl bromide. In such a coupling, two new stereogenic centers are created. It has been found that the *syn, syn* isomer is formed preferentially. To account for the *syn, syn* stereochemistry, chelation of the allylindium species with the hydroxyaldehyde function with intramolecular attack through a cyclic transition state is postulated.[143] The stereochemistry of the adduct is then dependent on the geometry of the attacking allylindium species. The In-mediated allylation of aldehydes with $BrCH_2CH=CHCF_3$ in water afforded α-trifluoromethylated homoallylic alcohols in high yields (Eq. 8.59).[144]

$$(8.59)$$

Allylation reactions of racemic and optically pure 4-oxoazetidine-2-carbaldehydes were investigated both under anhydrous conditions and in aqueous media. Indium-promoted allylation showed a reverse diastereofacial preference, although the observed selectivity is not synthetically useful.[145] Indium-mediated allylation of gem-diacetates gave excellent yields of the corresponding homoallylic acetates in aqueous media.[146] Allyl addition to α-diketones by treatment with allyl bromide and indium in water/THF gives diallyl diols such as $H_2C{:}CHCH_2CPh(OH)CH(OH)CH_2CH{:}CH_2$ with moderate stereoselectivities. Ring-closing metathesis of the diallyl diols with Grubbs' ruthenium olefin metathesis catalyst, followed by diol cleavage with lead tetraacetate, gives *cis*-alkenediones (Eq. 8.60).[147] The allylation was applied in the total asymmetric synthesis of the putative structure of the cytotoxic diterpenoid (−)-sclerophytin and of the natural sclerophytins.[148]

$$(8.60)$$

Allylation of the C-3 position of the cephem nucleus was accomplished by either indium-mediated or indium trichloride-promoted tin-mediated allylation reactions in aqueous media. Both methods gave 3-allyl-3-hydroxycephams in moderate to excellent yields.[149]

A new method has been developed for the synthesis of (E)-β-methyl Baylis-Hillman-type adducts with high E/Z (>93%) selectivity in modest to good yields. The process consists of two steps: an indium-mediated allylation reaction and a simple base-catalyzed isomerization step (Eq. 8.61). Various aldehydes were allylated with allyl bromides using indium under very mild conditions in aqueous media and thus converted to the Baylis-Hillman-type adducts.[150]

$$\begin{array}{c} \text{(structure: } R^1\text{-CHO + Br-CH=CH-CO}_2R^2 \xrightarrow[\text{THF/H}_2\text{O}]{\text{In, HCl}} R^1\text{-CH(OH)-CH(CH=CH}_2\text{)-CO-OR}^2 \xrightarrow{\text{DBU}} R^1\text{-CH(OH)-C(=CH-CH}_3\text{)-CO-OR}^2\text{)} \end{array} \tag{8.61}$$

C-branched sugars or C-oligosaccharides are obtainable through indium-promoted Barbier-type allylations in aqueous media.[151] Indium-mediated allylation of α-chlorocarbonyl compounds with various allyl bromides in aqueous media gave the corresponding homoallylic chlorohydrins, which could be transformed into the corresponding epoxides in the presence of a base (Eq. 8.62).[152]

$$\begin{array}{c} \text{(structure: } R^1\text{-CH(Cl)-CO-R}^2 + \text{CH}_2\text{=CH-CH}_2\text{Br} \xrightarrow[\text{THF/H}_2\text{O}]{\text{In}} R^1\text{-CH(Cl)-C(OH)(R}^2\text{)-CH}_2\text{-CH=CH}_2 \xrightarrow{\text{base}} R^1\text{-epoxide-C(R}^2\text{)-CH}_2\text{-CH=CH}_2\text{)} \end{array} \tag{8.62}$$

Linear α-homoallylic alcohol adducts were obtained with high regioselectivities in moderate-to-good yields using allylic indium reagents in the presence of 10 M water (Eq. 8.63).[153] A new mechanism is proposed for the α-regioselective indium-mediated allylation reaction in water. Based on the results and observations obtained from an NMR study, a cross-over experiment and the complete inversion of the stereochemistry of β,γ-adduct homoallylic sterols to the α,α-adduct homoallylic sterols, it is suggested that the initially formed γ-adduct undergoes a bond cleavage to generate the parent aldehyde in situ followed by a concerted rearrangement, perhaps a retro-ene reaction followed by a 2-oxonia [3,3]-sigmatropic rearrangement, to furnish the α-adduct (Scheme 8.17).[154]

$$\begin{array}{c} \text{(structure: } R\text{-CHO} \xrightarrow[\text{In, water (10M)}]{R_1\text{-CH=CH-CH}_2\text{Br}} R\text{-CH(OH)-CH=CH-R}_1 + R\text{-CH(OH)-CH(R}_1\text{)-CH=CH}_2\text{)} \end{array} \tag{8.63}$$

α-adduct   γ-adduct
>96:4 (α:γ)

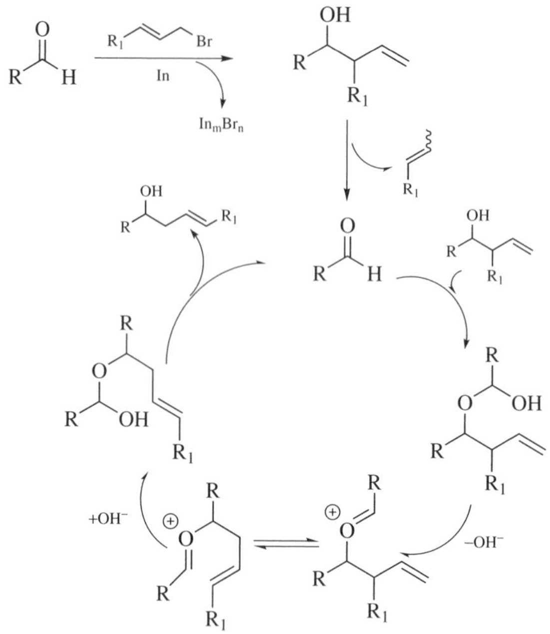

Scheme 8.17

Various nitrobenzaldehydes were simultaneously allylated and reduced using indium in the presence of HCl in aqueous media to give compounds having both homoallylic alcohol and aromatic amine functionalities.[155] Reactions of racemic as well as optically pure carbonyl-β-lactams with stabilized organo-indium reagents were investigated in aqueous media. The regio- and stereochemistry of the processes were generally good, offering a convenient asymmetric entry to densely functionalized hydroxy-β-lactams (Eq. 8.64).[156] A highly stereoselective In-mediated allylation was used for asymmetric synthesis of a highly functionalized THF-derivative (Eq. 8.65).[157]

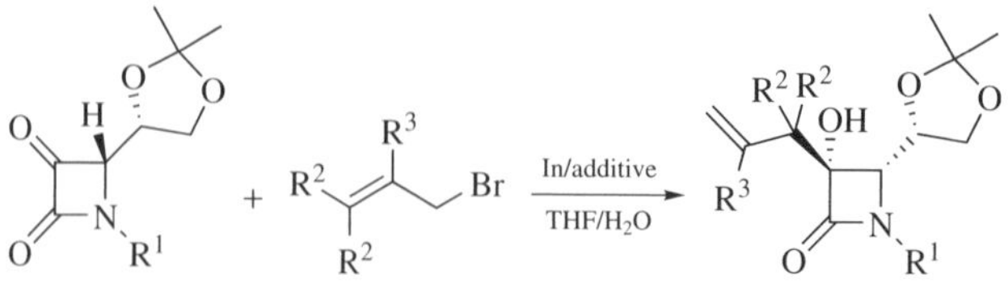

(8.64)

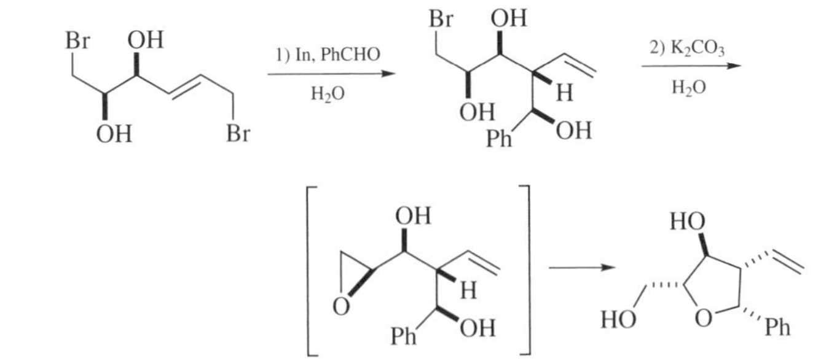

$$(8.65)$$

Palladium catalyzes allylation of carbonyl compounds with various allylic compounds using In-InCl$_3$ in aqueous media (Eq. 8.66).[158] Various allylic compounds can be effectively applied via the formation of π-allylpalladium(II) intermediates and their transmetalation with indium in the presence of indium trichloride in aqueous media.

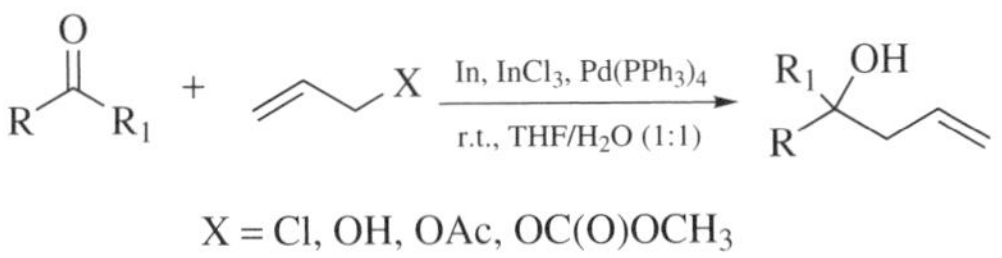

$$(8.66)$$

$$X = Cl, OH, OAc, OC(O)OCH_3$$

The indium-mediated allylation reaction in aqueous media has been applied to the studies on the total synthesis of dysiherbaine (Scheme 8.18).[159]

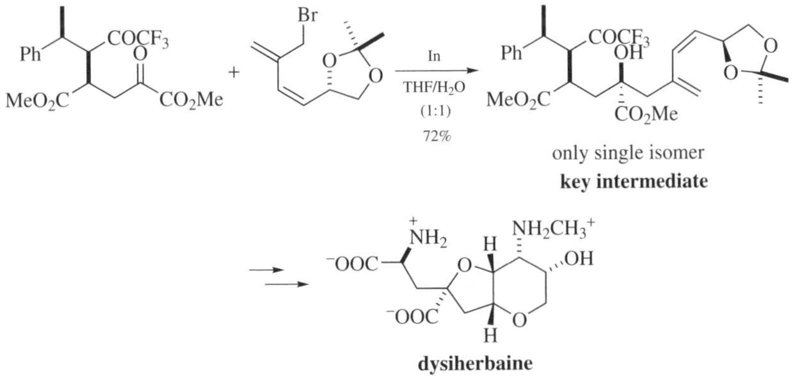

dysiherbaine

Scheme 8.18

Metal-mediated allylation of difluoroacetyltrialkylsilanes with various allyl bromides in aqueous media formed homoallylic alcohols exclusively. The Brook rearrangement, carbon to oxygen-silyl migration, was totally suppressed with no detectable formation of silyl ether (Eq. 8.67).[160] The reaction afforded high *syn* selectivity regardless of the allylic bromide geometry. The enantioselective indium-mediated allylation was attempted and found to give the desired products in moderate yields with high *syn* selectivity and enantioselectivity in organic solvent (Eq. 8.68).[161] Indium trichloride catalyzed indium-mediated allylation of dihydropyrans and dihydrofurans in water. This catalytic system afforded the allylated diols in moderate-to-high yields (Eq. 8.69).[162]

$$
\text{HF}_2\text{C}\overset{\text{O}}{\underset{}{\parallel}}\text{SiPh}_2{}^t\text{Bu} + \diagup\!\!\!\diagup\!\!\!\diagdown\text{Br} \xrightarrow[\substack{\text{THF/H}_2\text{O(1:1)} \\ 97\%}]{\text{In}} \text{HF}_2\text{C}\overset{\text{OH}}{\underset{\text{SiPh}_2{}^t\text{Bu}}{\vert}} \qquad (8.67)
$$

$$
\begin{array}{c}
\text{HO}\diagdown\!\!\diagup\diagup\text{Br} + \text{O}=\!\!\text{CH aldehyde (polyene)} \\
\xrightarrow[\substack{\text{THF-hexane (3:1)} \\ -78°\text{C, 2h}}]{\text{In, (-)-cinchonidine}} \\
\text{HO product} \\
36\% \ (85\% \ \text{ee, } syn{:}anti = 99{:}1)
\end{array} \qquad (8.68)
$$

$$
\text{dihydropyran} + \diagup\!\!\!\diagup\!\!\!\diagdown\text{Br} \xrightarrow[\substack{\text{In} \\ \text{H}_2\text{O, 16h, r.t.} \\ 83\%}]{\text{InCl}_3 \ (20 \ \text{mol}\%)} \text{HO}\diagdown\!\!\diagup\diagdown\!\!\diagup\diagdown\overset{\text{OH}}{\underset{}{\vert}}\diagup\!\!\diagdown \qquad (8.69)
$$

8.3.1.5 *Mechanistic Discussion.*

For the mechanism of the metal-mediated allylation reaction in aqueous media, Li proposed a carbanion-allylmetal-radical triad (Figure 8.1) in which the specific mechanism of the

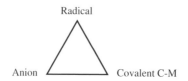

Figure 8.1   Mechanistic possibilities of metal-mediated allylation.

reaction is dependent on the metal being used. Recently, detailed mechanistic study of the indium-mediated allylation in water has been carried out. Using ^{1}H NMR studies, it was found that allyl bromide and indium in D_2O gave allylindium(I) as the reactive intermediate (Scheme 8.19).[163] The conclusion that allylindium(I) was formed in aqueous media was rather unexpected, as it was thought previously that allylindium(III) species were the more likely intermediate. Secondary deuterium kinetic isotope effects (SDKIE) in the metal-mediated allylation of benzaldehyde in aqueous media were examined.[164] The inverse SDKIE observed for the indium and tin cases are consistent with the polar addition mechanism. For magnesium and antimony, normal SDKIE was observed. These were interpreted as single electron transfer processes on metal surfaces in the case of magnesium as well as between the allylmetal and the carbonyl compound in the case of antimony.

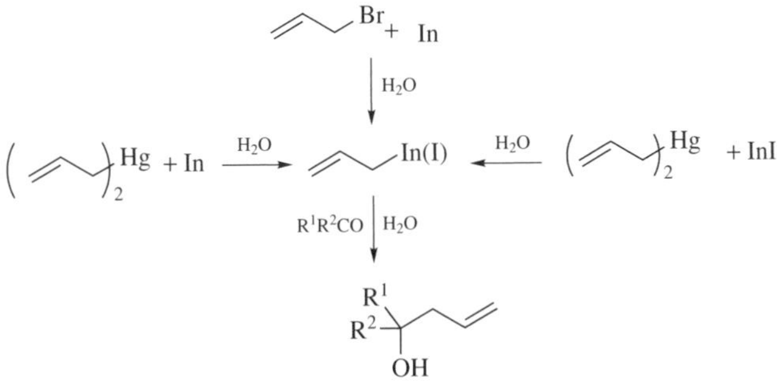

Scheme 8.19

8.3.1.6 *Mediated by Other Metals.*
In addition to the metals discussed above, other metals have been found to mediate the Barbier-Grignard-type conversions in water, but investigations on these metals are relatively limited.

B. Potassium allyl- and crotyltrifluoroborates undergo addition to aldehydes in biphasic media as well as water to provide homoallylic alcohol in high yields (>94%) and excellent diastereoselectivity (dr >98:2). The presence of a phase-transfer catalyst (e.g., Bu$_4$NI) significantly accelerates the rate of reaction, whereas adding fluoride ion retards the reaction (Eq. 8.70).[165] The method was applied to the asymmetric total synthesis of the antiobesity agent tetrahydrolipstatin (orlistat).[166]

$$\text{(8.70)}$$

Si. Tris(pentafluorophenyl)boron was found to be an efficient, air-stable, and water-tolerant Lewis-acid catalyst for the allylation reaction of allylsilanes with aldehydes.[167] Sc(OTf)$_3$-catalyzed allylations of hydrates of α-keto aldehydes, glyoxylates and activated aromatic aldehydes with allyltrimethylsilane in H$_2$O-CH$_3$CN were examined. α-Keto and α-ester homoallylic alcohols and aromatic homoallylic alcohols were obtained in good to excellent yields.[168] Allylation reactions of carbonyl compounds such as aldehydes and reactive ketones using allyltrimethoxysilane in aqueous media proceeded smoothly in the presence of 5 mol% of a CdF$_2$-terpyridine complex (Eq. 8.71).[169]

$$\text{(8.71)}$$

Ga. Gallium was used to mediate the allylation of carbonyl compounds in water.[170] The reaction can also be carried out by using preformed allylgallium reagents.[171] The corresponding homoallyl alcohols were obtained in high yields without the assistance of either acidic media or sonication (Eq. 8.72).

$$\text{(8.72)}$$

Mg. Li and co-worker first reported magnesium-mediated Barbier-Grignard allylation of benzaldehyde in water (Eq. 8.73).[172] Recently, a study was completed in which some water-tolerant allylating agents were prepared in situ from allylmagnesium chloride and various metallic salts reacted with aldehydes in THF-H$_2$O to afford the desired homoallylic alcohols.[173]

$$\text{(8.73)}$$

Co. In the presence of cobalt(II) chloride and metallic aluminum, allylic halides react with aldehydes at room temperature in tetrahydro-furan-water to afford the corresponding alcohols in high yields.[174]

Ge. Scandium(III) triflate-catalyzed allylation of carbonyl compounds with tetraallylgermane proceeded readily in aqueous nitromethane to afford homoallyl alcohols in excellent to good yields.[175] The presence of H_2O is indispensable for the allylation of aldehydes to proceed smoothly. Aldehydes were allylated exclusively in the presence of ketone moieties (Eq. 8.74).

$$\underset{Ph \quad H}{\overset{O}{\parallel}} + \underset{Ph}{\overset{O}{\parallel}} + \left(\diagup\!\!\!\diagup\right)_4 Ge \xrightarrow[CH_3NO_2/H_2O]{Sc(OTf)_3} \underset{Ph}{\overset{OH}{|}}\diagdown\!\!\!\diagup + \underset{Ph}{\overset{OH}{|}}\diagdown\!\!\!\diagup \quad (8.74)$$

$$96\% \qquad\qquad <1\%$$

Pb. Homoallylic alcohols can be obtained from allylation of alde-hydes and ketones with allyl bromide promoted by metallic lead in aqueous media.[176]

Cd. Cadmium perchlorate was found to catalyze allylation reactions using allyltributyltin in aqueous media very efficiently.[177] These cadmium-catalyzed allylation reactions are accelerated by ligands such as N, N, N', N'', N''-pentamethyldiethylenetriamine or 2,9-dimethylphenanthroline (Eq. 8.75). This accelerated the catalytic system to give allylation products of various aldehydes and ketones in high yields.

$$\underset{Me}{\underset{|}{Me_2N \diagup N \diagdown NMe_2}} \qquad \text{(phenanthroline structure)}$$

$$\text{ligand}$$

$$\underset{R^1 \quad R^2}{\overset{O}{\parallel}} + \diagup\!\!\!\diagup SnBu_3 \xrightarrow[EtOH/H_2O]{\underset{\text{ligand}}{cat.\ Cd(ClO_4)}} \underset{R^1 \quad R^2}{\overset{OH}{|}}\diagdown\!\!\!\diagup \quad (8.75)$$

Mn. Manganese is also effective for mediating aqueous carbonyl ally-lations and pinacol-coupling reactions. Manganese offers a higher reac-tivity and complete chemoselectivity toward allylation of aromatic aldehydes.[178]

Sb. Commercial antimony metal, in aqueous 1 M HCl or DCl solu-tion, reacts with allyl bromide and aldehydes to give the corresponding homoallylic alcohols in good yield. The reaction proceeds through the formation of allylstibine intermediates.[179] The same allylation can be performed in aqueous media with antimony using fluoride salt as the promoter instead of HCl.[180]

Bi.   Wada et al. reported[180] that metallic bismuth can also be used for allylation in an aqueous medium in a way similar to that of tin, in which aluminum powder and hydrobromic acid were used as the promoter. Again, the reaction is more effective than the same one conducted in an organic solvent. As a comparison, the allylation of phenylacetaldehyde carried out in a mixture of THF/water at room temperature gave the corresponding alcohol in 90% yield. Under the same conditions, the use of THF as solvent led to decreased yields and irreproducible results. Other metal promoters are also effective under the same conditions. Such combinations include Al(0)/BiCl$_3$, Zn(0)/BiCl$_3$, Fe(0)/BiCl$_3$, and Mg/BiCl$_3$.[181] Bismuth-mediated allylation was found to be promoted by the presence of fluoride ions[182] or sonication.[183] Katritzky et al. found that the bismuth(III)-aluminum system also mediated the allylation of imminium cations to give amines.[184] In this case, even methylation with iodomethane took place smoothly. Allylation of aldehydes carried out by electrochemically regenerated bismuth metal in an aqueous two-phase system was reported by Tsuji et al.[185]

Hg.   Allylmercury bromide and diallylmercury are both stable in water. They can allylate aldehydes in aqueous media but, in the case of allylmercury bromide, activation with tetrahexylammonium bromide was necessary.[186] The allylation reaction is chemoselective towards aldehyde; on the contrary, ketone compounds are unaffected.

Fe.   Iron metal was found to be able to mediate the allylation reactions of aryl aldehydes with allyl bromide using sodium fluoride as the promoter. The formation of an allyliron species was proposed as the reactive intermediate in the reaction (Scheme 8.20).[187] In view of

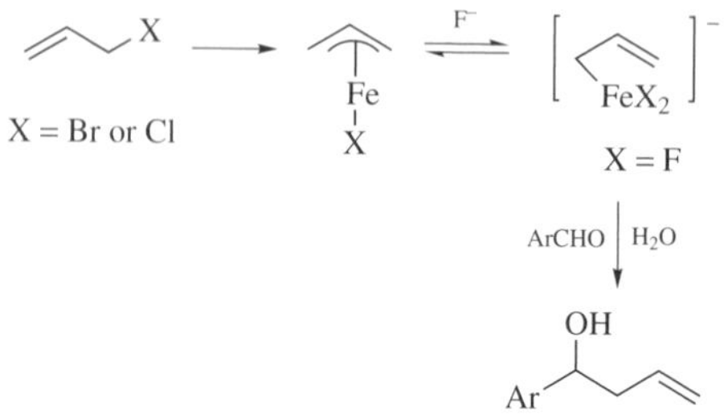

Scheme 8.20

its low cost and lack of toxicity, the use of iron may have potential practical applications.

8.3.1.7 Asymmetric Allylation.

One of the recent new developments on this subject is the asymmetric allylation reaction. It was found that native and trimethylated cyclodextrins (CDs) promote enantioselective allylation of 2-cyclohexenone and aldehydes using Zn dust and alkyl halides in 5:1 H_2O-THF. Moderately optically active products with ee up to 50% were obtained.[188] The results can be rationalized in terms of the formation of inclusion complexes between the substrates and the CDs and of their interaction with the surface of the metal.

An (S)-Tol-BINAP/AgNO$_3$ catalyst was successfully applied to a catalytic enantioselective allylation reaction of aldehydes in an aqueous system. The reactions with aromatic aldehydes afforded the desired products in high yields with good stereoselectivities and up to 81% ee (Eq. 8.76).[189] In the presence of the monothiobinaphthol (MTB) ligand, aryl ketones are allylated by a mixture of $Sn(CH_2CH=CH_2)_4$/$RSn(CH_2CH=CH_2)_3$ (R = Et, Bu) in high ee. The presence of water suppresses racemic background allylation. Allylation using the pure components alone was rather ineffective. The (1R)-2-mercapto[1,1′-binaphthalen]-2-ol-mediated allylation of acetophenone using a mixture of tetra(2-propenyl)stannane/ethyltri(2-propenyl)stannane/butyltri(2-propenyl)stannane gave (αR)-α-methyl-α-(2-propenyl)benzenemethanol in >98% yield and in 86–89% ee (Eq. 8.77). Aliphatic ketones gave complicated mixtures of products. A similar reaction using cyclohexyl methyl ketone gave α-methyl-α-(2-propenyl)cyclohexanemethanol with only 59% enantiomeric excess. tert-butyl methyl ketone did not react.[190] Catalytic asymmetric allylation of aldehydes with allyltributyltin in aqueous media has been realized using combinations of cadmium bromide and chiral diamine ligands. These ligands were found to accelerate the reactions significantly.[191]

$$(8.76)$$

yields up to 100%
ee up to 81%

$$Ph \overset{O}{\underset{}{\bigg\|}} + \left(\diagdown\diagdown Sn \right)_{\!4} \bigg/ \left(\diagdown\diagdown Sn - R \right)_{\!3} \xrightarrow[]{\underset{H_2O}{\text{cat.}}} R^1 \overset{OH}{\underset{*}{\bigg|}} \diagdown\diagdown (8.77)$$

yields up to 98%
ee up to 89%

A chemoenzymatic methodology has been developed using indium-mediated allylation (and propargylation) of heterocyclic aldehydes under aqueous conditions followed by *Pseudomonas cepacia* lipase-catalyzed enantioselective acylation of racemic homoallylic and homo-propargylic alcohols in organic media.[192]

8.3.2 Propargylation

The reaction of propargyl bromide with aldehydes mediated by tin in water generated a mixture of propargylation and allenylation products. The selectivity in the product formation is rather low.[193] Allenylations and propargylations of carbonyl compounds in aqueous media also could be carried out with the preformed organic tin reagents, instead of using metals.[194,195,196,197] The combination of $SnCl_2$ and KI was found to be more effective for the reaction (Eq. 8.78)[198] The zinc-mediated propargylation of 3-formylcephalosporins was also studied in aqueous media.[199] Zn-mediated propargylation (and allylation) of a chiral aldehyde in the presence of water proceeded efficiently with very high diastereoselectivity (>99%) to give homopropargylic alcohols (Eq. 8.79).[200] Chan et al. studied the behavior of aldehydes with propargyl bromides in aqueous medium mediated by indium.[201] They found that simple prop-2-yn-1-yl bromide reacted with both aliphatic and aromatic aldehydes in water, affecting mainly the homopropargyl alcohols (Scheme 8.21). In contrast, when propargyl bromide is γ-substituted the coupling products were predominantly or exclusively the allenylic alcohols. Such couplings also proceed with α-chloropropargyl phenyl sulfide.[202]

$$RCHO + R^1 \!\!\!\equiv\!\!\!\diagdown_{\!X} \xrightarrow[\text{water, } 35^\circ C, \, 15h]{SnCl_2(1.5 \text{ equiv}), \, KI \, (3 \text{ equiv})} R \overset{OH}{\underset{\|}{\bigg|}} R^1 (8.78)$$

X = Cl or Br

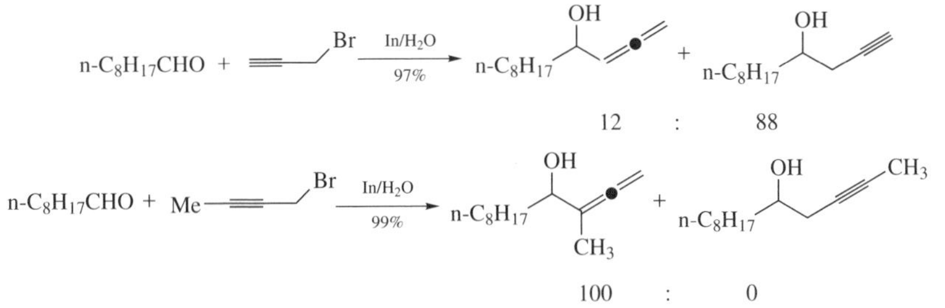

Scheme 8.21

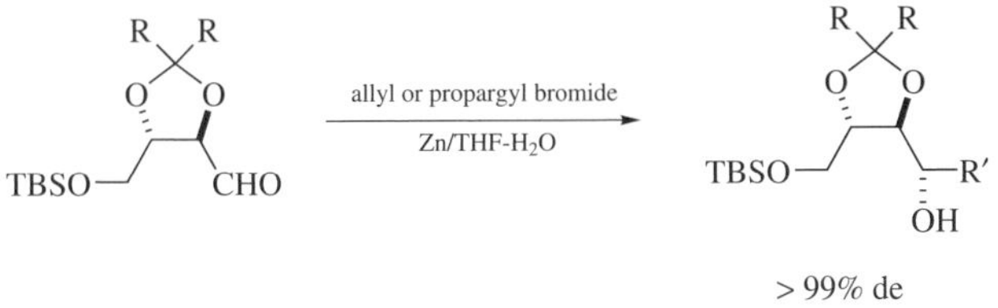

$$(8.79)$$

> 99% de

In synthetic applications, Li et al. examined the propargylation-allenylation of carbonyl compounds by using a variety of metals including Sn, Zn, Bi, Cd, and In.[203] By using the indium-mediated allenylation reaction, Li and co-workers developed the synthesis of the antiviral, antitumor compound (+)-goniofufurone (Scheme 8.22),[204] a key component isolated from the Asian trees of the genus *Goniothalamus*,[205] and other styryl lactone derivatives (Eq. 8.80).

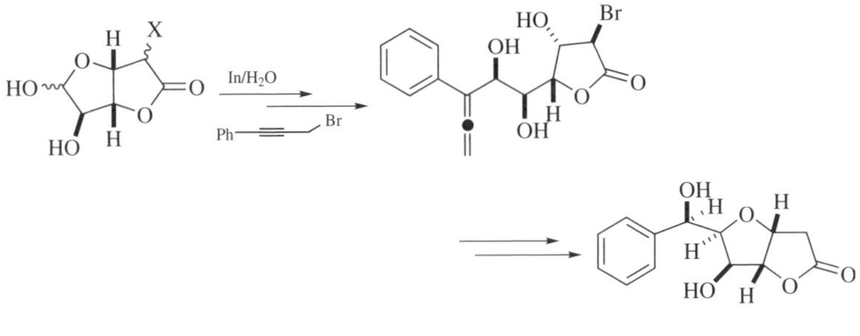

(+)-goniofufurone

Scheme 8.22

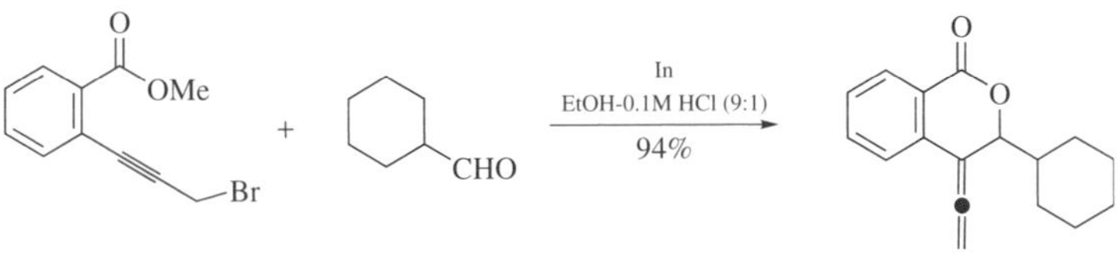

$$(8.80)$$

Propargylation (allylation) of diphenylmethyl 6-oxopenicillanate and 7-oxocephalosporanate was accomplished by reacting the corresponding bromides in the presence of indium or zinc in moderate yields in aqueous conditions.[206]

In-mediated propargylation of acetals and ketals with various allyl or propargyl bromides in aqueous media successfully provided the corresponding homoallylic or homopropargylic (and allenylic) alcohols (Eq. 8.81).[207]

$$R' = H, \text{ alkyl or aryl} \qquad\qquad R'' = \text{allyl, propargyl, allenyl} \tag{8.81}$$

Metal-mediated carbonyl allylation, allenylation, and propargylation of optically pure azetidine-2,3-diones were investigated in aqueous environments.[208] Different metal promoters showed varied regioselectivities on the product formation during allenylation/propargylation reactions of the keto-β-lactams. The stereochemistry of the new C3-substituted C3-hydroxy quaternary center was controlled by placing a chiral substituent at C4. The process led to a convenient entry to densely functionalized hydroxy-β-lactams (Eq. 8.82).

$$(8.82)$$

The transient organoindium intermediates formed in the reaction of propargyl bromide (**8.5a**) and indium in aqueous media have been investigated.[209] It was concluded that in D_2O, the reactive organoindium intermediate was allenylindium(I), whereas in THF-d_8, it was a mixture of allenylindium(I) and allenylindium(III) dibromide (Scheme 8.23). Theoretical calculations by either the HF or the B3LYP methods supported the experimental observation that between the propargylindium structures (**8.6a** and **8.7a**) and the allenylindium structures (**8.8a** and **8.9a**) the allenyl structures are the more stable irrespective of the oxidation sate of indium. The predominance of the allenylindium(I) (**8.8a**) structure in aqueous media is also compatible with the regioselection observed in its reaction with carbonyl compounds to give the homopropargyl alcohols **8.11a**.[201] For the γ-methylpropargyl/α-methylallenylindiums (**8.6–8.9b**) derived from 1-bromo-2-butyne (**8.5b**) and indium, the NMR studies indicated that the equilibria are in favor of the propargylic intermediates **8.6b** and **8.7b**. On the other hand, in the case of α-methylpropargyl/γ-methylallenylindiums (**8.8–8.9c**) derived from 3-bromo-1-butyne (**8.5c**) and indium, the equilibria are in favor of the allenyl structures **8.8c** and **8.9c**. These results are consistent with the previous conclusions regarding the regioselectivity in indium-mediated allenylation and propargylation reactions of carbonyl compounds. In all cases, the products **8.10** and **8.11** have the regioselection expected from the S_E2' pathway.[201,210]

Indium can effectively mediate the coupling of 1,4-dibromo-2-butyne with aldehydes in a 1:1 ratio to give 1,3-butadien-2-yl-methanols in aqueous media (Eq. 8.83).[211] When a 1:2 ratio was used, the reaction

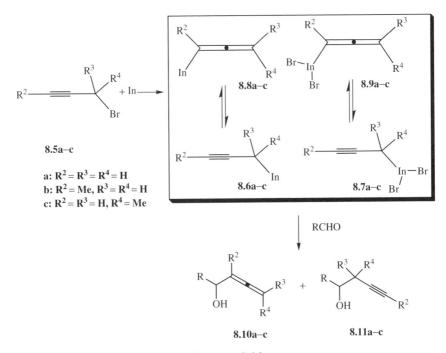

Scheme 8.23

gave a mixture of products (Scheme 8.24). Remarkably, compound **8.12** was formed as a single diastereomer with *anti*-stereochemistry. By using THF as the solvent, compound **8.12** can be obtained as the major product. The remarkable 1,6-diastereoselectivity was attributed to the intramolecular chelation of the allenylindium intermediate with the half alkoxide adduct.[212]

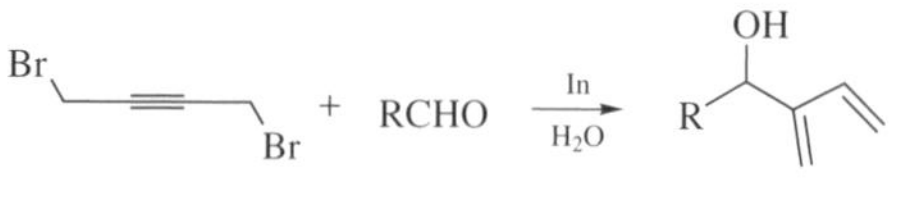

$$\text{(8.83)}$$

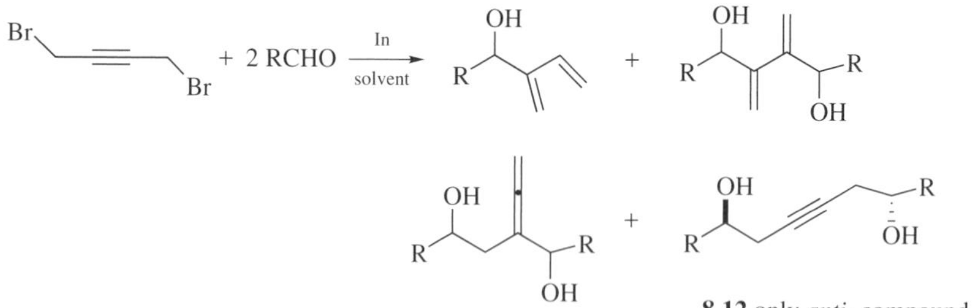

Scheme 8.24

8.12 only *anti-* compound

8.3.3 Cyclopentadienylation

Cyclopentadienylindium(I) reacts with aldehydes in aqueous THF to give the corresponding adducts as a mixture of isomers due to the facile 1,5-sigmatropic rearrangement. The alkylated cyclopentadienes react with dimethyl acetylenedicarboxylate in aqueous medium with ytterbium triflate as catalyst to give the Diels-Alder adducts **8.13** and **8.14** (Scheme 8.25).[213] Using aldehyde **8.15** as the substrate, the same sequence of cyclopentadienylation followed by intramolecular Diels-Alder reaction can be carried out in one pot in aqueous media to give the complex tricyclic adducts **8.16** in good overall yield (Scheme 8.26).

8.3.4 Benzylation

Zinc-mediated benzylation of carbonyl compounds in aqueous media was reported by Bieber.[214] The benzylation of 4-nitrobenzaldehyde

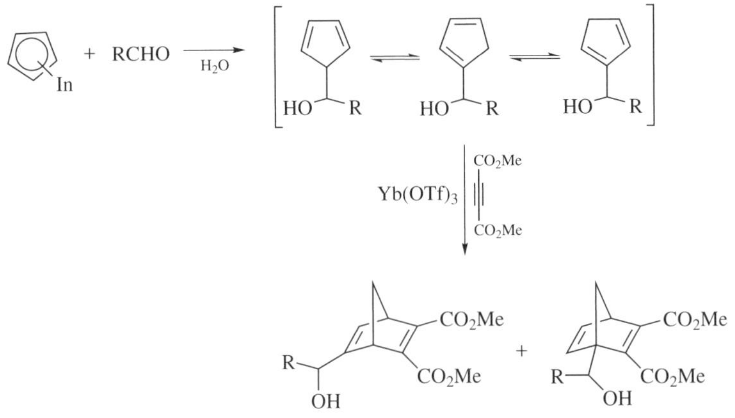

8.13   **8.14**

Scheme 8.25

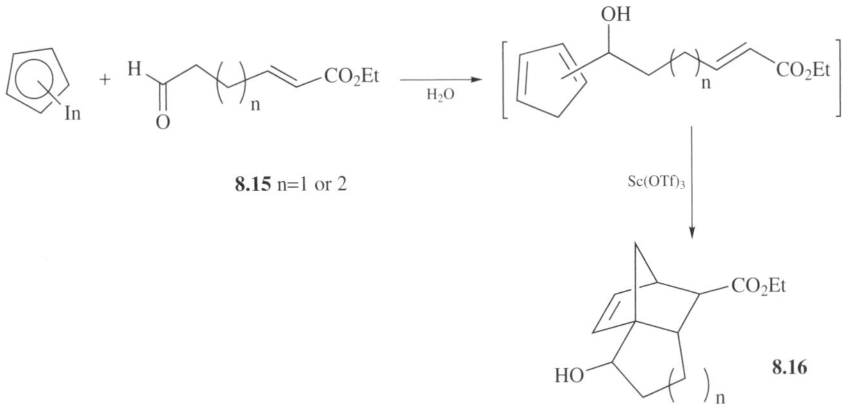

8.15 n=1 or 2

8.16

Scheme 8.26

could be controlled chemoselectively by using different phase-transfer catalysts and different metal reductants in water (Eq. 8.84).[215]

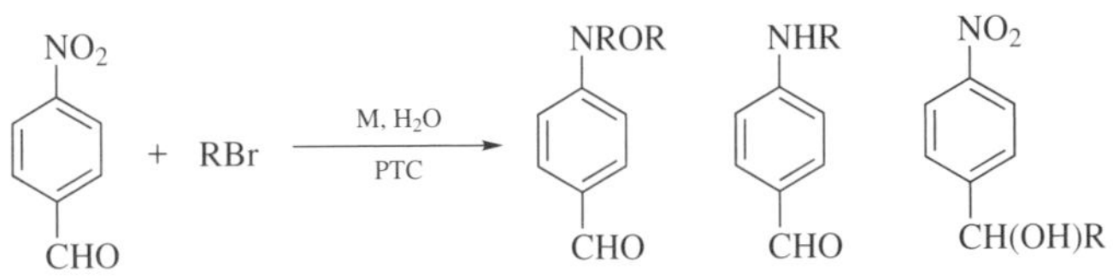

R = CH$_2$Ph or CH$_2$CH=CH$_2$

(8.84)

8.3.5 Arylation/Vinylation

In 1998, Miyaura reported a Rh(acac)(CO)$_2$/dppp-catalyzed addition of aryl or alkenylboronic acids to aldehydes in aqueous organic mixtures under an inert atmosphere (Eq. 8.85).[216] The use of electron-rich tri(*tert*-butyl)phosphine as ligand was found to be beneficial for obtaining good yields of the corresponding aldehyde addition products.[217]

$$\text{Ar-B(OH)}_2 \;+\; \text{RCHO} \xrightarrow[\text{50°C, organic/H}_2\text{O = (6:1)}]{\text{cat. Rh(acac)(CO)}_2\text{/dppb}} \quad \underset{\text{Ar} \qquad \text{R}}{\overset{\text{OH}}{\bigwedge}} \qquad (8.85)$$

On the other hand, after being unable to carry out a direct Grignard-type reaction of aryl halide with aldehyde, Li and co-workers attributed the failure to the addition of arylmetal onto the carbonyl. Subsequently, they studied the addition of various aryl- and vinylmetal reagents to aldehydes in air and water by using a variety of catalysts. It was found that aryl- and vinyltin compounds added to aldehydes smoothly when catalyzed by either Rh$_2$(COD)$_2$Cl$_2$ or Rh(COD)$_2$BF$_4$ (Eq. 8.86).[218] The carbonyl addition was found to be highly sensitive to both the metal and the groups attached to the metal. Except for organoarsenic and organoantimony reagents, aryl or vinyl derivatives of all other metals (and metalloids) examined were able to generate the desired carbonyl addition and conjugate addition products with variable efficiency. Among them, aryl and vinyltin, silicon, boron, lead, and bismuth derivatives were found to be the most effective. The corresponding indium and germanium reagents provided only low yields of the products. Taking the organotin reagents as an example, in the presence of a catalytic amount of Rh(COD)$_2$BF$_4$ at refluxing temperature in air and water, benzaldehyde underwent nucleophilic addition with trimethylphenyltin and dibutyldiphenyltin to give the corresponding nucleophilic addition product smoothly. On the other hand, under the same reaction conditions, no reaction was observed between benzaldehyde and phenyltin trichloride even after several days. When the reaction was carried out in the presence of potassium hydroxide, a smooth reaction again occurred to give the desired product overnight. A more dramatic effect was observed among triphenyltin chloride, triphenyltin hydroxide, and butyltriphenyltin. No reaction was observed with the chloride derivatives, but the reaction with either hydroxide or butyl derivatives proceeded smoothly. The use of different bases also affects the reaction progress. Various bases such as lithium hydroxide, sodium hydroxide, and potassium hydroxide were tested (showing the same trend as the relative basicity), and potassium hydroxide appeared to be the most effective for this reaction. A similar electronic effect was also observed with organobismuth, organolead, organoindium, and organoboron compounds. Li and co-workers explained the phenomena by *d-p* π bonding between these metals and the substituent and developed analogous chemistry using aromatic compounds (Figure 8.2).[219] Toward this end, aryltriethoxysilanes reacted with aldehydes in high yield in the presence of a Rh(I) catalyst and aqueous NaOH.[220] On the other hand, treatment of α, β-acetylenic ketones with chromium(II) in the presence of aldehydes, Me_3SiCl, and water in THF gives 2,5-disubstituted furans in good to excellent yields.[221] Under air and water conditions, as developed by Li and co-workers, carbonyl hydrates can also react. A highly diastereoselective rhodium-catalyzed addition of arylbismuth and aryllead reagents to chiral glyoxylate hydrate in air and water was reported (Eq. 8.87).[222]

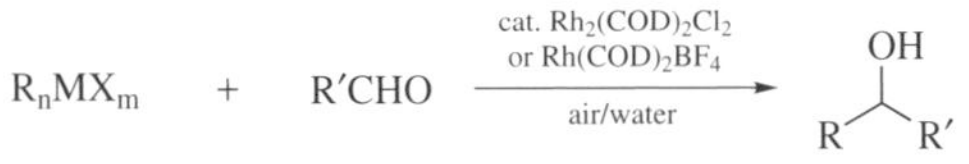
R$_n$MX$_m$
+
R′CHO
cat. Rh$_2$(COD)$_2$Cl$_2$
or Rh(COD)$_2$BF$_4$
air/water
OH
R
R′

(8.86)

R = aryl or vinyl
M = B, In, Si, Sn, Ge, Pb, As, Sb, Bi
X = alkyl, aryl, halogen, hydroxyl, and alkoxyl

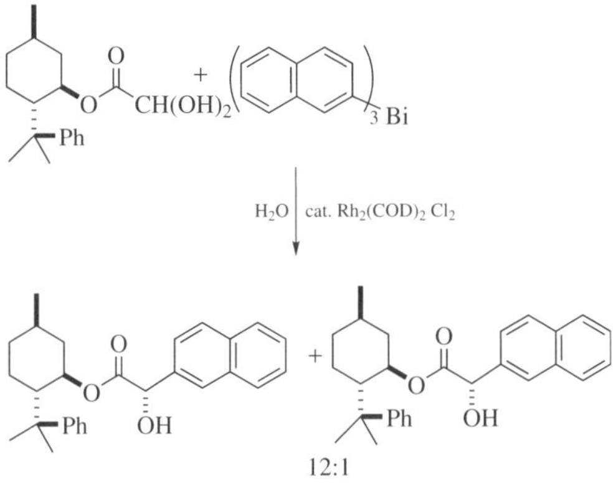
O
O
CH(OH)$_2$
+
3 Bi
Ph
H$_2$O
cat. Rh$_2$(COD)$_2$ Cl$_2$
O
O
Ph
OH
+
O
O
Ph
OH
12:1

(8.87)

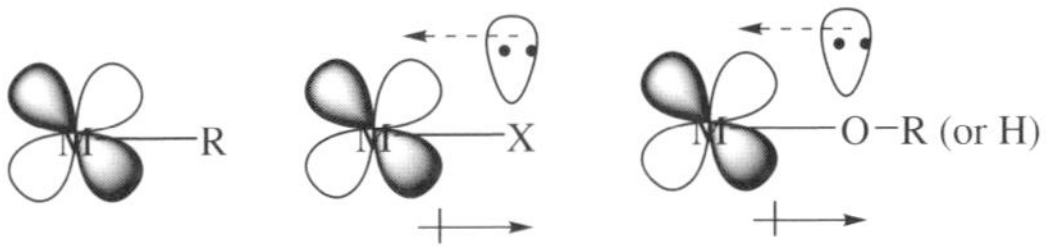
M
R
M
X
M
O−R (or H)

Figure 8.2 Postulated electronic effect on tin and other metals.

8.3.6 Alkynylation

An indium chloride–catalyzed coupling of alkynes to aldehydes was also possible in giving α, β-unsaturated carbonyl compounds in water in low yields (Eq. 8.88).[223] (Alkynylation is discussed in Chapter 4 on alkynes.)

$$\text{(8.88)}$$

8.3.7 Alkylation

The direct addition of simple alkyl groups onto aldehydes is the most challenging in water. Mitzel reported the indium-mediated alkylation of carbonyl compounds with α-sulfur stabilized systems.[224] Recently, Li and co-workers reported an efficient addition of simple alkyl halides to aldehydes mediated by Zn/CuI with a catalytic amount of InCl in water (Eq. 8.89).[225]

$$\text{(8.89)}$$

8.3.8 Reformatsky-Type Reaction

The reaction of an α-halo carbonyl compound with zinc, tin, or indium together with an aldehyde in water gave a direct cross-aldol reaction product (Eq. 8.90).[226,227] A direct Reformatsky-type reaction occurred when an aromatic aldehyde reacted with an α-bromo ester in water mediated by zinc in low yields. Recently, it was found that such a reaction mediated by indium was successful and was promoted by sonication (Eq. 8.91).[228] The combination of $BiCl_3$-Al,[229] $CdCl_2$-Sm,[230] and Zn-Et_3B-Et_2O[231] is also an effective mediator. Bismuth metal, upon activation by zinc fluoride, effected the crossed aldol reaction between α-bromo carbonyl compounds and aldehydes in aqueous media. The reaction was found to be regiospecific and *syn*-diastereoselective (Eq. 8.92).[232]

$$\text{(8.90)}$$

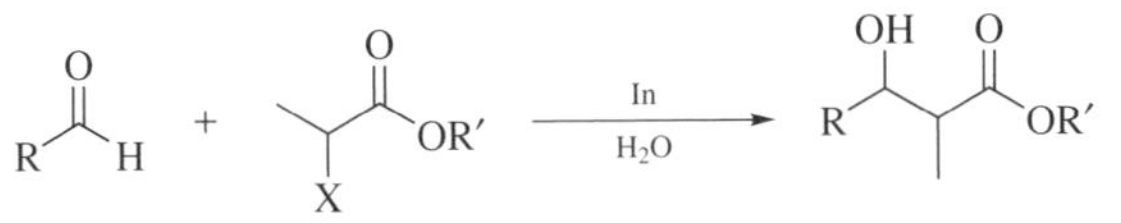

$$\text{(8.91)}$$

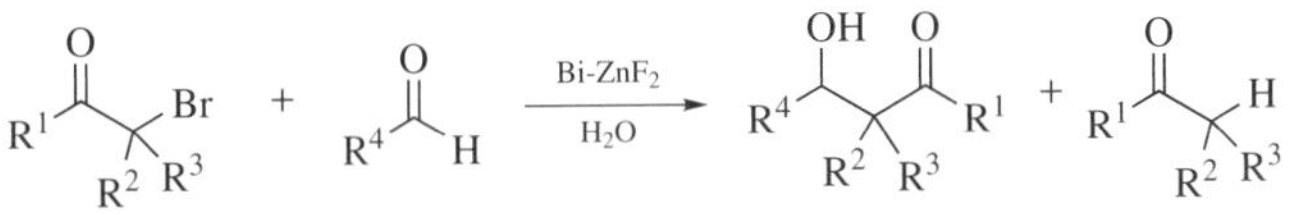

$$\text{(8.92)}$$

Bieber reported that the reaction of bromoacetates is greatly enhanced by catalytic amounts of benzoyl peroxide or peracids and gives satisfactory yields with aromatic aldehydes. A radical chain mechanism, initiated by electron abstraction from the organometallic Reformatsky reagent, is proposed (Scheme 8.27).[233] However, an alternative process of reacting aldehydes with 2,3-dichloro-1-propene and indium in water followed by ozonolysis provided the Reformatsky product in practical yields.[234] An electrochemical Reformatsky reaction in an aqueous medium and in the absence of metal mediator has also been reported.[235]

The indium-mediated aqueous Reformatsky reaction was used in the synthesis of α,α-difluoro-β-hydroxy ketones (Eq. 8.93).[236]

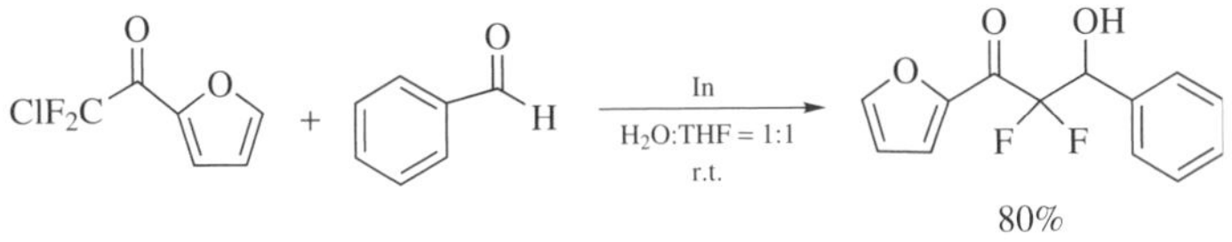

$$\text{(8.93)}$$

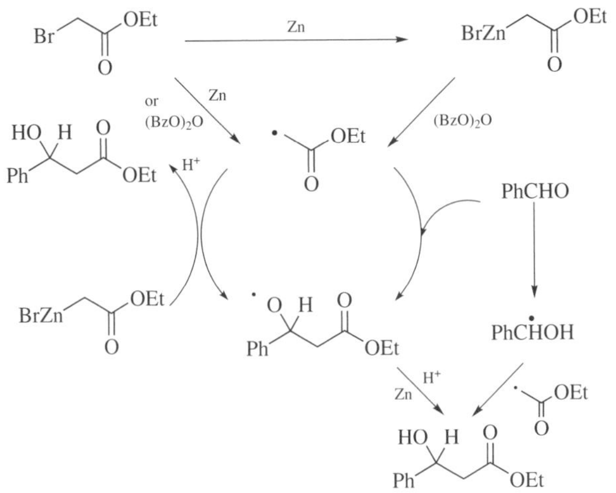

80%

Scheme 8.27

8.3.9 Direct Aldol Reaction

8.3.9.1 Classical Aldol.

Aldol reaction is an important reaction for creating carbon–carbon bonds. The condensation reactions of active methylene compounds such as acetophenone or cyclohexanone with aryl aldehydes under basic or acidic conditions gave good yields of aldols along with the dehydration compounds in water.[237] The presence of surfactants led mainly to the dehydration reactions. The most common solvents for aldol reactions are ethanol, aqueous ethanol, and water.[238] The two-phase system, aqueous sodium hydroxide–ether, has been found to be excellent for the condensation reactions of reactive aliphatic aldehydes.[239]

The Henry (nitroaldol) reaction was reported under very mild reaction conditions, in aqueous media using a stoichiometric amount of a nitroalkane and an aldehyde, in NaOH 0.025 M and in the presence of cetyltrimethylammonium chloride (CTACl) as cationic surfactant (Eq. 8.94).[240] Good to excellent yields of β-nitroalkanol are obtained. Under these conditions several functionalities are preserved, and side-reactions such as retro-aldol reaction or dehydration of 2-nitroalcohols are avoided.

$$(8.94)$$

Base-catalyzed aldol reactions have been carried out intramolecularly.[241] The aqueous acid-catalyzed intramolecular aldol condensation of 3-oxocyclohexaneacetaldehyde proceeded diastereoselectively (Eq. 8.95).[242]

$$(8.95)$$

85:15

Selective retro-aldol has also been reported by using aqueous HCl in THF.[243] Recently, catalytic aldol reactions in aqueous media have generated great interest due to the atom-economy related to the reaction. Reaction of 2-alkyl-1,3-diketones with the aqueous formaldehyde using aqueous 6–10 M potassium carbonate as base afforded aldol reaction products, which are cleaved by the base to give vinyl ketones (Eq. 8.96).[244]

$$(8.96)$$

78%

8.3.9.2 Lewis-Acid Catalyzed.

Recently, various Lewis acids have been examined as catalyst for the aldol reaction. In the presence of complexes of zinc with aminoesters or aminoalcohols, the dehydration can be avoided and the aldol addition becomes essentially quantitative (Eq. 8.97).[245] A microporous coordination polymer obtained by treating anthracene-*bis*(resorcinol) with $La(OiPr)_3$ possesses catalytic activity for ketone enolization and aldol reactions in pure water at neutral pH.[246] The La network is stable against hydrolysis and maintains microporosity and reversible substrate binding that mimicked an enzyme. Zn complexes of proline, lysine, and arginine were found to be efficient catalysts for the aldol addition of *p*-nitrobenzaldehyde and acetone in an aqueous medium to give quantitative yields and the enantiomeric excesses were up to 56% with 5 mol% of the catalysts at room temperature.[247]

$$(8.97)$$

8.3.9.3 Enzyme Catalyzed.

The enzyme aldolases are the most important catalysts for catalyzing carbon–carbon bond formations in nature.[248] A multienzyme system has also been developed for forming C–C bonds.[249] Recently, an antibody was developed by Schultz and co-workers that can catalyze the retro-aldol reaction and Henry-type reactions.[250] These results demonstrate that antibodies can stabilize the aldol transition state but point to the need for improved strategies for enolate formation under aqueous conditions.

8.3.9.4 Organic-Base Catalyzed.

Asymmetric direct aldol reactions have received considerable attention recently (Eq. 8.98).[251] Direct asymmetric catalytic aldol reactions have been successfully performed using aldehydes and unmodified ketones together with chiral cyclic secondary amines as catalysts.[252] L-proline and 5,5-dimethylthiazolidinium-4-carboxylate (DMTC) were found to be the most powerful amino acid catalysts for the reaction of both acyclic and cyclic ketones as aldol donors with aromatic and aliphatic aldehydes to afford the corresponding aldol products with high regio-, diastereo-, and enantioselectivities. Reactions employing hydroxyacetone as an aldol donor provide *anti*-1,2-diols as the major product with ee values up to >99%. The observed stereochemistry of the products was explained by a metal-free Zimmerman-Traxler-type transition state and involves an enamine intermediate. The reactions tolerate a small amount of water (<4 vol%), do not require inert reaction conditions or preformed enolate equivalents, and can be conveniently performed at room temperature in various solvents.

$$(8.98)$$

up to >99% ee

Quite recently, Janda reported that the aqueous direct aldol reaction can be catalyzed by nornicotine (a metabolite of nicotine). This is the only example of a metabolite capable of being used as a catalyst (Eq. 8.99).[253] Interestingly, catalysis was not observed in most common organic solvents including DMSO, chloroform, benzene, acetonitrile, THF, and DMF. This is in contrast to the (*S*)-proline catalyzed aldol reaction, which is usually carried out in DMSO or DMF. Direct cross-aldol can also be catalyzed by organoamines in buffered aqueous media (0.01 M phosphate, 2.7 mM KCl, 137 mM NaCl, pH = 7.4). Various aldehydes and ketones including carbohydrate derivatives can be chosen as the substrates. It is predicted that the synthetic method will get further attention as a prebiotic route to sugar.[254] Proline-catalyzed aldol reaction of nitrobenzaldehydes with various ketones was investigated in aqueous anionic micelles. Satisfactory reaction yields were obtained with SDS (sodium dodecylsulfonate), SDBS (sodium dodecylbenzenesulfonate), and SLS (sodium laurylsulfate), all anionic surfactants. Other surfactants such as Triton-100, CTAB (cetyltrimethylammonium bromide), and OTAC (octadecyltrimethylammonium chloride), which are either neutral or cationic surfactants, did not promote the reaction in aqueous media even with proline as catalyst.[255]

$$(8.99)$$

8.3.9.5 No Catalyst.
High-intensity ultrasound was employed to reinvestigate the aldol reaction in water without any catalyst.[256] Within

15–30 min, acetophenone reacted with non-enolizable aldehydes to afford the aldol exclusively, while under conventional heating conditions, the same compounds either failed to react or gave the enone often in complicated product mixture. Benzaldehyde reacted with a series of 1,3-dicarbonyl compounds to afford the corresponding *bis*(benzylidene) adducts.

8.3.9.6 Other Aldols.   Alternatively, Li and co-workers reported a $RuCl_2(PPh_3)_3$-catalyzed migration of functional groups of homoallylic alcohols in water (Eq. 8.100)[257] leading to an aldol-type reaction by reacting allyl alcohols with aldehyde (Scheme 8.28).[258] The presence of $In(OAc)_3$ promoted the aldol reaction of α-vinylbenzyl alcohol with aldehyde.[259] An indium hydride–promoted reductive aldol reaction of unsaturated ketones in aqueous media was developed.[260] The use of water and methanol as solvent dramatically reversed stereochemistry from *anti* to *syn* (Eq. 8.101). Boron-enolates have been used for aldol reactions in water using catalytic amounts of boron reagents.[261]

$$\text{(8.100)}$$

$$\text{(8.101)}$$

Solvent: THF   yield 86%   *syn:anti* =   4:96
MeOH   69%   *syn:anti* = >99:1
THF/H_2O 79%   *syn:anti* = 94:6

8.3.9.7 Synthetic Applications.   By the condensation of an arylaldehyde in alkaline aqueous medium with an arylmethylketone followed by oxidation with hydrogen peroxide, 7- and 3′,4′-substituted flavonols were synthesized under one-pot conditions.[262] An example is

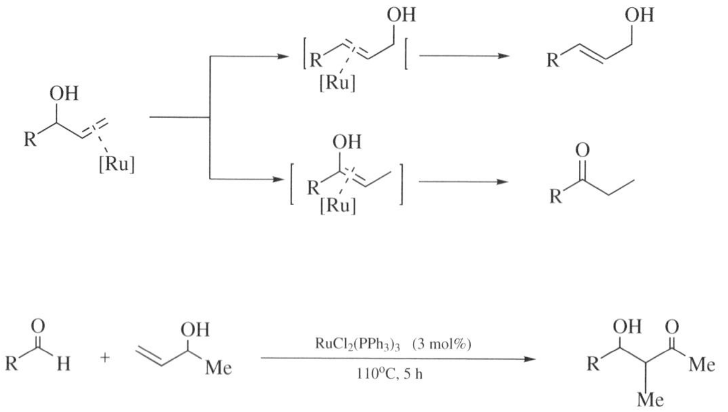

Scheme 8.28

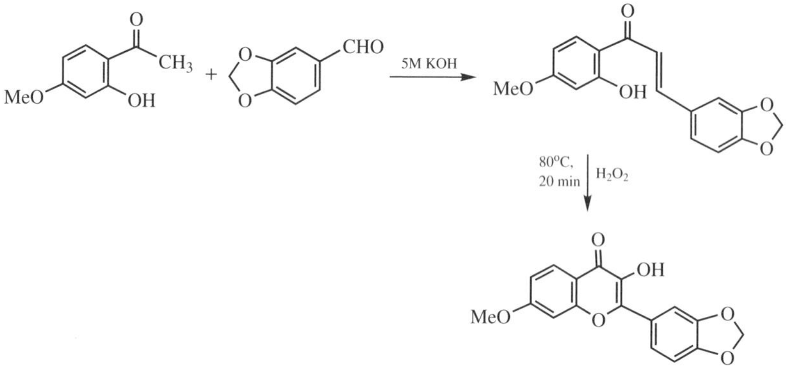

Scheme 8.29

illustrated in Scheme 8.29. The condensation of the hydroxyl-methoxy-substituted acetophenone with piperonal was carried out in a 5 M KOH alkaline aqueous solution. The chalcone was then oxidized with a 35% solution of H_2O_2 in the same pot to give the flavonol in 70% overall yield. Other examples of synthetic applications of aqueous aldol reactions include carminic acid.[263]

8.3.10 Mukaiyama Aldol Reaction

The crossed aldol reaction of silyl enol ethers with carbonyl compounds (Mukaiyama-aldol) was studied by Lubineau and co-workers in aqueous solvents without any acid catalyst. However, the reaction took several days to complete.[264] A major development is the use of water-tolerant Lewis acids for such reactions, pioneered by Kobayashi and co-workers.[265] Adding a catalytic amount of lanthanide triflate (a strong Lewis acid) greatly improved the rate and the yield of such reactions (Eq. 8.102).[266] Among the lanthanide triflates, ytterbium triflate ($Yb(OTf)_3$), gadolinium triflate ($Gd(OTf)_3$), and lutetium triflate ($Lu(OTf)_3$) generally gave better yields of the aldol condensation product; the diastereoselectivities of these reactions were moderate. Water-soluble aldehydes were applicable and the catalyst could be recovered and reused in this procedure. Since then, various Lewis acids have been used for such reactions. The catalytic activities of Lewis acids in water were related to hydrolysis constants and exchange rate constants for substitution of inner-sphere water ligands.[267]

$$\text{PhCHO} + \underset{\text{OSiMe}_3}{\overset{}{\bigcirc}} \xrightarrow[\text{H}_2\text{O/THF(1/4), r.t. 19 h}]{\text{Ln(III) salt, 10 mol\%}} \underset{\text{O OH}}{\overset{}{\bigcirc\!\!-\!\!\text{Ph}}} \qquad (8.102)$$

8–91% yield, $Yb(OTf)_3$: 91%

8.3.10.1 Lewis-Acid Catalyzed.
Other metals used with some success as catalysts in aqueous Mukaiyama aldol reactions are $Bi(OTf)_3$,[268] $Cu(OTf)_2$,[269] and $InCl_3$.[270] The formation of *syn*-aldol products is predominant in the water-based aldol reactions, which is in contrast with the analogous reactions run under anhydrous conditions where the *anti*-isomer is usually the major product. Boron has been shown to be an efficient mediator of stereoselective aldol reactions (Eq. 8.103).[271] The reaction between aldehydes and silyl enol ethers in a water/SDS mixture was catalyzed by 10 mol% Ph_2BOH to give *syn*-substituted β-hydroxy ketones in high diastereomeric excesses (80–94% de). In organic solvents (Et_2O, CH_2Cl_2) the reaction was almost completely thwarted. The accelerating effect of water was proposed via a boron enolate intermediate generated by silicon-metal exchange. $CeCl_3$ and $InCl_3$ were found to work best in i-$PrOH/H_2O$ (95:5) and no reaction was observed in pure water, which was attributed to the destruction of the starting silyl enol ether.[272]

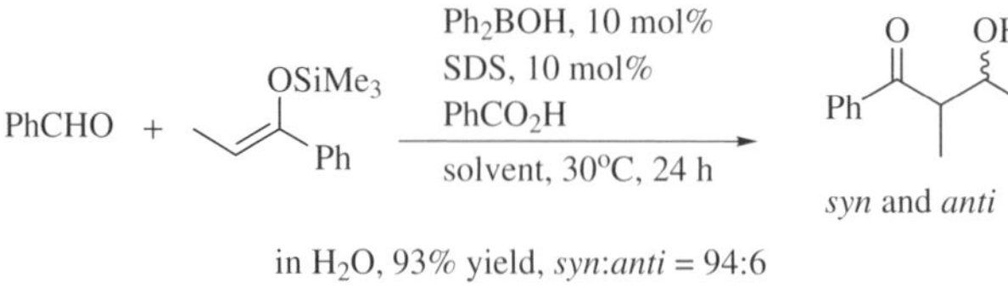
Ph2BOH, 10 mol%
SDS, 10 mol%
PhCO2H
solvent, 30°C, 24 h
PhCHO +
OSiMe3
Ph
O
OH
Ph
Ph
syn and anti
in H2O, 93% yield, syn:anti = 94:6
(8.103)

Scandium triflate-catalyzed aldol reactions of silyl enol ethers with aldehyde were successfully carried out in micellar systems[273] and encapsulating systems.[274] While the reactions proceeded sluggishly in water alone, strong enhancement of the reactivity was observed in the presence of a small amount of a surfactant. The effect of surfactant was attributed to the stabilization of enol silyl ether by it. Versatile carbon–carbon bond-forming reactions proceeded in water without using any organic solvents. Cross-linked Sc-containing dendrimers were also found to be effective and the catalyst can be readily recycled without any appreciable loss of catalytic activity.[275] Aldol reaction of 1-phenyl-1-(trimethylsilyloxy) ethylene and benzaldehyde was also conducted in a gel medium of fluoroalkyl end-capped 2-acrylamido-2-methylpropanesulfonic acid polymer.[276] A nanostructured, polymer-supported Sc(III) catalyst (NP-Sc) functions in water at ambient temperature and can be efficiently recycled. It also affords stereoselectivities different from isotropic solution and solid-state scandium catalysts in Mukaiyama aldol and Mannich-type reactions.[277]

LiOAc-catalyzed aldol reactions between trimethylsilyl enolates and aldehydes in a DMF-H_2O (50:1) solvent proceeded smoothly to afford the corresponding aldols in good to high yields.[278] In the presence of $CaCl_2$ in DMF, dimethylsilyl enolates reacted smoothly with aldehydes to give aldol adducts in good-to-high yields. This catalytic system was applicable to the aldol reaction with aqueous aldehydes such as formalin. The reaction is regiospecific as illustrated in Scheme 8.30,

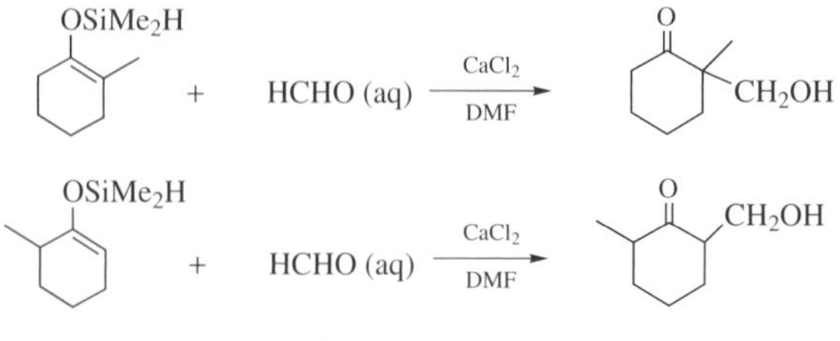
OSiMe2H
+
HCHO (aq)
CaCl2
DMF
O
CH2OH
OSiMe2H
+
HCHO (aq)
CaCl2
DMF
O
CH2OH

Scheme 8.30

suggesting that the Lewis base, $CaCl_2$, does not isomerize the enolate systems.[279]

Montmorillonite K10 was also used for aldol the reaction in water.[280] Hydrates of aldehydes such as glyoxylic acid can be used directly. Thermal treatment of K10 increased the catalytic activity. The catalytic activity is attributed to the structural features of K10 and its inherent Brönsted acidity. The aldol reactions of more reactive ketene silyl acetals with reactive aldehydes proceed smoothly in water to afford the corresponding aldol products in good yields (Eq. 8.104).[281]

$$RCHO \quad + \quad \underset{OMe}{\overset{OSiMe_3}{\diagup}} \quad \xrightarrow{\text{cat. K10, H}_2\text{O, r.t. 24h}} \quad \underset{MeO}{\overset{O \quad OH}{\diagdown}} R \qquad (8.104)$$

Polar polyoxyethylene-polyoxypropylene (POEPOP) resin, derivatized with a 4-hydroxymethyl phenoxy linker, was used as a solid support for lanthanide triflate–catalyzed Mukaiyama-type solid-phase aldol reactions.[282] The use of an aqueous solvent was found to be crucial. The reactions on an N-terminal peptide aldehyde substrate proceeded in very high yields.

8.3.10.2 Asymmetric Lewis-Acid Catalyzed.

Another important advance in aqueous Mukaiyama aldol reaction is the recent success of asymmetric catalysis.[283] In aqueous ethanol, Kobayashi and co-workers achieved asymmetric inductions by using $Cu(OTf)_2$/chiral *bis*(oxazoline) ligand,[284] $Pb(OTf)_2$/chiral crown ether,[285] and $Ln(OTf)_3$/chiral *bis*-pyridino-18-crown-6 (Eq. 8.105).[286]

$$PhCHO + \underset{Ph}{\overset{OSiMe_3}{\diagup}} \quad \xrightarrow[\substack{H_2O/EtOH = 1/9 \\ 0°C, 24 h}]{MX_n / L^*} \quad \underset{Ph}{\overset{OH \quad O}{\diagdown}} Ph \qquad (8.105)$$

syn/anti = 90/10, 55% ee(*syn*)   *syn/anti* = 93/7, 82% ee(*syn*)

On the other hand, Li and Wang recently developed a highly efficient asymmetric Mukaiyama reaction by using chiral gallium catalysts with Trost's chiral semicrown ligands (Eq. 8.106).[287] Such a system can achieve high enantioselectivity even in pure water. The combination of Ga(OTf)$_3$ and the semicrown ligand is able to make the aqueous reaction proceed smoothly with good yield (89%), diastereoselectivity (*syn/anti* 89:11), and enantioselectivity of *syn*-3a (ee 87%).

$$(8.106)$$

A chiral zirconium catalyst generated from Zr(O$_t$-Bu)$_4$ and (*R*)-3,3′-diiodo-1,1′-binaphthalene-2,2′-diol [(*R*)-3,3′-I$_2$BINOL] catalyzed the reaction in high yields under mild conditions in the presence of a small amount of water. *Anti*-adducts were obtained in high diastereo- and enantioselectivities (Eq. 8.107).[288] Sulfonate derivatives of chiral 1,1′-binaphthol were used as chiral anionic surfactants in asymmetric aldol-type reaction in water to give aldol adducts with moderate to good diastereo- and enantioselectivities; Ga(OTf)$_3$ and Cu(OTf)$_2$ provided better results than Sc(OTf)$_3$ as Lewis-acid catalysts in this system (Eq. 8.108).[289] The aldol reaction of trimethylsilyl enol ethers with aqueous CH$_2$O proceeded moderately well using tetrabutylammonium fluoride as an activator. Asymmetric hydroxymethylation of trimethoxysilyl enol ethers using (R)-BINAP-AgOTf as Lewis acid and KF as Lewis base has been achieved in aqueous media. (up to 57% ee) (Eq. 8.109).[290]

$$(8.107)$$

$$(8.108)$$

yield = 63-89%
syn:anti = 51:49–78:22
ee (*syn*) = 8-48%

cat.= Cu(OTf)$_2$, Ga(OTf)$_3$, Nd(OTf)$_3$, Sc(OTf)$_3$, Zn(OTf)$_2$

ligand =

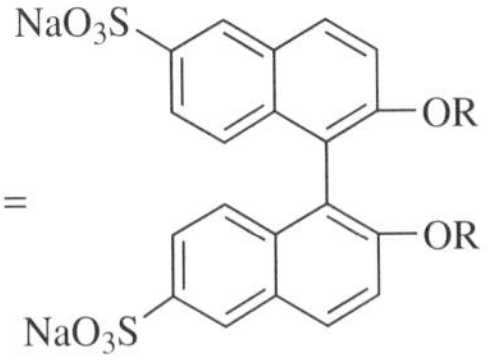

R=C$_8$H$_{17}$, C$_{12}$H$_{25}$

(*R*)-(+) or (*S*)-(−)

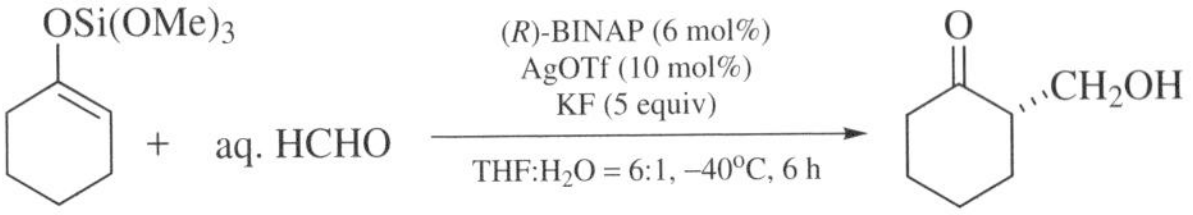

$$\text{(8.109)}$$

31% yield, 57% ee (*S*)

Chiral *bis*(oxazoline) ligands disubstituted at the carbon atom linking the two oxazolines by Frechet-type polyether dendrimers coordinated with copper(II) triflate were found to provide good yields and moderate enantioselectivities for Mukaiyama aldol reactions in water that are comparable with those resulting from the corresponding smaller catalysts.[291] AgPF$_6$-BINAP is very active in this reaction and the addition of a small amount of water enhanced the reactivity.[292]

A tin(II)-catalyzed asymmetric aldol reaction and lanthanide-catalyzed aqueous three-component reaction have been used as the key steps for the synthesis of febrifugine and isofebrifugine (Scheme 8.31).[293]

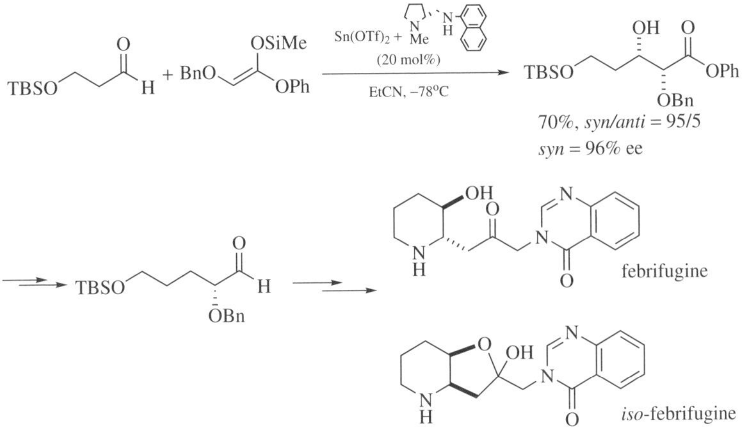

70%, *syn/anti* = 95/5
syn = 96% ee

febrifugine

iso-febrifugine

Scheme 8.31

8.3.11   Hydrogen Cyanide Addition

The addition reactions of HCN to carbon-hetero multiple bonds such as cyanohydrin formation reactions[294] and C=N or C≡N addition reactions[295] have all been performed in aqueous media. In particular, synthesis of higher sugars through cyanohydrin formation, the Kiliani-Fischer synthesis, is a classical reaction and probably is the earliest carbohydrate chemical synthesis without the use of a protecting group (Scheme 8.32).[296] The reaction is governed by the acidity of the media. The addition of cyanide to simple aldoses is essentially quantitative at pH=9.1, whereas the reaction is much slower at a lower pH. The cyanohydrins are not isolated, but are converted to the corresponding lactones. Reduction of the lactones by sodium amalgam, by catalytic hydrogenation, or by reduction with sodium borohydride at pH 3–4 in aqueous solution, generates the higher aldoses. By controlling the counter ion of the cyanide, one can change the proportions of diastereomers, originated from the creation of a new stereogenic center. The synthesis of, essentially, a single diastereomer is also possible.[297] Enzyme-catalyzed asymmetric formation of (S)-cyanohydrins derived from aldehydes and ketones has been reported in a biphasic solvent system (Eq. 8.110).[298]

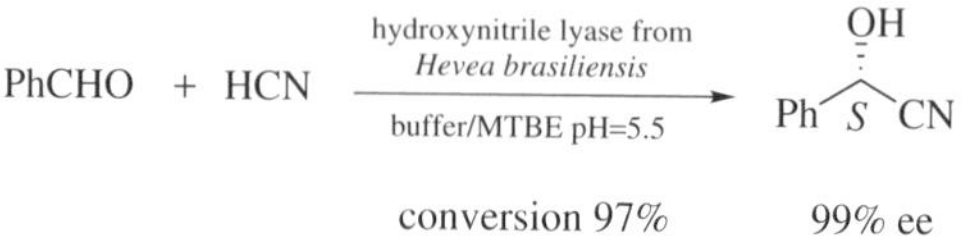

$$
PhCHO \;+\; HCN \xrightarrow[\substack{\textit{Hevea brasiliensis} \\ \text{buffer/MTBE pH=5.5}}]{\text{hydroxynitrile lyase from}} \overset{\displaystyle OH}{\underset{\substack{\text{Ph } S \text{ CN}}}{\bigwedge}}
\qquad (8.110)
$$

conversion 97%   99% ee

8.3.11.1   Benzoin Condensation.
The benzoin condensation is a related reaction consisting of treating an aromatic aldehyde with potassium cyanide or sodium cyanide usually in an aqueous ethanolic solution. Breslow studied the effects of inorganic salts on the rate

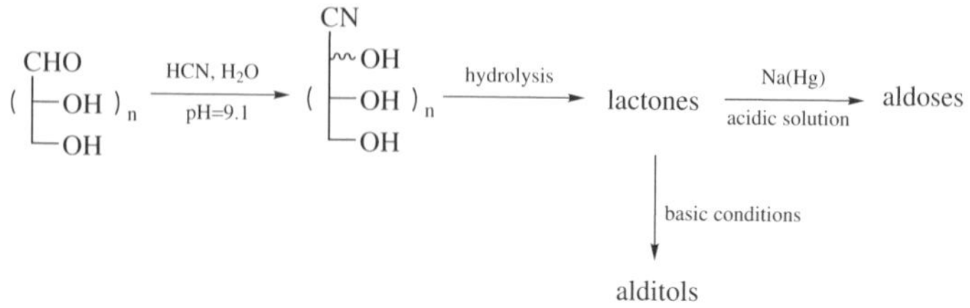

Scheme 8.32   Kiliani-Fischer synthesis of higher sugars.

of the cyanide-catalyzed benzoin condensation in aqueous media (Eq. 8.111).[299] The reaction is 200 times faster in water than in ethanol. Through the use of a quantitative antihydrophobic effect as probes for transition state structures, it was postulated that the acceleration of benzoin condensation in water was related to the amount of hydrophobic surfaces that are solvent accessible in the transition states compared with the initial state.[300] Thus, the addition of salts, which increased the hydrophobic effect, further increased the rate of the reaction. The addition of γ-cyclodextrin (in which both substrates can fit) also accelerates the reaction; whereas the addition of β-cyclodextrin (with a smaller cavity) inhibits the condensation. The cyanide-catalyzed benzoin condensation is comparable to the benzoin condensation catalyzed by thiamine in the biological systems, previously elucidated by Breslow.[301] Recently, Breslow has prepared several γ-cyclodextrin thiazolium salts that mimic the action of thiamine and catalyze the benzoin condensation very effectively.[302]

$$2PhCHO \quad \xrightarrow[\text{H}_2\text{O}]{\text{CN}^-} \quad Ph\overset{\displaystyle O}{\underset{\displaystyle OH}{\diagup}}Ph \qquad (8.111)$$

(R)-Benzoins and (R)-2-hydroxypropiophenone derivatives are formed on a preparative scale by benzaldehyde lyase (BAL)–catalyzed C–C bond formation from aromatic aldehydes and acetaldehyde in aqueous buffer/DMSO solution with remarkable ease in high chemical yield and high optical purity (Eq. 8.112).[303] Less-stable mixed benzoins were also generated via reductive coupling of benzoyl cyanide and carbonyl compounds by aqueous titanium(III) ions.[304]

$$ArCHO \ + \ MeCHO \quad \xrightarrow[\text{buffer/DMSO}]{\text{BAL}} \quad Ar\overset{\displaystyle O}{\underset{\displaystyle OH}{\diagup}} \quad 91\text{–}99\% \ ee \qquad (8.112)$$

8.3.12 Wittig Reactions

The Witting reaction has been investigated in aqueous conditions.[305] Wittig olefination reactions with stabilized ylides (known as the Wittig-Horner or Horner-Wadsworth-Emmons reaction) are sometimes performed in an organic/water biphase system.[306] Very often, a phase-transfer catalyst is used. Recently, the use of water alone as solvent has been investigated.[307] The reaction proceeds smoothly with a much weaker base such as K_2CO_3 or $KHCO_3$. No phase-transfer catalyst is required. Compounds with base- and acid-sensitive functional groups can be used directly. For example, under such conditions, β-dimethyl-hydrazoneacetaldehyde can be olefinated efficiently (Eq. 8.113).[308]

$$\text{(8.113)}$$

Recently, water-soluble phosphonium salts were synthesized and their Wittig reactions with substituted benzaldehydes were carried out in aqueous sodium hydroxide solution (Eq. 8.114).[309]

$$\text{(8.114)}$$

A tandem enzymatic aldol-intramolecular Horner-Wadsworth-Emmons reaction has been used in the synthesis of a cyclitol.[310] The key steps are illustrated in Scheme 8.33. The phosphonate aldehyde was condensed with dihydroxyacetone phosphate (DHAP) in water with FDP aldolase to give the aldol adduct, which cyclizes with an intramolecular Horner-Wadsworth-Emmons reaction to give the cyclo-pentene product. The one-pot reaction takes place in aqueous solution at slightly acidic (pH 6.1–6.8) conditions. The aqueous Wittig-type reaction has also been investigated in DNA-templated synthesis.[311]

The use of water-soluble reagents and catalysts allows reactions to be performed in aqueous buffered solutions. PEG-supported triarylphos-phine has been used in a Wittig reaction under mildly basic aqueous conditions (Eq. 8.115). The PEG-supported phosphine oxide byproduct can be easily recovered and reduced by alane to regenerate the starting reagent for reuse.[312] The aqueous Wittig reaction has also been used in

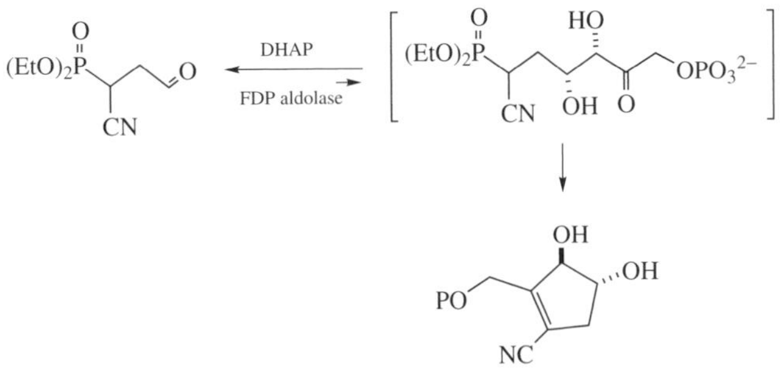

Scheme 8.33

the synthesis of picenadol and analogs.[313]

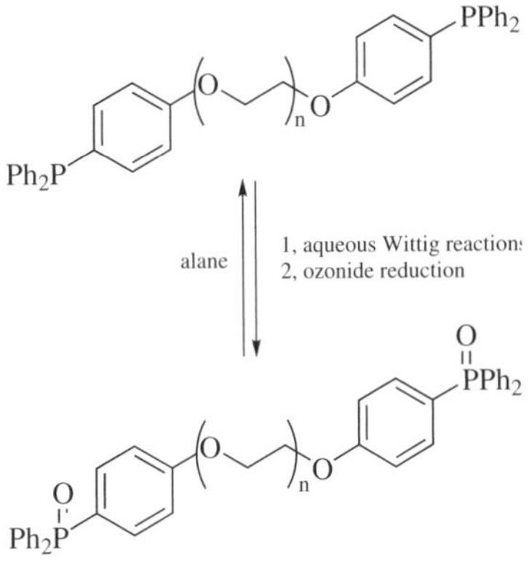

(8.115)

8.4 PINACOL COUPLING

The coupling of carbonyl compounds[314] to give 1,2-diols, known as the *pinacol coupling,* has been carried out in aqueous media.[315] Clerici and Porta extensively studied the aqueous pinacol coupling reactions mediated by Ti(III).[316] Schwartz reported a stereoselective pinacol coupling with a cyclopentadienetitanium complex.[317] Pinacol-type couplings were also developed by using a Zn-Cu couple,[318] Mg,[319] Mn,[320] Zn,[321] In,[322] Sm,[323] Al/NaOH,[324] Al/F⁻,[325] Ga,[326] Cd,[327] and other metals (Eq. 8.116). Under sonication, Kim and co-workers[328] found that aromatic aldehydes homocoupled to generate pinacol-type products. Pinacol coupling of benzaldehyde in an aqueous medium mediated by magnesium, zinc, iron, nickel, and tin was studied under the effect of ultrasound on these reactions with magnesium providing the best results.[329] The reaction occurred in neutral aqueous media over 8 to 22 hours. In the absence of sonication, the reaction was much slower and the yield of the product was decreased by a factor of 2 to 3. Interestingly, the reaction did not proceed under nitrogen protection! Water alone or a 1:1 mixture of water and t-BuOH were used in these reactions. Aliphatic aldehydes and ketones are inert under the reaction conditions. Solid aldehydes resulted in poor yield of the product or no product.

$$2 \quad \underset{R}{\overset{O}{\underset{\|}{\bigwedge}}} \overset{}{H} \quad \xrightarrow{\text{M in water}} \quad \underset{R \quad R}{\overset{HO \quad OH}{\bigwedge}} \qquad (8.116)$$

M = Zn-Cu, Mg, Mn, Zn, In, Sm, Al, Ga, C

A cross-coupling reaction of aldehydes with α-diketones proceeded in the presence of water to give the corresponding adducts in moderate to good yield. It is possible to use the substrates such as phenyl-glyoxal monohydrate, aqueous methylglyoxal, formalin, and aqueous α-chloroacetaldehyde for this reaction.[330]

8.5 OTHER REACTIONS (HALOGENATION AND OXIDATION OF α-H)

A mixture of 1,4-dioxane and water is often used as the solvent for the conversion of aldehydes and ketones by H_2SeO_3 to α-dicarbonyl compounds in one step (Eq. 8.117).[331] Dehydrogenation of carbonyl compounds with selenium dioxide generates the α, β-unsaturated carbonyl compounds in aqueous acetic acid.[332] Using water as the reaction medium, ketones can be transformed into α-iodo ketones upon treatment with sodium iodide, hydrogen peroxide, and an acid.[333] Interestingly, α-iodo ketones can be also obtained from secondary alcohol through a metal-free tandem oxidation-iodination approach.

$$\underset{}{\overset{O}{\bigvee}} \quad \xrightarrow[\text{1,4-dioxane/H}_2\text{O}]{\text{H}_2\text{SeO}_3} \quad \underset{}{\overset{O}{\bigvee}}\text{O} \qquad (8.117)$$

Protonic acid and Lewis acids can activate carbonyls to facilitate the addition of nucleophile attacks in aqueous media. The Prins reaction, reaction with alkyne, and Friedel-Crafts-type reactions have been discussed in related chapters in detail.

REFERENCES

1. Kukolev, V. P., Balyushina, N. A., Chukhadzhyan, G. A., *Arm. Khim. Zh.* **1982**, *35*, 688.

2. Joo, F., Benyei, A., *J. Organomet. Chem.* **1989**, *363*, C19.

3. Joo, F., Toth, Z., Beck, M. T., *Inorg. Chim. Acta* **1977**, *25*, L61.

4. Grosselin, J. M., Mercier, C., *J. Mol. Catal.* **1990**, *63*, L25.

5. Fache, E., Senocq, F., Santini, C., Basset, J. M., *J. Chem. Soc., Chem. Commun.* **1990**, 1776.

6. Darensbourg, D. J., Joo, F., Kannisto, M., Katho, A., Reibenspies, J. H., Daigle, D. J., *Inorg. Chem.* **1994**, *33*, 200.

7. Darensbourg, D. J., Joo, F., Kannisto, M., Katho, A., Reibenspies, J. H., *Organometallics* **1992**, *11*, 1990; Benyei, A., Joo, F., *J. Mol. Catal.* **1990**, *58*, 151.

8. Rhyoo, H. Y., Park, H.-J., Suh, W. H., Chung, Y. K., *Tetrahedron Lett.* **2002**, *43*, 269.

9. Schlatter, A., Kundu, M. K., Woggon, W.-D., *Angew. Chem. Int. Ed.* **2004**, *43*, 6731.

10. Lamaty, G., Riviere, M. H., Roque, J. P., *Bull. Soc. Chim. Fr.* **1983**, 33.

11. Boyer, B., Betzer, J. F., Lamaty, G., Leydet, A., Roque, J. P., *New J. Chem.* **1995**, *19*, 807.

12. Kondo, S., Nakanishi, M., Yamane, K., Miyagawa, K., Tsuda, K., *J. Macromol. Sci. Chem.* **1990**, *A27*, 391.

13. Luo, H.-M., Li, Y.-Q., *Chin. J. Chem.* **2005**, *23*, 345.

14. Bordoloi, M., *Tetrahedron Lett.* **1993**, *34*, 1681.

15. Singh, A. K., Bakshi, R. K., Corey, E. J., *J. Am. Chem. Soc.* **1987**, *109*, 6187; Hasegawa, E., Curran, D. P., *J. Org. Chem.* **1993**, *58*, 5008; Talukdar, S., Fang, J.-M., *J. Org. Chem.* **2001**, *66*, 330.

16. Krapcho, A. P., Seidman, D. A., *Tetrahedron Lett.* **1981**, *22*, 179; Singh, J., Kad, G. L., Sharma, M., Dhillon, R. S., *Synth. Commun.* **1998**, *28*, 2253.

17. Oller-Lopez, J. L., Campana, A. G., Cuerva, J. M., Oltra, J. E., *Synthesis* **2005**, 2619.

18. Desai, D. G., Swami, S. S., Nandurdikar, R. S., *Synth. Commun.* **2002**, *32*, 931.

19. Swami, S. S., Desai, D. G., Bhosale, D. G., *Synth. Commun.* **2000**, *30*, 3097.

20. Kuo, L. Y., Weakley, T. J. R., Awana, K., Hsia, C., *Organometallics* **2001**, *20*, 4969.

21. Baruah, R. N., *Ind. J. Chem. Sect. B*: **1994**, 33*B*, 1823.

22. Wang, W., Shi, L., Huang, Y., *Tetrahedron Lett.* **1990**, *31*, 182.

23. Satagopan, V., Chandalia, S. B., *Synth. Commun.* **1989**, *19*, 1217.

24. Petrier, C., Luche, J. L., *Tetrahedron Lett.* **1987**, *28*, 2347; Baruah, R. N., *Tetrahedron Lett.* **1992**, *33*, 5417.

25. Bhar, S., Guha, S., *Tetrahedron Lett.* **2004**, *45*, 3775.

26. Wiemann, J., *Compt. Rend.* **1945**, *220*, 606; Kapron, J., Wiemann, J., *Bull. Soc. Chim.* **1945**, *12*, 945.

27. Kamiura, K., Wada, M., *Tetrahedron Lett.* **1999**, *40*, 9059.

28. Petragnani, N., Comasseto, J. V., *Synthesis* **1986**, 1.

29. Boyer, S. K., Bach, J., McKenna, J., Jagdmann, E. Jr., *J. Org. Chem.* **1985**, *50*, 3408.

30. Libot, C., Pletcher, D., *Electrochem. Commun.* **2000**, *2*, 141; Reishakhrit, L. S., Argova, T. B., Ivanova, E. A., *Khim.* **1987**, 110.

31. Bryson, T. A., Jennings, J. M., Gibson, J. M., *Tetrahedron Lett.* **2000**, *41*, 3523.

32. Mizugaki, T., Ebitani, K., Kaneda, K., *Tetrahedron Lett.* **1997**, *38*, 3005.

33. Biscoe, M. R., Breslow, R., *J. Am. Chem. Soc.* **2003**, *125*, 12718.

34. Groger, H., Hummel, W., Rollmann, C., Chamouleau, F., Husken, H., Werner, H., Wunderlich, C., Abokitse, K., Drauz, K., Buchholz, S., *Tetrahedron* **2004**, *60*, 633.

35. Doiuchi, T., Minoura, Y., *Isr. J. Chem.* **1977**, *15*, 84.

36. Yadav, A. K., Singh, A., *Bull. Chem. Soc. Jpn.* **2002**, *75*, 587.

37. Reddy, M. A., Bhanumathi, N., Rao, K. R., *Chem. Commun.* **2001**, 1974.

38. Schmitzer, A. R., Franceschi, S., Perez, E., Rico-Lattes, I., Lattes, A., Thion, L., Erard, M., Vidal, C., *J. Am. Chem. Soc.* **2001**, *123*, 5956.

39. Rico-Lattes, I., Schmitzer, A., Perez, E., Lattes, A., *Chirality* **2001**, *13*, 24.

40. Sato, S., Sakamoto, T., Miyazawa, E., Kikugawa, Y., *Tetrahedron* **2004**, *60*, 7899.

41. Gross, T., Seayad, A. M., Ahmad, M., Beller, M., *Org. Lett.* **2002**, *4*, 2055.

42. Ishimoto, K., Mitoma, Y., Nagashima, S., Tashiro, H., Prakash, G. K. S., Olah, G. A., Tashiro, M., *Chem. Commun.* **2003**, 514.

43. Chaubey, G. S., Das, S., Mahanti, M. K., *Bull. Chem. Soc. Jpn.* **2002**, *75*, 2215.

44. Venturello, C., Gambaro, M., *J. Org. Chem.* **1991**, *56*, 5924.

45. Biella, S., Prati, L., Rossi, M., *J. Mol. Catal. A: Chem.* **2003**, *197*, 207.

46. Tanaka, M., Hirokane, Y., *J. Biosci. Bioeng.* **2000**, *90*, 341.

47. Meziane, S., Lanteri, P., Longeray, R., Arnaud, C., *Compt. Rend. l'Acad. Sci. Serie IIc: Chim.* **1998**, *1*, 91.

48. Pande, C. S., Gupta, N., *Monat. Chem.* **1995**, *126*, 647.

49. Corma, A., Fornes, V., Iborra, S., Mifsud, M., Renz, M., *J. Cat.* **2004**, *221*, 67.

50. For a recent review on the reaction of α-hydroxy hydroperoxide, see: Ganeshpure, P. A., Adam, W., *Synthesis* **1996**, 179.

51. Minisci, F., *Gazz. Chim. Ital.* **1959**, *89*, 1910.

52. Peters, W., *Ber.* **1905**, *38*, 2567.

53. Shishido, K., Takeda,Y., Kinugawa, J., *J. Am. Chem. Soc.* **1961**, *83*, 538; Shishido, K., Kozima, S., Hanada, T., *J. Organomet. Chem.* **1967**, *9*, 99; Shishido, K., Kozima, S., *J. Organomet. Chem.* **1968**, *11*, 503.

54. Nosek, J., *Collect. Czech. Chem. Commun.* **1964**, *29*, 597.

55. Killinger, T. A., Boughton, N. A., Runge, T. A., Wolinsky, J., *J. Organomet. Chem.* **1977**, *124*, 131.

56. Petrier, C., Luche, J. L., *J. Org. Chem.* **1985**, *50*, 910.

57. Einhorn, C., Luche, J. L., *J. Organomet. Chem.* **1987**, *322*, 177; Petrier, C., Einhorn, J., Luche, J. L., *Tetrahedron Lett.* **1985**, *26*, 1449.

58. Mattes, H., Benezra, C., *Tetrahedron Lett.* **1985**, *26*, 5697; Zhou, J. Y., Lu, G. D., Wu, S. H., *Synth. Commun.* **1992**, *22*, 481.

59. Wilson, S. R., Guazzaroni, M. E., *J. Org. Chem.* **1989**, *54*, 3087.

60. Kunz, T., Reißig, H. U., *Liebigs Ann. Chem.* **1989**, 891.

61. Chan, T.-H., Li, C.-J., *Organometallics* **1990**, *9*, 2649.

62. Li, C.-J., Chan, T.-H., *Organometallics* **1991**, *10*, 2548.

63. Trost, B. M., King, S. A., *J. Am. Chem. Soc.* **1990**, *112*, 408.

64. Oda, Y., Matsuo, S., Saito, K., *Tetrahedron Lett.* **1992**, *33*, 97.

65. Cripps, H. N., Kiefer, E. F., *Org. Syn.* **1962**, *42*, 12.

66. Durant, A., Delplancke, J. L., Winand, R., Reisse, J., *Tetrahedron Lett.* **1995**, *36*, 4257.

67. Marton, D., Stivanello, D., Tagliavini, G., *J. Org. Chem.* **1996**, *61*, 2731.

68. Sjoholm, R., Rairama, R., Ahonen, M., *Chem. Commun.* **1994**, 1217.

69. Ahonen, M., Sjöholm, R., *Chem. Lett.* **1995**, 341.

70. Hanessian, S., Park, H., Yang, R. Y., *Synlett.* **1997**, 351; Hanessian, S., Park, H., Yang, R. Y., *Synlett.* **1997**, 353.

71. Zha, Z., Xie, Z., Zhou, C., Chang, M., Wang, Z., *New J. Chem.* **2003**, *27*, 1297.

72. Archibald, S. C., Hoffmann, R. W., *Chemtracts-Organic Chemistry,* **1993**, *6*, 194.

73. Marquez, F., Montoro, R., Llebaria, A., Lago, E., Molins, E., Delgado, A., *J. Org. Chem.* **2002**, *67*, 308.

74. Nokami, J., Otera, J., Sudo, T., Okawara, R., *Organometallics* **1983**, *2*, 191.

75. Nokami, J., Wakabayashi, S., Okawara, R., *Chem. Lett.* **1984**, 869.

76. Zhou, J. Y., Chen, Z. G., Wu, S. H., *J. Chem. Soc., Chem Commun.* **1994**, 2783.

77. Uneyama, K., Kamaki, N., Moriya, A., Torii, S., *J. Org. Chem.* **1985**, *50*, 5396.

78. Wu, S. H., Huang, B. Z., Zhu, T. M., Yiao, D. Z., Chu, Y. L., *Act. Chim. Sinica.* **1990**, *48*, 372; *ibid*, **1987**, *45*, 1135.

79. Uneyama, K., Matsuda, H., Torii, S., *Tetrahedron Lett.* **1984**, *25*, 6017.

80. Mandai, T., Nokami, J., Yano, T., Yoshinaga, Y., Otera, J. *J. Org. Chem.* **1984**, *49*, 172.

81. Chan, T.-H., Yang, Y., Li, C.-J., *J. Org. Chem.* **1999**, *64*, 4452.

82. Petrier, C., Luche, J. L., *J. Org. Chem.* **1985**, *50*, 910; Petrier, C., Einhorn, J., Luche, J. L., *Tetrahedron Lett.* **1985**, *26*, 1449.

83. Einhorn, C., Luche, J. L., *J. Organomet. Chem.* **1987**, *322*, 177.

84. Masuyama, Y., Takahara, J. P., Kurusu, Y., *Tetrahedron Lett.* **1989**, *30*, 3437; Masuyama, Y., Nimura, Y., Kurusu, Y., *Tetrahedron Lett.* **1991**, *32*, 225.

85. Safi, M., Sinou, D., *Tetrahedron Lett.* **1991**, *32*, 2025.

86. Boaretto, A., Marton, D., Tagliavini, G., Gambaro, A., *J. Organomet. Chem.* **1985**, *286*, 9.

87. Boaretto, A., Marton, D., Tagliavini, G., *J. Organomet. Chem.* **1985**, *297*, 149.

88. Furlani, D., Marton, D., Tagliavini, G., Zordan, M., *J. Organomet. Chem.* **1988**, *341*, 345.

89. Hachiya, I., Kobayashi, S., *J. Org. Chem.* **1993**, *58*, 6958; Kobayashi, S., Wakabayashi, T., Oyamada, H., *Chem. Lett.* **1997**, 831; McCluskey, A., *Green Chemistry.* **1999**, *1*, 167.

90. Zha, Z., Wang, Y., Yang, G., Zhang, L., Wang, Z., *Green Chemistry* **2002**, *4*, 578.

91. Wang, Z., Zha, Z., Zhou, C., *Org. Lett.* **2002**, *4*, 1683.

92. Kumaraswamy, S., Nagabrahmanandachari, S., Kumara Swamy, K. C., *Synth. Commun.* **1996**, *26*, 729.

93. Yanagisawa, A., Morodome, M., Nakashima, H., Yamamoto, H., *Synlett.* **1997**, 1309.

94. Marshall, R. L., Muderawan, I. W., Young, D. J., *J. Chem. Soc. Perkin 2*, **2000**, 957.

95. Shibata, I., Yoshimura, N., Yabu, M., Baba, A., *Eur. J. Org. Chem.* **2001**, 3207.

96. Manabe, K., Mori, Y., Wakabayashi, T., Nagayama, S., Kobayashi, S., *J. Am. Chem. Soc.* **2000**, *122*, 7202.

97. Nagayama, S., Kobayashi, S., *Angew. Chem., Int. Ed.* **2000**, *39*, 567.

98. Masuyama, Y., Kishida, M., Kurusu, Y., *Chem. Commun.* **1995**, 1405; Ito, A., Kishida, M., Kurusu, Y., Masuyama, Y., *J. Org. Chem.* **2000**, *65*, 494.

99. Kundu, A., Prabhakar, S., Vairamani, M., Roy, S., *Organometallics* **1997**, *16*, 4796; Tan, X.-H., Shen, B., Liu, L., Guo, Q.-X., *Tetrahedron Lett.* **2002**, *43*, 9373; Debroy, P., Roy, S., *J. Organomet. Chem.* **2003**, *675*, 105.

100. Okano, T., Kiji, J., Doi, T., *Chem. Lett.* **1998**, 5.

101 Tan, X.-H., Shen, B., Deng, W., Zhao, H., Liu, L., Guo, Q. X., *Org. Lett.* **2003**, *5*, 1833.

102. Carde, L., Llebaria, A., Delgado, A., *Tetrahedron Lett.* **2001**, *42*, 3299.

103. Samoshin, V. V., Gremyachinskiy, D. E., Smith, L. L., Bliznets, I. V., Gross, P. H., *Tetrahedron Lett.* **2002**, *43*, 6329; Gremyachinskiy, D. E., Smith, L. L., Gross, P. H., Samoshin, V. V., *Green Chemistry* **2002**, *4*, 317.

104. Wang, J., Yuan, G., Dong, C.-Q., *Chem. Lett.* **2004**, *33*, 286.

105. Chang, H.-M., Cheng, C.-H., *Org. Lett.* **2000**, *2*, 3439.

106. Schmid, W., Whitesides, G. M., *J. Am. Chem. Soc.* **1991**, *113*, 6674.

107. Li, C.-J., *Ph. D. Thesis*, McGill University, 1992.

108. Li, C.-J., Chan, T.-H., *Tetrahedron Lett.* **1991**, *32*, 7017.

109. Araki, S., Jin, S. J., Idou, Y., Butsugan, Y., *Bull. Chem. Soc. Jpn.* **1992**, *65*, 1736.

110. Kim, E., Gordon, D. M., Schmid, W., Whitesides, G. M., *J. Org. Chem.* **1993**, *58*, 5500.

111. Li, C.-J., *Tetrahedron Lett.* **1995**, *36*, 517.

112. Chen, D. L., Li, C.-J., *Tetrahedron Lett.* **1996**, *37*, 295.

113. Marshall, J. A., Hinkle, K. W., *J. Org. Chem.* **1995**, *60*, 1920.

114. Yi, X.H., Meng, Y., Li, C.-J., *Tetrahedron Lett.* **1997**, *38*, 4731.

115. Chan, T.-H., Li, C.-J., *J. Chem. Soc., Chem. Commun.* **1992**, 747.

116. Chan, T.-H., Li, C.-J., *203rd National Meeting of the American Chemical Society*, San Francisco, CA, April 1992, Abstract ORGN435.

117. Dondoni, A., Merino, P., Orduna, J., *Tetrahedron Lett.* **1991**, *32*, 3247.

118. Gordon, D. M., Whitesides, G. M., *J. Org. Chem.* **1993**, *58*, 7937; Casiraghi, G., Rassu, G., *Chemtracts: Organic Chemistry* **1993**, *6*, 336.

119. Gao, J., Haerter, R., Gordon, D. M., Whitesides, G. M., *J. Org. Chem.* **1994**, *59*, 3714.

120. Binder, W. H., Prenner, R. H., Schmid, W., *Tetrahedron* **1994**, *50*, 749.

121. Prenner, R. H., Binder, W. H., Schmid, W., *Libigs Ann. Chem.*, **1994**, 73.

122. Chan, T.-H., Li, C.-J., Lee, M.-C., Wei, Z.-Y., *Can. J. Chem.* **1994**, *72*, 1181.

123. Chan, T.-H., Lee, M.-C., *J. Org. Chem.* **1995**, *60*, 4228.

124. Chan, T.-H., Xin, Y.-C., *J. Chem. Soc. Chem. Comm.* **1996**, 905; Chan, T.-H., Xin, Y.-C., Von Itzstein, M., *J. Org. Chem.* **1997**, *62*, 3500; Gao, J., Martichonok, V., Whitesides, G. M., *J. Org. Chem.* **1996**, *61*, 9538.

125. Wang, R., Lim, C. M., Tan, C. H., Lim, B. K., Sim, K. Y., Loh, T. P., *Tetrahedron: Asymmetry* **1995**, *6*, 1825.

126. Li, C.-J., Lu, Y.-Q., *Tetrahedron Lett.* **1995**, *36*, 2721.

127. Li, C.-J., Lu, Y.-Q., *Tetrahedron Lett.* **1996**, *37*, 471.

128. Yang, Z. Y., Burton, D. J., *J. Org. Chem.* **1991**, *56*, 1037.

129. Li, C.-J., Chen, D.-L., Lu, Y.-Q., Haberman, J. X., Mague, J. T., *J. Am. Chem. Soc.* **1996**, *118*, 4216; Li, C.-J., Chen, D.-L., Lu, Y.-Q., Haberman, J. X., Mague, J. T., *Tetrahedron* **1998**, *54*, 2347.

130. Li, C.-J., Chen, D.-L., *Synlett.* **1999**, 735.

131. Haberman, J. X., Li, C.-J., *Tetrahedron Lett.* **1997**, *38*, 4735.

132. Loh, T.-P., Cao, G.-Q., Pei, J., *Tetrahedron Lett.* **1998**, *39*, 1453.

133. Orjala, J., Nagle, G. D., Hsu, L. V., Gerwick, W. H., *J. Am. Chem. Soc.* **1995**, *117*, 8281.

134. Loh, T.-P., Cao, G. Q., Pei, J., *Tetrahedron Lett.* **1998**, *39*, 1457; Loh, T.-P., Song, H.-Y., *Synlett* **2002**, 2119.

135. Loh, T. P., Li, X. R., *J. Chem. Soc., Chem. Commun.* **1996**, 1929.

136. Loh, T. P., Wang, R. B., Sim, K. Y., *Main Group Met. Chem.* **1997**, *20*, 237.

137. Bryan, V. J., Chan, T.-H., *Tetrahedron Lett.* **1996**, *37*, 5341.

138. Hao, J., Aiguade, J., Forsyth, C. J., *Tetrahedron Lett.* **2001**, *42*, 821.

139. Chan, T.-H., Isaac, M., *Tetrahedron Lett.* **1995**, *36*, 8957.

140. Chan, T.-H., Li, C.-J., *Can. J. Chem.* **1992**, *70*, 2726.

141. Waldmann, H., *Synlett.* **1990**, 627.

142. Cram, D. J., Kopecky, K. R., *J. Am. Chem. Soc.* **1959**, *81*, 2748; Reetz, M. T., *Angew. Chem. Int. Ed. Eng.* **1984**, *23*, 556; Midland, M. M., Koops, R. W., *J. Org. Chem.* **1990**, *55*, 5058.

143. Paquette, L. A., in *Green Chemistry: Frontiers in Benign Chemical Syntheses and Processes*, eds. Anastas, P. A., Williamson, T. C., Oxford University Press, **1998;** Isaac, M. B., Chan, T.-H., *Tetrahedron Lett.* **1995**, *36*, 8957; Chan, T.-H., Li, C.-J., *Can. J. Chem.* **1992**, *70*, 2726; Paquette, L., Mitzel, T. M., *J. Am. Chem. Soc.* **1996**, *118*, 1931.

144. Loh, T.-P., Li, X.-R., *Eur. J. Org. Chem.* **1999**, 1893; Loh, T.-P., Li, X.-R., *Tetrahedron* **1999**, *55*, 5611.

145. Alcaide, B., Almendros, P., Salgado, N. R., *J. Org. Chem.* **2000**, *65*, 3310; Alcaide, B., Almendros, P., Aragoncillo, C., Rodriguez-Acebes, R., *J. Org. Chem.* **2001**, *66*, 5208.

146. Yadav, J. S., Reddy, B. V. S., Reddy, G. S., Kiran, K., *Tetrahedron Lett.* **2000**, *41*, 2695.

147. Mendez-Andino, J., Paquette, L. A., *Org. Lett.* **2000**, *2*, 1263.

148. Bernardelli, P., Moradei, O. M., Friedrich, D., Yang, J., Gallou, F., Dyck, B. P., Doskotch, R. W., Lange, T., Paquette, L. A., *J. Am. Chem. Soc.* **2001**, *123*, 9021.

149. Lee, J. E., Cha, J. H., Pae, A. N., Choi, K. I., Koh, H. Y., Kim, Y., Cho, Y. S., *Synth. Commun.* **2000**, *30*, 4299.

150. Cha, J. H., Pae, A. N., Choi, K. I., Cho, Y. S., Koh, H. Y., Lee, E., *J. Chem. Soc., Perkin 1*, **2001**, 2079.

151. Canac, Y., Levoirier, E., Lubineau, A., *J. Org. Chem.* **2001**, *66*, 3206; Lubineau, A., Canac, Y., Le Goff, N., *Adv. Synth. Catal.* **2002**, *344*, 319.

152. Shin, J. A., Choi, K. I., Pae, A. N., Koh, H. Y., Kang, H.-Y., Cho, Y. S., *J. Chem. Soc. Perkin 1*, **2001**, 946.

153. Loh, T.-P., Tan, K.-T., Yang, J.-Y., Xiang, C.-L., *Tetrahedron Lett.* **2001**, *42*, 8701.

154. Loh, T.-P., Tan, K.-T., Hu, Q.-Y., *Tetrahedron Lett.* **2001**, *42*, 8705; Tan, K.-T., Chng, S.-S., Cheng, H.-S., Loh, T.-P., *J. Am. Chem. Soc.* **2003**, *125*, 2958.

155. Cho, Y. S., Kang, K. H., Cha, J. H., Choi, K. I., Pae, A. N., Koh, H. Y., Chang, M. H., *Bull. Kor. Chem. Soc.* **2002**, *23*, 1285.

156. Alcaide, B., Almendros, P., Aragoncillo, C., Rodriguez-Acebes, R., *Synthesis* **2003**, 1163.

157. Hidestål, O., Ding, R., Almesåker, A., Lindström, U. M., *Green Chemistry* **2005**, *7*, 259.

158. Jang, T.-S., Keum, G., Kang, S. B., Chung, B. Y., Kim, Y., *Synthesis* **2003**, 775.

159. Huang, J.-M., Xu, K.-C., Loh, T.-P., *Synthesis* **2003**, 755.

160. Chung, W. J., Higashiya, S., Oba, Y., Welch, J. T., *Tetrahedron* **2003**, *59*, 10031.

161. Loh, T.-P., Yin, Z., Song, H.-Y., Tan, K.-L., *Tetrahedron Lett.* **2003**, *44*, 911.

162. Song, J., Hua, Z.-H., Qi, S., Ji, S.-J., Loh, T.-P., *Synlett* **2004**, 829.

163. Chan, T. H., Yang, Y., *J. Am. Chem. Soc.* **1999**, *121*, 3228.

164. Gajewski, J. J., Bocian, W., Brichford, N. L., Henderson, J. L., *J. Org. Chem.* **2002**, *67*, 4236; Lucas, P., Gajewski, J. J., Chan, T. H., *Canad. J. Anal. Sci. Spectr.* **2003**, *48*, 1.

165. Thadani, A. N., Batey, R. A., *Org. Lett.* **2002**, *4*, 3827.

166. Thadani, A. N., Batey, R. A., *Tetrahedron Lett.* **2003**, *44*, 8051.

167. Ishihara, K., Hanaki, N., Funahashi, M., Miyata, M., Yamamoto, H., *Bull. Chem. Soc. Jpn* **1995**, *68*, 1721.

168. Wang, M.-W., Chen, Y.-J., Liu, L., Wang, D., Liu, X.-L., *J. Chem. Res., Synop.* **2000**, 80.

169. Aoyama, N., Hamada, T., Manabe, K., Kobayashi, S., *Chem. Commun.* **2003**, 676.

170. Wang, Z., Yuan, S., Li, C.-J., *Tetrahedron Lett.* **2002**, *43*, 5097.

171. Tsuji, T., Usugi, S.-I., Yorimitsu, H., Shinokubo, H., Matsubara, S., Oshima, K., *Chem. Lett.* **2002**, 2.

172. Li, C.-J., Zhang, W.-C., *J. Am. Chem. Soc.* **1998**, *120*, 9102; Zhang, W.-C., Li, C.-J., *J. Org. Chem.* **1999**, *64*, 3230.

173. Fukuma, T., Lock, S., Miyoshi, N., Wada, M., *Chem. Lett.* **2002**, 376.

174. Khan, R. H., Prasada Rao, T. S. R., *J. Chem. Res., Synop.* **1998**, 202.

175. Akiyama, T., Iwai, J., *Tetrahedron Lett.* **1997**, *38*, 853.

176. Zhou, J.-Y., Jia, Y., Sun, G.-F., Wu, S.-H., *Synth. Commun.* **1997**, *27*, 1899.

177. Kobayashi, S., Aoyama, N., Manabe, K., *Synlett* **2002**, 483.

178. Li, C.-J., Meng, Y., Yi, X.-H., Ma, J., Chan, T.-H., *J. Org. Chem.* **1997**, *62*, 8632.

179. Li, L.-H., Chan, T.-H., *Tetrahedron Lett.* **2000**, *41*, 5009; Li, L.-H., Chan, T.-H., *Can. J. Chem.* **2001**, *79*, 1536.

180. Wada, M., Ohki, H., Akiba, K. Y., *Bull. Chem. Soc. Jpn.* **1990**, *63*, 1738; *J. Chem. Soc., Chem. Commun.* **1987**, 708; Fukuma, T., Lock, S., Miyoshi, N., Wada, M., *Chem. Lett.* **2002**, 376.

181. Wada, M., Fukuma, T., Morioka, M., Takahashi, T., Miyoshi, N., *Tetrahedron Lett.* **1997**, *38*, 8045.

182. Smith, K., Lock, S., El-Hiti, G. A., Wada, M., Miyoshi, N., *Org. Biomol. Chem.* **2004**, *2*, 935.

183. Matsumura, N., Doi, T., Mishima, K., Kitagawa, Y., Okumura, Y., Mizuno, K., *ITE Lett. Batter., New Technol. Med.* **2003**, *4*, 473.

184. Katritzky, A. R., Shobana, N., Harris, P. A., *Organometallics* **1992**, *11*, 1381.

185. Minato, M., Tsuji, J., *Chem. Lett.* **1988**, 2049.

186. Chan, T.-H., Yang, Y., *Tetrahedron Lett.* **1999**, *40*, 3863.,

187. Chan, T.-C., Lau, C.-P., Chan, T.-H., *Tetrahedron Lett.* **2004**, *45*, 4189.

188. Fornasier, R., Marcuzzi, F., Piva, M., Tonellato, U., *Gazz. Chim. Ital.* **1996**, *126*, 633; Fornasier, R., Marcuzzi, F., Marton, D., *Main Gr. Met. Chem.* **1998**, *21*, 65; Abele, E., Lukevics, E., *Latv. Kim. Zur.* **1998**, 73.

189. Loh, T.-P., Zhou, J.-R., *Tetrahedron Lett.* **2000**, *41*, 5261.

190. Cunningham, A., Woodward, S., *Synlett* **2002**, 43.

191. Kobayashi, S., Aoyama, N., Manabe, K., *Chirality* **2003**, *15*, 124.

192. Singh, S., Kumar, S., Chimni, S. S., *Tetrahedron: Asymmetry* **2002**, *13*, 2679.

193. Wu, S. H., Huang, B. Z., Gao, X., *Synth. Commun.* **1990**, *20*, 1279.

194. Boaretto, A., Marton, D., Tagliavini, G., Gambaro, A., *J. Organomet. Chem.* **1985**, *286*, 9.

195. Boaretto, A., Marton, D., Tagliavini, G., *J. Organomet. Chem.* **1985**, *297*, 149.

196. Furlani, D., Marton, D., Tagliavini, G., Zordan, M., *J. Organomet. Chem.* **1988**, *341*, 345.

197. Hachiya, I., Kobayashi, S., *J. Org. Chem.* **1993**, *58*, 6958.

198. Houllemare, D., Outurquin, F., Paulmier, C., *J. Chem. Soc. Perkin 1*, **1997**, 1629.

199. Chattopadhyay, A., *J. Org. Chem.* **1996**, *61*, 6104; Keltjens, R., Bieber, L. W., da Silva, M. F., da Costa, R. C., Silva, L. O. S., *Tetrahedron Lett.* **1998**, *39*, 3655; Vadivel, S. K., de Gelder, R., Klunder, A. J. H., Zwanenburg, B., *Eur. J. Org. Chem.* **2003**, 1749.

200. Chattopadhyay, A., Dhotare, B., *Tetrahedron: Asymmetry* **1998**, *9*, 2715.

201. Isaac, M. B., Chan, T.-H., *J. Chem. Soc., Chem. Commun.* **1995**, 1003.

202. Mitzel, T. M., Palomo, C., Jendza, K., *J. Org. Chem.* **2002**, *67*, 136.

203. Yi, X.-H., Meng, Y., Hua, X.-G., Li, C.-J., *J. Org. Chem.* **1998**, *63*, 7472.

204. Yi, X.-H., Meng, Y., Li, C.-J., *Chem. Commun.* **1998**, 449; Hua, X.-G., Li, C.-J., *Main Gr. Met. Chem.* **1999**, *22*, 533; Hua, X.-G., Mague, J. T., Li, C.-J., *Tetrahedron Lett.* **1998**, *39*, 6837; Mague, J. T., Hua, X.-G., Li, C.-J., *Acta Crystallogr., Sect. C: Cryst. Struct. Commun.* **1998**, *C54*, 1934.

205. Fang, X. P., Anderson, J. E., Chang, C. J., Fanwick, P. E., McLaughlin, J. L., *J. Chem. Soc., Perkin Trans. 1*, **1990**, 1655.

206. Cho, Y. S., Lee, J. E., Pae, A. N., Choi, K. I., Koh, H. Y., *Tetrahedron Lett.* **1999**, *40*, 1725.

207. Kwon, J. S., Pae, A. N., Choi, K. I., Koh, H. Y., Kim, Y., Cho, Y. S., *Tetrahedron Lett.* **2001**, *42*, 1957.

208. Alcaide, B., Almendros, P., Aragoncillo, C., Rodriguez-Acebes, R., *J. Org. Chem.* **2001**, *66*, 5208; Alcaide B., Almendros P., Aragoncillo C., *Chemistry* **2002**, *8*, 1719; *Chem. Eur. J.* **2002**, *8*, 1719; Alcaide, B., Almendros, P., Aragoncillo, C., Rodriguez-Acebes, R., *Synthesis* **2003**, 1163.

209. Miao, W., Chung, L.-W., Wu, Y.-D., Chan, T.-H., *J. Am. Chem. Soc.* **2004**, *126*, 13326.

210. Marshall, J. A., Chobanian, H. R., *J. Org. Chem.* **2000**, *65*, 8357.

211. Lu, W., Ma, J., Yang, Y., Chan, T.-H., *Org. Lett.* **2000**, *2*, 3469.

212. Miao, W., Chan, T.-H., *J. Am. Chem. Soc.* **2003**, *125*, 2412.

213. Yang, Y., Chan, T. H., *J. Am. Chem. Soc.***2000**, *122*, 402.

214. Bieber, L. W., Storch, E. C., Malvestiti, I., da Sila, M. F., *Tetrahedron Lett.* **1998**, *39*, 9393.

215. Zha, Z.-G., Xie, Z., Zhou, C.-L., Wang, Z.-Y., Wang, Y.-S., *Chin. J. Chem.* **2002**, *20*, 1477.

216. Sakai, M., Ueda, M., Miyaura, N., *Angew. Chem. Int. Ed.* **1998**, *37*, 3279.

217. Ueda, M., Miyaura, N., *J. Org. Chem.* **2000**, *65*, 4450.

218. Li, C.-J., Meng, Y., *J. Am. Chem. Soc.* **2000**, *120*, 9538; Huang, T. S., Venkatraman, S., Nguyen, T., Meng Y., Kort, D., Ding, R. Wang, D., Li, C.-J., *Pure Appl. Chem.* **2001**, *73*, 1315.

219. Huang, T., Meng, Y., Venkatraman, S., Wang, D., Li, C.-J., *J. Am. Chem. Soc.* **2001**, *123*, 7451.

220. Murata, M., Shimazaki, R., Ishikura, M., Watanabe, S., Masuda, Y., *Synthesis* **2002**, 717.

221. Takai, K., Morita, R., Sakamoto, S., *Synlett.* **2001**, *10*, 1614.

222. Ding, R., Ge, C. S., Chen, Y. J., Wang, D., Li, C.-J., *Tetrahedron Lett.* **2002**, *43*, 7789.

223. Viswanathan, G. S., Li, C.-J., *Tetrahedron Lett.* **2002**, *43*, 1613.

224. Engstrom, G., Morelli, M., Palomo, C., Mitzel, T., *Tetrahedron Lett.* **1999**, *40*, 5967.

225. Keh, C. C. K., Wei, C., Li, C.-J., *J. Am. Chem. Soc.***2003**, *125*, 4062.

226. Chan, T.-H., Li, C.-J., Wei, Z.-Y., *J. Chem. Soc., Chem. Commun.* **1990**, 505.

227. Zhou, J. Y., Jia, Y., Sha, Q. Y., Wu, S. H., *Synth. Commun.* **1996**, *26*, 769.

228. Chan, T.-H., Li, C.-J., Lee, M.-C., Wei, Z.-Y., *Can. J. Chem.* **1994**, *72*, 1181.

229. Shen, Z., Zhang, J., Zou, H., Yang, M., *Tetrahedron Lett.* **1997**, *38*, 2733.

230. Xu, X. L., Lu, P., Zhang, Y. M., *Chin. Chem. Lett.* **1999**, *10*, 729.

231. Chattopadhyay, A., Salaskar, A., *Synthesis* **2000**, 561.

232. Lee, Y. J., Chan, T.-H., *Can. J. Chem.* **2003**, *81*, 1406.

233. Bieber, L. W., Malvestiti, I., Storch, E. C., *J. Org. Chem.* **1997**, *62*, 9061.

234. Yi, X.-H., Meng, Y., Li, C.-J., *Tetrahedron Lett.* **1997**, *38*, 4731.

235. Areias, M. C. C., Bieber, L. W., Navarro, M., Diniz, F. B., *J. Electroanal. Chem.* **2003**, *558*, 125.

236. Chung, W. J., Higashiya, S., Welch, J. T., *J. Flu. Chem.* **2001**, *112*, 343.

237. For a recent example, see: Ben Ayed, T., Amri, H., *Synth. Commun.* **1995**, *25*, 3813.

238. For an excellent review on classical Aldol reactions, see: Nielsen, A. T., Houlihan, W. J., *Org. React.* **1968**, *16*, 1.

239. Grignard, V., Dubien, M., *Ann. Chim. (Paris)*, **1924**, *[10]2*, 282.

240. Ballini, R., Bosica, G., *J. Org. Chem.* **1997**, *62*, 425.

241. Bowden, K., Brownhill, A., *J. Chem. Soc. Perkin 2* **1997**, 997.

242. De Santis, B., Iamiceli, A. L., Marini Bettolo, R., Migneco, L. M., Scarpelli, R., Cerichelli, G., Fabrizi, G., Lamba, D., *Helv. Chim. Acta* **1998**, *81*, 2375.

243. Peseke, K., Aldinger, S., Reinke, H., *Liebigs Annal.* **1996**, 953.

244. Ben Ayed, T., Amri, H., *Synth. Commun.* **1995**, *25*, 3813.

245. Buonora, P. T., Rosauer, K. G., Dai, L., *Tetrahedron Lett.* **1995**, *36*, 4009.

246. Dewa, T., Saiki, T., Aoyama, Y., *J. Am. Chem. Soc.* **2001**, *123*, 502.

247. Darbre, T., Machuqueiro, M., *Chem. Commun.* **2003**, 1090.

248. For recent reviews, see: Machajewski, T. D., Wong, C.-H., Lerner, R. *Angew. Chem., Int. Ed.* **2000**, *39*, 1352; Fessner, W.-D., Helaine, V., *Curr. Opin. Biotechnol.* **2001**, *12*, 574; Breuer, M., Hauer, B., *Curr. Opin. Biotechnol.* **2003**, *14*, 570.

249. Sanchez-Moreno, I., Garcia-Garcia, J. F., Bastida, A., Garcia-Junceda, E., *Chem. Commun.* **2004**, 1634.

250. Flanagan, M. E., Jacobsen, J. R., Sweet, E., Schultz, P. G., *J. Am. Chem. Soc.* **1996**, *118*, 6078.

251. List, B., Pojarliev, P., Biller, W. T., Martin, H. J., *J. Am. Chem. Soc.* **2002**, *124*, 827, Trost, B. M., Ito, H., *J. Am. Chem. Soc.* **2000**, *122*, 12003.

252. Sakthivel, K., Notz, W., Bui, T., Barbas, C. F. III, *J. Am. Chem. Soc.* **2001**, *123*, 5260.

253. Dickerson, T. J., Janda, K. D., *J. Am. Chem. Soc.* **2002**, *124*, 3220.

254. Cordova, A., Notz, W., Barbas C. F. III, *Chem. Commun.* **2002**, 3024.

255. Peng, Y.-Y., Ding, Q.-P., Li, Z., Wang, P. G., Cheng, J.-P., *Tetrahedron Lett.* **2003**, *44*, 3871.

256. Cravotto, G., Demetri, A., Nano, G. M., Palmisano, G., Penoni, A., Tagliapietra, S., *Eur. J. Org. Chem.* **2003**, 4438.

257. Li, C.-J., Wang, D., Chen, D.-L., *J. Am. Chem. Soc.* **1995**, *117*, 12867.

258. Wang, M., Li, C.-J., *Tetrahedron Lett.* **2002**, *43*, 3589.

259. Wang, M., Yang, X.-F., Li, C.-J., *Eur. J. Org. Chem.* **2003**, 998.

260. Inoue, K., Ishida, T., Shibata, I., Baba, A., *Adv. Synth. Catal.* **2002**, *344*, 283.

261. Mori, Y., Kobayashi, J., Manabe, K., Kobayashi, S., *Tetrahedron* **2002**, *58*, 8263.

262. Fringuelli, F., Pani, G., Piermatti, O., Pizzo, F., *Tetrahedron* **1994**, *50*, 11499.

263. Bingham, S. J., Tyman, J. H. P., Malik, K. M. A., Hibbs, D. E., Hursthouse, M. B., *J. Chem. Res. Synop.* **1998**, *546*.

264. Lubineau, A., *J. Org. Chem.* **1986**, *51*, 2142; Lubineau, A., Meyer, E., *Tetrahedron* **1988**, *44*, 6065.

265. Kobayashi, S., Hachiya, I., *Tetrahedron Lett.* **1992,** *33*, 1625; Hachiya, I., Kobayashi, S., *J. Org. Chem.* **1993**, *58*, 6958; Kobayashi, S., Sugiura, M., Kitagawa, H., Lam, W. W. L., *Chem. Rev.* **2002**, *102*, 2227.

266. Kobayashi, S., Hachiya, I., *J. Org. Chem.* **1994**, *59*, 3590; For a review on lanthanides catalyzed organic reactions in aqueous medium, see: Kobayashi, S., *Synlett.* **1994**, 689.

267. Kobayashi, S., Nagayama, S., Busujima, T., *J. Am. Chem. Soc.* **1998**, *120*, 8287.

268. Le Roux, C., Ciliberti, L., Laurent-Robert, H., Laporterie, A., Dubac, J., *Synlett* **1998**, 1249.

269. Kobayashi, S., Nagayama, S., Busujima, T., *Chem. Lett.* **1997**, 959.

270. Loh, T.-P., Chua, G.-L., Vittal, J. J., Wong, M.-W., *Chem. Commun.* **1998**, 861; Loh, T.-P., Pei, J., Cao, G.-Q., *Chem. Commun.* **1996**, 1819; Loh, T.-P., Pei, J., Koh, K. S.-V., Cao, G.-Q., Li, X.-R., *Tetrahedron Lett.* **1997**, *38*, 3465.

271. Mori, Y., Manabe, K., Kobayashi, S., *Angew. Chem., Int. Ed.* **2001**, *40*, 2815.

272. Munoz-Muniz, O., Quintanar-Audelo, M., Juaristi, E., *J. Org. Chem.* **2003**, *68*, 1622.

273. Manabe, K., Mori, Y., Kobayashi, S., *Tetrahedron* **1999**, *55*, 11203; Manabe, K., Kobayashi, S., *Tetrahedron Lett.* **1999**, *40*, 3773; Manabe, K., Kobayashi, S., *Synlett.* **1999**, 547; Kobayashi, S., Wakabayashi, T., Nagayama, S., Oyamada, H., *Tetrahedron Lett.* **1997**, *38*, 4559; Kobayashi, S., Wakabayashi, T., *Tetrahedron Lett.* **1998**, *39*, 5389.

274. Tian, H.-Y., Chen, Y.-J., Wang, D., Zeng, C.-C., Li, C.-J., *Tetrahedron Lett.* **2000**, *41*, 2529; Tian, H.-Y., Chen, Y.-J., Wang, D., Bu, Y.-P., Li, C.-J., *Tetrahedron Lett.* **2001**, *42*, 1803; Tian, H.-Y., Li, H.-J., Chen, Y.-J., Wang, D., Li, C.-J., *Ind. Eng. Chem. Res.* **2002**, *41*, 4523; Nishikido, J., Nanbo, M., Yoshida, A., Nakajima, H., Matsumoto, Y., Mikami, K., *Synlett.* **2002**, 1613.

275. Reetz, M. T., Giebel, D., *Angew. Chem., Int. Ed.* **2000**, *39*, 2498.

276. Sawada, H., Kurachi, J., Maekawa, T., Kawase, T., Oharu, K., Nakagawa, H., Ohira, K., *Polym. J.* **2002**, *34*, 858.

277. Gu, W., Zhou, W.-J., Gin, D. L., *Chem. Mat.* **2001**, *13*, 1949.

278. Nakagawa, T., Fujisawa, H., Mukaiyama, T., *Chem. Lett.* **2003**, *32*, 696.

279. Miura, K., Nakagawa, T., Hosomi, A., *J. Am. Chem. Soc.* **2002**, *124*, 536.

280. Loh, T.-P., Li, X.-R., *Tetrahedron* **1999**, *55*, 10789.

281. Loh, T.-P., Feng, L.-C., Wei, L.-L., *Tetrahedron* **2000**, *56*, 7309.

282. Graven, A., Grotli, M., Meldal, M., *J. Chem. Soc., Perkin 1*, **2000**, 955.

283. For a review, see: Correa, I. R., Jr., Pilli, R. A., *Quim. Nova* **2003**, *26*, 531; Kobayashi, S., Hamada, T., Nagayama, S., Manabe, K., *J. Braz. Chem. Soc.* **2001**, *12*, 627.

284. Kobayashi, S., Nagayama, S., Busujima, T., *Chem. Lett.* **1999**, 71; Kobayashi, S., Mori, Y., Nagayama, S., Manabe, K., *Green Chem.* **1999**, *1*, 175; Kobayashi, S., Nagayama, S., Busujima, T., *Tetrahedron* **1999**, *55*, 8739.

285. Nagayama, S., Kobayashi, S., *J. Am. Chem. Soc.* **2000**, *122*, 11531.

286. Kobayashi, S., Hamada, T., Nagayama, S., Manabe, K., *Org. Lett.* **2001**, *3*, 165.

287. Li, H. J., Tian, H. Y., Chen, Y. J., Wang, D., Li, C. J., *Chem. Commun.* **2002**, 2994.

288. Yamashita, Y., Ishitani, H., Shimizu, H., Kobayashi, S., *J. Am. Chem. Soc.* **2002**, *124*, 3292.

289. Li, H.-J., Tian, H.-Y., Chen, Y.-J., Wang, D., Li, C.-J., *J. Chem. Res., Synop.* **2003**, 153.

290. Ozasa, N., Wadamoto, M., Ishihara, K., Yamamoto, H., *Synlett.* **2003**, 2219.

291. Yang, B.-Y., Chen, X.-M., Deng, G.-J., Zhang, Y.-L., Fan, Q.-H., *Tetrahedron Lett.* **2003**, *44*, 3535.

292. Ohkouchi, M., Yamaguchi, M., Yamagishi, T., *Enantimer.* **2000**, *5*, 71.

293. Kobayashi, S., Ueno, M., Suzuki, R., Ishitani, H., *Tetrahedron Lett.* **1999**, *40*, 2175.

294. Friedrich, K., in eds. Patai, S., Rappoport, Z., *The Chemistry of Functional Groups, Supplement C*, pt. 2, 1345, Wiley, New York, 1983; Guthrie, J. P., *J. Am. Chem. Soc.* **1998**, *120*, 1688.

295. Taillades, J., Commeyras, A., *Tetrahedron* **1974**, *30*, 2493.

296. Webber, J. M., in eds. Wolfrom, M. L., Tipson, R. S., *Advances in Carbohydrate Chemistry*, Vol. 17, Academic Press, 1962.

297. Fischer, E., Passmore, F., *Ber.* **1890**, *23*, 2226.

298. Griengl, H., Klempier, N., Pochlauer, P., Schmidt, M., Shi, N., Zabelinskaja-Mackova, A. A., *Tetrahedron* **1998**, *54*, 14477.

299. Kool, E. T., Breslow, R., *J. Am. Chem. Soc.* **1988**, *110*, 1596.

300. Breslow, R., Conners, R. V., *J. Am. Chem. Soc.* **1995**, *117*, 6601.

301. Breslow, R. D., *J. Am. Chem. Soc.* **1958**, *80*, 3719.

302. Breslow, R., Kool, E., *Tetrahedron Lett.* **1988**, *29*, 1635; Gao, G., Xiao, R., Yuan, Y., Zhou, C.-H., You, J., Xie, R.-G., *J. Chem. Res., Synop.* **2002**, *6*, 262.

303. Demir, A. S., Sesenoglu, O., Eren, E., Hosrik, B., Pohl, M., Janzen, E., Kolter, D., Feldmann, R., Dunkelmann, P., Muller, M., *Adv. Synth.*

Catal. **2002**, *344*, 96; Demir, A. S., Dunnwald, T., Iding, H., Pohl, M., Muller, M., *Tetrahedron: Asymmetry* **1999**, *10*, 4769.

304. Clerici, A., Porta, O., *J. Org. Chem.* **1993**, *58*, 2889.

305. Maerkl, G., Merz, A., *Synthesis*, **1973**, 295; Hwang, J.-J., Lin, R.-L., Shieh, R.-L., Jwo, J.-J., *J. Mol. Cat. A: Chem.* **1999**, *142*, 125.

306. Piechucki, C., *Synthesis*, **1976**, 187; Mikolajczyk, M., Grzejszczak, S., Midura, W., Zatorski, A., *Synthesis*, **1976**, 396.

307. Rambaud, M., de Vecchio, A., Villieras, J., *Synth. Commun.* **1984**, *14*, 833.

308. Schimitt, M., Bourguignon, J. J., Wermuth, C. G., *Tetrahedron Lett.* **1990**, *31*, 2145.

309. Russell, M. G., Warren, S., *J. Chem. Soc. Perkin 1* , **2000**, *4*, 505; Russell, M. G., Warren, S., *Tetrahedron Lett.* **1998**, *39*, 7995.

310. Gijsen, H. J. M., Wong, C. H., *Tetrahedron Lett.* **1995**, *36*, 7057.

311. Gartner, Z. J., Kanan, M. W., Liu, D. R., *Angew. Chem., Int. Ed.* **2002**, *41*, 1796.

312. Sieber, F., Wentworth, P. Jr., Toker, J. D., Wentworth, A. D., Metz, W. A., Reed, N. N., Janda, K. D., *J. Org. Chem.* **1999**, *64*, 5188.

313. Martinelli, M. J., Peterson, B. C., Hutchinson, D. R., *Heterocycles* **1993**, *36*, 2087.

314. For recent reviews, see: Kahn, B. E., Rieke, R. D., *Chem. Rev.* **1988**, *88*, 733; Pons, J. M., Santelli, M., *Tetrahedron* **1988**, *44*, 4295.

315. Grinder, M. G., *Ann. Chim. Phys.* **1892**, *26*, 369.

316. Clerici, A., Porta, O., *J. Org. Chem.* **1989**, *54*, 3872.

317. Barden, M. C., Schwartz, J., *J. Am. Chem. Soc.* **1996**, *118*, 5484.

318. Delair, P., Luche, J. L., *J. Chem. Soc., Chem. Commun.* **1989**, 398.

319. Zhang, W.-C., Li, C.-J., *J. Chem. Soc. Perkin Trans. I*, **1998**, 3131; Zhang, W.-C., Li, C.-J., *J. Org. Chem.* **1999**, *64*, 3230; Li, J.-T., Bian, Y.-J., Zang, H.-J., Li, T.-S., *Synth. Commun.* **2002**, *32*, 547.

320. Li, C.-J., Meng, Y., Yi, X.-H., Ma, J., Chan, T.-H., *J. Org. Chem.* **1997**, *62*, 8632; Li, C.-J., Meng, Y., Yi, X.-H., Ma, J., Chan, T.-H., *J. Org. Chem.* **1998**, *63*, 7498; Li, J.-T., Bian, Y.-J., Liu, S.-M., Li, T.-S., *Youji Huaxue* **2003**, *23*, 479.

321. Tsukinoki, T., Kawaji, T., Hashimoto, I., Mataka, S., Tashiro, M., *Chem. Lett.* **1997**, 235; Wang, L., Sun, X., Zhang, Y., *J. Chem. Res., Synop.* **1998**, 336; Hekmatshoar, R., Yavari, I., Beheshtiha, Y. S., Heravi, M. M., *Monatsh. Chem.* **2001**, *132*, 689; Yang, J.-H., Li, J.-T., Zhao, J.-L., Li, T.-S., *Synth. Commun.* **2004**, *34*, 993.

322. Lim, H. J., Keum, G., Nair, V., Ros, S., Jayan, C. N., Rath, N. P., *Tetrahedron Lett.* **2002**, *43*, 8967.

323. Wang, L., Zhang, Y., *Tetrahedron* **1998**, *54*, 11129; Wang, L., Zhang, Y.-M., *Chin. J. Chem.* **1999**, *17*, 550; Wang, L., Zhang, Y., *Tetrahedron* **1998**, *54*, 11129.

324. Bhar, S., Panja, C., *Green Chem.* **1999**, *1*, 253; Bian, Y.-J., Liu, S.-M., Li, J.-T., Li, T.-S., *Synth. Commun.* **2002**, *32*, 1169.

325. Li, L.-H., Chan, T. H., *Org. Lett.* **2000**, *2*, 1129.

326. Wang, Z.-Y., Yuan, S.-Z., Zha, Z.-G., Zhang, Z.-D., *Chin. J. Chem.* **2003**, *21*, 1231.

327. Zheng, Y., Bao, W., Zhang, Y., *Synth. Commun.* **2000**, *30*, 3517.

328. Kang, S. B., Chung, B. Y., Kim, Y., *Tetrahedron Lett.* **1998**, *39*, 4367.

329. Meciarova, M., Toma, S., Babiak, P., *Chem. Papers* **2001**, *55*, 302.

330. Miyoshi, N., Takeuchi, S., Ohgo, Y., *Chem. Lett.* **1993**, 2129.

331. Nabjohn, N., *Org. React.* **1976**, *24*, 261.

332. Leonard, N. J., Hay, A. S., Fulmer, R. W., Gash, V. W., *J. Am. Chem. Soc.* **1955**, *77*, 439.

333. Barluenga, J., Marco-Arias, M., Gonzalez-Bobes, F., Ballesteros, A., Gonzalez, J. M., *Chem. Commun.* **2004**, 2616.

CHAPTER 9

CARBOXYLIC ACIDS AND DERIVATIVES

9.1 GENERAL

9.1.1 Reaction of α-Hydrogen

Like aldehydes and ketones, the α-hydrogens of acid and acid derivatives are acidic and can be abstracted with base to generate the carbanions, which can then react with various electrophiles such as halogens, aldehydes, ketones, unsaturated carbonyl compounds, and imines, to give the corresponding products. Many of these reactions can be performed in aqueous conditions. These have been covered in related chapters.

9.1.2 Reduction

Compared with aldehydes and ketones, carboxylic acids and their derivatives are less reactive toward reduction. Nevertheless, it is still possible to reduce various acid derivatives in aqueous conditions. Aromatic carboxylic acids, esters, amides, nitriles, and chlorides (and ketones and nitro compounds) were rapidly reduced by the SmI_2-H_2O system to the corresponding products at room temperature in good yields

Comprehensive Organic Reactions in Aqueous Media, Second Edition, by Chao-Jun Li and Tak-Hang Chan
Copyright © 2007 John Wiley & Sons, Inc.

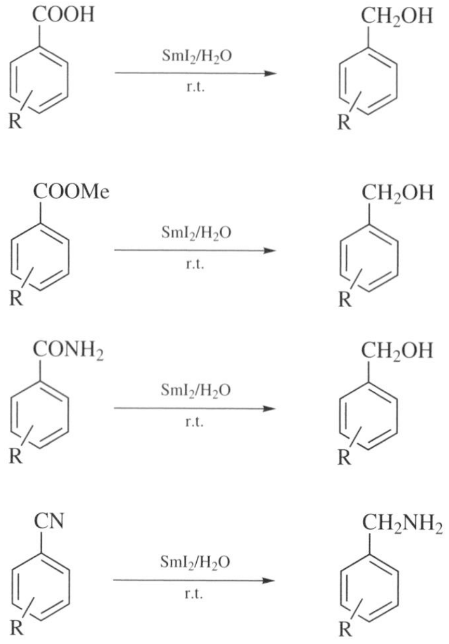

Scheme 9.1

(Scheme 9.1).[1] For example, $PhCO_2H$ can be reduced to $PhCH_2OH$ in 89% yield; $PhCO_2Me$ afforded $PhCH_2OH$ in 93% yield; $PhCONH_2$ afforded $PhCH_2OH$ in 94% yield; whereas $2\text{-}MeC_6H_4CN$ was reduced to $2\text{-}MeC_6H_4CH_2NH_2$ in 94% yield.

9.2 CARBOXYLIC ACIDS

9.2.1 Esterification/Amidation

The most common reactions of carboxylic acids are esterifications and amidations, which are accomplished by reacting with alcohols and amines respectively. In nature, such reactions occur via the action of enzymes in aqueous media, which generates an equilibrium mixture of the starting materials and the product. In the laboratory, such reactions are generally carried out in anhydrous organic solvent with acylating agents. However, studies have shown that such esterification reactions and amidation reactions can also be carried out in aqueous media. A convenient method for the preparation of esters was developed on the basis of the reaction of cage-like polycyclic olefins with carboxylic acid anhydrides and water (Eq. 9.1).[2] Mixed anhydrides were found to give rise to the corresponding low-molecular-weight acid esters. Among the obtained esters, acetates possess a pleasant odor and they can be used as components of synthetic fragrant substances. A benzoyl group can be introduced onto the 3′-hydroxyl group of 6-chloropurine riboside by treatment with benzoic anhydride in the presence of a base in aqueous solution (Eq. 9.2).[3] The reaction can be applied to the synthesis of nucleoside derivatives. Reactions between dodecyl succinic anhydride and starch slurries under aqueous alkaline conditions give the esterification product.[4] Esterification of sucrose occurs with octanoyl chloride in a discontinuous reactor in aqueous media in the presence of 4-dimethylaminopyridine as catalyst (Eq. 9.3).[5]

$$\text{(9.1)}$$

$$\text{(9.2)}$$

$$\text{(9.3)}$$

Because the reaction of an amine with an acyl chloride is much faster than the hydrolysis of the acyl chloride, the reaction can usually be carried out in an aqueous alkali solution. This is well known as the Schotten-Baumann procedure.[6] For example, a number of N-acyl taxol analogs have been prepared under Schotten-Baumann conditions by the reaction of N-debenzoyltaxol with various acid chlorides (Eq. 9.4).[7] Highly purified N-long-chain-acyl neutral amino acids such as potassium N-lauroyl-γ-aminobutyrate, useful as surfactants for detergent formulations, can be prepared in high yield by reacting a neutral amino acid such as glycine, γ-aminobutyric acid, and alanine with a saturated or unsaturated fatty acid halide such as lauroyl chloride in a mixture of water and a hydrophilic organic solvent (acetone, acetonitrile, or alcohol) in the presence of a base (e.g., potassium hydroxide).[8] The reaction of an amino acid with a carboxylic acid anhydride in biphasic aqueous media can generate amides. The presence of water aids in product isolation by phase separation and recycle of the reactants.[9]

R = benzoyl, 95%
R = isovaleryl, 89%
R = t-butyacetyl, 92%
R = trimethylacetyl, 93%
R = 4-azido-2,3,5,6-tetrafluorobenzoyl, 93%

$$(9.4)$$

Amines can be efficiently acylated with cyclic and acyclic anhydrides, by dissolving them in an aqueous medium, in the presence of a surfactant such as sodium dodecyl sulfate (SDS) under neutral conditions (Eq. 9.5).[10] Cyclic and acyclic anhydrides reacted with various amines chemoselectively in the presence of phenols and thiols. No acidic or basic reagents were used during the reaction. The acylated products can be purified by simple phase separation. Diamines and their salts are acetylated easily at room temperature with Ac_2O in aqueous media giving high yields (82–98%).[11] Acylation of phenols, acids, and (or) amines can be performed with the water-soluble acylating agents, of 2-acylthio-1-alkylpyridinium salts, in an aqueous phase (Eq. 9.6, R = Ph).[12] Reaction of phenols, amines, and acids with 2-benzoylthio-1-methylpyridinium chloride, prepared in situ from benzoyl chloride and 1-methyl-2(1H)-pyridinethione, afforded the corresponding benzoyl deriveratives in good yields. Thiocarbamates were prepared by adding RNHCOCl or RNCO to an aqueous alkaline solution (or emulsion) of thiophenols in high yields.[13] The trifluoroacetylation of amino acids under aqueous conditions can be carried out with S-ethyltrifluoro-thioacetate to give the N-trifluoroacetyl derivatives.[14] However, a major drawback of this procedure is that the strongly odoriferous ethanethiol is released in the course of the reaction. Recently, this problem has been circumvented by the use of S-dodecyltrifluorothioacetate as the acylating agent under aqueous conditions since the liberated dodecanethiol is non-odoriferous (Eq. 9.7).[15] The pH optimization of the Schotten-Baumann reaction in aqueous media has been analyzed and can be used to obtain selective acylation of two amine sites. An example is the selective monoacetylation of 4-aminobenzylamine to give either 4-acetylaminobenzylamine or N-(4-aminobenzyl)acetamide (Eq. 9.8).[16] Matsumura and co-workers found that the addition of a small amount of water dramatically increased the enantioselectivity of a copper-catalyzed enantioselective monoacylation of diols in THF (Eq. 9.9).[17] It was proposed that a thin aqueous phase formed on the surface of the Na_2CO_3 promoted both the reactivity and enantioselectivity through hydrogen bondings.

$$R-NH_2 \quad + \quad (R'CO)_2O \quad \xrightarrow[\substack{R'=CH_3, \\ CH(CH_3)_2, \\ C_6H_5, \\ COCH_2CH_2COOH, \\ COCH=CHCOOH}]{SDS/H_2O} \quad R-NHC(O)R' \qquad (9.5)$$

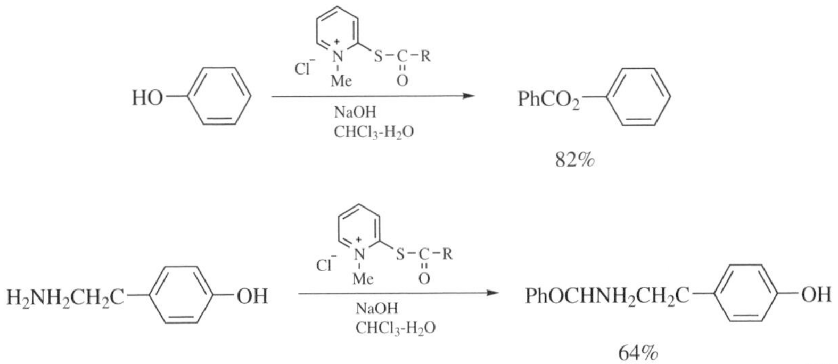

82%

64%

$$(9.6)$$

86–92%

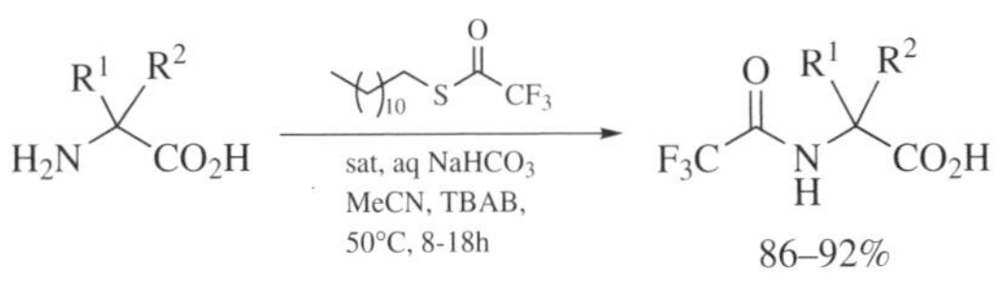

$$(9.7)$$

(9.8)

(84%)

pKa's 3.74, 9.70

(85%)

(9.9)

	yield%	ee%
without water	31	32
with 100μl H$_2$O	42	58

cat =

The reaction of AcOH and EtOH showed that esterification in aqueous solution takes place to a very slight extent and reaches equilibrium with the esterification of about 1.2% AcOH.[18] The addition of NaCl to the solution accelerates the rate of esterification and also changes the equilibrium to 7% ester. The addition of HCl and H$_2$SO$_4$ accelerates the esterification considerably, which also reaches equilibrium quite quickly with 19% of the substance being esterified. The addition of more acid increases the rate of esterification still more but the final equilibrium remains at 19% esterified. NaCl in the same solution increases and accelerates the esterification by the separation of the ester as a separate layer and reaches equilibrium with 45% esterified. Esters, ethers, and thioethers are prepared selectively by condensation reactions in water in the presence of a surfactant, dodecylbenzenesulfonic acid. Dehydrative esterification of carboxylic acids with alcohol in water was efficiently catalyzed by hydrophobic polystyrene-supported sulfonic acids as recoverable and reusable catalysts. In these reactions, esters were obtained in high yields without using any dehydrating agents (Figure 9.1).[19] As illustrated in Figure 9.1, it is proposed that the polystyrene-supported sulfonic acid forms a micelle. The reaction of the acid and the alcohol occurs inside the hydrophobic interior of the micelle followed by expulsion of water into the aqueous surrounding. The sulfonic acid contents of the catalysts and the presence of long

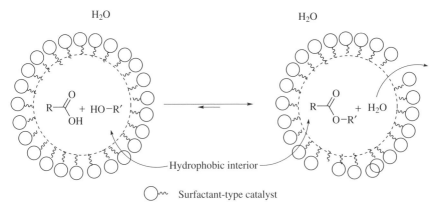

Figure 9.1 Direct esterification by dehydration in the presence of a surfactant-type catalyst in water.

alkyl chains on the benzene rings of polystyrene significantly affected the catalytic activity.

Alternatively, esterification of carboxylic acid can be carried out in aqueous media by reacting carboxylic acid salts with alkyl halides through nucleophilic substitutions (Eq. 9.10).[20] The reaction rate of alkyl halides with alkali metal salts of carboxylic acids to give esters increases with the increasing concentration of catalyst, halide, and solvent polarity and is reduced by water. Various thymyl ethers and esters can be synthesized by the reactions of thymol with alkyl halides and acid chlorides, respectively, in aqueous medium under microwave irradiation (Eq. 9.11).[21] Such an esterification reaction of poly(methacrylic acid) can be performed readily with alkyl halides using DBU in aqueous solutions, although the rate of the reaction decreases with increasing water content.[22]

$$RCO_2^-M^{n+} \ + \ R'X \ \xrightarrow[\text{H}_2\text{O}]{\text{phase-transfer catalyst}} \ RCO_2R' \qquad (9.10)$$

$$RCOCl \ + \ R'OH \ \xrightarrow[\substack{\text{microwave} \\ \text{2–5 min}}]{\text{1\% aq NaOH}} \ \underset{97–99\%}{RCO_2R'} \qquad (9.11)$$

Direct amide formation in aqueous solution between carboxylic acids and amines can occur and the rates are first order in the anion of the acid and the basic form of the amine (Eq. 9.12).[23] *The second-order rate constant is independent of the acidity of the medium.* The condensation reaction of glycine to form di- and triglycine occurs in aqueous solution in the presence of a higher concentration of sodium chloride and copper ions, and forms the corresponding peptides.[24] An aqueous solution of 1 M amino acid and 0.1 M trimetaphosphate maintained at chosen pH values of 8.0–10.5 and at room temperature in the presence of imidazole or 1,2,4-triazole gives rise, after a few days, to the corresponding peptides.[25] Yields are highest when the pH is adjusted with concentrated NaOH or KOH instead of NH_3. In some cases, glycine is quantatively transformed within 10 to 15 days into peptides, mainly di- and tripeptides. Synthesis of peptides in aqueous solutions can be carried out efficiently by means of water-soluble carbodiimides (Eq. 9.13).[26] A variety of peptides containing C-terminal aromatic amino acids were prepared through water-soluble carbodiimides in a reaction in aqueous media at 5–7°C.[27] When serine was used as the amino component, no protection of the HO groups proved necessary. Solid-phase peptide synthesis has been carried out using water-soluble coupling reagents in aqueous media.[28] Resin-bound active esters such as 1-hydroxybenzotriazole (HOBt) esters showed high hydrolysis rates relative to peptide coupling while active esters derived from resin-bound 4-hydroxy-2,3,5,6-tetrafluorobenzoic acid produced mainly dipeptide coupling products when treated with amino acids in water (Eq. 9.14).[29] (p-Hydroxyphenyl)dimethylsulfonium methyl sulfate (Figure 9.2) was found to be an excellent peptide coupling reagent having a water-soluble property and a high reactivity.[30] Interestingly, the formation of peptide bonds in water can be performed by coupling the condensation reaction with the oxidation of carbon monoxide by dioxygen.[31] The reaction takes place on a simple palladium catalyst and is aided by the addition of ionic salts to the solution.

$$R^1COO^- \ + \ H_3\overset{+}{N}R^2 \ \rightleftharpoons \ R^1CONHR^2 + H_2O \qquad (9.12)$$

$$Z{-}Ala{-}OH \ + \ Amines \ \xrightarrow[\substack{EDC \ hydrochloride \\ MeOH}]{0.1 \ eq \ HOAt \ (or \ HOBt)} \ Z\text{-alanine amide} \qquad (9.13)$$

$$(9.14)$$

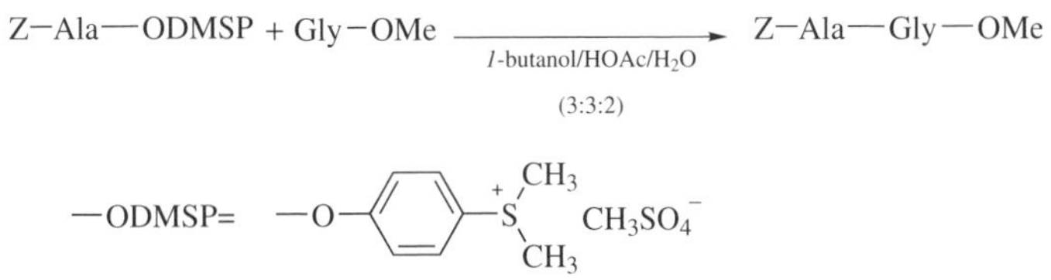

Z—Ala—ODMSP + Gly—OMe $\xrightarrow[\text{(3:3:2)}]{\textit{l}\text{-butanol/HOAc/H}_2\text{O}}$ Z—Ala—Gly—OMe

$$-\text{ODMSP} = \quad -\text{O}-\!\!\!\left\langle\!\!\bigcirc\!\!\right\rangle\!\!-\!\!\overset{+}{\underset{\text{CH}_3}{\overset{\text{CH}_3}{\text{S}}}}\;\;\text{CH}_3\text{SO}_4^-$$

Figure 9.2   Preparation of protected peptide.

9.2.2   Decarboxylation

The decarboxylation of carboxylic acid in the presence of a nucleophile is a classical reaction known as the Hunsdiecker reaction. Such reactions can be carried out sometimes in aqueous conditions. Manganese(II) acetate catalyzed the reaction of α,β-unsaturated aromatic carboxylic acids with NBS (1 and 2 equiv) in MeCN/water to afford haloalkenes and α-(dibromomethyl)benzenemethanols, respectively (Eq. 9.15).[32] Decarboxylation of free carboxylic acids catalyzed by Pd/C under hydrothermal water (250°C/4 MPa) gave the corresponding hydrocarbons (Eq. 9.16).[33] Under the hydrothermal conditions of deuterium oxide, decarbonylative deuteration was observed to give fully deuterated hydrocarbons from carboxylic acids or aldehydes.

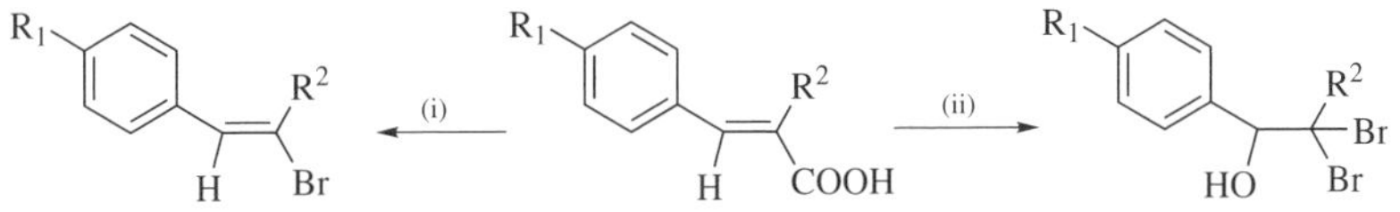

(i) NBS(1 eq)/Mn(OAc)$_2$(0.1 eq)/MeCN-H$_2$O(1:1)/16 h/r.t.
(ii)NBS(2 eq)/Mn(OAc)$_2$(0.1 eq)/MeCN-H$_2$O(1:1)/16 h/r.t.

$$(9.15)$$

$$\text{R}-\text{CO}_2\text{H} \xrightarrow[\text{H}_2\text{O, 250°C, 4–5 MPa}]{\text{10 wt\% Pd/C (3–5 mmol\%)}} \text{R}-\text{H} \qquad (9.16)$$

9.3   CARBOXYLIC ACID DERIVATIVES

9.3.1   Esters and Thiol Esters

The hydrolysis of esters and thiol esters is a classical reaction. In nature, these reactions are catalyzed by a variety of enzymes in an aqueous environment. Chemically, hydrolysis of esters and thiol esters is catalyzed both by acids and bases (Scheme 9.2). There has been

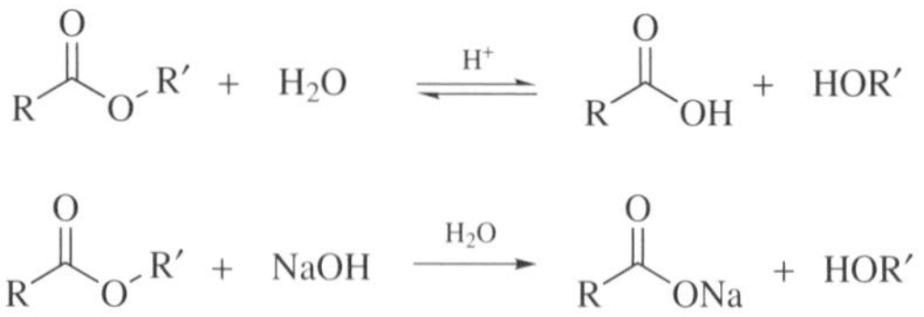

Scheme 9.2

Scheme 9.3

extensive coverage of the subject in various organic chemistry text-books.

On the other hand, the reactions of esters with amines generate the aminolysis products. A theoretical study[34] on ester aminolysis reaction mechanisms in aqueous solution shows that the formation of a tetra-hedral zwitterionic intermediate (Scheme 9.3) plays a key role in the aminolysis process. The rate-determining step is the formation or break-down of such an intermediate, depending on the pH of the medium. Stepwise and concerted processes have been studied by using compu-tation methods. Static and dynamic solvent effects have been analyzed by using a dielectric continuum model in the first case and molecular dynamics simulations together with the QM/MM method in the second case. The results show that a zwitterionic structure is always formed in the reaction path although its lifetime appears to be quite dependent on solvent dynamics.

9.3.2   Acid Halides and Anhydrides

Compared with esters, acid halides and anhydrides are more reactive and are hydrolyzed more readily. It is interesting to note that there is a substantial lifetime for these acid derivatives in aqueous media. Acid halides dissolved in PhCl or in PhBr shaken at a constant rate with water shows that hydrolysis occurs at the boundary between the two liquid phases.[35] The reaction of benzoyl chloride (PhCOCl) and benzoate ion with pyridine N-oxide (PNO) as the inverse phase-transfer catalyst yields both the substitution product (benzoic anhydride) and the

$$RCOCl \ + \ PNO \ \xrightarrow[CH_2Cl_2]{} \ RCOONP^+Cl^-$$

$$RCOONP^+ \ + \ R'CO_2^- \ \xrightarrow[H_2O]{} \ RCOOCOR' \ + \ PNO$$

Scheme 9.4

hydrolysis product (benzoic acid) in a two-phase water/dichloromethane medium (Scheme 9.4).[36] The hydrolysis of benzoic anhydride enjoys two mechanisms of acid catalysis in 60% (vol/vol) dioxane-water as solvent, one at low and another at high values of $[H_3O^+]$; the former becomes effectively undetectable in a purely aqueous solvent.[37]

9.3.3 Amides

As in the case of esters, hydrolysis of amides is also a fundamental reaction in organic chemistry and plays a key role in biological systems. The reaction has been covered extensively in organic chemistry and biochemistry textbooks.

For conversion of amide to other acid derivatives, a novel synthesis of urea glycosides in aqueous media has been reported via the reaction of Steyermark's glucosyl carbamate with amines in good yields (Eq. 9.17).[38] This method was successfully applied to develop a new route to the synthesis of urea-tethered neo-glycoconjugates and pseudo-oligosaccharides.

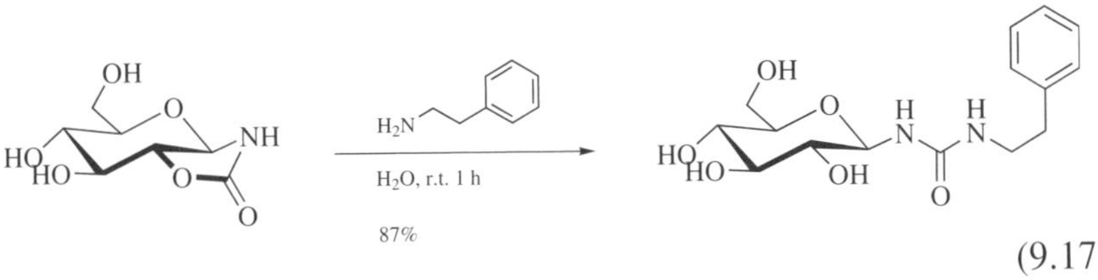

(9.17)

9.3.4 Nitriles

The reaction of aliphatic, aromatic, heterocyclic, conjugated, and poly-hydroxy aldehydes with NBS and ammonia gave the corresponding nitriles in high yields at 0°C in water (Eq. 9.18).[39] Ceric ammonium nitrate (CAN)[40] and iodine[41] are also effective as the oxidizing reagents.

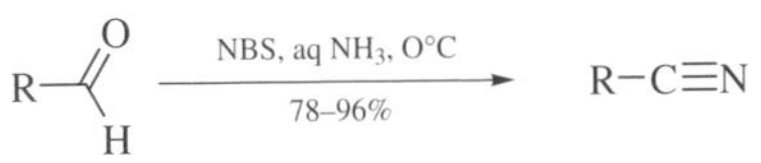

(9.18)

Partial hydrolysis of nitrile gives amides. Conventionally, such reactions occur under strongly basic or acidic conditions.[42] A broad range of amides are accessed in excellent yields by hydration of the corresponding nitriles in water and in the presence of the supported ruthenium catalyst $Ru(OH)_x/Al_2O_3$ (Eq. 9.19).[43] The conversion of acrylonitrile into acrylamide has been achieved in a quantitative yield with better than 99% selectivity. The catalyst was reused without loss of catalytic activity and selectivity. This conversion has important industrial applications.

$$R-C\equiv N \ + \ H_2O \ \xrightarrow[\text{97–99\%}]{Ru(OH)_x/Al_2O_3} \ \underset{R\quad NH_2}{\overset{O}{\parallel}} \qquad (9.19)$$

R = alkyl, aryl, vinyl

REFERENCES

1. Kamochi, Y., Kudo, T., *Chem. Lett.* **1993**, 1495.

2. Mamedov, M. K., Nabieva, E. K., Rasulova, R. A., *Russ. J. Org. Chem.* **2005**, *41*, 974.

3. Maruyama, T., Kozai, S., Takamatsu, S., Izawa, K., *Nucleic Acids Symposium Series* **2000**, *44* 103.

4. Jeon, J. S., Viswanathan, A., Gross, R. A., *Polymer Preprints (ACS Division of Polymer Chemistry)* **1998**, *39*, 142.

5. Braga, C. F., Abreu, C. A. M., Lima Filho, N. M., Queneau, Y., Descotes, G., *React. Kinet. Catal. Lett.* **2002**, *77*, 91.

6. Schotten, C., *Ber.* **1884**, *17*, 2544; Baumann, E., *Ber.* **1886**, *19*, 3218.

7. Georg, G. I., Boge, T. C., Cheruvallath, Z. S., Harriman, G. C. B., Hepperle, M., Park, H., Himes, R. H., *Bioorg. Med. Chem. Lett.* **1994**, *4*, 335.

8. Hattori, T., Kitamura, N., Yamato, N, Yokota, H. (Ajinomoto Co., Inc., Japan), Eur. Pat. Appl. (2003), 13 pp. EP: 1314717, A1: 20030528.

9. Zima, G. C., Williams, T. H. (Eastman Chemical Company, USA), U.S. (1995), 5 pp. US: 5391780, A: 19950221.

10. Naik, S., Bhattacharjya, G., Talukdar, B., Patel, B. K., *Eur. J. Org. Chem.* **2004**, 1254.

11. Lapin, V. V., Tishchenko, D. V., *Zh. Prik.Khim.* **1972**, *45*, 929.

12. Sakakibara, T., Watabe, Y., Yamada, M. Sudoh, R., *Bull. Chem. Soc. Jpn* **1988**, *61*, 247.

13. Akiyoshi, S., Mori, T., Shigematsu, T. (Mitsubishi Chemical Industries Co., Ltd.), Jpn. Tokkyo Koho (1973), 6 pp. JP: 48007423, 19730305.

14. Weygand, F., Spiess, B., *Chem. Ber.* **1964**, *97*, 3456.

15. Hickey, M. R., Nelson, T. D., Secord, E. A., Allwein, S. P., Kress, M. H., *Synlett*, **2005**, 255.

16. King, J. F., Rathore, R., Lam, J. Y. L., Guo, Z. R., Klassen, D. F., *J. Am. Chem. Soc.* **1992**, *114*, 3028.

17. Iwasaki, F., Maki, T., Nakashima, W., Onomura, O., Matsumura, Y., *Org. Lett.* **1999**, *1*, 969; Iwasaki, F., Maki, T., Onomura, O., Nakashima, W., Matsumura, Y., *J. Org. Chem.* **2000**, *65*, 996.

18. Purgotti, A., *Gazz. Chim. Ital.* **1918**, *48*, 58.

19. Manabe, K., Iimura, S., Sun, X.-M., Kobayashi, S., *J. Am. Chem. Soc.* **2002**, *124*, 11971; Manabe, K., Kobayashi, S., *Adv. Synth. Catal.* **2002**, *344*, 270; Manabe, K., Sun, X.-M., Kobayashi, S., *J. Am. Chem. Soc.* **2001**, *123*, 10101.

20. Trautmann, H., Schoellner, R., *J. Prakt. Chem.* **1971**, *313*, 561; Bodroux, F., *Compt. Rend.* **1914**, *157*, 938.

21. More, D. H., Pawar, N. S., Dewang, P. M., Patil, S. L., Mahulikar, P. P., *Russ. J. Gen. Chem.* **2004**, *74*, 217.

22. Nishikubo, T., Kameyama, A., Yamada, Y., Yoshida, Y., *J. Polym. Sci. Part A: Polym. Chem.* **1996**, *34*, 3531.

23. Morawetz, H., Otaki, P. S., *J. Am. Chem. Soc.* **1963**, *85*, 463.

24. Rode, B. M., Schwendinger, M. G., *Orig. Life Evol. Biosph.* **1990**, *20*, 401.

25. Rabinowitz, J., Hampai, A., *J. Mol. Evol.* **1985**, *21*, 199.

26. Nozaki, S., *J. Pept. Res.* **1999**, *54*, 162; Ranganathan, D., Singh, G. P., Ranganathan, S., *J. Am. Chem. Soc.* **1989**, *111*, 1144.

27. Knorre, D. G., Shubina, T. N., *Zhurnal Obshchei Khimii* **1966**, *36*, 656.

28. Hojo, K., Maeda, M., Tanakamaru, N., Mochida, K., Kawasaki, K., *Protein Pept. Lett.* **2006**, *13*, 189.

29. Corbett, A. D., Gleason, J. L., *Tetrahedron Lett.* **2002**, *43*, 1369.

30. Kouge, K., Koizumi, T., Okai, H., Kato, T., *Bull. Chem. Soc. Jpn.* **1987**, *60*, 2409.

31. Remias, J. E., Elia, C. N., Sen, A. Abstracts of Papers, *224th ACS National Meeting*, Boston, MA, August 18–22, 2002, INOR-594.

32. Chowdhury, S., Roy, S., *Tetrahedron Lett.* **1996**, *37*, 2623.

33. Matsubara, S., Yokota, Y., Oshima, K., *Org. Lett.* **2004**, *6*, 2071.

34. Chalmet, S., Harb, W., Ruiz-Lopez, M. F., *J. Phys. Chem. A* **2001**, *105*, 11574.

35. Karve, D., Dole, K. K., *J. Univ. Bomby Sci.* **1938**, *7(Pt. 3)*, 108; *Chem. Abstract.* **1939**, *33*, 37340.

36. Kuo, C. S., Jwo, J. J., *J. Org. Chem.* **1992**, *57*, 1991.

37. Satchell, D. P. N., Wassef, W. N., *J. Chem. Soc. Perkin Trans. 2*, **1992**, 1855.

38. Ichikawa, Y., Matsukawa, Y., Isobe, M., *Synlett* **2004**, 1019.

39. Bandgar, B. P., Makone, S. S., *Synth. Commun.* **2006**, *36*, 1347.

40. Bandgar, B. P., Makone, S. S., *Synlett* **2003**, 262.

41. Talukdar, S., Hsu, J.-L., Chou, T.-C., Fang, J.-M., *Tetrahedron Lett.* **2001**, *42*, 1103; Shie, J.-J., Fang, J.-M., *J. Org. Chem.* **2003**, *68*, 1158.

42. For reviews, see: Beckwith, A. L. J., in *The Chemistry of Amides*, Zabicky, J., ed. Wiley, New York, 1970; Cacchi, S., Misiti, D., La Torre, F., *Synthesis*, **1980**, 243.

43. Yamaguchi, K., Matsushita, M., Mizuno, N., *Angew. Chem. Int. Ed.* **2004**, *43*, 1576.

CHAPTER 10

CONJUGATED CARBONYL COMPOUNDS

10.1 REDUCTION

10.1.1 Hydrogenation

α, β-Unsaturated carbonyl compounds are more electrophilic. Usually, the C=C and C≡C bonds in these compounds are more readily reduced than simple alkenes and alkynes. Furthermore, it is also possible to reduce such electron-deficient C=C and C≡C bonds selectively without affecting the nonconjugated ones. The most common methods for reducing such unsaturated bonds in aqueous media are by catalytic hydrogenation. Many transition-metal catalysts can be used to hydrogenate both activated and unactivated alkenes and alkynes.[1] Some can catalyze the hydrogenation of conjugated alkene and alkyne bonds selectively. For example, half-sandwich Ru-cyclopentadienyl derivatives containing the cage-like water-soluble monodentate phosphine 1,3,5-triaza-7-phosphaadamantane (PTA) show high selectivity to C=C double-bond hydrogenation in aqueous phase or biphasic homogeneous hydrogenation of activated olefins via hydrogen transfer or under hydrogen pressure at mild conditions (Eq. 10.1).[2] A similarly selective hydrogenation of benzylidene acetone was accomplished under biphasic

Comprehensive Organic Reactions in Aqueous Media, Second Edition, by Chao-Jun Li and Tak-Hang Chan
Copyright © 2007 John Wiley & Sons, Inc.

conditions (water/Et$_2$O) using CpRu(PTA)$_2$H as the catalyst. The presence of anions such as BF$_4^-$ and PF$_6^-$ significantly improves the catalytic activity, while Cl$^-$ shuts down the catalysis due to the formation of the catalytically inactive CpRu(PTA)$_2$Cl. The reaction was proposed to proceed via a 1,4-conjugate addition mechanism (Eq. 10.2).[3] Other studies include the use of water-soluble ruthenium-TPPTS complexes.[4]

$$(10.1)$$

$$(10.2)$$

In addition to the use of hydrogen directly, hydrogen generated from CO and water (water–gas shift reaction) is also very effective in hydrogenating activated alkenes under basic conditions (Eq. 10.3).[5]

Interestingly, selenium can also catalyze such a reaction with[6] or without (Eq. 10.4)[7] using a base.

$$RHC{=}CHCOR' \ + \ CO \ + \ H_2O \ \xrightarrow[\text{30°C, 20h}]{\text{Rh}_6(\text{CO})_{16},\ \text{Et}_3\text{N}} \ RCH_2CH_2COR' \ + \ CO_2 \quad (10.3)$$

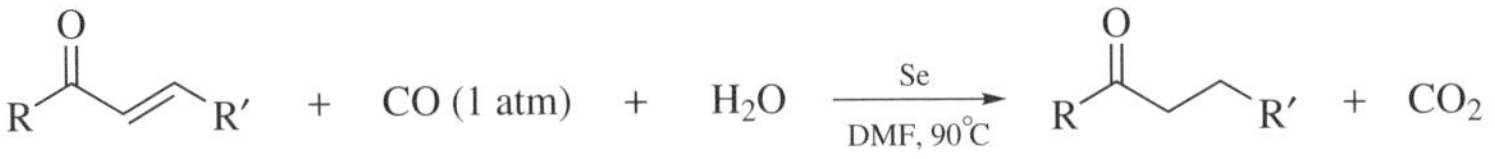

+ CO (1 atm) + H_2O $\xrightarrow[\text{DMF, 90°C}]{\text{Se}}$ + CO_2 (10.4)

10.1.2 Asymmetric Hydrogenation

By using chiral phosphine ligands, hydrogenation of prochiral alkenes can provide optically active products. Asymmetric hydrogenation in an aqueous medium with several water-soluble chiral catalysts has been intensively investigated using a variety of substrates. For example, by using a water-soluble Rh catalyst, an α-acetamidoacrylic ester can be hydrogenated with up to 94% ee (Eq. 10.5).[8] The hydrogenation of a prochiral imine gives a product with up to 96% ee (Eq. 10.6).[9]

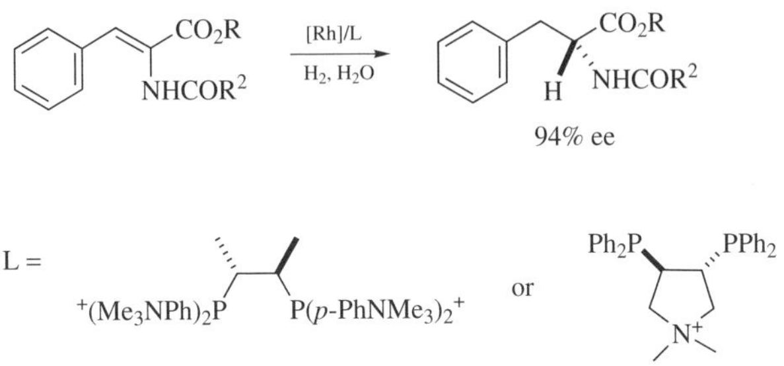

(10.5)

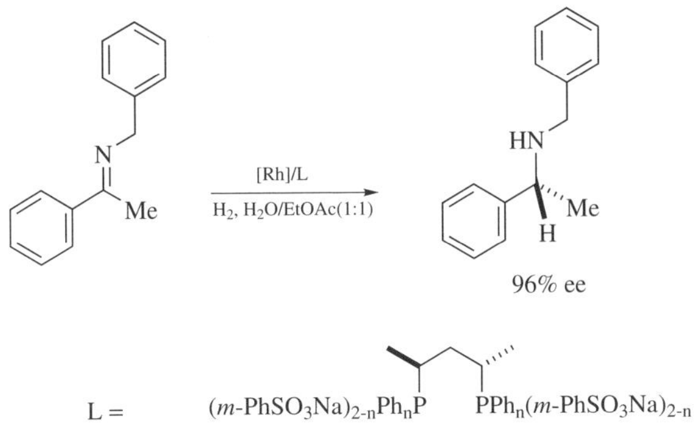

(10.6)

However, the optical yield of the product for asymmetric hydrogenation in an aqueous medium is generally lower than those obtained

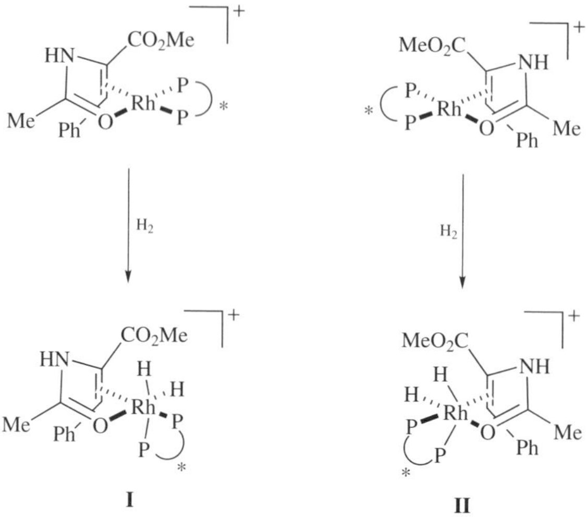

Scheme 10.1

in organic solvents.[10] Often, an increase in water content reduces the enantioselectivity. The decrease of enantioselectivity caused by the addition of water was explained by Amrani and Sinou[11] through the use of Halpern's model[12] for asymmetric hydrogenation (Scheme 10.1). The two enantiomers are generated from two different pathways and the optical yield is determined by the transition-state energy difference. The lower ee in water was attributed to the smaller energy difference between the transition states.

By using a mixture of ethyl acetate and D_2O as solvent for hydrogenation, up to 75% deuterium is incorporated in the reduced product.[13] This result indicates that the role of water here is not only as a solvent. Research on asymmetric hydrogenation in an aqueous medium is still actively being pursued. The method has been applied extensively in the synthesis of various amino acid derivatives.[14]

10.1.3 Reduction by Other Methods

When the conditions are controlled properly, Zn can mediate the reduction of the C–C double bond of α, β-unsaturated carbonyl compounds in the presence of a nickel catalyst in aqueous ammonium chloride (Eq. 10.7). The use of ultrasonication enhances the rate of the reaction.[15] Sodium hydrogen telluride, (NaTeH), prepared in situ from the reaction of tellurium powder with an aqueous ethanol solution of sodium borohydride, is an effective reducing reagent for many functionalities, including activated C=C bonds. Water is a convenient solvent for these transformations.[16] Other reagents effective for the selective reduction of activated alkenes and alkynes include Mg-ZnCl$_2$,[17] Zn/CH$_3$COONH$_4$,[18] SmI$_2$,[19] Co$_2$(CO)$_8$,[20] and sodium dithionite/β-cyclodextrin.[21]

$$(10.7)$$

10.2 EPOXIDATION, DIHYDROXYLATION, HYDROXYAMINATION

Epoxidation of compounds in which a double bond is conjugated to electron-withdrawing groups occurs only very slowly or not at all with peroxy acids or alkyl peroxides. On the other hand, epoxidation with hydrogen peroxide under basic biphase conditions, known as the Weitz-Scheffer epoxidation (Eq. 10.8),[22] is an efficient method for conversion into epoxides. This reaction has been applied to many α, β-unsaturated aldehydes, ketones, nitriles, esters, sulfones, and so on. The reaction is first order in both unsaturated ketone and $^-$O$_2$H through a Michael-type addition of the hydrogen peroxide anion to the conjugated system, followed by ring closure of the intermediate enolate with expulsion of $^-$OH. The epoxidation of electron-deficient olefins can also be performed with hydrogen peroxide in the presence of sodium tungstate as catalyst.[23] Such epoxidation of α, β-unsaturated carbonyl compounds also occurs in ionic liquid/water biphasic conditions.[24]

$$(10.8)$$

Besides the basic hydrogen peroxide reaction, epoxidation of α, β-unsaturated carbonyl compounds can be carried out by using dimethyl-dioxirane in dry acetone (Eq. 10.9).[25] The addition of water, however, resulted in a measurable increase in the rate of the reaction. On the other hand, aziridination of α-bromo-2-cyclopenten-1-one via a conjugated addition–S_N2 substitution generates bicyclic α-keto aziridines efficiently by using aliphatic primary amines mediated by phase-transfer catalysts (PTCs) in water at room temperature (Eq. 10.10).[26] Bicyclic α-keto aziridine derivatives are highly strained and reactive compounds that can be used in the synthesis of biologically active compounds.

(10.9)

(10.10)

91–98%

Dihydroxylation and asymmetric dihydroxylation of electronically deficient conjugate alkenes have been developed in aqueous media. These reactions were discussed in Chapter 3.

10.3   CONJUGATE ADDITION: HETEROATOM

In α, β-unsaturated carbonyl compounds and related electron-deficient alkenes and alkynes, there exist two electrophilic sites and both are prone to be attacked by nucleophiles. However, the conjugated site is considerably "softer" compared with the unconjugated site, based on the Frontier Molecular Orbital analysis.[27] Consequently, softer nucleophiles predominantly react with α, β-unsaturated carbonyl compounds through conjugate addition (or Michael addition). Water is a "hard" solvent. This property of water has two significant implications for conjugate addition reactions: (1) Such reactions can tolerate water since the nucleophiles and the electrophiles are softer whereas water is hard; and (2) water will not compete with nucleophiles significantly in such reactions. Hence, various conjugate additions have been successfully carried out in aqueous media. Soft heteroatom-based nucleophiles such as amines, thiols, phosphines, and some halides are quite effective for such conjugate additions. The use of amines and phosphines as catalysts for Baylis-Hillman reactions is based on such conjugate additions as the key step. The phosphine-catalyzed γ-addition of alkynoates also proceeds via the initial conjugate addition of phosphines. These reactions lead to C–C bond formations and will be discussed in the next section.

The kinetics, products, and stereochemistry of the addition of HCl, HBr, and HI to propiolic acid in water have been studied.[28] The addition is predominantly *trans* to give the *cis*-3-haloacrylic acid. Both the rate of addition and the selectivity giving *trans*-addition increase with the nucleophilicity of the halide in water (i.e., $I^- > Br^- > Cl^-$). The order of reactivity is also consistent with the order of the softness of the nucleophiles. The reaction is first order in propiolic acid and the halide anion. It was proposed that the addition involves two mechanistic pathways: a major *trans*-addition via a transient carbanion formed with specific geometry and a minor *cis*-addition process (Scheme 10.2).

Thiolates, generated in situ by the action of ammonium tetrathiomolybdate on alkyl halides, thiocyanates, and disulfides, undergo conjugate addition to α, β-unsaturated esters, nitriles, and ketones in water under neutral conditions (Eq. 10.11).[29] Conjugate addition of thiols was also carried out in a hydrophobic ionic liquid [bmim]PF$_6$/water-solvent system (2:1) in the absence of any acid catalyst to afford the corresponding Michael adducts in high to quantitative yields with excellent 1,4-selectivity under mild and neutral conditions (Eq. 10.12). The use of ionic liquids helps to avoid the use of either acid or base catalysts

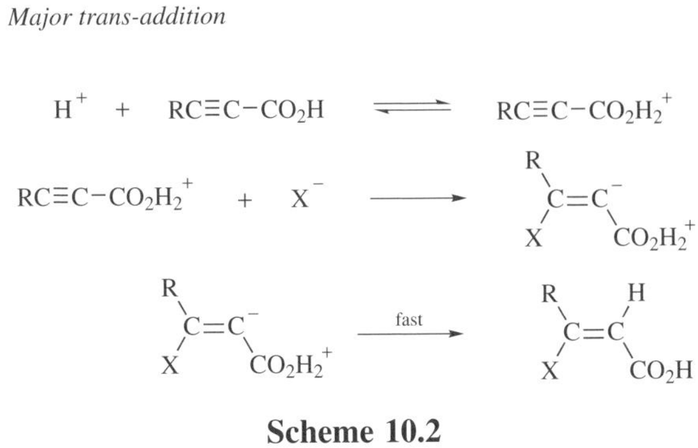

Major trans-addition

Scheme 10.2

for this conversion.[30] An intramolecular Michael addition of thiols from the secondary side of β-cyclodextrin to α, β-unsaturated compounds at the primary side in water was reported (Eq. 10.13).[31] Products of undesirable side reactions resulting from polymerization were not observed; the use of cyclodextrin precluded the use of either acid or base and the catalyst can be recovered and reused.

$$(10.11)$$

73%

$$(10.12)$$

n = 1, 2   R = aryl, naphthyl, benzyl

R = aryl, cyclohexyl   90–98%
X = CHO, COMe, CN, CO$_2$Me, CONH$_2$

$$(10.13)$$

Michael addition of amines to α, β-unsaturated compounds catalyzed by β-cyclodextrin in water generates the corresponding β-amino compounds in excellent yield under mild conditions (Eq. 10.14).[32] The β-cyclodextrin can be recovered and reused in subsequent reactions without loss of activity. Aqueous ionic liquids have also been used as solvent.[33] Boric acid[34] and copper salts[35] were found to be effective catalysts for such additions with aliphatic amines in water (Eq. 10.15). Aromatic amines do not participate effectively in the reaction. A rapid Michael addition of secondary amines has been achieved in good-to-excellent yields in the presence of water under microwave irradiation (Eq. 10.16).[36] In the absence of water and under conventional heating, the reaction does not proceed or takes place in very low yield after a long reaction time. A new route for synthesizing sulfonate-containing amines has been developed by using the Michael-type addition reaction of primary or secondary amines with sodium vinylsulfonate in water (Eq. 10.17)[37] No bases or acids were used in the reaction. The water-soluble amines can be used as ligands in aqueous catalysis.

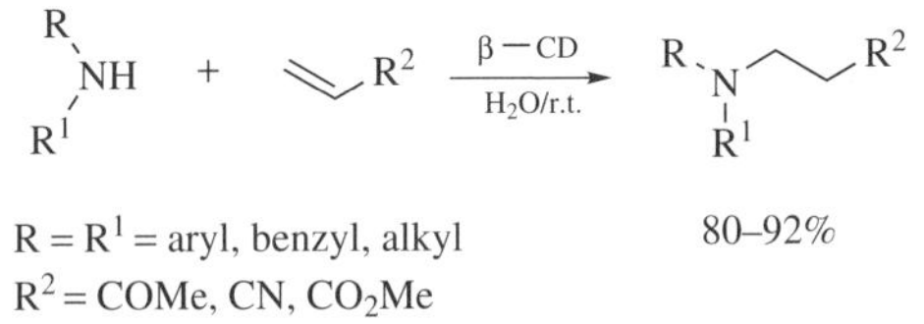
β−CD
H2O/r.t.
80–92%
R = R1 = aryl, benzyl, alkyl
R2 = COMe, CN, CO2Me

(10.14)

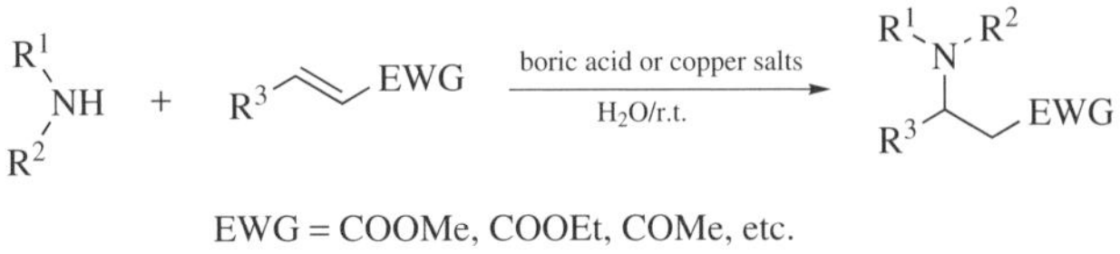
boric acid or copper salts
H2O/r.t.
EWG = COOMe, COOEt, COMe, etc.

(10.15)

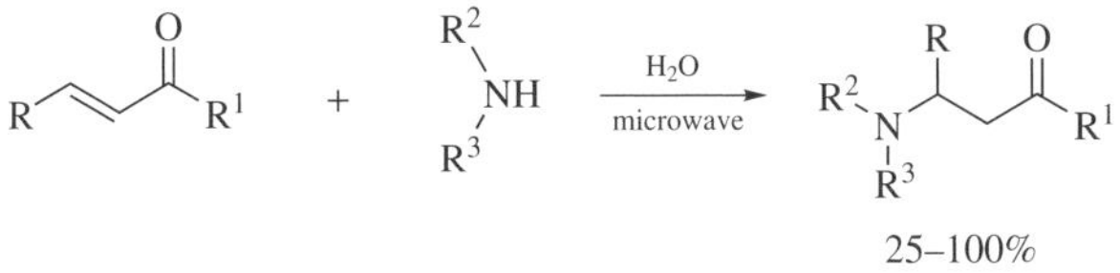
H2O
microwave
25–100%

(10.16)

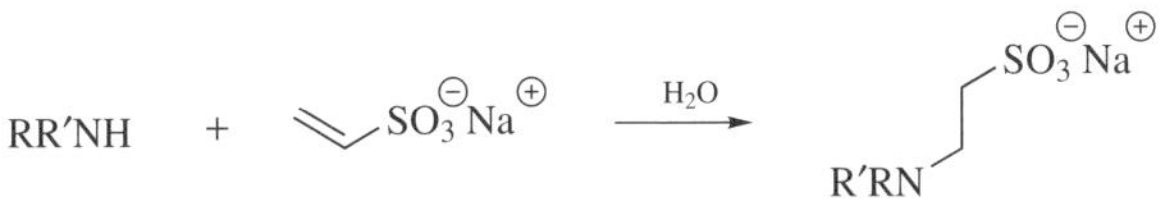
RR′NH
H2O
R′RN

(10.17)

Addition of azide ion to conjugated systems can be carried out by using NaN_3 and acetic acid in water (Eq. 10.18).[38] Some reactions were very rapid while others took 1 to 3 days to complete. Lewis base was found to catalyze such conjugate additions of azide ion to cyclic enones in water.[39]

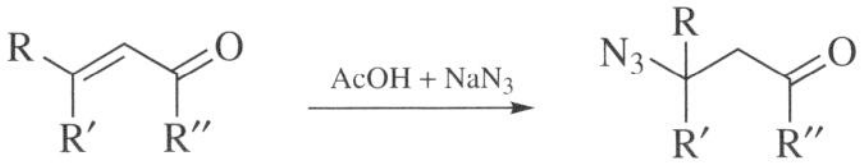
AcOH + NaN3

(10.18)

Highly stabilized phosphorus ylides are prepared from acetylenic esters, a carbon-based nucleophile, and triphenylphosphine in aqueous media.[40] In acetone-water (2:1) solvent, the reaction proceeds via the conjugate addition of triphenylphosphine to dialkyl acetylenedicarboxylates; the resulting vinyl triphenylphosphonium salts undergo Michael addition reaction with a carbon-nucleophile to give the corresponding highly stabilized phosphorus ylides.

10.4 C–C BOND FORMATION

Delocalized carbanions are soft and can undergo conjugate additions readily with unsaturated carbonyl compounds in water. For nonstabilized carbanions, the change of counter-cation can significantly affect their softness and hardness. The use of soft metals as counter-ions greatly enhances the softness of such carbanions, which makes it possible to carry out the conjugate additions in aqueous conditions.

10.4.1 Addition of Hydrogen Cyanide

The conjugate addition of hydrogen cyanide, generated in situ from KCN and acetic acid to β-mesityl ketones, gives high yields of the corresponding oxo nitriles in aqueous ethanol (Eq. 10.19).[41]

$$(10.19)$$

yield up to 92%

10.4.2 Addition of α-Carbonyl Compounds

In the 1970s, Hajos[42] and Wiechert[43] independently reported that the Michael addition of 2-methylcyclopentane-1,3-dione to vinyl ketone in water gives the corresponding conjugate addition product without the use of a basic catalyst. The Michael addition product further cyclizes to give a 5–6 fused-ring system. The use of water as solvent is significantly superior both in yield and the purity of the product in comparison with the same reaction in methanol with base. A similar enhancement of reactivity was found in the Michael addition of 2-methyl-cyclohexane-1,3-dione to vinyl ketone, which eventually led to optically pure Wieland-Miescher ketone.[44] The reaction, however, proceeds under more drastic conditions. Deslongchamps extended the aqueous Michael addition to acrolein.[45] The study has been applied to the total synthesis of 13-α-methyl-14α-hydroxysteroid. The addition of ytterbium triflate (Yb(OTf)$_3$) further enhances the rate of the Michael addition reactions in water (Eq. 10.20).[46]

$$(10.20)$$

A significant acceleration of Michael addition was reported by Lubineau in the reaction of nitroalkanes with buten-2-one when the reaction media was changed from nonpolar organic solvents to water

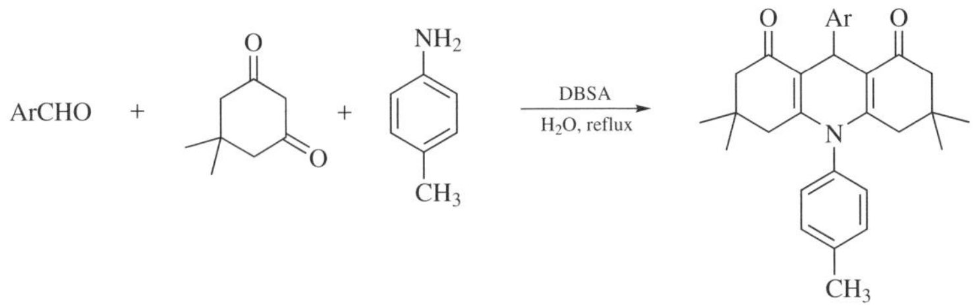

ArCHO +

Scheme 10.3

10.1   Ar = p-ClC$_6$H$_4$–

(Eq. 10.21).[47] Additives, such as glucose and saccharose, further increased the rate of the reaction. The Michael addition reaction between ascorbic acid and cyclohexenone was effectively carried out in water in the presence of an inorganic acid rather than a base as catalyst (Eq. 10.22).[48] A one-pot clean synthesis of 1,8-dioxohexahydroacridines was carried out in aqueous media by using p-dodecylbenzenesulfonic acid as a Brönsted acid-surfactant-combined catalyst (Scheme 10.3).[49] Thus, the Knoevenagel reaction/conjugate addition/cyclization/elimination reaction of 5,5-dimethyl-1,3-cyclohexanedione with 4-methylaniline and 4-chlorobenzaldehyde gave 9-(4-chlorophenyl)-3,4,6,7,9,10-hexahydro-3,3,6,6-tetramethyl-10-(4-methylphenyl)-1,8(2H,5H)-acridinedione (**10.1**).

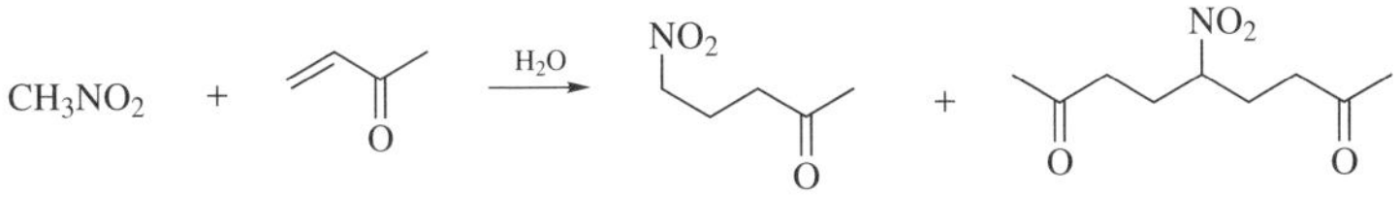

$$(10.21)$$

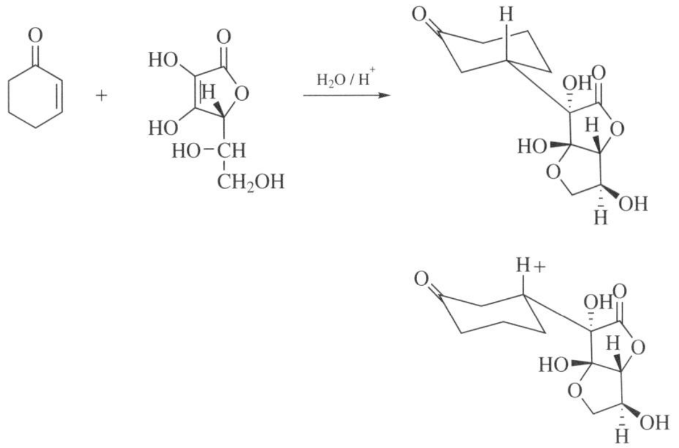

$$(10.22)$$

The Michael reaction of benzylidene acetophenone and benzylidene acetone with ethyl acetoacetate, nitromethane, and acetylacetone was studied by Musaliar and co-workers in the presence of a cetyltrimethylammonium bromide–containing aqueous micellar medium.[50] The Michael reaction of various nitro alkanes with electrophilic alkenes is performed in NaOH (0.025–0.1 M) in the presence of cetyltrimethylammonium chloride (CTACl) without any organic solvent (Eq. 10.23).[51]

$$\text{EWG} = \text{COMe, CN, SO}_2\text{Ph} \tag{10.23}$$

Jenner investigated the kinetic pressure effect on some specific Michael and Henry reactions and found that the observed activation volumes of the Michael reaction between nitromethane and methyl vinyl ketone are largely dependent on the magnitude of the electrostriction effect, which is highest in the lanthanide-catalyzed reaction and lowest in the base-catalyzed version. In the latter case, the reverse reaction is insensitive to pressure.[52] Recently, Kobayashi and co-workers reported a highly efficient Lewis-acid-catalyzed asymmetric Michael addition in water.[53] A variety of unsaturated carbonyl derivatives gave selective Michael additions with α-nitrocycloalkanones in water, at room temperature without any added catalyst or in a very dilute aqueous solution of potassium carbonate (Eq. 10.24).[54]

$$n = 1,2,3,5,7$$
$$Z = \text{COMe, COEt, CHO, CO}_2\text{Me, CN, SO}_2\text{Ph}$$

yield up to 99%

$$\tag{10.24}$$

10.4.3 Addition of Allyl Groups

It was reported that the indium-mediated Michael addition of allyl bromide to 1,1-dicyano-2-arylethenes proceeded well in an aqueous medium.[55] Similarly, cyclopentadienylindium(I) was reported to add in a Michael reaction fashion to electron-deficient alkenes giving substituted cyclopentadienes (Eq. 10.25).[56]

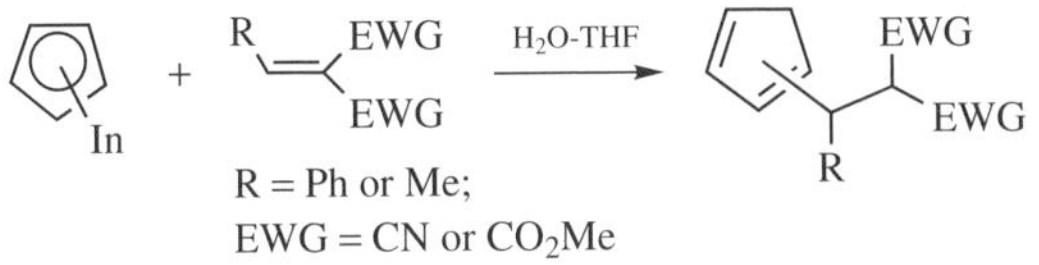

$$(10.25)$$

R = Ph or Me;
EWG = CN or CO$_2$Me

10.4.4 Addition of Alkyl Groups

Luche reported that when a zinc-copper couple was used, alkyl halides reacted with conjugated carbonyl compounds and nitriles to give 1,4-addition products in good yields under sonication conditions (Eq. 10.26).[57]

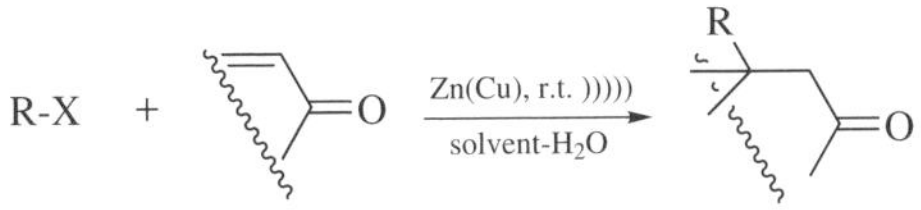

$$(10.26)$$

A moderate diastereoselectivity was observed in these reactions where a mixture of diastereomers could be generated.[58] The reactivity of the halides followed the order of tertiary > secondary ≫ primary and iodide > bromide (chlorides did not react). The preferred solvent system was aqueous ethanol. The process was suggested to proceed by a free radical mechanism occurring on the metal surface under sonochemical conditions. Efforts to trap the intermediate [A] intramolecularly gave only a very low yield of the cyclization product (Scheme 10.4).[59]

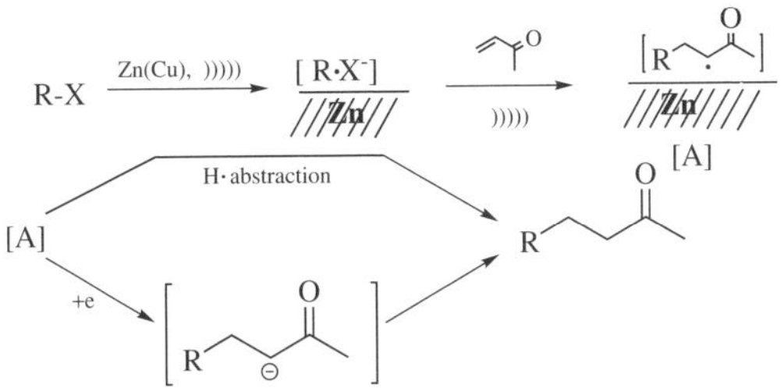

Scheme 10.4

Similar additions also occurred on vinylphosphine oxides. When the optically active vinylphosphine oxide was used, P-chiral alkylphosphine oxide was obtained with retention of the configuration (Eq. 10.27)[60]

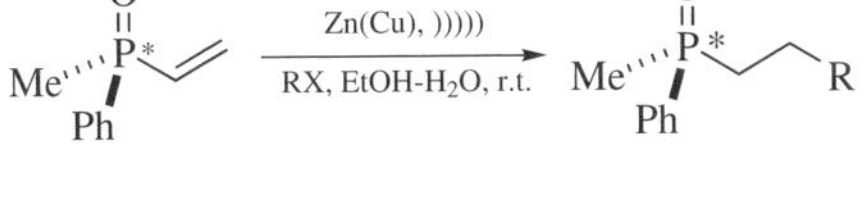

$$(10.27)$$

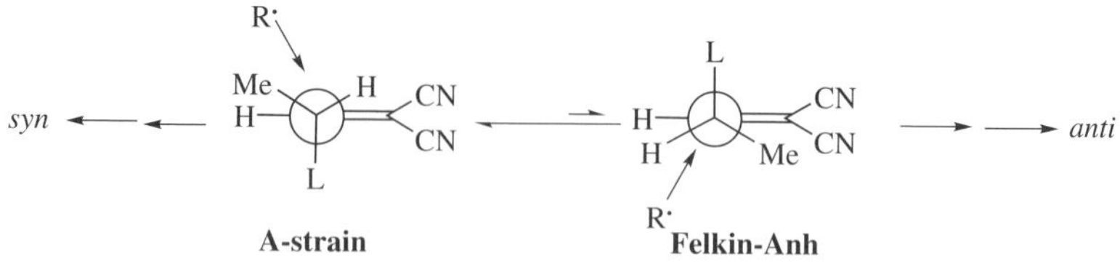

Scheme 10.5

Giese studied the diastereoselectivity associated with such a conjugated addition in water (Eq. 10.28).[61] *Anti*-isomer was found as the major product if the attacking radical was bulky, which can be explained by the argument that the more stable "A-strain" conformer of the alkene reacts more slowly with the bulky alkyl radical than the less stable "Felkin-Anh" conformer (Scheme 10.5). A new silica-supported zinc-copper matrix dramatically promotes conjugate addition of alkyl iodides to alkenenitriles in water.[62] Conjugate additions with ω-chloroalkyl iodides generate cyclic nitriles primed for cyclization, providing one of the few annulation methods for cyclic alkenenitriles.

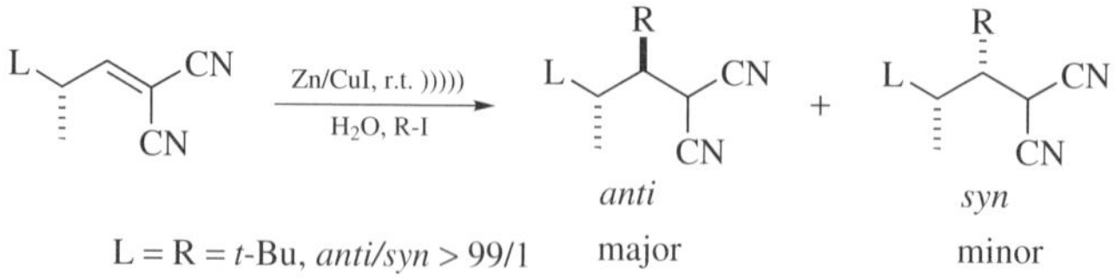

$$L = R = t\text{-Bu}, \, anti/syn > 99/1 \quad \text{major} \qquad \qquad \text{minor} \tag{10.28}$$

The diastereoselective ultrasonically induced zinc-copper 1,4-addition of alkyl iodides to chiral α, β-unsaturated systems in aqueous media was studied by Suares and co-workers: the *Z*-isomer gives good diastereoselectivities while reactions with the *E*-isomer are nonstereoselective.[63] The 1,4-addition to chiral γ, α-dioxolanyl-α, β-unsaturated esters also proceeds with good yields (51–99%) (Eq.10.29).[64]

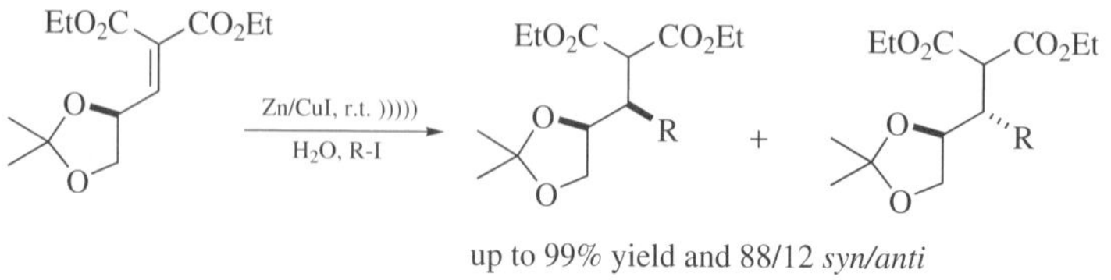

up to 99% yield and 88/12 *syn/anti*

Li and co-workers reported the conjugate addition of an alkyl group to enamides mediated by zinc in aqueous NH_4Cl to generate α-amino acid derivatives (Eq. 10.30).[65] Miyabe et al.[66] as well as Jang and

Cho[67] reported the addition of alkyl radicals from alkyl iodide to α,β-unsaturated ketones, esters, and nitriles mediated by indium in aqueous media. Recently, enantiomerically pure natural and unnatural α-amino acids have been synthesized from a chiral methyleneoxazolidinone by such a highly diastereoselective 1,4-conjugate addition of alkyl iodides in aqueous media (Eq. 10.31).[68] The zinc-copper conjugate addition reaction exhibits high chemoselectivity with the possibility of using functionalized iodides to afford a single diastereomer in short reaction times with good yields.

$$(10.30)$$

38–94%   (20–82% de)

$$(10.31)$$

10.4.5 Addition of Vinyl and Aryl Groups

Miyaura and co-workers[69] reported the Rh(I)-catalyzed conjugate addition of aryl- or 1-alkenylboronic acids, $RB(OH)_2$ to enones in high yields at 50°C in an aqueous solvent. A combination of $(acac)Rh(CO)_2$ and dppb was highly effective for the addition to acyclic and cyclic enones. For example, a 96% yield of 2-phenyl-4-octanone was obtained from $PhB(OH)_2$ and 2-octen-4-one in aqueous MeOH in the presence of $(acac)Rh(CO)_2$ and dppb. Since then, extensive studies have been carried out on the boronic acid chemistry largely related to conjugated additions including asymmetric conjugated additions, most noticeably by Hayashi and co-workers.[70] For example, reactions of α,β-unsaturated ketones with excess arylboronic acids in the presence of a rhodium catalyst generated in situ from $Rh(acac)(C_2H_4)$ and $(3S)$-4,4'-bis-(diphenylphosphino)-2,2',6,6'-tetramethoxy-3,3'-bipyridine $((S)$-P-Phos) in dioxane/water at 100°C gave high yields of the corresponding products in up to 99% ee (Eq. 10.32).[71] Arylboronate esters bearing a pendant Michael-acceptor alkene can add to norbonene and cyclize

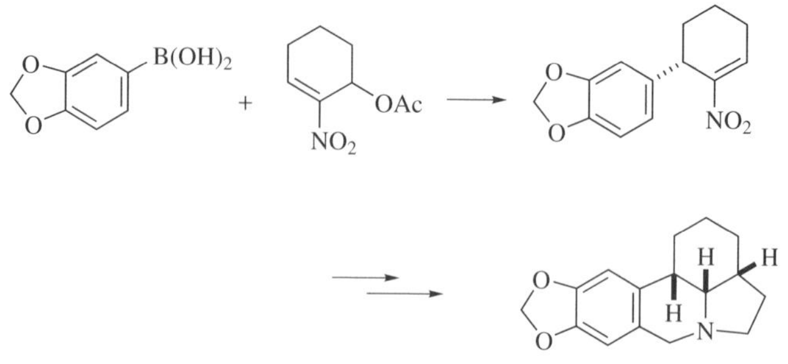

(+)-γ-lycorane

Scheme 10.6

to give indane systems in yields ranging from 62% to 95% with high diastereomeric excess (>20:1).[72] The reaction is accelerated by bases and ligands.[73] An interesting application of the reaction is the pseudo-asymmetric allylic arylation of 2-nitro-cyclohex-2-enols ester which occurs by asymmetric conjugate addition followed by elimination (Eq. 10.33).[74] Application of this reaction has led to an elegant synthesis of the alkaloid (+)-γ-lycorane (Scheme 10.6).

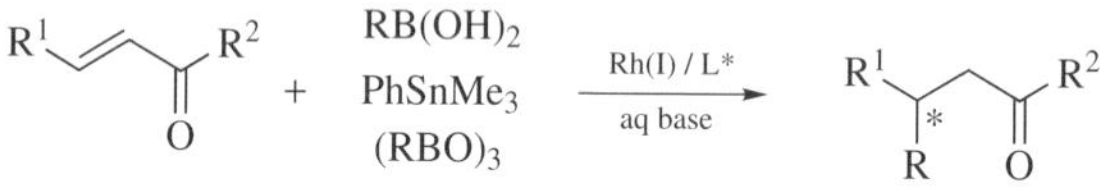

$$(10.32)$$

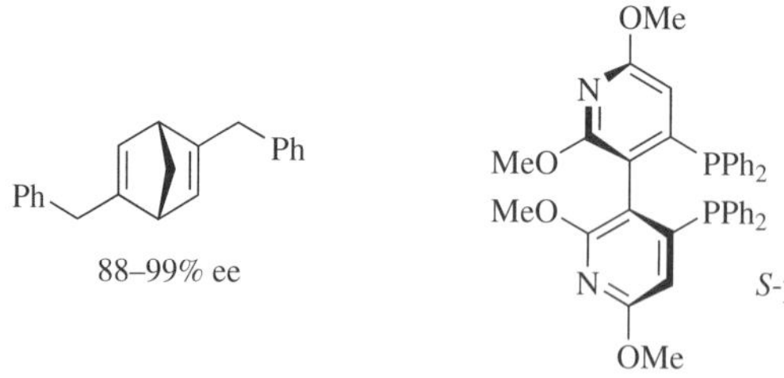

88–99% ee

S-phos, up to 99% ee

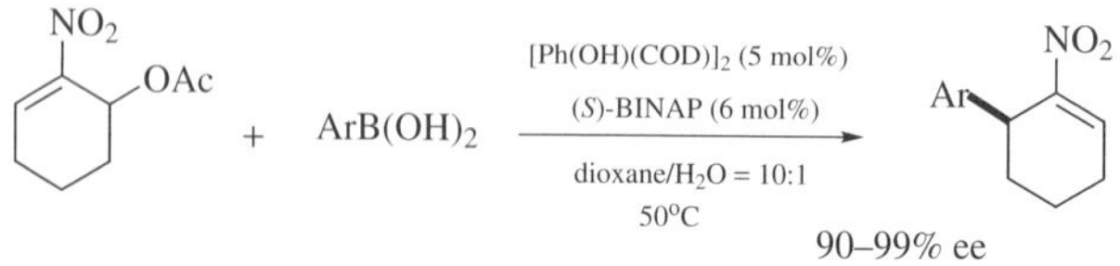

$$(10.33)$$

90–99% ee

Li and co-workers examined the addition of various aryl and vinyl organometallic reagents to α,β-unsaturated carbonyl compounds in air and water. It was found that both $Rh_2(COD)_2Cl_2$ and $Rh(COD)_2BF_4$ are effective.[75] The organometallic reagents include: organotin,[76] organo-indium, organobismuth,[77] organolead,[78] and organosilicon compounds (arylhalosilanes and arylsilanols) (Eq. 10.34)[79] in addition to organo-boron compounds. For the conjugated addition, both ketones (linear and cyclic) and esters were effective as the electron-withdrawing functional groups. When either a mono- or disubstituted unsaturated C=C was involved, the reaction proceeded rapidly. In some cases, a mixture of several products including both the conjugated addition and Heck-type reaction products was observed for the monosubstituted derivatives. Either no reaction was observed or very low yields of the products were obtained with trisubstituted derivatives. A novel synthesis of α-amino acids was developed by this method in air and water using α-amidoacrylates (Eq. 10.35).[80] The rhodium-catalyzed addition of boronic acids to α-substituted activated alkenes proceeds smoothly in water, resulting in a useful synthesis of both racemic 2-substituted succinic esters and β-amino acid derivatives. The catalyst [Rh(COD)Cl$_2$] (COD = cycloocta-1,5-diene) was used with sodium dodecyl sulfate (SDS) as a phase-transfer catalyst. The conjugate additions of organo-siloxanes to α,β-unsaturated carbonyl compounds catalyzed by a cationic rhodium complex (2 mol% [Rh(cod)(MeCN)$_2$]BF$_4$) in water-containing solvent (dioxane/H$_2$O = 10:1) give arylation products in yields of 72–97%.[81]

$$R_nMX_m \ + \quad \underset{R'}{\overset{O}{\diagdown}}\!\!\!\diagup\!\!\!\underset{R''}{\diagdown} \quad \xrightarrow[\text{air / water}]{\text{cat. Rh(I)}} \quad \underset{R'}{\overset{R \quad O}{\diagup\!\!\!\diagdown\!\!\!\diagup\!\!\!\diagdown R''}} \qquad (10.34)$$

R = aryl or vinyl
M = B, In, Si, Sn, Ge, Pb, As, Sb, Bi
X = alkyl, aryl, halogen, hydroxyl, and alkoxyl

$$\text{(structure)} \quad \xrightarrow[\substack{\text{cat. Rh(I)/H}_2\text{O/r.t./sonication} \\ \text{air atmosphere}}]{\text{ArSnR}'_3} \quad \text{(structure)} \qquad (10.35)$$

Rhodium-catalyzed Heck-type coupling of boronic acids with activated alkenes was carried out in an aqueous emulsion.[82] The couplings between arylboronic acids and activated alkenes catalyzed by a water-soluble *tert*-butyl amphosrhodium complex were found to progress at room temperature to generate Heck-type products with high yields and excellent selectivity. It was necessary to add two equivalents of the alkene component, whereby one equivalent is believed to act as a sacrificial hydride acceptor in this reaction.

10.4.6   Addition of Alkynyl Groups

The addition of alkynyl groups is discussed in Chapter 4, Alkynes.

10.4.7   Other Conjugate Additions

The hydrocyanation of conjugated carbonyl compounds is a related reaction.[83] Very often such a conjugated addition is carried out in aqueous conditions. For example, in the pioneer work of Lapworth, hydrocyanation of activated olefins was carried out with KCN or NaCN in aqueous ethanol in the presence of acetic acid (Eq. 10.36).[84]

$$PhCH=C(CO_2Et)_2 \xrightarrow[\text{aq EtOH}]{\text{NaCN-HOAc}} PhHC-CH(CO_2Et)_2$$
$$| \atop CN$$

93–96% yield

$$(10.36)$$

10.5   OTHER REACTIONS

10.5.1   Reductive Coupling

The production of adiponitrile is an important industrial process involving the electrohydrodimerization (EHD) of acrylonitrile. Adiponitrile is used as an important precursor for hexamethylenediamine and adipic acid, the monomers required for the manufacture of the nylon-66 polymer. The annual production of adiponitrile is about a million tons.[85] Although it was initially studied in the 1940s,[86] the electroreductive coupling of acrylonitrile to adiponitrile was only commercialized more than a decade later after Baizer (at Monsanto) developed the supporting electrolyte. It was found that a 90% yield of adiponitrile could be achieved when a concentrated solution of certain quaternary ammonium salts (QAS), such as tetraethylammonium p-toluenesulfonate, is used together with lead or mercury cathodes (Eq. 10.37).[87] Initially, Monsanto employed a divided cell for the EHD process, which was soon replaced by an undivided-cell process because of several shortcomings with the former process.[88] The undivided-cell system involves electrolysis of a dilute solution of acrylonitrile in a mixed sodium phosphate–borate electrolyte using a cadmium cathode and a carbon steel anode. The presence of a quaternary ammonium salt is essential for the adiponitrile selectivity.

$$2 \; \diagup CN + 2H_2O \; \xrightarrow{2e^-} \; NC \diagdown\diagup\diagdown\diagup_{CN} \; + \; 2\,{}^-OH \qquad (10.37)$$

Zinc-mediated reductive dimerization cyclization of 1,1-dicyano-alkenes occurs to give functionalized cyclopentenes in good yields under saturated aqueous NH_4Cl-THF solution at room temperature. The *trans* isomers are the major products (Eq. 10.38).[89]

total yield of *cis* and *trans* 54–90%
trans:cis up to 84:16

$$(10.38)$$

10.5.2 Baylis-Hillman Reactions

For Michael addition with nitrogen nucleophiles, a quantitative study of the Michael addition of activated olefins by using substituted pyridines as nitrogen nucleophiles in water was also reported.[90] In this report, the rate-determining step was investigated. A related reaction between activated olefins with aldehydes in the presence of tertiary amines, the so-called Baylis-Hillman reaction, generates synthetically useful allyl alcohols.[91] In some cases, an aqueous medium was used for the reaction. The reaction, however, is generally very slow, requiring several days for completion. Recently, Augé et al. studied the reaction in an aqueous medium in detail.[92] A significant increase in reactivity has been observed when the reaction is carried out in water (Eq. 10.39). The addition of lithium or sodium iodide further increases the reactivity. Most of the studies were done with DABCO as the catalyst.

$$(10.39)$$

A practical and efficient set of conditions were developed using a stoichiometric base catalyst, 1,4-diazabicyclo[2,2,2]octane (DABCO) and an aqueous medium to overcome problems commonly associated with the Baylis-Hillman reaction such as low reaction yields and long reaction time.[93] Acrylamide reacted with aromatic aldehydes to give the corresponding 3-hydroxy-2-methylenepropionamides in 61–99% yield (Eq. 10.40).[94]

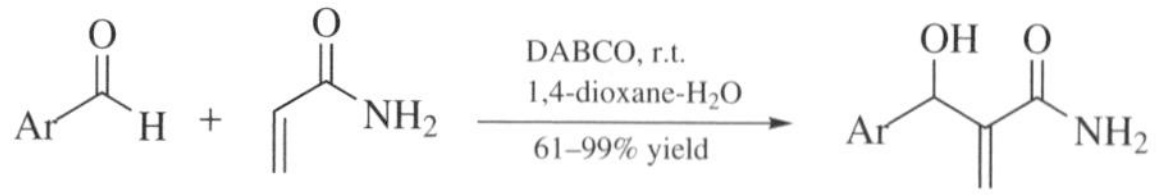

$$(10.40)$$

A convenient synthesis of 2-methylenealkanoates and alkanenitriles is accomplished via the regioselective nucleophilic addition of hydride ion from NaBH$_4$ to (2Z)-2-(bromomethyl)alk-2-enoates and 2-(bromomethyl)alk-2-enenitriles respectively in the presence of DABCO in aqueous media (Eq. 10.41).[95] The reaction of the DABCO [1,4-diazabicyclo[2.2.2]octane] salt of Baylis-Hillman acetate with tosylamide in aqueous THF gave the Baylis-Hillman adduct of N-tosylimine in good yield.[96] Similarly, the reaction of the DABCO salts, generated in situ from the Baylis-Hillman acetates, and KCN in aqueous THF gave ethyl 3-cyano-2-methylcinnamates or 3-cyano-2-methylcinnamonitriles in good yields (Scheme 10.7).

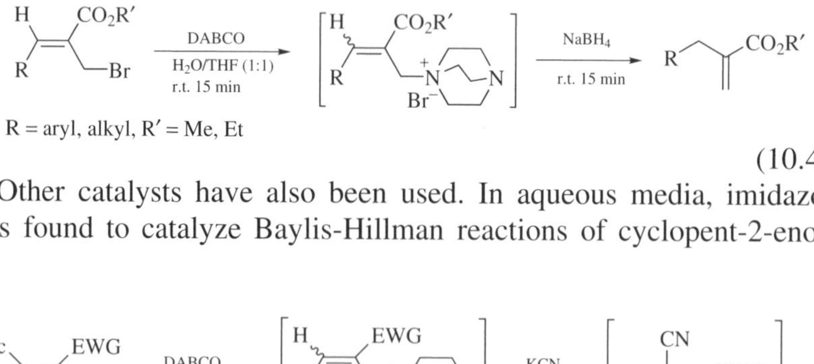

R = aryl, alkyl, R' = Me, Et

$$(10.41)$$

Other catalysts have also been used. In aqueous media, imidazole was found to catalyze Baylis-Hillman reactions of cyclopent-2-enone

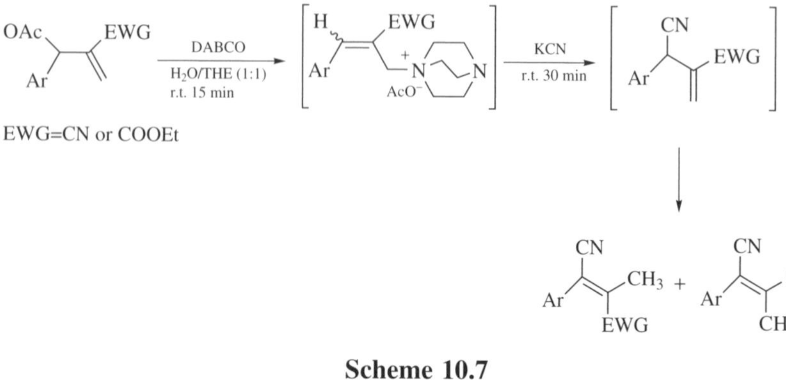

EWG=CN or COOEt

Scheme 10.7

with various aldehydes to afford the desired adducts in high yields (Eq. 10.42).[97] Reaction of 2-cyclohexenones with aqueous formaldehyde, catalyzed by DMAP in THF, affords the corresponding 2-(hydroxymethyl)-2-cyclohexenone in good yields (Eq. 10.43).[98] Trimethylamine-mediated Baylis-Hillman coupling of alkyl acrylates with aldehydes proceeds in aqueous media (Eq. 10.44).[99] In homogeneous H_2O/solvent medium, the reaction rate of aromatic aldehydes and acrylonitrile or acrylate was greatly accelerated, which led to a shorter reaction time, a lower reaction temperature, and a higher yield.[100] High pressure (ca. 200 MPa), generated by freezing H_2O in a sealed autoclave, was successfully applied to the Baylis-Hillman reaction, in which an efficient rate enhancement was observed (Eq. 10.45).[101] The dominant role in the Baylis-Hillman reaction is attributed to hydrogen bonding rather than the hydrophobic effect.[102] The reactivity of a variety of quinuclidine-based catalysts in the Baylis-Hillman reaction has been examined and a straightforward correlation between the basicity of the base and reactivity has been established, without any exceptions. The following order of reactivity was established with the pKa of the conjugate acids (measured in water) given in parentheses: quinuclidine (11.3), 3-hydroxyquinuclidine (9.9), DABCO (8.7), 3-acetoxyquinuclidine (7.2) (Eq. 10.46).[103]

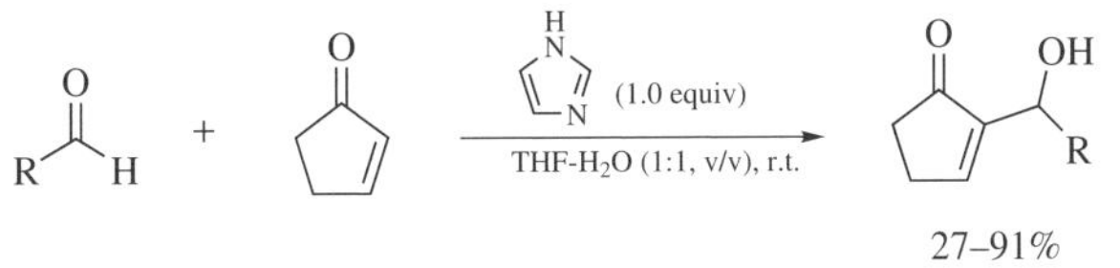

$$(10.42)$$

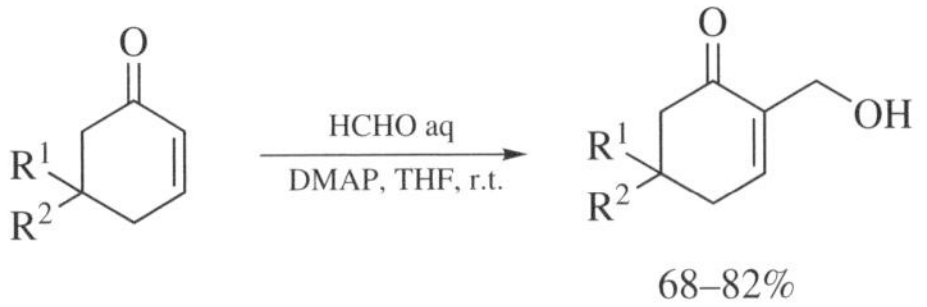

27–91%

$$(10.43)$$

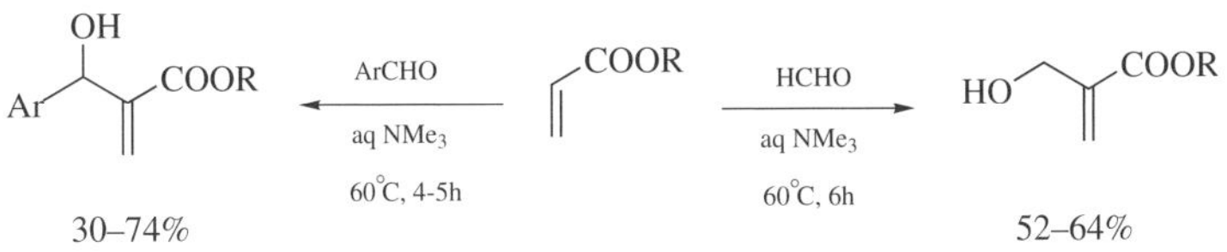

68–82%

$$(10.44)$$

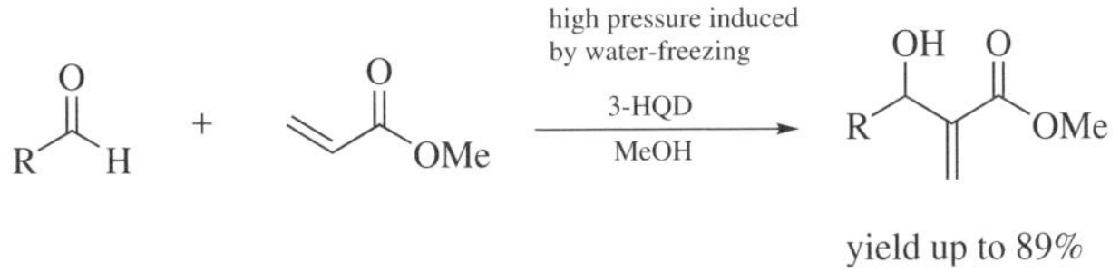

30–74%   52–64%

$$(10.45)$$

yield up to 89%

$$(10.46)$$

The asymmetric Baylis-Hillman reaction of sugar-derived aldehydes as chiral electrophiles with an activated olefin in dioxane:water (1:1) proceeded with 36–86% de and in good yields of the corresponding glycosides (Eq. 10.47).[104] The use of chiral N-methylprolinol as a chiral base catalyst for the Baylis-Hillman reaction of aromatic aldehydes with ethyl acrylate or methyl vinyl ketone gave the adducts in good yields with moderate-to-good enantioselectivities in 1,4-dioxane:water (1:1, vol/vol) under ambient conditions.[105]

$$(10.47)$$

It should be noted that catalytic amounts of *bis*-arylureas and *bis*-arylthioureas greatly accelerated the DABCO-promoted Baylis-Hillman reaction of aromatic aldehydes with methyl acrylate in the absence of solvent. These robust organocatalysts were better mole-per-mole promoters of the reaction than either methanol or water and they were recovered in higher yields.[106]

10.5.3 γ-Addition of Alkynoates

While the Michael-addition of α,β-unsaturated carbonyl and related compounds constitutes one of the most important fundamental synthetic reactions in organic chemistry, the γ-*addition* developed by Trost and Li of nucleophiles to 2-alkynoates catalyzed by a phosphine provides a complementary C–C bond formation method.[107] Recently, Li and co-workers reported an efficient γ–addition of carbon nucleophiles to alkynoates bearing an electron-withdrawing group in aqueous media in the presence of either phosphine or polymer-supported triphenylphosphine and without using any co-catalyst (Eq. 10.48).[108] In this system, water serves the role of both solvent and a proton shuttle. The polymer-supported triphenylphosphine catalyst can be easily recovered by simple filtration and reused as a catalyst for subsequent reactions. Furthermore, inert gas atmosphere was not required for this catalytic reaction.

$$(10.48)$$

REFERENCES

1. Joo, F., *Aqueous Organometallic Catalysis*, Springer-Verlag, 2001.

2. Bolano, S., Gonsalvi, L., Zanobini, F., Vizza, F., Bertolasi, V., Romerosa, A., Peruzzini, M., *J. Mol. Catal. A: Chem.* **2004**, *224*, 61.

3. Mebi, C. A., Frost, B. J., *Organometallics* **2005**, *24*, 2339.

4. Hernandez, M., Kalck, P., *J. Mol. Catal. A: Chem.* **1997**, *116*, 131.

5. Joh, T., Fujiwara, K., Takahashi, S., *Bull. Chem. Soc. Jpn.* **1993**, *66*, 978.

6. Nishiyama, Y., Makino, Y., Hamanaka, S., Ogawa, A., Sonoda, N., *Bull. Chem. Soc. Jpn.* **1989**, *62*, 1682.

7. Tian, F., Lu, S., *Synlett* **2004**, 1953.

8. Toth, I., Hanson, B. E., Davis, M. E., *Tetrahedron Asymmetry* **1990**, *1*, 895.

9. Bakos, J., Orosz, A., Heil, B., Laghmari, M., Lhoste, P., Sinou, D., *J. Chem. Soc., Chem. Commun.* **1991**, 1684; Lensink, C., de Vries, J. G., *Tetrahedron Asymmetry* **1992**, *3*, 235.

10. Lecomte, L., Sinou, D., Bakos, J., Toth, I., Heil, B., *J. Organomet. Chem.* **1989**, *370*, 277.

11. Sinou, D., Amrani, Y., *J. Mol. Catal.* **1986**, *36*, 319.

12. Halpern, J., in Morrison, J. D. ed., *Asymmetric Synthesis,* Vol. 5, Academic Press, Orlando, FL, 1985.

13. Laghmari, M., Sinou, D., *J. Mol. Catal.* **1991**, *66*, L15.

14. RajanBabu, T. V., Yan, Y.-Y., Shin, S., *Current Organic Chemistry*, **2003**, *7*, 1759; Dwars, T., Oehme, G., *Adv. Synth. Catal.* **2002**, *344*, 239.

15. Petrier, C., Luche, J. L., Lavaitte, S., Morat, C., *J. Org. Chem.* **1989**, *54*, 5313.

16. Petragnani, N., Comasseto, J. V., *Synthesis*, **1986**, 1; Kambe, N., Kondo, K., Morita, S., Murai, S., Sonoda, N., *Angew. Chem.* **1980**, *92*, 1041.

17. Saikia, A., Barthakur, M. G., Boruah, R. C., *Synlett* **2005**, 523.

18. Zhou, Y.-B., Wang, Y.-L., Wang, J. Y., *J. Chem. Res.* **2004**, 118.

19. Dahlen, A., Hilmersson, G., *Tetrahedron Lett.* **2003**, *44*, 2661.

20. Lee, H.-Y., An, M., *Tetrahedron Lett.* **2003**, *44*, 2775.

21. Fornasier, R., Marcuzzi, F., Tonellato, U. J., *Inclus. Phenom. Mol. Recog. Chem.* **1994**, *18*, 81.

22. Berti, G., in Allinger, N. L., Eliel, E. L. eds., *Topics in Stereochemistry*, Vol. 7, p. 97, Wiley, New York, NY, 1973.

23. Kirshenbaum, K. S., Sharpless, K. B., *J. Org. Chem.* **1985**, *50*, 1979.

24. Wang, B., Kang, Y.-R., Yang, L.-M., Suo, J.-S., *J. Mol. Catal. A: Chem.* **2003**, *203*, 29.

25. Baumstark, A. L., Harden, D. B., Jr., *J. Org. Chem.* **1993**, *58*, 7615.

26. Mekonnen, A., Carlson, R., *Tetrahedron* **2006**, *62*, 852.

27. Fleming, I., *Frontier Orbitals and Organic Chemical Reactions*, John Wiley & Sons, Chichester, 1976.

28. Bowden, K., Price, M. J., *J. Chem. Soc. B:* **1970**, 1466.

29. Devan, N., Sureshkumar, D., Beadham, I., Prabhu, K. R., Chandrasekaran, S., *Ind. J. Chem. Sect. B* **2002**, 41*B*, 2112; Zhang, Y., Fu, C., Lu, P., Fan, W., *Gaodeng Xuexiao Huaxue Xuebao* **1989**, *10*, 1208.

30. Yadav, J. S., Reddy, B. V. S., Baishya, G., *J. Org. Chem.* **2003**, *68*, 7098.

31. Krishnaveni, N. S., Surendra, K., Rao, K. R., *Chem. Commun.* **2005**, 669.

32. Surendra, K., Krishnaveni, N. S., Sridhar, R., Rao, K. R., *Tetrahedron Lett.* **2006**, *47*, 2125.

33. Xu, L.-W., Li, J.-W., Zhou, S.-L., Xia, C.-G., *New J. Chem.* **2004**, *28*,183.

34. Chaudhuri, M. K., Hussain, S., Kantam, M. L., Neelima, B., *Tetrahedron Lett.* **2005**, *46*, 8329.

35. Xu, L.-W., Li, J.-W., Xia, C.-G., Zhou, S.-L., Hu, X.-X., *Synlett* **2003**, 2425.

36. Moghaddam, F. M., Mohammadi, M., Hosseinnia, A., *Synth. Commun.* **2000**, *30*, 643.

37. Liang, H.-C., Das, S. K., Galvan, J. R., Sato, S. M., Zhang, Y., Zakharov, L. N., Rheingold, A. L., *Green Chemistry* **2005**, *7*, 410.

38. Boyer, J. H., *J. Am. Chem. Soc.* **1951**, *73*, 5248.

39. Xu, L.-W., Xia, C.-G., Li, J.-W., Zhou, S.-L., *Synlett* **2003**, 2246.

40. Ramazani, A., Dolatyari, L., Kazemizadeh, A. R., Ahmadi, E., *Asian J. Chem.* **2005**, *17*, 297.

41. Fuson, R. C., Bannister, R. G., *J. Am. Chem. Soc.* **1952**, *74*, 1631.

42. Hajos, Z. G., Parrish, D. R., *J. Org. Chem.* **1974**, *39*, 1612.

43. Eder, U., Sauer, G., Wiechert, R., *Angew. Chem. Int. Ed. Engl.* **1971**, *10*, 496.

44. Harada, N., Sugioka, T., Uda, H., Kuriki, T., *Synthesis* **1990**, 53.

45. Lavallee, J.-F., Deslongchamps, P., *Tetrahedron Lett.* **1988**, *29*, 6033.

46. Keller, E., Feringa, B. L., *Tetrahedron Lett.* **1996**, *37*, 1879.

47. Lubineau, A., Auge, J., *Tetrahedron Lett.* **1992**, *33*, 8073.

48. Sussangkarn, K., Fodor, G., Karle, I., George, C., *Tetrahedron* **1988**, *44*, 7047.

49. Jin, T.-S., Zhang, J.-S., Guo, T.-T., Wang, A.-Q., Li, T.-S., *Synthesis* **2004**, 2001.

50. Mudaliar, C. D., Nivalkar, K. R., Mashraqui, S. H., *Org. Prep. Proc. Int.* **1997**, *29*, 584.

51. Ballini, R., Bosica, G., *Eur. J. Org. Chem.* **1998**, *2*, 355.

52. Jenner, G., *New J. Chem.* **1999**, *23*, 525.

53. Kobayashi, S., Kakumoto, K., Mori, Y., Manabi, K., *Isr. J. Chem.* **2002**, *41*, 247.

54. Miranda, S., Lopez-Alvarado, P., Giorgi, G., Rodriguez, J., Avendano, C., Menendez, J., *Synlett* **2003**, 2159.

55. Wang, L., Sun, X., Zhang, Y., *Synth. Commun.* **1998**, *28*, 3263.

56. Yang, Y., Chan, T. H., *J. Am. Chem. Soc.* **2000**, *122*, 402.

57. Petrier, C., Dupuy, C., Luche, J. L., *Tetrahedron Lett.* **1986**, *27*, 3149; Luche, J. L., Allavena, C., *Tetrahedron Lett.* **1988,** *29*, 5369; Dupuy, C., Petrier, C., Sarandeses, L. A., Luche, J. L., *Synth. Commun.* **1991**, *21*, 643.

58. Giese, B., Damm, W., Roth, M., Zehnder, M., *Synlett.* **1992**, 441; Erdmann, P., Schafer, J., Springer, R., Zeitz, H. G., Giese, B., *Helv. Chim. Act.* **1992**, *75*, 638.

59. Luche, J. L., Allavena, C., Petrier, C., Dupuy, C., *Tetrahedron Lett.* **1988,** *29*, 5373.

60. Pietrusiewicz, K. M., Zabtocka, M., *Tetrahedron Lett.* **1988,** *29*, 937.

61. Roth, M., Damm, W., Giese, B., *Tetrahedron Lett.* **1996**, *37*, 351.

62. Fleming, F. F., Gudipati, S., *Org. Lett.* **2006**, *8*, 1557.

63. Suarez, R. M., Sestelo, J. P., Sarandeses, L. A., *Synlett.* **2002**, 1435.

64. Suarez, R. M., Sestelo, J. P., Sarandeses, L. A., *Chem. Eur. J.* **2003**, 9, 4179.

65. Huang, T., Keh, C. C. K., Li, C.-J., *Chem. Commun.* **2002**, 2440.

66. Miyabe, H., Ueda, M., Nishimura, A., Naito, T., *Org. Lett.* **2002**, 4, 131; Itooka, R., Iguchi, Y., Miyaura, N., *Chem. Lett.* **2001**, 722.

67. Jang, D. O., Cho, D. H., *Synlett.* **2002**, 631.

68. Suarez, R. M; Sestelo, J. P., Sarandeses, L. A *Org. Biomol. Chem.* **2004**, 2, 3584; Suarez, R. M., Perez, S. J., Sarandeses, L. A., *Chem. Eur. J.* **2003**, 9, 4179.

69. Sakai, M., Hayashi, H., Miyaura, N., *Organometallics* **1997**, 16, 4229; Itooka, R., Iguchi, Y., Miyaura, N., *Chem. Lett.* **2001**, 722.

70. For an excellent review on this subject, see: Hayashi, T., *Synlett.* **2001**, 879; for additional information, see Hayashi, T., Ueyama, K., Tokunaga, N., Yoshida, K., *J. Am. Chem. Soc.* **2003**, 125, 11508; Shintani, R., Tokunaga, N., Doi, H., Hayashi, T., *J. Am. Chem. Soc.* **2004**, 126, 6240: PS-supported BINAP as ligand; Otomaru, Y., Senda, T., Hayashi, T., *Org. Lett.* **2004**, 6, 3357; Uozumi, Y., Kobayashi, Y., *Heterocycles* **2003**, 59, 71; Uozumi, Y., Tanaka, H., Shibatomi, K., *Org. Lett.* **2004**, 6, 281.

71. Shi, Q., Xu, L., Li, X., Jia, X., Wang, R., Au-Yeung, T. T.-L., Chan, A. S. C., Hayashi, T., Cao, R., Hong, M., *Tetrahedron Lett.* **2003**, 44, 6505.

72. Lautens, M., Mancuso, J., *Org. Lett.* **2002**, 4, 2105.

73. Itooka, R., Iguchi, Y., Miyaura, N., *J. Org. Chem.* **2003**, 68, 6000.

74. Dong, L., Xu, Y.-J., Cun, L.-F., Cui, X., Mi, A.-Q., Jiang, Y.-Z., Gong, L.-Z., *Org. Lett.* **2005**, 7, 4285.

75. Huang, T., Venkatraman, S., Meng, Y., Nguyen, T. V., Kort, D., Wang, D., Ding, R., Li, C.-J., *Pure Appl. Chem.* **2001**, 73, 1315.

76. Huang, T.-S., Meng, Y., Venkatraman, S., Wang, D., Li, C.-J., *J. Am. Chem. Soc.* **2001**, 123, 7451; Venkatraman, S., Meng, Y., Li, C.-J., *Tetrahedron Lett.* **2001**, 42, 4459; Oi, S., Moro, M., Ito, H., Honma, Y., Miyano, S., Inoue, Y., *Tetrahedron* **2002**, 58, 91.

77. Venkatraman, S., Li, C.-J., *Tetrahedron Lett.* **2001**, 42, 781.

78. Ding, R., Chen, Y.-J., Wang, D., Li, C.-J., *Synlett* **2001**, 1470.

79. Huang, T.-S., Li, C.-J., *Chem. Commun.* **2001**, 2348.

80. Huang, T.-S., Li, C.-J., *Org. Lett.* **2001**, 3, 2037; Chapman, C. J., Frost, C. G., *Adv. Synth. Catal.* **2003**, 345, 353; Wadsworth, K.J., Wood, F. K., Chapman, C. J., Frost, C. G., *Synlett* **2004**, 2022.

81. Oi, S., Honma, Y., Inoue, Y., *Org. Lett.* **2002**, 4, 667; Murata, M., Shimazaki, R., Ishikura, M., Watanabe, S., Masuda, Y., *Synthesis* **2002**, 717.

82. Lautens, M., Mancuso, J., Grover, H., *Synthesis* **2004**, 2006; de la Herran, G., Murcia, C., Csaky, A. G., *Org. Lett.* **2005**, *7*, 5629.

83. For a review, see: Nagata, W., Yoshioka, M., *Org. React.* **1977**, *25*, 255.

84. Lapworth, A., Wechsler, E., *J. Chem. Soc.* **1910**, *97*, 38.

85. Danly, D. E., King, C. J. H. in Lund, H., Baizer, M. M. eds., Organic Electrochemistry, 3rd ed., Marcel Dekker, 1991.

86. Baizer, M. M., *Tetrahedron Lett.* **1963**, *4*, 973; Moncelli, M. R., Guidelli, R., *J. Electroanal. Chem. Interf. Electrochem.* **1981**, *129*, 373.

87. Baizer, M. M., Carla, M., *Chemtech.* **1980**, *10(3)*, 161.

88. Danly, D. E. *AIChE Symp. Series* **1981**, *77(204)*, 39.

89. Wang, L., Zhang, Y., *Synth. Commun.* **1998**, *28*, 3991.

90. Heo, C. K., Bunting, J. W., *J. Org. Chem.* **1992**, *57*, 3570.

91. For a recent review, see: Basavaiah, D., Rao, P. D., Hyma, R. S., *Tetrahedron*, **1996**, *52*, 8001.

92. Augé, J., Lubin, N., Lubineau, A., *Tetrahedron Lett.* **1994**, *35*, 7947.

93. Yu, C., Liu, B., Hu, L., *J. Org. Chem.* **2001**, *66*, 5413.

94. Yu, C., Hu, L., *J. Org. Chem.* **2002**, *67*, 219.

95. Basavaiah, D., Kumaragurubaran, N., *Tetrahedron Lett.* **2001**, *42*, 477.

96. Chung, Y. M., Gong, J. H., Kim, T. H., Kim, J. N., *Tetrahedron Lett.* **2001**, *42*, 9023; Kim, J. N., Lee, H. J., Lee, K. Y., Gong, J. H., *Synlett* **2002**, 173.

97. Luo, S., Zhang, B., He, J., Janczuk, A., Wang, P. G., Cheng, J.-P., *Tetrahedron Lett.* **2002**, *43*, 7369; Luo, S., Wang, P. G., Cheng, J.-P., *J. Org. Chem.* **2004**, *69*, 555.

98. Rezgui, F., El Gaied, M. M., *Tetrahedron Lett.* **1998**, *39*, 5965; Lee, K. Y., Gong, J. H., Kim, J. N., *Bull. Kor. Chem. Soc.* **2002**, *23*, 659.

99. Basavaiah, D., Krishnamacharyulu, M., Rao, J., *Synth. Commun.* **2000**, *30*, 2061.

100. Cai, J., Zhou, Z., Zhao, G., Tang, C., *Org. Lett.* **2002**, *4*, 4723.

101. Hayashi, Y., Okado, K., Ashimine, I., Shoji, M., *Tetrahedron Lett.* **2002**, *43*, 8683.

102. Aggarwal, V. K., Dean, D. K., Mereu, A., Williams, R., *J. Org. Chem.* **2002**, *67*, 510.

103. Aggarwal, V. K., Emme, I., Fulford, S. Y., *J. Org. Chem.* **2003**, *68*, 692.

104. Krishna, P. R., Kannan, V., Sharma, G. V. M., Rao, M. H. V. R., *Synlett* **2003**, 888.

105. Krishna, P. R., Kannan, V., Reddy, P. V. N., *Adv. Synth. Catal.* **2004**, *346*, 603.

106. Maher, D. J., Connon, S. J., *Tetrahedron Lett.* **2004**, *45*, 1301.

107. Trost, B. M., Li, C.-J., *J. Am. Chem. Soc.* **1994**, *116*, 3167; Trost, B. M., Li, C.-J., *J. Am. Chem. Soc.* **1994**, *116*, 10819; Trost, B. M., Dake, G. R., *J. Org. Chem.* **1997**, *62*, 5670; Alvarez-Ibarra, C., Csaky, A. G., de la Oliva, C. G., *J. Org. Chem.* **2000**, *65*, 3544.

108. Skouta, R., Varma, R. S., Li, C.-J., *Green Chemistry*, **2005**, *7*, 571.

CHAPTER 11

NITROGEN COMPOUNDS

11.1 AMINES

Amines are widely used as organic bases in organic chemistry. In aqueous media, amines exist in equilibria with their conjugated acids, the ammonium ions, and hydroxide ion. In addition to serving as bases, common reactions of amines include serving as the nucleophile to react with various organic electrophiles. Because the nitrogen atom in amine is softer than the oxygen atom in water, such nucleophilic reactions with softer electrophiles can occur readily in water. For example, nucleophilic substitution reactions and nucleophilic conjugate-addition reactions occur readily in aqueous media. Organic amines are also excellent ligands for various transition-metals in water because of the softness of both amines and transition-metals. In addition, the electron-lone pair on the nitrogen of amines renders them prone to oxidation reactions.

11.1.1 Alkylation

Because the S_N2 nucleophilic substitution of uncharged amines with uncharged aliphatic organic halides involves a transition state that is more polar than that of the starting materials, such substitution reactions

Comprehensive Organic Reactions in Aqueous Media, Second Edition, by Chao-Jun Li and Tak-Hang Chan
Copyright © 2007 John Wiley & Sons, Inc.

occur readily in aqueous media. The use of polar water as solvent can stabilize the transition state and lower the activation energy. These reactions have been discussed in Chapter 6, on organic halides.

Selective N-alkylation of primary amines with chloroacetamides proceeds in aqueous conditions under controlled pH by using stoichiometric amounts of sodium iodide in acetonitrile/water (Eq. 11.1).[1] It has been found that α-(alkylamino)acetamides are less basic than the corresponding primary alkylamines by about 2 pKa units, which makes the selective monoalkylation possible. Reductive amination also proceeds under aqueous conditions.[2] Stirring primary and secondary amines with an aldehyde and wet EtOH at 180°C under 70 atm CO with RhCl$_3$ gives the corresponding N-alkylation products. For example, morpholine reacts with aqueous HCHO and RhCl$_3$·3H$_2$O as catalyst to give 90% N-methylmorpholine (Eq. 11.2).

$$(11.1)$$

$$(11.2)$$

11.1.2   Diazotization and Nitrosation[3]

Primary aromatic amines can be readily converted to diazonium salts by treatment with nitrous acid in aqueous media (Eq. 11.3).[4] The reaction also occurs with aliphatic primary amines. However, aliphatic diazonium salts are extremely unstable and decompose to give a complicated mixture of substitution, elimination, and rearrangement products. Aromatic diazonium ions are more stable because of the resonance interaction between the nitrogen and the ring (Figure 11.1). These salts are generally prepared in situ and used directly. Various groups such as halogen, nitro, alkyl, aldehyde, and sulfonic acid do not interfere with the preparation. It is also possible to diazotize an aromatic amine without disturbing an aliphatic amine group in the same molecule by

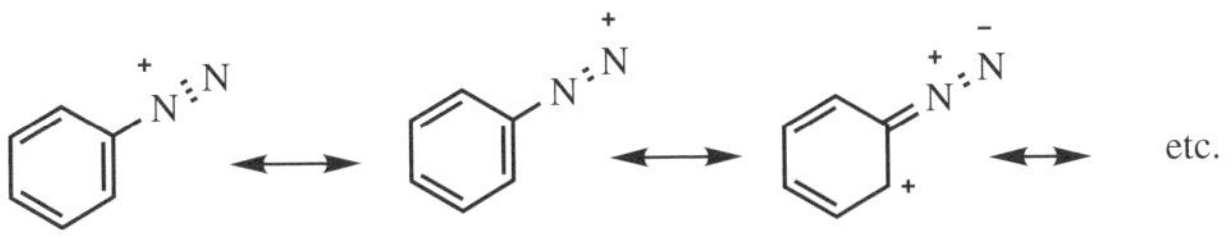

Figure 11.1 Resonance structures of aromatic diazonium ion.

working at approximately pH = 1 because aliphatic amines do not react with nitrous acid at pH < 3.[5]

$$Ar-NH_2 \xrightarrow{\text{aq HONO}} Ar-\overset{+}{N}\equiv N \qquad (11.3)$$

If an aliphatic amino group is next to an electron-withdrawing group such as CO_2R, CN, CHO, COR and has a hydrogen, reaction with aqueous nitrous acid gives a diazo compound (Eq. 11.4). Such compounds are used widely in 1,3-dipolar cycloaddition reactions, which will be covered in Chapter 12.

$$EtO_2C-CH_2-NH_2 \xrightarrow{\text{aq HONO}} EtO_2C-CH=\overset{+}{N}=\overset{-}{N} \qquad (11.4)$$

The reaction of secondary amines with aqueous nitrous acid generates N-nitroso compounds (Eq. 11.5). Such a nitrosation occurs with both aromatic and aliphatic secondary amines.[6]

$$R_2NH \xrightarrow{\text{aq HONO}} R_2N-NO \qquad (11.5)$$

11.1.3 Oxidation

The oxidation of amines can be carried out by a variety of oxidizing reagents. The one-electron oxidation of aromatic amines and diamines by hydroxyl radicals in water initially gives radical adducts that decay by first-order kinetics and have lifetimes of approximately $5-50$ μsec.[7] The decay products are cation radicals and are long-lived in the absence of oxygen. Mono- and polyaliphatic amines are oxidized to the corresponding nitro derivatives by HOF·CH_3CN made directly from F_2, acetonitrile, and water (Eq. 11.6).[8] Salts of the amines are nonreactive toward the oxidizer. Oxidation of alkylamines with a tertiary alkyl group gives the corresponding nitro compounds without undesired rearrangement. An example is the reaction of 1,3-adamantanediamine with $H_2{}^{18}O$/MeCN, which gives labeled 1,3-dinitroadamantane with both oxygens of the nitro group being ^{18}O. Other reagents used for oxidation of aliphatic amines in water include dimethyldioxirane (Eq. 11.7),[9] bromamine-T/RuCl$_3$ (Eq. 11.8),[10] ozone,[11] (batho)$_2$Cu(II) (Eq. 11.9),[12] and nitrous oxide with platinum particles (Eq. 11.10)[13] to give various products. A hydroxyapatite-bound ruthenium complex efficiently catalyzed the aerobic oxidation of various primary amines to nitriles, which were further hydrated to amides in the presence of water (Eq. 11.11).[14] Using a recyclable tungstate-exchanged Mg-Al layered double hydroxide heterogenized catalyst and aqueous H$_2$O$_2$ as oxidant, tertiary amines were oxidized into N-oxides in quantitative yields at a high rate at rt (Eq. 11.12).[15] The oxidation of a series of tertiary amines with ClO$_2$ in buffered water or aqueous NaCN yielded iminium ions, which were trapped in situ with either internal ($^-$OH) or external (CN$^-$) nucleophiles (Eq. 11.13).[16] Direct transformations of amines to ketones were achieved by using Pd/C catalyst in water under microwave irradiation using water as the oxygen source (Eq. 11.14).[17]

$$RNH_2 \xrightarrow[-HF]{HOF \cdot CH_3CN} [RNHOH] \xrightarrow[-HF; -H_2O]{HOF \cdot CH_3CN} [RNO] \xrightarrow[-HF]{HOF \cdot CH_3CN} RNO_2 \qquad (11.6)$$

$$H_3C-NH_2 \xrightarrow{O-O} H_3C-NO_2 \qquad (11.7)$$

$$\underset{R}{\overset{R}{R}}N \xrightarrow[\substack{CH_3CN/H_2O \ (1:1), \ 80°C \\ pH \ 8.4}]{RuCl_3/bromamine-T} \underset{R}{\overset{R}{R}}\overset{+}{N}-O^- \qquad (11.8)$$

$$50–92\%$$

$$R_2NCH_2R' \xrightarrow[H_2O \ -2e]{(batho)_2CuSO_4} R_2N=CHR' \xrightarrow{H_2O}$$

$$R_2NH + R'CHO \xrightarrow[H_2O \ -2e]{(batho)_2CuSO_4} R'COOH \qquad (11.9)$$

$$R\diagup NH_2 \xrightarrow[r.t.]{N_2O/Pt} R\diagup CHO$$

$$\underset{R \quad R'}{\overset{NH_2}{\diagdown}} \xrightarrow[r.t.]{N_2O/Pt} \underset{R \quad R'}{\overset{O}{\diagdown}} \qquad (11.10)$$

$$R\diagup NH_2 \xrightarrow[\substack{toluene \\ O_2 \ atmosphere \\ 110°C \\ 12 \ h}]{RuHAP} R\diagdown CN \xrightarrow[\substack{H_2O \\ N_2 \ atmosphere \\ 150°C \\ 24 \ h}]{RuHAP} \underset{R \quad NH_2}{\overset{O}{\diagdown}} \qquad (11.11)$$

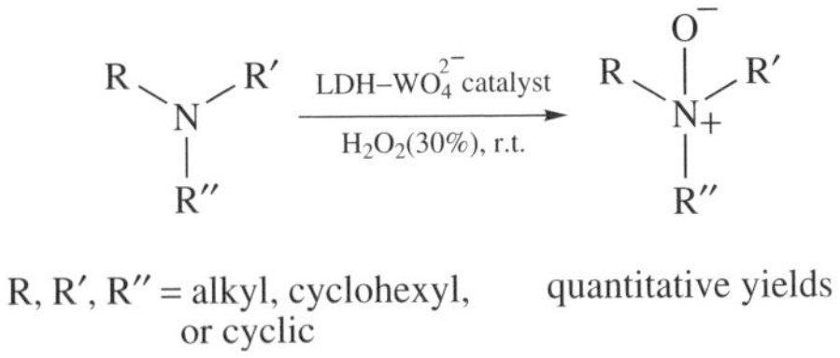

$$(11.12)$$

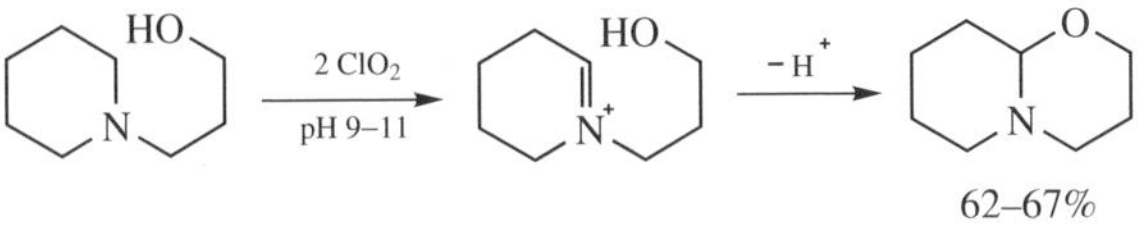

R, R′, R″ = alkyl, cyclohexyl, or cyclic   quantitative yields

62–67%

$$(11.13)$$

67%

quantitative yield

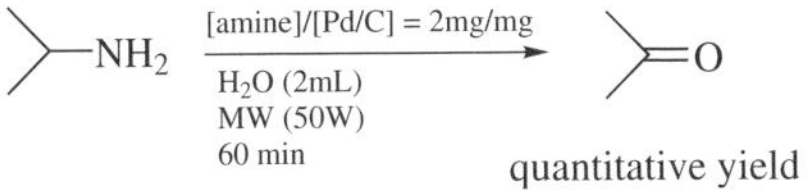

$$(11.14)$$

In the presence of a copper catalyst and an oxidizing reagent, the C–H bonds at the α-position of nitrogen in amines can be activated readily. Li and co-workers developed various cross-dehydrogenative couplings (CDC) via the oxidative removal of two different C–H bonds (Eq. 11.15).[18] All these reactions can tolerate water as solvent. On the other hand, without using additional nucleophiles, $CuBr/H_2O_2$ or cerium(IV) ammonium nitrate (CAN) mediates the oxidative coupling of N,N-dialkylarylamines in water to give benzidine derivatives (Eq. 11.16).[19]

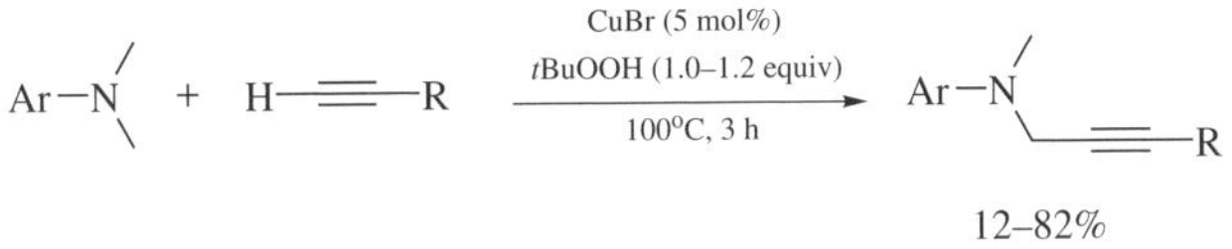

12–82%

$$(11.15)$$

53–85%

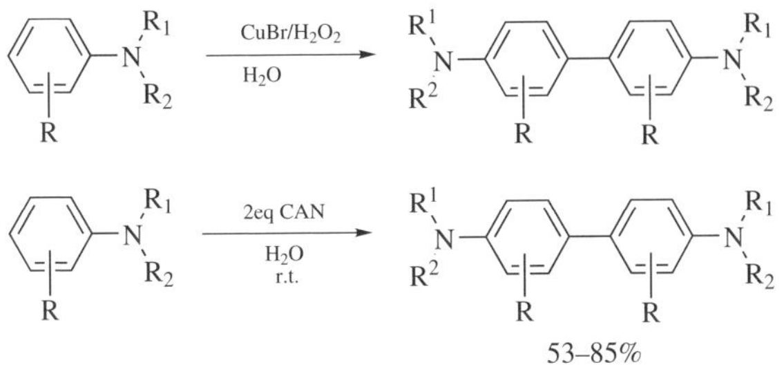

$$(11.16)$$

Oxidation of aromatic amines (anilines) is also accomplished by a variety of oxidizing reagents[20] including enzymes in aqueous media. Polyanilines have been produced by this method. Oxidative cleavage of aryl aziridines involving β-cyclodextrin-aziridine complexes in water with NBS gives the corresponding α-amino ketones in high yields (Eq. 11.17).[21] Reaction of 2-iodoxybenzoic acid (IBX) with 2-aryl epoxides and 2-aryl aziridines as their β-cyclodextrin complexes in water affords the corresponding substituted α-hydroxyacetophenones and α-aminoacetophenones, respectively (Eq. 11.18).[22]

$$
\underset{\beta-\text{CD}}{\text{R}}\begin{array}{c}\text{Ts}\\\text{N}\\\triangle\end{array} \xrightarrow[50°C, 12 h]{NBS/H_2O} \underset{80–90\%}{\text{R}}\begin{array}{c}\text{O}\\\|\\\end{array}\text{NHTs}
$$

$$(11.17)$$

R = H, p–Me, p–Cl, p–Br, p–OMe, and p–COMe

$$
\underset{R}{\overset{X}{\triangle}}\underset{\beta-\text{CD}}{} \xrightarrow[H_2O]{IBX} \underset{R}{\overset{O}{\underset{82–92\%}{}}}X
$$

$$(11.18)$$

R = H, Br, Cl, Me, OMe, and NO$_2$

1. X = O   3. X = OH
2. X = NTs   4. X = NHTs

11.1.4   Reaction with Carbonyl Compounds

Amines can react with various carbonyl compounds and their derivatives in aqueous media to give the corresponding imine derivatives. These reactions have been discussed in related chapters. The synthetically most useful reaction of this type is the formation of imines and imine derivatives from the condensation of amines with aldehydes and ketones. Water is an excellent solvent for such condensation reactions. For example, water was found to be an ideal solvent for a high-yield, fast preparation of easily hydrolyzable 2-pyrrolecarbaldimines.[23] In the presence of Cu^{2+}, the reaction afforded the corresponding Cu(II) chelates (Eq. 11.19).

$$
\begin{array}{c}\text{pyrrole-CHO}\end{array} + \text{RNH}_2 \xrightarrow[Cu(OAc)_2]{H_2O} \left[\begin{array}{c}\text{imine}\end{array}\right] \longrightarrow \text{Cu(II) chelate}
$$

$$(11.19)$$

The corresponding imine products undergo a wide range of reactions to form various synthetically and biologically important products (discussed in the following).

11.2 IMINES

11.2.1 Reductions

Imines (and immonium ions) can be reduced to the corresponding amines by using a variety of reagents in aqueous media. Dichloro*bis*-(1,4-diazabicyclo[2.2.2]octane)-(tetrahydroborato)zirconium(IV), which is stable under mild aqueous acidic conditions (pH 4–6) and survives in H_2O for several days, reduces imines and enamines in aqueous media.[24] Imines were reduced to amines in good yields with zinc powder in 5% aqueous NaOH solution without any organic solvents under mild conditions (Eq. 11.20).[25] Electrochemical methods were also used to reduce imines in aqueous media.[26] The most common reagents used for such reductions are sodium borohydride (NaBH₄) and sodium cyanoborohydride (NaBH(CN)₃).[27] It is possible to carry out one-pot reductive amination of carbonyl compounds by reaction of a primary or secondary amine in aqueous MeCN or methanol at controlled pH with these reagents (Eq. 11.21).

$$\underset{R^2}{\overset{R^1}{>}}{=}N{\overset{R^3}{\diagdown}} \quad \xrightarrow[\text{5\% aq NaOH, r.t.}]{\text{Zn}} \quad H{-}\underset{R^2}{\overset{R^1}{\diagup}}{-}\underset{H}{\overset{R^3}{N}} \qquad (11.20)$$

53–88%

$$\xrightarrow[\text{NaCNBH}_3/\text{aq MeCN, pH (6–8)}]{\text{aq CH}_2\text{O (excess)}} \qquad (11.21)$$

The deoxygenation of aldehydes and ketones to the corresponding hydrocarbons via the hydrazones is known as the Wolff-Kishner reduction.[28] Various modifications of the original protocols have been suggested. One of the most useful is the Huang-Minlon modification, which substituted hydrazine hydrate as a safer and less expensive replacement of anhydrous hydrazine. In addition, diethylene glycol together with sodium hydroxide was used to increase the reaction temperature by driving off water and excess hydrazine hydrate to complete the reaction (Eq. 11.22).[29] It has been recently reported that the multiple-batch Huang-Minlon procedure has been improved by recycling excess hydrazine hydrate instead of distilling it.[30] The reaction of the carbonyl compounds with hydrazine hydrate under microwave conditions has also been described.[31] The subsequent conversion of the hydrazones to the hydrocarbons under microwave conditions can also be accomplished.

$$RCHO \xrightleftharpoons[KOH]{H_2NNH_2.H_2O} H_2O + RCH=NNH_2 \longrightarrow RCH_3 + N_2 \qquad (11.22)$$

In addition to chemical reduction, catalytic hydrogenations of imines and iminium ions have also been carried out in aqueous media. For example, Wilkinson and co-workers reported the hydrogenation of imines using $RhCl(PPh_3)_3$ and $[Rh(PPh_3)_2(diene)]PF_6$ as catalyst precursors at room temperature under 1 atm of hydrogen, using alcohol as solvents. For a solid imine, a water-Et_2O phase-transfer system was used (Eq. 11.23).[32] Chiral rhodium(I) catalysts formed with chiral sulfonated diphosphines catalyzed the asymmetric hydrogenation of carbon–nitrogen double bonds to give 58% ee of the amine product in aqueous-organic two-phase solvent systems (Eq. 11.24).[33] Recently, an efficient enantioselective transfer hydrogenation of imines and iminium ions in water was reported by using a chiral (p-cymene)ruthenium(II) chloride dimer complex in aqueous media using sodium formate as the hydrogen source (Eq. 11. 25).[34]

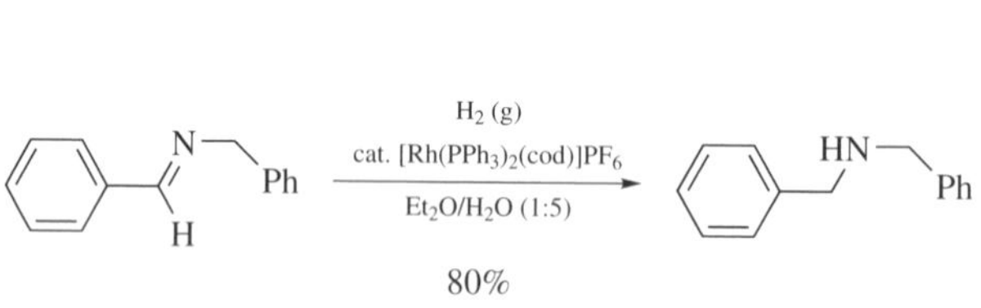

(11.23)

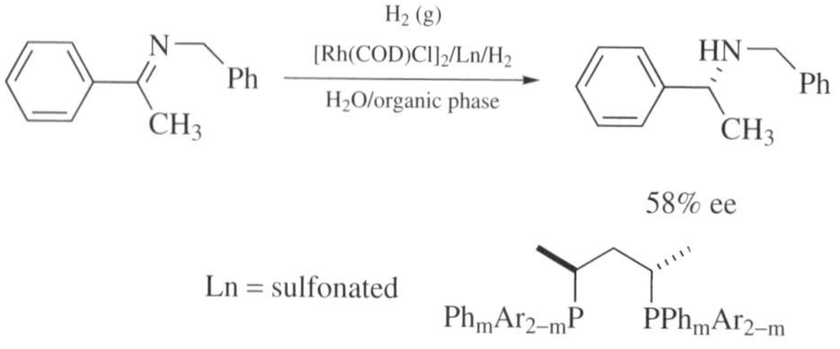

(11.24)

$$(11.25)$$

95% ee

11.2.2 Nucleophilic Additions

11.2.2.1 Mannich-Type Reactions.

The Mannich-type reaction is useful for the synthesis of β-aminocarbonyl compounds. The rate of Mannich-type reaction of phenols and of ketones with secondary amines is greatly increased in an aqueous medium compared with alcoholic or hydrocarbon solvents.[35] Recently, Mannich-type reactions catalyzed by various Lewis acids have been studied extensively. Under an aqueous condition of THF-water (9:1), Kobayashi reported the reaction of an aldehyde with an amine and a vinyl ether to give the Mannich-type product, in the presence of 10 mol% of ytterbium triflate, Yb(OTf)$_3$.[36] Additionally, Loh reported a one-pot Mannich-type reaction between aldehydes, amines, and silyl enol ethers catalyzed by indium trichloride in water to give β-amino ketones and esters in moderate-to-good yields.[37] Suitable silyl enol ethers were MeOC(=CMe$_2$)OSiMe$_3$ and CH$_2$=CPhOSiMe$_3$. The catalyst can be recycled when the reaction is complete. In the presence of a catalytic amount of Ln(OTf)$_3$ or Cu(OTf)$_2$ (OTf = O$_3$SCF$_3$), Kobayashi found that the three-component Mannich-type reactions of aldehydes, amines, and silyl enolates proceeded smoothly in micellar systems to afford the corresponding β-amino ketones and esters in high yields (Eq.11.26).[38] Kobayashi also found that a Lewis-acid-surfactant-combined catalyst (LASC), composed of water-stable Lewis-acidic metal cations such as scandium (III) and copper (II) and anionic surfactants such as dodecyl sulfate and dodecanesulfonate, are highly successful for the Mannich-type reaction and other Lewis-acid-catalyzed C–C bond formations in water (Eq. 11.27).[39] Furthermore, the results of aldol reactions in various solvents show that water is the best for the LASC-catalyzed reactions. The LASCs were found to form stable colloidal dispersions rapidly in the presence of reaction substrates in water, even when the substrates are solid. The Mannich-type reaction between N-pyrrolecarboxylates, CH$_2$O, and hydrochlorides of primary amines is catalyzed by

$Y(O_3SCF_3)_3$ to afford a monoaminoalkylation product in good yield in aqueous media.[40] Zinc tetrafluoroborate is also highly effective for such couplings in aqueous THF.[41] Kobayashi also reported a Mannich-type reaction of imines with silyl enolates catalyzed by neutral salts such as sodium triflate in water as a suspension medium. Unusual kinetic behavior indicates that the presence of the Mannich adduct facilitates the rate of its formation.[42]

$$(11.26)$$

$$(11.27)$$

Brönsted acids have also been quite effective in catalyzing the Mannich-type reactions in aqueous media. Following their studies on the $BF_3 \cdot OEt_2$-catalyzed Mannich-type reaction in aqueous media,[43] Akiyama reported that HBF_4-catalyzed Mannich-type reactions of aldehydes, amines, and silyl enolates took place smoothly in water in the presence of a surfactant to afford β-amino carbonyl compounds in high yields.[44] One-pot synthesis of β-amino carbonyl compounds from aldehydes and amines also worked well (Eq. 11.28).[45] The diastereoselectivity on the HBF_4-catalyzed Mannich-type reaction of ketene silyl acetal derived from α-oxy esters with aldimines has been studied. It was shown that the use of ketene silyl acetal derived from aryl ester in aqueous 2-propanol gave *anti* β-amino-α-siloxy ester with excellent stereoselectivity and the use of ketene silyl acetal derived from methyl ester in water in the presence of sodium dodecyl sulfate gave the *syn* isomer preferentially.[46]

$$(11.28)$$

On the other hand, Kobayashi has developed a Brönsted acid–combined catalyst for aqueous Mannich-type reactions. Three-component Mannich-type reactions of aldehydes, amines, and ketones (e.g., benzaldehyde, *p*-anisidine, and cyclohexanone) were efficiently catalyzed by dodecylbenzenesulfonic acid at ambient temperature in water to give various β-amino ketones in good yields; whereas the same reaction proceeded sluggishly in organic solvents (Eq. 11.29).[47] The catalyst is also effective for the reactions of aldehydes, amines, and silyl enolates in water.[48] The Brönsted acid and surfactant can be used separately.[49] For example, the HBF$_4$ (0.1 equiv)-catalyzed Mannich-type reactions of ketene silyl acetals with aldimines proceeded smoothly in water in the coexistence of as low as 1 mol% of SDS. Hydrophobic polystyrene-supported sulfonic acid (PS-SO$_3$H) has also been used for such couplings in water.[50]

$$
\text{(11.29)}
$$

in H$_2$O: up to 100%
in MeOH or CH$_2$Cl$_2$:   < 10%

It has also been shown that dimethylsilyl enolates can be activated by diisopropylamine and water and exhibit a high reactivity toward N-tosyl imines to give Mannich-type reaction products in the absence of a Lewis acid or a Brönsted acid.[51] For example, the reaction of [(1-cyclohexen-1-yl)oxy]dimethylsilane with 4-methyl-N-(phenylmethylene)benzene sulfonamide gave rel-4-methyl-N-{(R)-[(1S)-(2-oxocyclohexyl)phenyl-methyl]}benzenesulfonamide ($anti$-isomer) in 91% yield stereoselectively (99:1 $anti$:syn) (Eq. 11.30). On the other hand, Li and co-workers reported a ruthenium-catalyzed tandem olefin migration/aldol and Mannich-type reactions by reacting allyl alcohol and imine in protic solvents.[52]

$$
\text{(11.30)}
$$

91% $anti$:$anti$
             99:1

More recently, asymmetric Mannich-type reactions have been studied in aqueous conditions. Barbas and co-worker reported a direct amino acid catalyzed asymmetric aldol and Mannich-type reactions that can tolerate small amounts of water (<4 vol%).[53] Kobayashi found that a diastereo- and enantioselective Mannich-type reaction of a hydrazono ester with silyl enol ethers in aqueous media has been successfully achieved with ZnF$_2$, a chiral diamine ligand, and trifluoromethanesulfonic acid (Eq. 11.31).[54] The diastereoselective Mannich-type reaction of chiral aldimines with 2-silyloxybutadienes in the presence of zinc triflate and water led to the products with 74–90% de.[55] Cordova and Barbas reported a direct organocatalytic asymmetric Mannich-type reaction in aqueous media by using L-proline as catalyst. The reaction between protected α-imino ethyl glyoxylate and aldehydes provided β-formyl substituted α-amino acid derivatives with excellent diastereoselectivities (up to 19:1, *syn/anti*) and high enantioselectivities (ee between 72 and >99%). By combining the proline-catalyzed Mannich-type reactions with indium promoted allylation in aqueous media, a one-pot asymmetric synthesis of cyclic γ-allyl substituted α-amino acid derivatives (up to >99% ee) was accomplished (Eq. 11.32).[56]

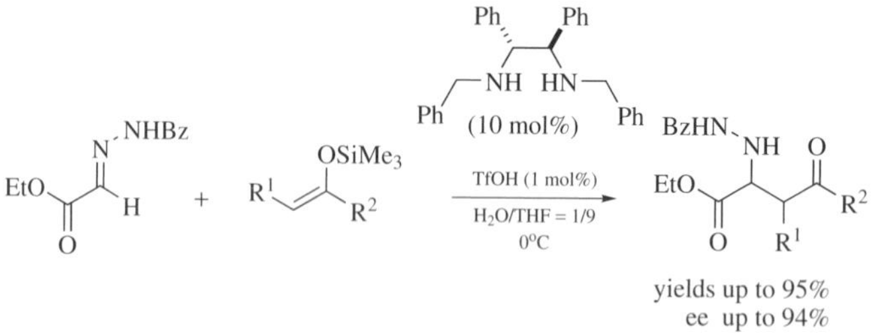

(11.31)

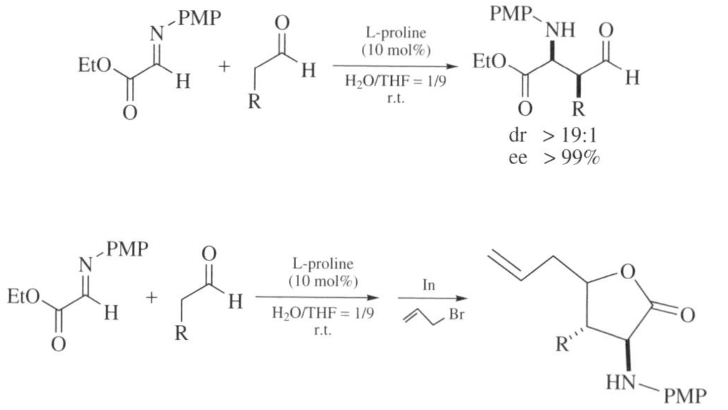

yields up to 95%
ee up to 94%

dr > 19:1
ee > 99%

(11.32)

11.2.2.2   *Addition of Allyl Groups.*

As reported by Grieco et al.,[57] iminium salts generated in situ from primary amines and formaldehyde can be allylated with preformed allylstannane under aqueous conditions. Chan reported that the allylation of sulfonimines in water can be mediated by either indium[58] or zinc (Eq. 11.33).[59] It should be noted that the corresponding aldimines cannot be allylated under such conditions as the aldimines would be hydrolyzed first. Crotylation of sulfonimines bearing a proximal chelating group in aqueous media have also been examined. The reaction gave diastereoselectively the *syn*-isomers (Eq. 11.34).[60]

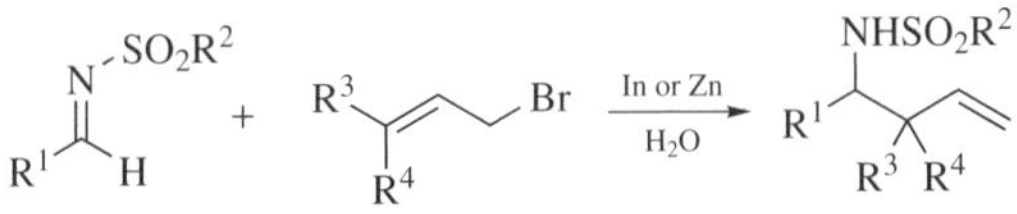

(11.33)

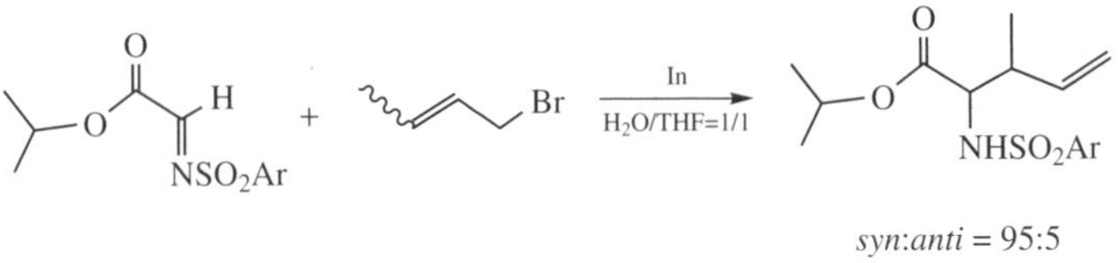

(11.34)

syn:anti = 95:5

Allylation of acyloyl-imidazoles and pyrazoles[61] with allyl halide mediated by indium in aqueous media provides a facile regioselective synthesis of β, γ-unsaturated ketones (Scheme 11.1), which has been applied to the synthesis of the monoterpene artemesia ketone. The same product can be obtained by indium-mediated allylation of acyl cyanide (Eq. 11.35).[62] Samarium, gallium, and bismuth can be used as a mediator for the allylation of nitrones and hydrazones to give homoallylic hydroxylamine and hydrazides in aqueous media in the presence of Bu_4NBr (Scheme 11.2).[63] The reaction with gallium and bismuth can be increased dramatically under microwave activation.

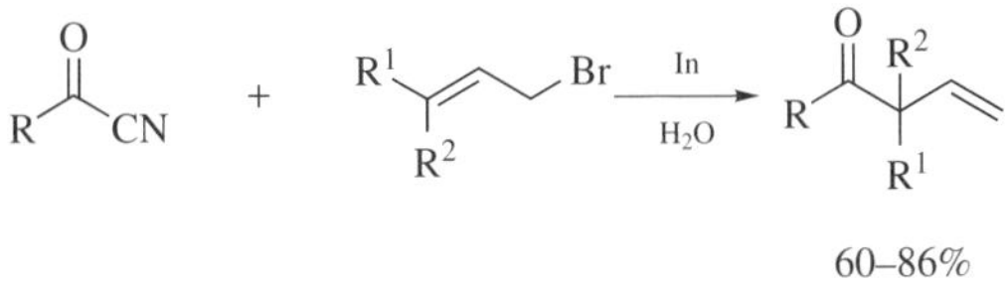

(11.35)

60–86%

Allylation of the nitro group on nitrobenzene derivatives proceeded under similar reaction conditions (Eq. 11.36).[64] Allylation reactions of various benzoylhydrazones with tetraallyltin were carried out in

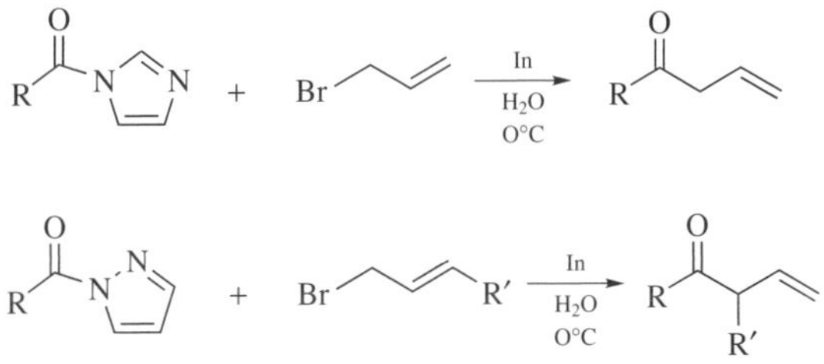

Scheme 11.1

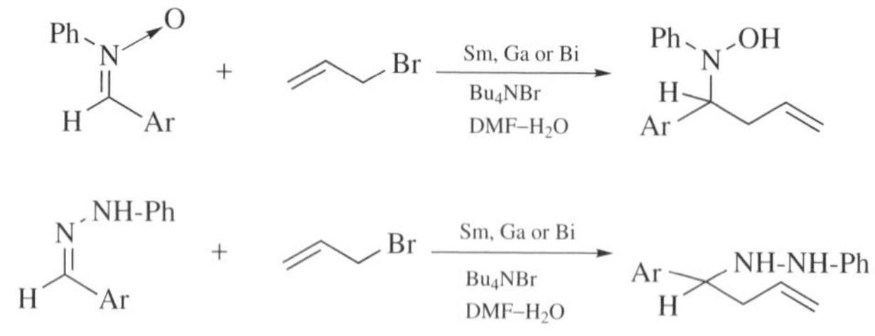

Scheme 11.2

the presence of scandium triflate as a Lewis-acid catalyst in aqueous media (Eq. 11.37).[65] Three-component reactions of aldehydes, benzoyl-hydrazine, and tetraallyltin were also catalyzed by scandium triflate in the same media. The reaction was used to prepare oxazolidinone derivatives. A three-component synthesis of homoallylic amines start-ing from aldehyde, amine, and allyltributylstannane were realized by Sn(II) chloride dihydrate in H_2O in the presence of SDS surfactant (Eq. 11.38).[66]

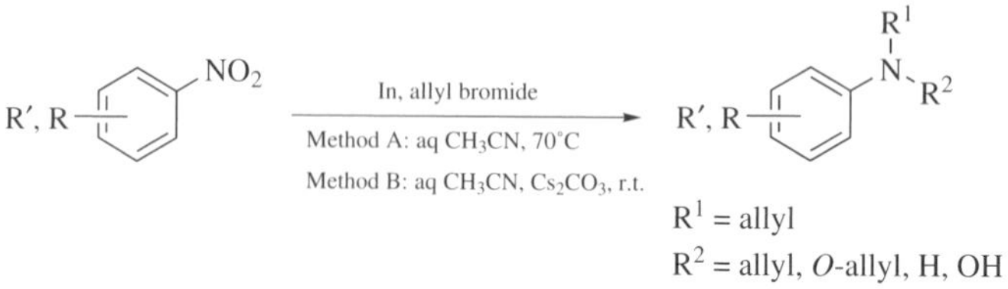

$$(11.36)$$

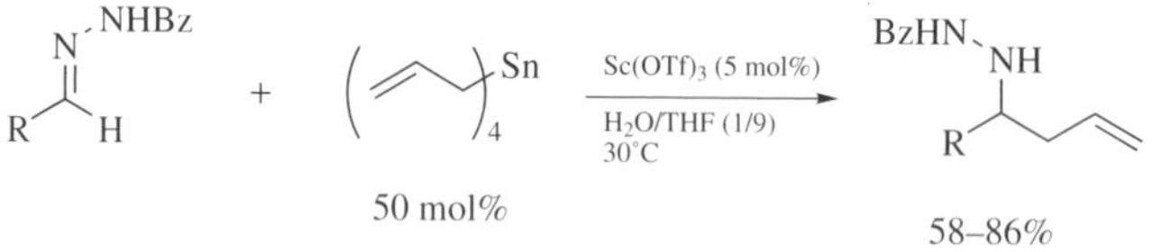

$$(11.37)$$

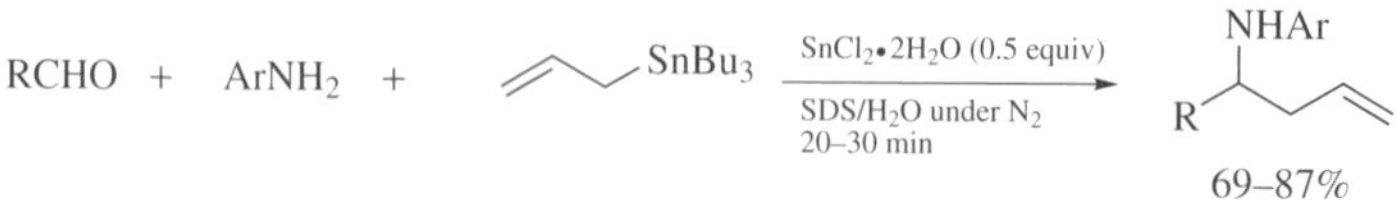

$$(11.38)$$

Iminium ions, generated in aqueous solution from secondary amines and formaldehyde, undergo a Barbier-type allylation mediated by tin, aluminum, and zinc. The reaction is catalyzed by copper and produces tertiary homoallylamines in up to 85% yield.[67] The imines generated in situ from 2-pyridinecarboxaldehyde/2-quinolinecarboxaldehyde and aryl amines undergo indium-mediated Barbier allylation in aqueous media to provide homoallylic amines.[68] Crotyl and cinnamyl bromides lead to diastereoselective allylation with dr up to 98:2 (Eq. 11.39).

$$(11.39)$$

$$\begin{array}{c} syn{:}anti \\ 98{:}2 \end{array}$$

The allylation reaction of electron-deficient imines with allylic alcohol derivatives in the presence of a catalytic amount of palladium(0) complex and indium(I) iodide was studied in the presence of water (Eq. 11.40).[69] α-Substituted homoallylic piperidines are prepared by indium-mediated addition of allylic bromides to α-methoxypiperidine derivatives in water (Scheme 11.3). γ-Substituted allylic bromides undergo bond formation at the most substituted termini; when the allylic bromide is γ-substituted, the *syn* stereoisomers of the products predominate. The reaction does not occur if the substrate does not possess a chelating group in close proximity. Water as a solvent was found to accelerate the indium-mediated Barbier-type allylation and benzylation of β,γ-unsaturated piperidinium ion, which was generated from β,γ-unsaturated α-methoxy-N-methoxycarbonylpiperidine, while a ring-opened allylated product was obtained in one case using β,γ-saturated α-methoxy-N-methoxycarbonylpiperidine. Solvents other than water resulted in low yield of the allylated and benzylated products, suggesting that water is essential to generate the piperidinium ion intermediate from β,γ-saturated α-methoxy-N-methoxycarbonyl-piperidine (Scheme 11.3).[70]

$$(11.40)$$

X = (*1S*)–camphorsultam   in H₂O–THF   in THF

The stereoselective allylation of carbon–nitrogen multiple bonds have also been studied. The addition of allylzinc bromide to aromatic imines derived from (*S*)-valine esters was affected by reversibility, which caused the lowering of the diastereoisomeric ratio with increasing reaction time. The retroallylation reaction could be avoided by performing the reaction in the presence of trace amounts of water or by using CeCl₃· 7H₂O as the catalyst with a decreased reaction rate.[71]

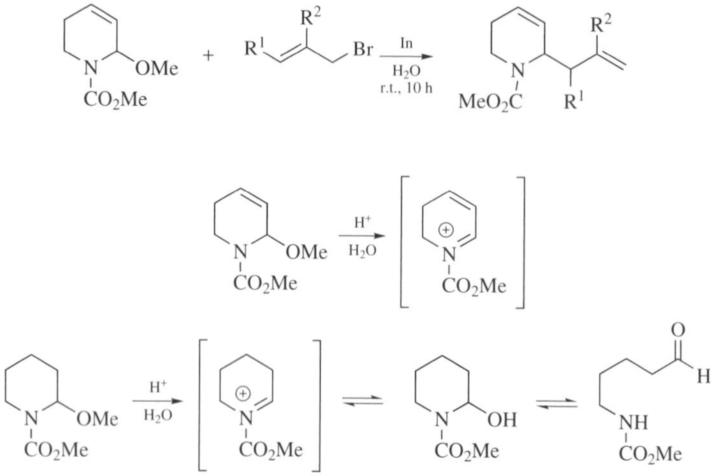

Scheme 11.3

Hanessian reported the synthesis of enantiomerically pure or highly enriched allylglycine and its chain-substituted analogs from the reaction of the sultam derivatives of O-benzyl glyoxylic acid oxime with allylic bromides in the presence of zinc powder in aqueous ammonium chloride (Eq. 11.41).[72] Brown noticed the critical importance of water in the asymmetric allylboration of N-trimethylsilylbenzaldimines with B-allyldiisopinocampheylborane.[73] The reaction required one equivalent of water to proceed (Eq. 11.42).

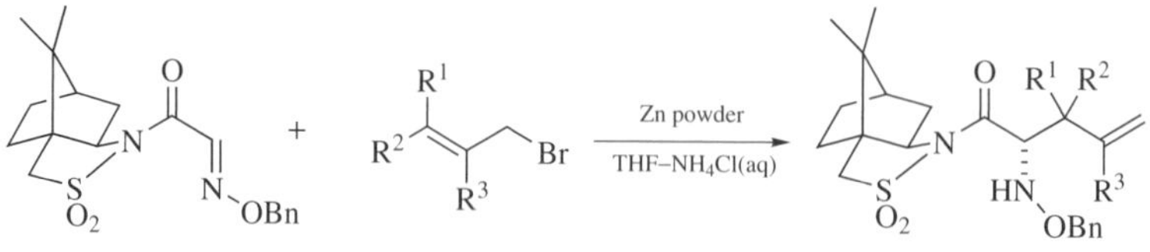

$$(11.41)$$

yield up to 99%
ee up to 99:1

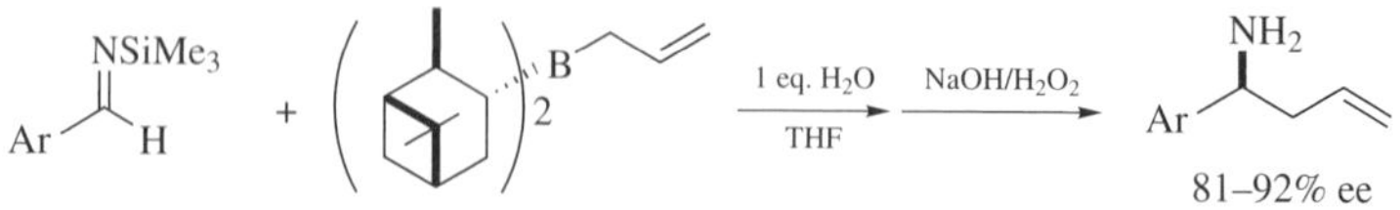

$$(11.42)$$

81–92% ee

Indium-mediated allylation reactions of α-keto imides derived from Oppolzer's sultam proceeded in aqueous THF in good yields and excellent diastereoselectivity (Eq. 11.43).[74] The indium-mediated allylation of the Oppolzer camphorsultam derivatives of glyoxylic oxime ether proceeded with excellent diastereoselectivity in aqueous media, providing a variety of enantiomerically pure α-amino acids (Eq. 11.44).[75] Interestingly, the indium-mediated alkylation of these substrates can also be performed in water (see Eq. 11.44 below).[87]

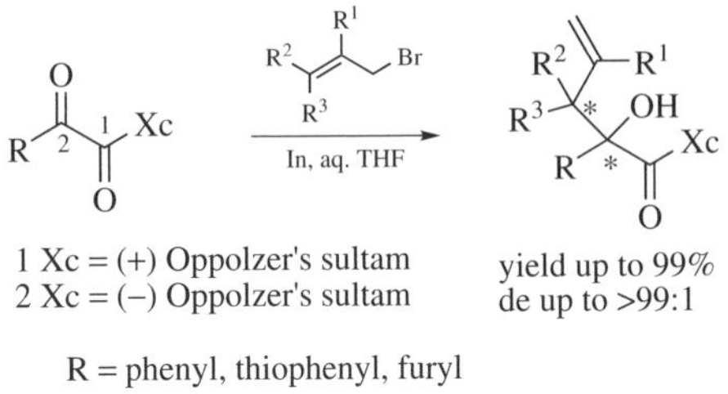

O
R
2
1
Xc
O
R1
R2
Br
R3
In, aq. THF
R2
R1
R3
OH
Xc
R
O
1 Xc = (+) Oppolzer's sultam
2 Xc = (−) Oppolzer's sultam
yield up to 99%
de up to >99:1
R = phenyl, thiophenyl, furyl

(11.43)

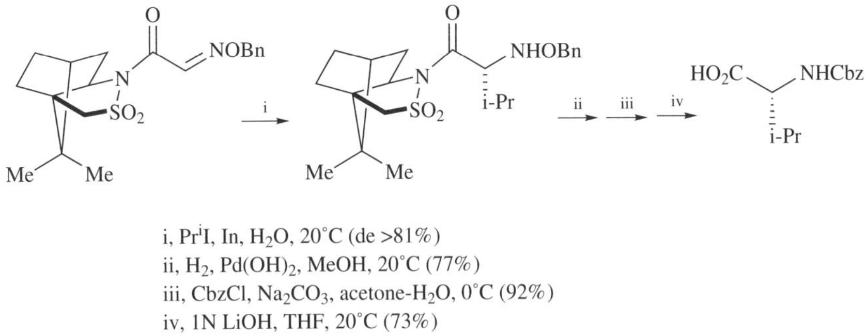

O
NOBn
N
SO2
Me
Me
i
NHOBn
i-Pr
ii
iii
iv
HO2C
NHCbz
i-Pr
i, PriI, In, H2O, 20°C (de >81%)
ii, H2, Pd(OH)2, MeOH, 20°C (77%)
iii, CbzCl, Na2CO3, acetone-H2O, 0°C (92%)
iv, 1N LiOH, THF, 20°C (73%)

(11.44)

More recently, catalytic asymmetric allylations of imines and imine derivatives in aqueous media have been studied. An *N*-spiro C_2-symmetrical chiral quaternary ammonium salt (*S*,*S*)-I-Br [(*S*,*S*)-β-Np-NAS-Br] has been evaluated in the allylation of glycine *tert*-Bu ester benzophenone Schiff base [$Ph_2C{=}NCH_2COOCMe_3$] for synthesis of both natural and unnatural α-amino acids (Eq. 11.45).[76]

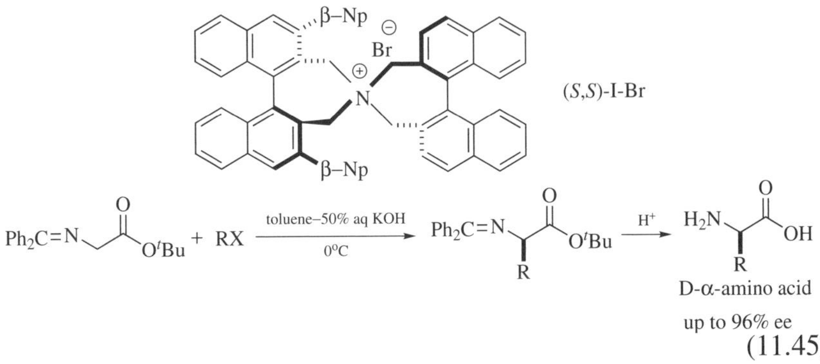

β-Np
Br
N
β-Np
(S,S)-I-Br
Ph2C=N
O
OtBu
+ RX
toluene–50% aq KOH
0°C
Ph2C=N
O
OtBu
R
H+
H2N
O
OH
R
D-α-amino acid
up to 96% ee

(11.45)

A formal enantioselective synthesis of the antibiotic L-azatyrosine was developed. The asymmetric allylation of hydrazono esters with allylsilanes in the presence of a catalytic amount of ZnF_2-chiral diamines in aqueous media generated (benzoyl)hydrazino-4-pentenoates in high enantioselectivity (Eq. 11.46).[77]

$$yield\ 60–92\%$$
$$ee\ 65–86\%$$

$$(11.46)$$

11.2.2.3 Reaction with Propargyl Halides.

The indium-mediated coupling of propargyl bromide with a variety of imines and imine oxides afforded homo-propargylamine derivatives in aqueous media under mild conditions.[78] Propargylation of glyoxylic oxime ether in the presence of a catalytic amount of palladium(0) complex and indium(I) iodide in aqueous media was also studied (Eq.11.47).[79]

$$(11.47)$$

11.2.2.4 Addition of Alkyl Groups.

Katrizky[80] reported that in the presence of bismuth(III) chloride-metallic aluminum, alkyl (as well as allyl) halides react with N-(alkylamino)benzotriazoles at 20°C in THF-water to give the corresponding homoalkylated amines in high yields (Eq. 11.48). Competitive addition of $^-CCl_3$ anions to N-alkyl-pyridinium salts was studied in a two-phase system of chloroform/concentrated aqueous NaOH and in a homogeneous medium.[81] Aminomethylation of 1-alkylpyrroles by aqueous formaldehyde and dimethylamine hydrochloride, followed by reaction with iodomethane, affords the 1-alkyl-2,5-bis[(trimethylammonio)methyl]pyrrole diiodide

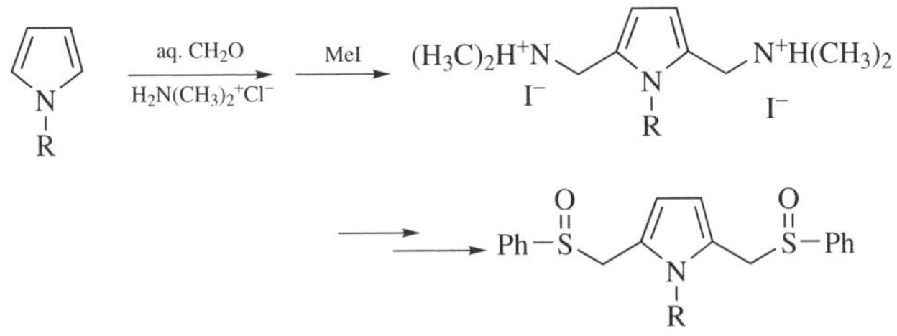

Scheme 11.4

(Scheme 11.4).[82]

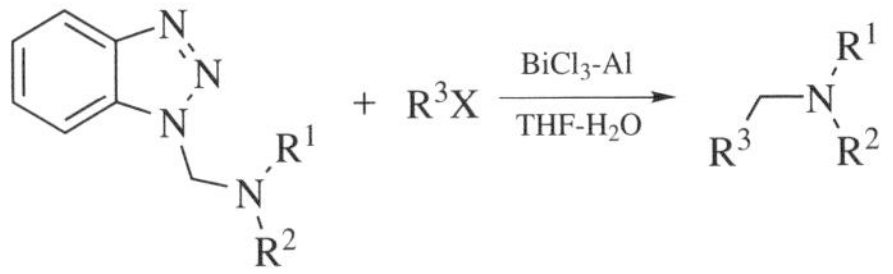

$$(11.48)$$

Clerici and Porta reported that phenyl, acetyl and methyl radicals add to the C_α atom of the iminium ion, $PhN^+Me=CHMe$, formed in situ by the titanium-catalyzed condensation of N-methylaniline with acetaldehyde to give PhNMeCHMePh, PhNMeCHMeAc, and PhNMeCHMe$_2$ in 80% overall yield.[83] Recently, Miyabe and co-workers studied the addition of various alkyl radicals to imine derivatives. Alkyl radicals generated from alkyl iodide and triethylborane were added to imine derivatives such as oxime ethers, hydrazones, and nitrones in an aqueous medium.[84] The reaction also proceeds on solid support.[85] N-sulfonylimines are also effective under such reaction conditions.[86] Indium is also effective as the mediator (Eq. 11.49).[87] A tandem radical addition-cyclization reaction of oxime ether and hydrazone was also developed (Eq. 11.50).[88] Li and co-workers reported the synthesis of α-amino acid derivatives and amines via the addition of simple alkyl halides to imines and enamides mediated by zinc in water (Eq. 11.51).[89] The zinc-mediated radical reaction of the hydrazone bearing a chiral camphorsultam provided the corresponding alkylated products with good diastereoselectivities that can be converted into enantiomerically pure α-amino acids (Eq. 11.52).[90]

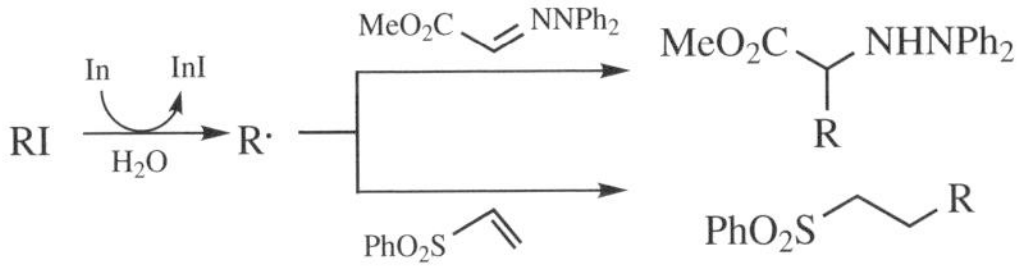

$$(11.49)$$

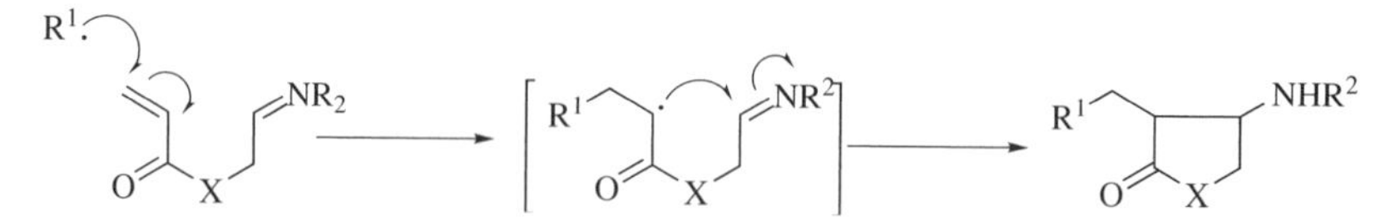

(11.50)

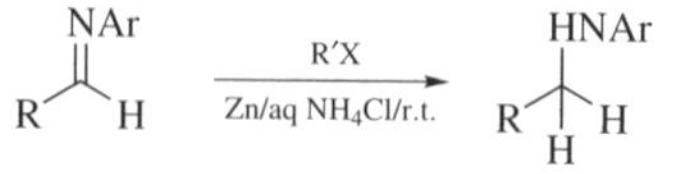

(11.51)

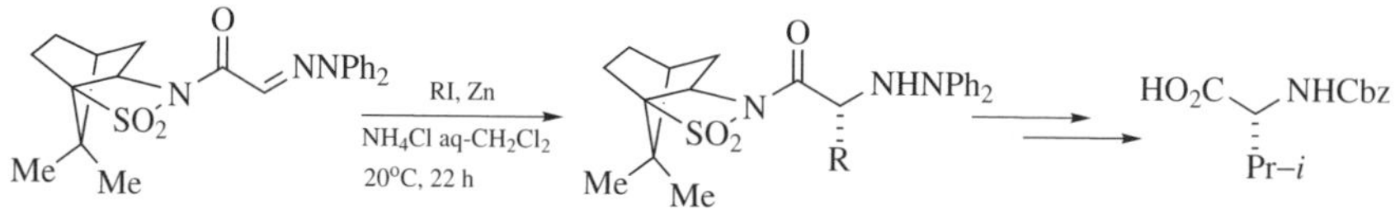

(11.52)

11.2.2.4 Addition of Vinyl and Aryl Groups.

The reaction of aromatic radicals, generated by decomposition of diazonium salts, with iminium salts in the presence of $TiCl_3$ in aqueous media produces secondary amines (Eq. 11.53).[91] The iminium salts are formed in situ from aromatic amines and aldehydes.

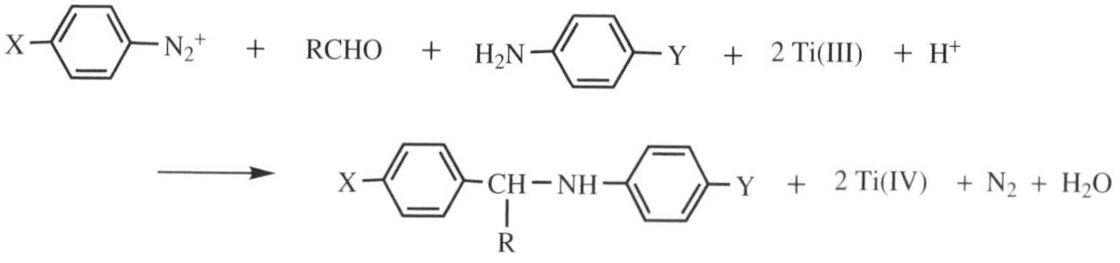

X=H, Cl, Me, OMe
Y=H, OMe
R=Me, Et, p-MePh

(11.53)

Petasis reported an efficient addition of vinyl boronic acid to iminium salts.[92] While no reaction was observed when acetonitrile was used as solvent, the reaction went smoothly in water to give allyl amines (Eq. 11.54). The reaction of the boron reagent with iminium ions generated from glyoxylic acid and amines affords novel α-amino acids (Eq. 11.55). Carboalumination of alkynes in the presence of catalytic Cp_2ZrCl_2 and H_2O affords vinylalane intermediates, which serve as nucleophiles in the subsequent addition to enantiomerically enriched (*tert*-butyl)- and (*p*-tolyl)sulfinimines. Chiral allylic sulfinamides are obtained in high diastereoselectivity and in good yield.[93]

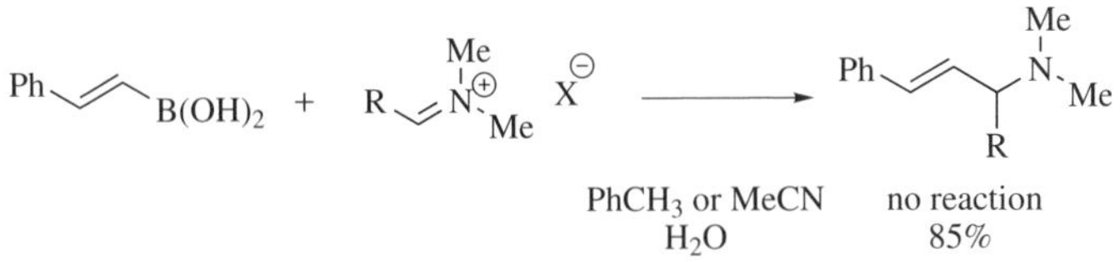

$$(11.54)$$

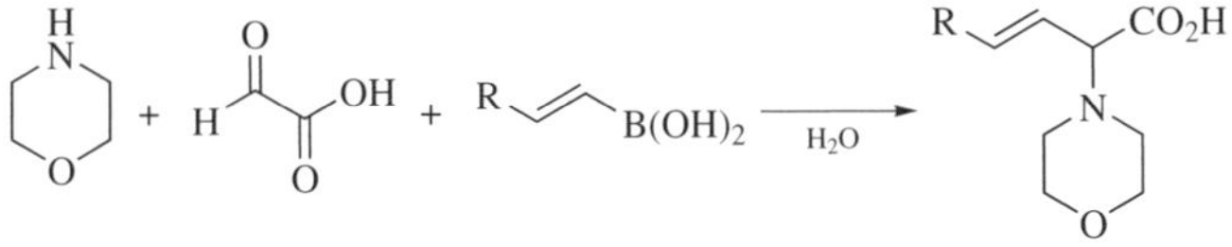

PhCH$_3$ or MeCN   no reaction
H$_2$O   85%

$$(11.55)$$

Miyaura and co-workers reported the rhodium-catalyzed reaction of arylboronic esters with *N*-sulfonylaldimines under aqueous conditions and found that hydrolysis of the *N*-sulfonylaldimines occurred first. The reaction had to be performed in anhydrous organic solvents (Eq. 11.56).[94] In contrast, Wang and Li[95] recently reported that in the presence of a rhodium catalyst, imines react with phenyltrimethyltin or phenyltrimethyllead in water and air under ultrasonic irradiation at 35°C to give the corresponding diarylmethylamines in good yields (Eq. 11.57).

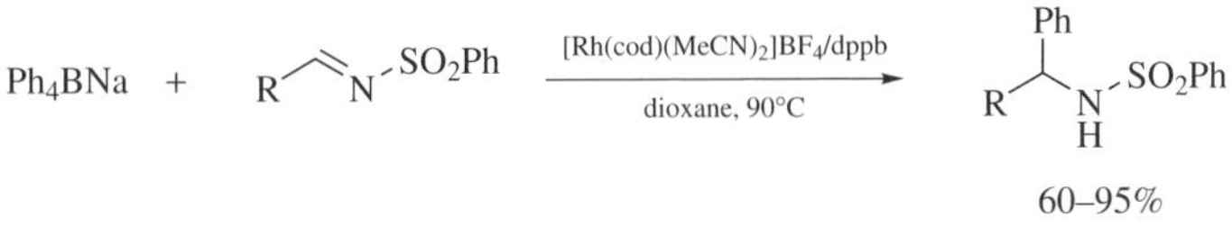

$$(11.56)$$

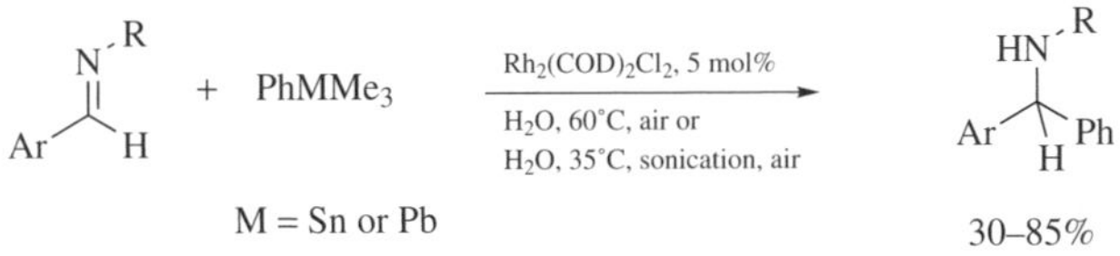

60–95%

$$(11.57)$$

M = Sn or Pb   30–85%

11.2.2.5 *Other Nucleophilic Additions.*

Besides the above addition reactions, the addition of alkynyl groups to imines has been recently investigated extensively in water and the subject was covered in Chapter 4, on alkynes. In the presence of catalytic amounts of Yb(O$_3$SCF$_3$)$_3$, imines reacted with Me$_3$SiCN to afford α-amino nitriles in excellent yields. Although some imines are difficult to prepare and purify, three-component reactions of aldehydes, amines, and Me$_3$SiCN proceeded smoothly.[96] By using Sc(O$_3$SCF$_3$)$_3$ as catalyst, Strecker-type reactions were successfully carried out by simply mixing aldehydes, amines, and tributyltin cyanide in aqueous media (Eq. 11.58).[97] The presence of a small amount of water was found to be crucial for the asymmetric addition of Reformatsky-type reagent to imines utilizing diisopropyl tartrate as a chiral auxiliary.[98]

$$R^1CHO \ + \ R^2NH_2 \ + \ Bu_3SnCN \ \xrightarrow{Sc(OTf)_3} \ \underset{R^1}{\overset{Bu_3Sn_\diagdown NR^2}{\underset{CN}{\bigg|}}} \ \xrightarrow{H_2O} \ \underset{R^1}{\overset{NHR^2}{\underset{CN}{\bigg|}}}$$

$$(11.58)$$

11.2.3 Reductive Coupling

Reductive coupling of aldimines obtained from aromatic aldehydes and aromatic amines to generate vicinal diamines mediated by indium was carried out in aqueous ethanol (Eq. 11.59).[99] Small indium rods were used in this study. No side-product was observed due to unimolecular reduction. The presence of NH_4Cl was found to accelerate the reaction. The reaction fails completely in CH_3CN, DMF, or wet DMF. The use of nonaromatic substrates also resulted in the failure of the reaction.

$$Ar^1CH=NAr^2 \ \xrightarrow[NH_4Cl]{In, \ H_2O\text{-}EtOH} \ \underset{Ar^2HN \ \ NHAr^2}{\overset{Ar^1HC-CHAr^1}{\underset{|\ \ \ \ \ |}{}}} \qquad (11.59)$$

$$meso + dl$$

Reductive coupling of aldimines into vicinal diamines has been performed by the action of zinc powder and 10% NaOH solution without using any organic solvents at ambient temperature in high yields.[100] Additives such as NH_4Cl and L-tyrosine can be used in lieu of 10% NaOH.[101] Vicinal disulfonamides were generated by the reductive coupling of N-sulfonylimines in Sm/HCl/THF;[102] whereas reductive coupling of aldimines and ketimine was examined by Sm(II)-based reagents (SmI_2, SmI_2-HMPA, $SmBr_2$, $Sm\{N[Si(CH_3)_3]_2\}_2$, SmI_2, triethylamine, and water).[103] Nitrones (e.g., $2\text{-}HOC_6H_4CH=N(O)Ph$) undergo deoxygenative reductive coupling and subsequent cyclization to 3-arylamino-2,3-dihydrobenzofuran derivatives in the presence of indium under aqueous conditions at ambient temperature in good yields (Eq. 11.60).[104] The reductive coupling and cyclization of 1,1-dicyanoalkenes were performed with metallic samarium in saturated aqueous NH_4Cl-THF solution at room temperature to give the amino compounds (Eq. 11.61).[105]

$$ (11.60) $$

$$ (11.61) $$

11.3 DIAZO COMPOUNDS

11.3.1 Substitution

Diazonium salts react with various nucleophiles in water (Eq. 11.62).[106] In acidic aqueous solution, p-phenylene*bis*diazonium ion reacts with alcohols more rapidly than it does with water.[107] In the presence of nucelophiles such as halides, the substitution products are obtained. Furthermore, diazonium salts of aromatic compounds are excellent substrates for palladium-catalyzed coupling reactions such as the Heck-type reactions in water.

$$ NO_2-C_6H_4-\overset{+}{\underset{\underset{\cdot\cdot}{\overset{\|}{N}}}{N}}\Big\}^-OH \rightleftharpoons NO_2-C_6H_4-N\!=\!N\!-\!OH \rightleftharpoons NO_2-C_6H_4-\overset{-}{\underset{NO}{N}}\Big\}^+Na $$
$$ + NaOH \qquad\qquad + H_2O $$

$$ (11.62) $$

11.3.2 Reduction

Aromatic diazo compounds can be reduced in water via a radical process (Scheme 11.5).[108] The reduction mechanism of arenediazonium salts by hydroquinone was studied in detail.[109] Arenediazonium tetrafluoroborate salts undergo facile electron-transfer reactions with hydroquinone in aqueous phosphate-buffered solution containing the hydrogen donor solvent acetonitrile. Reaction rates are first order in a

$$e_{aq}^{-} + X-C_6H_4N_2^{+} \longrightarrow X-C_6H_4N_2^{\bullet}$$

$$X-C_6H_4N_2^{\bullet} + X-C_6H_4N_2^{+} \rightleftharpoons (X-C_6H_4)_2N_4^{\bullet+}$$

$$X-C_6H_4N_2^{\bullet} \longrightarrow X-C_6H_4^{\bullet} + N_2$$

$$X-C_6H_4^{\bullet} + X-C_6H_4N_2^{+} \longrightarrow (X-C_6H_4)_2N_2^{\bullet+}$$

$$(X-C_6H_4)_2N_2^{\bullet+} + H_2O \rightleftharpoons (X-C_6H_4)_2N_2OH^{\bullet} + H^+$$

$$X = CO_2C_2H_5, F, H, CH_3, OCH_3, N(CH_3)_2$$

Scheme 11.5

hydroquinone and arenediazonium ion concentration and exhibit inverse first-order dependence on the hydrogen ion concentration at pH 1.0–9.5. Benzoquinone and arene are the principal products, but arylation of acetonitrile and benzoquinone are competitive in a reaction process that exhibits a 2:1 stoichiometric relation between the arenediazonium ion salt and hydroquinone. Reactions performed in deuterium oxide show kinetic isotope effects that reflect the solvent isotope effect on the acidity constant for hydroquinone. The reaction was proposed to proceed via an outer-sphere electron transfer mechanism. Free-radical reduction of arenediazonium ions with KI via γ-ray photolysis in aqueous solution generates the iodo-dediazonization product.[110]

11.3.3 Cyclopropanation

The reaction of diazoacetates and olefins efficiently gives cyclopropanation products. A cyclopropanation reaction involving ethyl diazoacetate with simple olefins proceeds with high efficiency in aqueous media using Rh(II) carboxylates as catalysts (Eq. 11.63).[111] Highly enantioselective cyclopropanation of alkenes was independently developed by using cobalt and ruthenium catalysts in water. An example is the enantioselective cyclopropanation of styrene in THF-H$_2$O or MeOH-H$_2$O catalyzed by an optically active β-ketoiminato cobalt(II) complex (Eq. 11.64).[112] Relative to THF, water and alcohol accelerated the reaction and improved the diastereo- and enantioselectivities. Even α,α-disubstituted styrenes were smoothly cyclopropanated with high enantiomeric excesses. Asymmetric cyclopropanation of styrene with chiral bis(hydroxymethyl-dihydrooxazolyl)pyridine-ruthenium catalyst in aqueous/organic biphasic media or homogeneous alcoholic media in the combination of toluene resulted in high enantiomeric excess up to 96–97% and *trans:cis* stereoselectivity to 97:3 (Eq. 11.65).[113] In the case of the intramolecular cyclopropanation reaction of *trans*-cinnamyl diazoester using Ru(II)(pybox-he) complex in biphasic medium, it gave the corresponding cyclopropane ring-fused lactone in 52% ee. The active catalyst in the water phase can be reused several times for the cyclopropanation reaction (Eq. 11.66).[114]

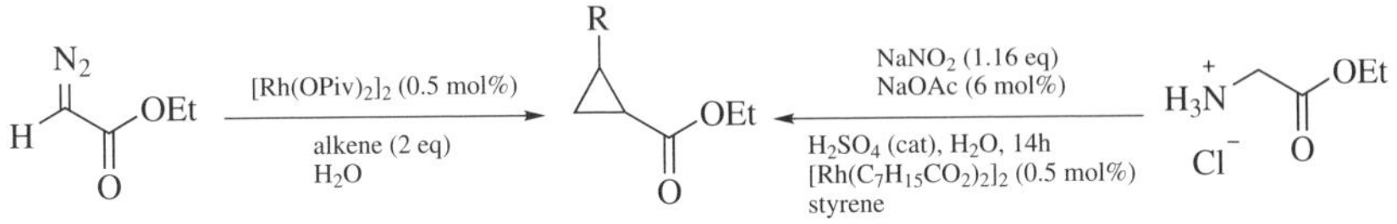

$$(11.63)$$

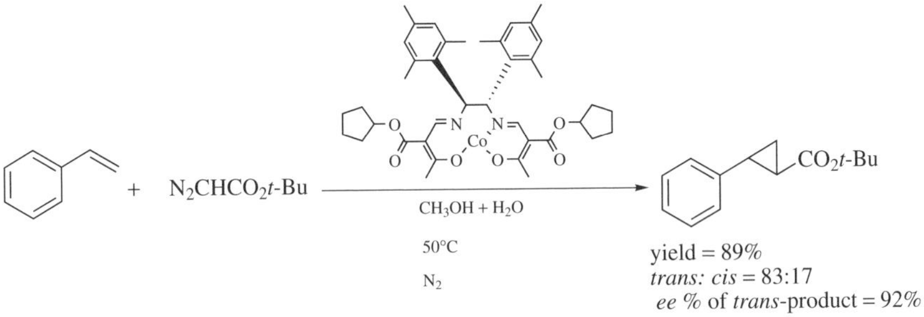

yield = 89%
trans: cis = 83:17
ee % of *trans*-product = 92%

$$(11.64)$$

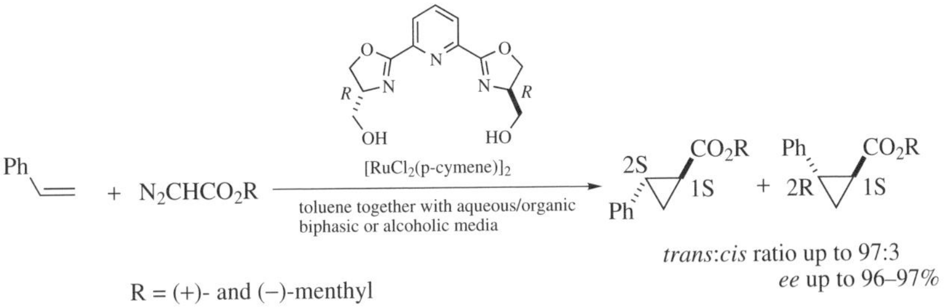

trans:cis ratio up to 97:3
ee up to 96–97%

R = (+)- and (–)-menthyl

$$(11.65)$$

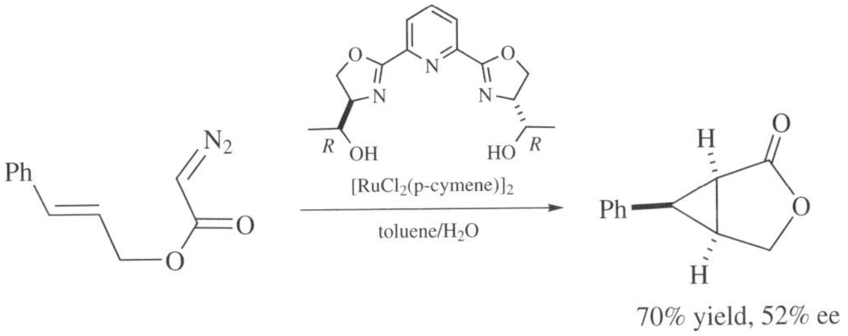

$$(11.66)$$

70% yield, 52% ee

11.3.4   Coupling Reaction

In the presence of metal salts, aryldiazo compounds undergo various coupling reactions in aqueous media. Such reactions have been used extensively in the dye and pigment industry[115] and in analytical chemistry.[116] For example, the $AlCl_3$-catalyzed diazo coupling of three different 3-substituted-1H-pyrazol-2-in-5-ones in water with seven different aryldiazonium salts yielded 21 different (5-hydroxy-3-substituted-1H-pyrazol-4-yl)-azobenzene derivatives as colored products with high yields (Eq. 11.67).[117]

$$(11.67)$$

11.4   AZIDES

11.4.1   Substitution

Simple azide is a good nucleophile that can undergo a variety of reactions such as the nucleophilic substitution reactions and conjugate addition reactions in aqueous media to give the corresponding organic azide compounds (Eq. 11.68).[118] Oxidative coupling of the azide ion with anions of primary nitroalkanes generates 1-azido-1-nitroalkanes (Eq. 11.69).[119] Baylis-Hillman acetates undergo smooth nucleophilic displacement with sodium azide in water under mild conditions to afford the corresponding 2-[(azido)methyl]-3-(substituted)propenoic acid ethyl ester in excellent yields (Eq. 11.70).[120]

$$(11.68)$$

$$RCH_2NO_2 \quad + \quad NaN_3 \quad \xrightarrow[\text{CH}_2\text{Cl}_2/\text{H}_2\text{O}]{\text{NaOH, (NH}_4)_2\text{S}_2\text{O}_8} \quad RCH(NO_2)N_3 \qquad (11.69)$$

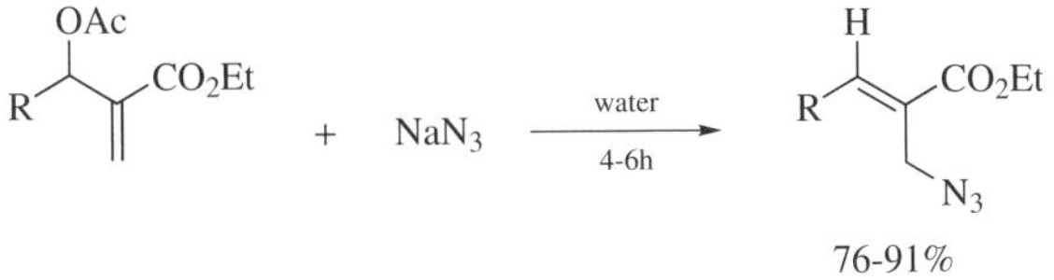

$$(11.70)$$

76-91%

11.4.2 Click Chemistry

As was mentioned in Chapter 5, "click" chemistry refers to a set of chemical reactions and processes that are easy to carry out, give high yield, and require readily available starting materials and reagents, with simple reaction conditions in no solvent or in a solvent that is benign (such as water) and simple product purification. A number of nitrogen compounds can participate in click chemistry. Useful among them are aziridines, which are the aza analogs of epoxides. While N-H or *N*-alkyl aziridines are stable under basic conditions, they can be readily opened by various heteronucleophiles such as azide ion under buffered conditions in various solvents including water (Scheme 11.6).[121]

Another important click reaction is the cycloaddition of azides. The addition of sodium azide to nitriles to give 1*H*-tetrazoles is shown to proceed readily in water with zinc salts as catalysts (Eq. 11.71).[122] The scope of the reaction is quite broad; a variety of aromatic nitriles, activated and nonactivated alkyl nitriles, substituted vinyl nitriles, thiocyanates, and cyanamides have all been shown to be viable substrates for this reaction. The reaction of an arylacetylene with an azide in hot water gave 1,4-disubstituted 1,2,3-triazoles in high yields,[123] while a similar reaction between a terminal aliphatic alkyne and an azide (except *m*-nitroazidobenzene) afforded a mixture of regioisomers with

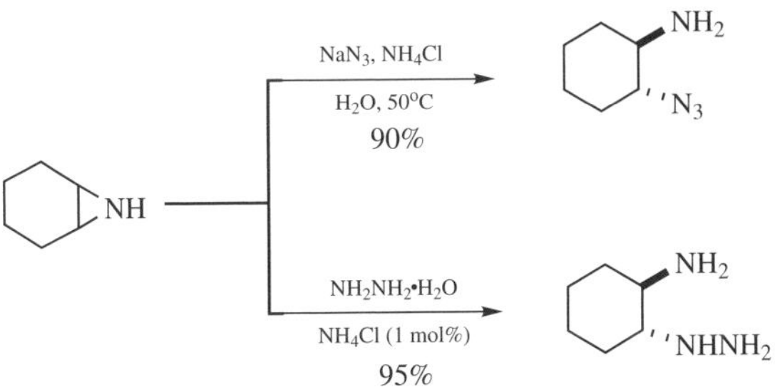

Scheme 11.6

the ratio of 1,4- to 1,5-isomers ranging from 3:1 to 28.6:1. Reactions of *m*-nitroazidobenzene with either arylalkynes or aliphatic alkynes formed only 1,4-disubstituted derivatives in excellent yields. A copper(I) catalyst in a mixture of ionic liquid [bmim][BF$_4$] and water can effect the three-component reaction of halides, sodium azide, and alkynes to form 1,4-disubstituted 1,2,3-triazoles in good-to-high yields (Eq. 11.72).[124] The reaction has been extensively discussed in Chapter 4.

$$R-C{\equiv}N \quad \xrightarrow[\substack{\text{water} \\ \text{reflux}}]{\substack{\text{1.1 eq. NaN}_3 \\ \text{1.0 eq. ZnBr}_2}} \quad \underset{R}{\overset{N{=}N}{\diagup}}NH \qquad (11.71)$$

R = Ar, Alk, Vinyl, SR, NR$_2$

52–96%

$$R{-}{\equiv} \; + \; R'{-}X \quad \xrightarrow[\text{CuI, Na}_2\text{CO}_3]{\text{NaN}_3,\,[\text{bmim}][\text{BF}_4]/\text{H}_2\text{O}} \quad \underset{R}{\overset{N{=}N}{\diagup}}N{-}R' \qquad (11.72)$$

X = Br, I

68–99%

N-sulfonylamides can be efficiently prepared by the hydrative reaction between terminal alkynes, sulfonyl azides, and water in the presence of copper catalyst and amine base under very mild conditions (Eq. 11.73).[125] The method is quite general and can tolerate a variety of labile functional groups. A wide range of alkynes and sulfonyl azides are readily coupled catalytically with water to furnish amides in high yields. The reaction is regioselective in that only terminal alkynes react while double or internal triple bonds are intact. The reaction is also adaptable to a solid-phase synthesis. A copper(I)-catalyzed three-component reaction of amine, propargyl halide, and azide in water forms (1-substituted-1*H*-1,2,3-triazol-4-ylmethyl)dialkylamines in good yields (Eq. 11.74).[126]

$$R^1{-}{\equiv} \; + \; R^2{-}SO_2N_3 \; + \; H_2O \quad \xrightarrow[\text{CH}_3\text{Cl, 25°C, 12h}]{\text{CuI (cat), Et}_3\text{N}} \quad R^1\underset{O}{\overset{\displaystyle }{\diagdown}}\overset{H}{\underset{}{N}}{\diagdown}SO_2R^2 \qquad (11.73)$$

74–97%

$$\underset{R^2}{\overset{R^1}{\diagdown}}NH \; + \; {\equiv}{-} \; + \; R{-}\overset{n}{\bigcirc}O\underset{}{-}N_3 \quad \xrightarrow[\text{H}_2\text{O, r.t. air}]{\text{CuI (10 mol\%), Et}_3\text{N}} \quad \underset{R^2}{\overset{R^1}{\diagdown}}N{-}\diagup\overset{N{=}N}{\underset{}{\diagdown}}N{-}O{-}R$$

X = Cl, Br   n = 0, 1

70–98%

(11.74)

11.5 NITRO COMPOUNDS

Decomposition and oxidation of nitromethane, nitroethane, and 1-nitropropane in supercritical water (SCW) near the critical point have been studied in a flow reactor.[127] Reduction of the nitro group can be carried out with various reducing reagents in water. For example, using indium-wire in water, nitro compounds are reduced to amines in water at room temperature under sonication (Eq. 11.75).[128] Preparation of aromatic amines can be performed by reduction of aromatic nitro compounds with metallic tellurium in near-critical water.[129] On the other hand, conjugated nitroalkenes undergo efficient conjugate additions with various nucleophiles and Diels-Alder-type reactions in water.[130] The conjugate addition reaction has been discussed together with other conjugated systems, and the Diels-Alder reaction will be discussed in Chapter 12.

$$R-NO_2 \xrightarrow[\text{H}_2\text{O, r.t. sonication}]{\text{In-wire, pH= 1~2}} R-NH_2 \qquad (11.75)$$

74–100%

REFERENCES

1. Loeser, E., Prasad, K., Repic, O., *Synth. Commun.* **2002**, *32*, 403.

2. Watanabe, Y., Yamamoto, M., Mitsudo, T., Takegami, Y., *Tetrahedron Lett.* **1978**, *19*, 1289.

3. Smith, M. B., March, J., *March's Advanced Organic Chemistry: Reactions, Mechanisms, and Structure,* 5th ed., John Wiley & Sons, New York, 2001.

4. Butler, A. R., *Chem. Rev.* **1975**, *75*, 241, Patai, S., *The Chemistry of Diazonium and Diazo Groups*, Wiley, New York, 1978.

5. Kornblum, N., Iffland, D. C., *J. Am. Chem. Soc.* **1949**, *71*, 2137.

6. Williams, D. L. H., *Nitrosation*, Cambridge University Press, Cambridge, 1988.

7. Rao, P. S., Hayon, E., *J. Phys. Chem.* **1975**, *79*, 1063.

8. Rozen, S., Kol, M., *J. Org. Chem.* **1992**, *57*, 7342.

9. Miaskiewicz, K., Teich, N. A., Smith, D. A., *J. Org. Chem.* **1997**, *62*, 6493.

10. Sharma, V. B., Jain, S. L., Sain, B., *Tetrahedron Lett.* **2004**, *45*, 4281.

11. Cocheci, V., Gerasimou, E., Csunderlik, C., Cotarca, L., Novac, A., *Rev. Roum. Chim.* **1989**, *34*, 749.

12. Wang, F., Sayre, L. M., *J. Am. Chem. Soc.* **1992**, *114*, 248.

13. Ohtani, B., Takamiya, S., Hirai, Y., Sudoh, M., Nishimoto, S. I., Kagiya, T., *J. Chem. Soc. Perkin Trans.* 2, **1992**, 175.

14. Mori, K., Yamaguchi, K., Mizugaki, T., Ebitani, K., Kaneda, K., *Chem. Commun.* **2001**, 461.

15. Choudary, B. M., Bharathi, B., Reddy, C. V., Kantam, M. L., Raghavan, K. V., *Chem. Commun.* **2001**, 1736.

16. Chen, C. K., Hortmann, A. G., Marzabadi, M. R., *J. Am. Chem. Soc.* **1988**, *110*, 4829.

17. Miyazawa, A., Tanaka, K., Sakakura, T., Tashiro, M., Tashiro, H., Prakash, G. K. S., Olah, G. A., *Chem. Commun.* **2005**, 2104.

18. Li, Z., Li, C.-J., *J. Am. Chem. Soc.*, **2004**, *126*, 11810.

19. Jiang, Y., Xi, C., Yang, X., *Synlett* **2005**, 1381; Xi, C., Jiang, Y., Yang, X., *Tetrahedron Lett.* **2005**, *46*, 3909

20. Laha, S., Luthy, R. G., *Envir. Sci. Techn.* **1990**, *24*, 363; Kaushik, R. D., Arya, R. K., Kumar, S., *Asian J. Chem.* **2000**, *12*, 1229; Kaushik, R. D., Joshi, R., *Asian J. Chem.* **1997**, *9*, 746.

21. Reddy, M. S., Narender, M., Rao, K. R., *Tetrahedron Lett.* **2005**, *46*, 1299.

22. Surendra, K., Krishnaveni, N. S., Reddy, M. A., Nageswar, Y. V. D., Rao, K. R., *J. Org. Chem.* **2003**, *68*, 9119.

23. Grushin, V. V., Marshall, W. J., *Adv. Synth. Cat.* **2004**, *346*, 1457.

24. Firouzabadi, H., Iranpoor, N., Alinezhad, H., *Bull. Chem. Soc. Jpn.* **2003**, *76*, 143.

25. Tsukinoki, T., Mitoma, Y., Nagashima, S., Kawaji, T., Hashimoto, I., Tashiro, M., *Tetrahedron Lett.* **1998**, *39*, 8873.

26. Stradins, J., Turovska, B., Glezer, V., Markava, E., Gustina, D., *J. Electroanal. Chem. Interfac. Electrochem.* **1991**, *317*, 243.

27. Seyden-Penne, J., *Reductions by the Alumino- and Borohydrides in Organic Synthesis*, VCH, New York, 1991.

28. Todd, D., *Organic Reactions*, **1948**, *4*, 378.

29. Huang, M., *J. Am. Chem. Soc.* **1946**, *68*, 2487.

30. Eisenbraun, E. J., Payne, K. W., Bymaster, J. S., *Ind. Eng. Chem. Res.* **2000**, *39*, 1119.

31. Gadhwal, S., Baruah, M., Sandhu, J. S., *Synlett* **1999**, 1573.

32. Longley, C. J., Goodwin, T. J., Wilkinson, G., *Polyhedron* **1986**, *5*, 1625.

33. Amrani, Y., Lecomte, L., Sinou, D., Bakos, J., Toth, I., Heil, B., *Organometallics* **1989**, *8*, 542.

34. Wu, J., Wang, F., Ma, Y., Cui, X., Cun, L., Zhu, J., Deng, J., Yu, B., *Chem. Commun.* **2006**, 1766.

35. Tychopoulos, V., Tyman, J. H. P., *Synth. Commun.* **1986**, *16*, 1401.

36. Kobayashi, S., Ishitani, H., *J. Chem. Soc., Chem. Commun.* **1995**, 1379.

37. Loh, T.-P., Wei, L.-L., *Tetrahedron Lett.* **1998**, *39*, 323; Loh, T.-P., Liung, S. B. K. W., Tan, K.-L., Wei, L.-L., *Tetrahedron* **2000**, *56*, 3227.

38. Kobayashi, S., Busujima, T., Nagayama, S., *Synlett* **1999**, 545.

39. Manabe, K., Mori, Y., Wakabayashi, T., Nagayama, S., Kobayashi, S., *J. Am. Chem. Soc.* **2000**, *122*, 7202.

40. Zhang, C., Dong, J., Cheng, T., Li, R., *Tetrahedron Lett.* **2001**, *42*, 461.

41. Ranu, B. C., Samanta, S., Guchhait, S. K., *Tetrahedron* **2002**, *58*, 983.

42. Loncaric, C., Manabe, K., Kobayashi, S., *Adv. Synth. Catal.* **2003**, *345*, 1187.

43. Akiyama, T., Takaya, J., Kagoshima, H., *Chem. Lett.* **1999**, *9*, 947.

44. Akiyama, T., Takaya, J., Kagoshima, H., *Synlett* **1999**, 1426.

45. Akiyama, T., Takaya, J., Kagoshima, H., *Synlett* **1999**, 1045.

46. Akiyama, T., Takaya, J., Kagoshima, H., *Tetrahedron Lett.* **2001**, *42*, 4025.

47. Manabe, K., Kobayashi, S., *Org. Lett.* **1999**, *1*, 1965.

48. Manabe, K., Mori, Y., Kobayashi, S., *Synlett* **1999**, 1401; Manabe, K., Mori, Y., Kobayashi, S., *Tetrahedron* **2001**, *57*, 2537.

49. Akiyama, T., Itoh, J., Fuchibe, K., *Synlett* **2002**, 1269; Akiyama, T., Takaya, J., Kagoshima, H., *Adv. Synth. Catal.* **2002**, *344*, 338.

50. Iimura, S., Nobutou, D., Manabe, K., Kobayashi, S., *Chem. Commun.* **2003**, 1644.

51. Miura, K., Tamaki, K., Nakagawa, T., Hosomi, A., *Angew. Chem., Int. Ed.* **2000**, *39*, 1958.

52. Wang, M., Yang, X.-F., Li, C.-J., *Eur. J. Org. Chem.* **2003**, 998.

53. Sakthivel, K., Notz, W., Bui, T., Barbas, C. F. III., *J. Am. Chem. Soc.* **2001**, *123*, 5260.

54. Kobayashi, S., Hamada, T., Manabe, K., *J. Am. Chem. Soc.* **2002**, *124*, 5640.

55. Ishimaru, K., Kojima, T., *J. Org. Chem.* **2003**, *68*, 4959.

56. Cordova, A., Barbas, C. F., *Tetrahedron Lett.* **2003**, *44*, 1923.

57. Grieco, P. A., Bahsas, A., *J. Org. Chem.* **1987**, *52*, 1378.

58. Chan, T.-H., Lu, W., *Tetrahedron Lett.* **1998**, *39*, 8605; Sampath Kumar, H. M., Anjaneyulu, S., Jagan Reddy, E., Yadav, J. S., *Tetrahedron Lett.* **2000**, *41*, 9311.

59. Lu, W., Chan, T.-H., *J. Org. Chem.* **2000**, *65*, 8589.

60. Lu, W., Chan, T.-H., *J. Org. Chem.* **2001**, *66*, 3467.

61. Bryan, V. J., Chan, T. H., *Tetrahedron Lett.* **1997**, *38*, 6493.

62. Yoo, B. W., Choi, K. H., Lee, S. J., Nam, G. S., Chang, K. Y., Kim, S. H., Kim, J. H., *Synth. Commun.* **2002**, *32*, 839.

63. Laskar, D. D., Prajapati, D., Sandhu, J. S., *Tetrahedron Lett.* **2001**, *42*, 7883; Laskar, D. D., Gohain, M., Prajapati, D., Sandhu, J. S., *New J. Chemistry* **2002**, *26*, 193.

64. Kang, K. H., Choi, K. I., Koh, H. Y., Kim, Y., Chung, B. Y., Cho, Y. S., *Synth. Commun.* **2001**, *31*, 2277.

65. Kobayashi, S., Hamada, T., Manabe, K., *Synlett* **2001**, 1140.

66. Akiyama, T., Onuma, Y., *J. Chem. Soc., Perkin 1* **2002**, 1157.

67. Estevam, I. H. S., Bieber, L. W., *Tetrahedron Lett.* **2003**, *44*, 667.

68. Kumar, S., Kaur, P., *Tetrahedron Lett.* **2004**, *45*, 3413.

69. Miyabe, H., Yamaoka, Y., Naito, T., Takemoto, Y., *J. Org. Chem.* **2003**, *68*, 6745.

70. Matsumura, Y., Onomura, O., Suzuki, H., Furukubo, S., Maki, T., Li, C.-J., *Tetrahedron Lett.* **2003**, *44*, 5519.

71. Basile, T., Bocoum, A., Savoia, D., Umani-Ronchi, A., *J. Org. Chem.* **1994**, *59*, 7766.

72. Hanessian, S., Yang, R.-Y., *Tetrahedron Lett.* **1996**, *37*, 5273.

73. Chen, G.-M., Ramachandran, P. V., Brown, H. C., *Angew. Chem., Int. Ed.* **1999**, *38*, 825.

74. Shin, J. A., Cha, J. H., Pae, A. N., Choi, K. I., Koh, H. Y., Kang, H.-Y., Cho, Y. S., *Tetrahedron Lett.* **2001**, *42*, 5489.

75. Miyabe, H., Nishimura, A., Ueda, M., Naito, T., *Chem. Commun.* **2002**, 1454.

76. Ooi, T., Uematsu, Y., Maruoka, K., *Adv. Synth. Catal.* **2002**, *344*, 288.

77. Hamada, T., Manabe, K., Kobayashi, S., *Angew. Chem., Int. Ed.* **2003**, *42*, 3927.

78. Prajapati, D., Laskar, D. D., Gogoi, B. J., Devi, G., *Tetrahedron Lett.* **2003**, *44*, 6755.

79. Miyabe, H., Yamaoka, Y., Naito, T., Takemoto, Y., *J. Org. Chem.* **2004**, *69*, 1415.

80. Katritzky, A. R., Shobana, N., Harris, P. A., *Tetrahedron Lett.* **1991**, *32*, 4247.

81. Makosza, M., Kryklowa, I., *Tetrahedron* **1999**, *55*, 6395.

82. Kim, I. T., Elsenbaumer, R. L., *Tetrahedron Lett.* **1998**, *39*, 1087.

83. Clerici, A., Porta, O., *Gazz. Chim. Ital.* **1992**, *122*, 165.

84. Miyabe, H., Ueda, M., Naito, T., *J. Org. Chem.* **2000**, *65*, 5043.

85. Miyabe, H., Nishimura, A., Fujishima, Y., Naito, T., *Tetrahedron* **2003**, *59*, 1901.

86. Miyabe, H., Ueda, M., Naito, T., *Chem. Commun.* **2000**, 2059.

87. Miyabe, H., Ueda, M., Nishimura, A., Naito, T., *Org. Lett.* **2002**, *4*, 131.

88. Miyabe, H., Ueda, M., Fujii, K., Nishimura, A., Naito, T., *J. Org. Chem.* **2003**, *68*, 5618.

89. Huang, T., Keh, C. C. K., Li, C.-J., *Chem. Commun.* **2002**, 2440.

90. Ueda, M., Miyabe, H., Nishimura, A., Sugino, H., Naito, T., *Tetrahedron: Asymmetry* **2003**, *14*, 2857.

91. Clerici, A., Porta, O., *Tetrahedron Lett.* **1990**, *31*, 2069.

92. Petasis, N. A., Zavialov, I. A., *J. Am. Chem. Soc.* **1997**, *119*, 445.

93. Wipf, P., Nunes, R. L., Ribe, S., *Helv. Chim. Acta* **2002**, *85*, 3478.

94. Ueda, M., Miyaura, N., *J. Organomet. Chem.* **2000**, *595*, 31.

95. Ding, R., Zhao, C.-H., Chen, Y.-J., Liu, L., Wang, D., Li, C.-J., *Tetrahedron Lett.* **2004**, *45*, 2995.

96. Kobayashi, S., Ishitani, H., Ueno, M., *Synlett* **1997**, 115.

97. Kobayashi, S., Busujima, T., *Chem. Commun.* **1998**, 981.

98. Ukaji, Y., Takenaka, S., Horita, Y., Inomata, K., *Chem. Lett.* **2001**, 254.

99. Kalyanam, N., Rao, G. V., *Tetrahedron Lett.* **1993**, *34*, 1647.

100. Dutta, M. P., Baruah, B., Boruah, A., Prajapati, D., Sandhu, J. S., *Synlett* **1998**, 857.

101. Tsukinoki, T., Nagashima, S., Mitoma, Y., Tashiro, M., *Green Chem.* **2000**, *2*, 117.

102. Liu, X., Liu, Y., Zhang, Y., *Tetrahedron Lett.* **2002**, *43*, 6787.

103. Kim, M., Knettle, B. W., Dahlen, A., Hilmersson, G., Flowers, R. A., *Tetrahedron* **2003**, *59*, 10397.

104. Jeevanandam, A., Ling, Y.-C., *Tetrahedron Lett.* **2001**, *42*, 4361.

105. Wang, L., Zhang, Y., *Tetrahedron Lett.* **1998**, *39*, 5257.

106. Hodgson, H. H., *J. Chem. Soc.* **1948**, 348.

107. Lewis, E. S., Chalmers, D. J., *J. Am. Chem. Soc.* **1971**, *93*, 3267; Lewis, E. S., Hartung, L. D., McKay, B. M., *J. Am. Chem. Soc.* **1969**, *91*, 419.

108. Daasbjerg, K., Sehested, K., *J. Phys. Chem. A* **2003**, *107*, 4462.

109. Brown, K. C., Doyle, M. P., *J. Org. Chem.* **1988**, *53*, 3255.

110. Packer, J. E., Monig, J., Dobson, B. C., *Austr. J. Chem.* **1981**, *34*, 1433.

111. Wurz, R. P., Charette, A. B., *Org. Lett.* **2002**, *4*, 4531.

112. Ikeno, T., Nishizuka, A., Sato, M., Yamada, T., *Synlett* **2001**, 406.

113. Iwasa, S., Takezawa, F., Tuchiya, Y., Nishiyama, H., *Chem. Commun.* **2001**, 59.

114. Iwasa, S., Tsushima, S., Nishiyama, K., Tsuchiya, Y., Takezawa, F., Nishiyama, H., *Tetrahedron: Asym.* **2003**, *14*, 855.

115. Iwamoto, H., Kobayashi, H., Murer, P., Sonoda, T., Zollinger, H., *Bull. Chem. Soc. Jpn* **1993**, *66*, 2590.

116. Al-Hatim, A. A., Ibraheem, B.B., *Anal. Lett.* **1989**, *22*, 2091.

117. Khalil, A. K., Hassan, M. A., Mohamed, M. M., El-Sayed, A. M., *Dyes Pigm.* **2005**, *66*, 241.

118. Richard, J. P., Lin, S.-S., Williams, K. B., *J. Org. Chem.* **1996**, *61*, 9033; Buckley, N., Oppenheimer, N. J., *J. Org. Chem.* **1996**, *61*, 7360.

119. Tselinskii, I. V., Mel'nikova, S. F., Fedotov, S. A., *Russ. Chem. Bull.* **2002**, *51*, 1466.

120. Yadav, J. S., Gupta, M. K., Pandey, S. K., Reddy, B. V. S., Sarma, A. V. S., *Tetrahedron Lett.* **2005**, *46*, 2761.

121. Kolb, H. C., Finn, M. G., Sharpless, K. B., *Angew. Chem. Int. Ed.* **2001**, *40*, 2004.

122. Demko, Z. P., Sharpless, K. B., *J. Org. Chem.* **2001**, *66*, 7945.

123. Wang, Z.-X., Qin, H.-L., *Chem. Commun.* **2003**, 2450.

124. Zhao, Y.-B., Yan, Z.-Y., Liang, Y.-M., *Tetrahedron Lett.* **2006**, *47*, 1545.

125. Cho, S. H., Yoo, E. J., Bae, I., Chang, S., *J. Am. Chem. Soc.* **2005**, *127*, 16046.

126. Yan, Z.-Y., Zhao, Y.-B.; Fan, M.-J., Liu, W.-M., Liang, Y.-M., *Tetrahedron* **2005**, *61*, 9331.

127. Anikeev, V., Yermakova, A., Goto, M., *Ind. Eng. Chem. Res.* **2004**, *43*, 8141.

128. Cho, Y. S., Jun, B. K., Kim, S., Cha, J. H., Pae, A. N., Koh, H. Y., Chang, M. H., Han, S.-Y., *Bull. Kor. Chem. Soc.* **2003**, *24*, 653.

129. Wang, L., Li, P.-H., Jiang, Z.-Q., *Chin. J. Chem.* **2003**, *21*, 222.

130. Fringuelli, F., Matteucci, M., Piermatti, O., Pizzo, F., Burla, M. C., *J. Org. Chem.* **2001**, *66*, 4661.

CHAPTER 12

PERICYCLIC REACTIONS

12.1 INTRODUCTION

Pericyclic reactions are reactions that occur by a concerted process through a cyclic transition state. Such reactions include cycloadditions, sigmatropic rearrangements, and electrocyclizations.[1] Based on Klopman's equation of chemical reactivity, pericyclic reactions occur predominantly via the interaction of frontier molecular orbitals.[2] Unlike the charge–charge interaction, orbitals–orbital interactions are relatively soft and thus can tolerate "hard" water as solvent. However, the special properties of water can provide additional benefits to such reactions. During such reactions, there is usually a large decrease in the volume of activation ($\Delta V^{\neq}$) in forming the transition state from the starting materials in order to have an effective interaction of the frontier molecular orbitals. For example, the values of $\Delta V^{\neq}$ are between -20 and -45 cm^3 mol^{-1} for Diels-Alder reactions and are about -10 cm^3 mol^{-1} typically for Cope and Claisen rearrangements.[3] This large-volume contraction can be translated into a subtraction of several kcal/mol from the free energy of activation under high pressure. Thus, the rates of these reactions are particularly susceptible to a change in pressure. For example, a Diels-Alder cycloaddition reaction under a pressure of 9–10 kbar at room temperature has approximately the same rate as one at

Comprehensive Organic Reactions in Aqueous Media, Second Edition, by Chao-Jun Li and Tak-Hang Chan
Copyright © 2007 John Wiley & Sons, Inc.

about 100°C at atmospheric pressure.[4] In addition, whereas an increase in temperature increases the rate of both forward and reverse reactions, the increase in pressure in a Diels-Alder reaction accelerates the forward reaction only.

An analogy can be drawn between pericyclic reactions in water and under high pressure. Water's internal pressure on hydrophobic substrates acts on the volume of activation of a reaction in the same way as an externally applied pressure does. Thus, the internal pressure of water influences the rates of pericyclic reactions in water in the same direction as external pressures. The use of salting-out salts will further increase the rate of pericyclic reactions. Recently, Kumar quantified the relationship between internal pressure and the rate of the aqueous Diels-Alder reaction. A linear relationship between the two was observed.[5]

On the other hand, many pericyclic reactions are accelerated by Lewis-acid catalysts. The acceleration has been attributed to a complex formation between the Lewis acid and the polar groups of the reactants that brings about changes in the energies and orbital coefficients of the frontier orbitals.[6] The complex formation also stabilizes the enhanced polarized transition state.

The effect of water molecules on pericyclic reactions can also be compared with the effects of Lewis acids on these reactions. The enhanced polarization of the transition state in these reactions would lead to stronger hydrogen bonds at the polar groups of the reactants, which will result in a substantial stabilization of the transition states in the same way Lewis acids do. A computer-simulation study on the Diels-Alder reaction of cyclopentadiene by Jorgensen indicated that this effect contributes about a factor of 10 to the rates.[7]

For intermolecular pericyclic reactions, the aggregation of nonpolar reactants in water (hydrophobic effect) results in a net gain in free energy. This gain in free energy would contribute to the increase in the rate of reactions. Breslow has studied the influence of the hydrophobic effect on the aqueous Diels-Alder reactions in detail.[8,9] Schneider has reported a quantitative correlation between solvophobicity and the rate enhancement of aqueous Diels-Alder reactions.[10,11] Enforced hydrophobic interactions between diene and dienophile have also been suggested by Engberts to account for the acceleration in water.[12] A pseudothermodynamic analysis of the rate acceleration in water relative to 1-propanol and 1-propanol-water mixtures indicates that hydrogen-bond stabilization of the polarized activated complex and the decrease of the hydrophobic surface area of the reactants during the activation process are the two main causes of the rate enhancement in water.[13] Studies of Diels-Alder reactions under high pressure by Jenner revealed that water can alter kinetics, chemo-, and enantioselectivity through polarity and hydrophobic effects.[14] The weight of these two, however, depends on the use of specific reaction partners. By studying the reaction in dilute aqueous ethanol, Smith[15] and Griesbeck[16] pointed out that the concentration of the reaction substrates is important for rate enhancement. The rate shows a maximum at 0.5 M under the aqueous ethanol condition.

The influence of the hydrophobic effect on the aqueous pericyclic reactions can be compared with the effect of catalytic antibodies. Antibodies have been found to catalyze Diels-Alder reactions,[17] hetero-Diels-Alder reactions,[18] and Claisen[19] and oxy-Cope[20] rearrangements. It is suggested that antibodies catalyze these reactions by acting as an entropy trap, primarily through binding and orienting the substrates in the cyclic conformations.

12.2 DIELS-ALDER REACTIONS

The Diels-Alder reaction is one of the most important methods used to form cyclic structures and is one of the earliest examples of carbon–carbon bond formation reactions in aqueous media.[21] Diels-Alder reactions in aqueous media were in fact first carried out in the 1930s, when the reaction was discovered,[22] but no particular attention was paid to this fact until 1980, when Breslow[23] made the dramatic observation that the reaction of cyclopentadiene with butenone in water (Eq. 12.1) was more than 700 times faster than the same reaction in isooctane, whereas the reaction rate in methanol is comparable to that in a hydrocarbon solvent. Such an unusual acceleration of the Diels-Alder reaction by water was attributed to the "hydrophobic effect,"[24] in which the hydrophobic interactions brought together the two nonpolar groups in the transition state.

$$\text{(12.1)}$$

In addition, cyclodextrins, because of their hydrophobic cavity, are capable of hydrophobic binding of the diene and/or dienophile into the cyclodextrin cavity in water. Therefore, cyclodextrins with the appropriate cavity size can function as catalysts for the Diels-Alder reactions of cyclopentadiene and butenone (Eq. 12.1) by mutual binding of the reagents in the cyclodextrin cavity relative to unassociated molecules. The use of β-cyclodextrin, which simultaneously forms an inclusion complex with the diene and dienophile, and the use of 4.86 M LiCl aqueous solution as solvent, which salts out nonpolar materials dissolved in water,[25] further enhanced the rate of aqueous Diels-Alder reactions.

On the other hand, the use of α-cyclodextrin decreased the rate of the reaction. This inhibition was explained by the fact that the relatively smaller cavity can only accommodate the binding of cyclopentadiene, leaving no room for the dienophile. Similar results were observed between the reaction of cyclopentadiene and acrylonitrile. The reaction between hydroxymethylanthracene and N-ethylmaleimide in water at 45°C has a second-order rate constant over 200 times larger than in acetonitrile (Eq. 12.2). In this case, the β-cyclodextrin became an inhibitor rather than an activator due to the even larger transition state, which cannot fit into its cavity. A slight deactivation was also observed with a salting-in salt solution (e.g., quanidinium chloride aqueous solution).

$$\qquad\qquad\qquad\qquad\qquad\qquad\qquad\qquad\qquad (12.2)$$

The stereoselectivity of some Diels-Alder reactions was also strongly affected in water.[26] At low concentrations, in which both components were completely dissolved, the reaction of cyclopentadiene with butenone gave a 21.4:1 ratio of endo/exo products when they were stirred at 0.15 M concentration in water, compared to only a 3.85:1 ratio in excess cyclopentadiene and an 8.5:1 ratio with ethanol as the solvent. Aqueous detergent solution had no effect on the product ratio. The stereochemical changes were explained by the need to minimize the transition-state surface area in water solution, thus favoring the more compact endo stereochemistry. The results are also consistent with the effect of polar media on the ratio.[27]

The catalytic behavior exhibited by β-cyclodextrin was also observed by Sternbach in the intramolecular Diels-Alder reaction of a furan-ene in water (Eq. 12.3).[28]

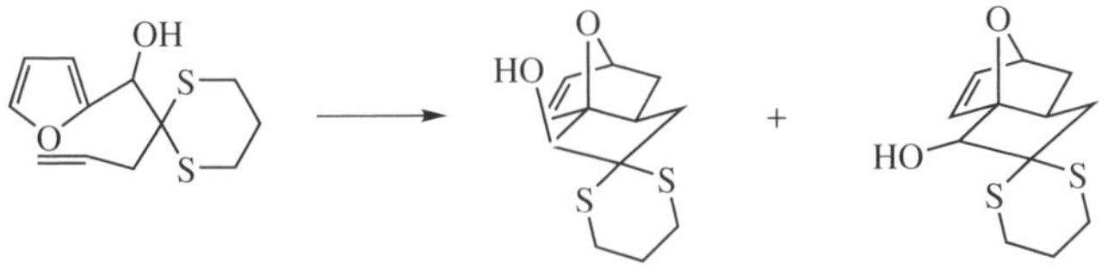

(12.3)

A   B

In water alone, the cyclization proceeded in 20% yield with an epimeric selectivity of 1:2 (**A:B**) at 89°C after 6 h. The same reaction gave 91% of the cyclized product when one equivalent of ß-cyclodextrin was present. In this case, the epimeric selectivity was also changed to 1:1.5 (**A:B**). However, no significant change of reactivity was observed with either α-cyclodextrin or the nonionic detergent, Brij-35, present. A similar enhancement of reactivity by β-cyclodextrin was observed in the cyclization of the amine derivative.

Roskamp reported[29] a similar intramolecular Diels-Alder reaction accelerated by silica gel saturated with water. The reaction led to the ready construction of the 11-oxabicyclo [6,2,1] ring systems (Eq. 12.4). The intramolecular Diels-Alder reaction has also been investigated by Keay.[30] The Diels-Alder reaction of 2,5-dimethylpyrrole derivatives with dimethyl acetylenedicarboxylate in water generated the corresponding cyclization products.[31]

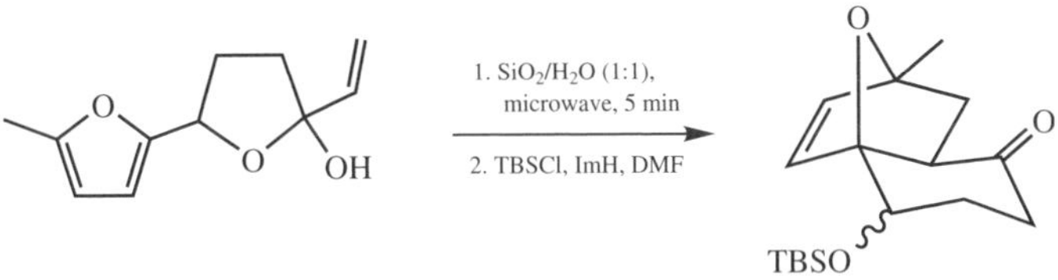

1. SiO$_2$/H$_2$O (1:1), microwave, 5 min

2. TBSCl, ImH, DMF

(12.4)

The retro-Diels-Alder (RDA) reaction of anthracenedione (Eq 12.5) proceeds considerably faster in aqueous solution than in organic solvents.[32] The addition of organic solvents to water retards the reaction, whereas glucose induces a modest acceleration. The results suggest that the origin of rate acceleration involves mainly enhanced hydrogen bonding of water to the activation complex for the RDA reaction.

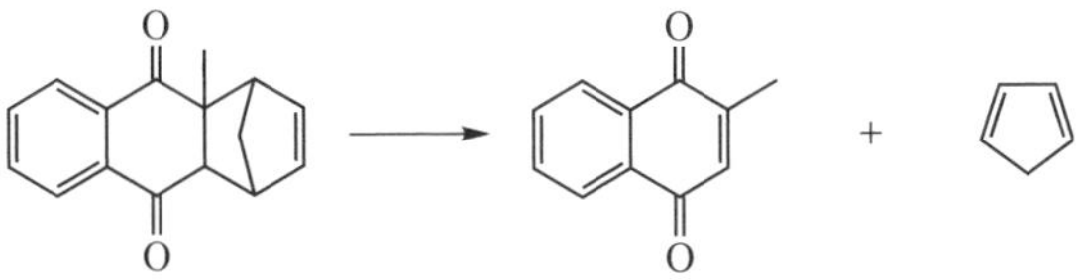

(12.5)

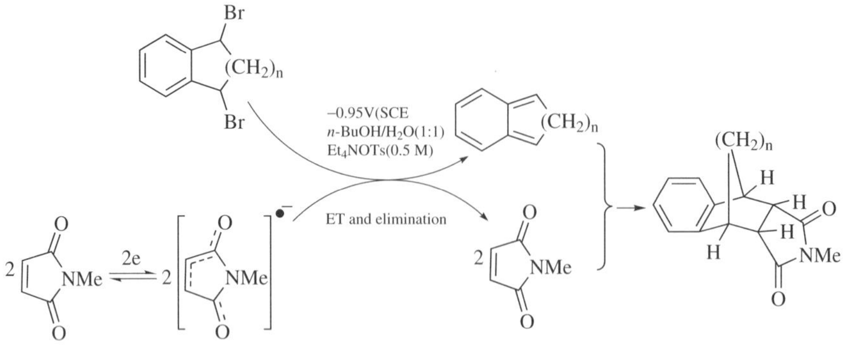

Scheme 12.1

Holt studied the Diels-Alder reaction in a mixture of water, 2-propanol, and toluene as microemulsions.[33] The endo/exo ratio between the reaction of cyclopentadiene and methyl methacrylate was enhanced with increasing amount of water in the presence of a surfactant.

Utley et al. were able to perform Diels-Alder reactions in aqueous solution via electrogenerated *ortho*-quinodimethanes.[34] They cathodically generated the *ortho*-quinodimethanes in aqueous electrolyte in the presence of N-methylmaleimide, which is both the redox mediator and the dienophile. Competition from the electrohydrodimerization of N-methylmaleimide is suppressed, allowing for the efficient formation of the *endo*-adduct (Scheme 12.1).

12.2.1 Lewis-Acid Catalysis

Recently, water-tolerating Lewis acid has been used to catalyze various Diels-Alder reactions in aqueous media. An important aspect of the Diels-Alder reaction is the use of Lewis acids for the activation of the substrates. While most Lewis acids are decomposed or deactivated in water, Bosnich reported that $[Ti(Cp^*)_2(H_2O)_2]^{2+}$ is an air-stable, water-tolerant Diels-Alder catalyst.[35] A variety of different substrates were subjected to the conditions to give high yields and selectivity (Eq. 12.6).

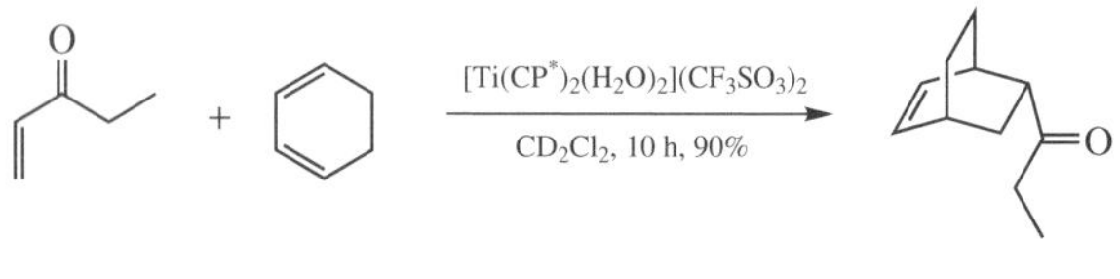

(12.6)

Kobayashi has found that scandium triflate, $Sc(OTf)_3$,[36] and lanthanide triflate, $Ln(OTf)_3$, are stable and can be used as Lewis catalysts under aqueous conditions. Many other Lewis acids have also been reported to catalyze Diels-Alder reactions in aqueous media. For example, Engberts reported[37] that the cyclization reaction in Eq. 12.7 in an aqueous solution containing 0.010 M $Cu(NO_3)_2$ is 250,000 times faster than that in acetonitrile and about 1,000 times faster than that in water alone. Other salts, such as Co^{2+}, Ni^{2+}, and Zn^{2+}, also catalyze the reaction, but not as effectively as Cu^{2+}. However, water has no effect on the endo-exo selectivity for the Lewis-acid catalyzed reaction.

X = NO$_2$, Cl, H, Me, OMe   (12.7)

Tris(pentafluorophenyl)boron was found to be an efficient, air-stable, and water-tolerant catalyst for Diels-Alder reactions.[38] Other Lewis acids[39] effective for catalyzing Diels-Alder reactions in aqueous conditions include $InCl_3$,[40] methylrhenium trioxide (MTO),[41] $In(OTf)_3$,[42] and $Bi(OTf)_3$.[43] A comparative study of specific-acid catalysis and Lewis-acid catalysis of Diels-Alder reactions between dienophiles and cyclopentadiene in water and mixed aqueous media has been carried out. At equimolar amounts of copper(II) nitrate as the Lewis-acid catalyst and hydrochloric acid (0.01 M) as the specific-acid catalyst, the reaction rate of a dienophile with cyclopentadiene is about 40 times faster with copper catalysis than with specific-acid catalysis under the same reaction conditions when the dienophile is capable of forming chelation (Eq. 12.7). On the other hand, when the dienophile is not capable of forming chelate with copper(II) ion, copper catalysis does not occur (Eq. 12.8).[44] In such a case, specific-acid catalysis with hydrochloric acid still occurs and the catalyzed reaction is 6 times faster than the uncatalyzed reaction in water. The stereoselectivity of the Diels-Alder reaction of (E)-γ-oxo-α, β-unsaturated thioesters with cyclopentadiene is greatly enhanced in the presence of Lewis acids favoring the endo acyl isomers. In the absence of a Lewis acid, the reaction at 25° gave two adducts; endo acyl isomers and exo acyl isomers in a ratio of 1:1 respectively. In the presence of Lewis acids, the reaction gave the two products in ratios of 75–94:25–6 respectively. The stereoselectivity was enhanced to ratios of 95–98:5–2 when lowering the reaction temperature (Eq. 12.9).[45]

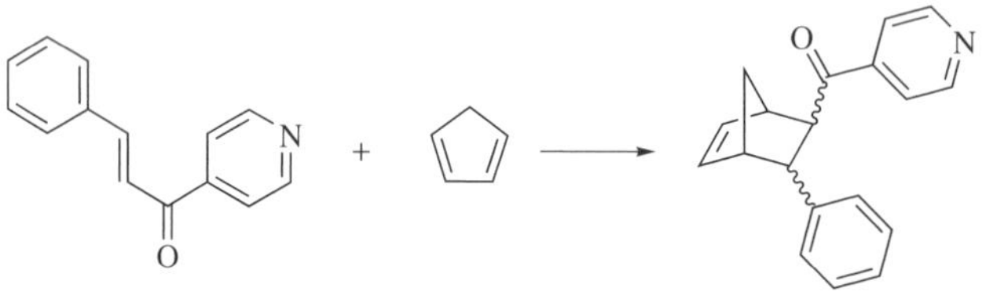

$$(12.8)$$

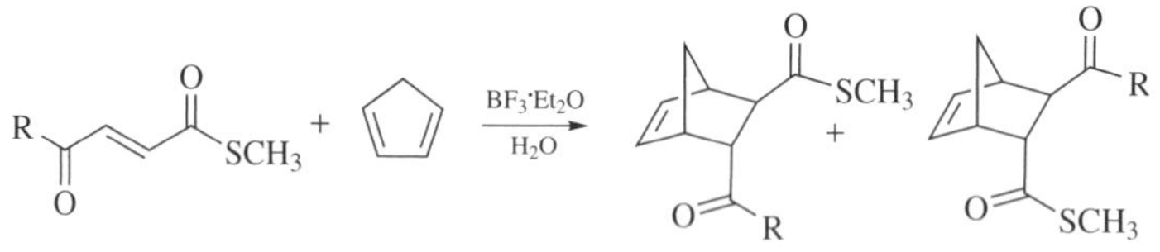

$$(12.9)$$

95:5

It has been found that the combination of Lewis acids and surfactants is particularly effective for catalyzing Diels-Alder reactions in water. The effect of micelles of SDS, CTAB, dodecyl heptaoxyethylene ether ($C_{12}E_7$), and copper and zinc didodecyl sulfate [M(DS)$_2$] on the Diels-Alder reaction of 3-(p-substituted phenyl)-1-(2-pyridyl)-2-propen-1-ones (Figure 12.1) with cyclopentadiene was studied.

In the absence of catalytically active transition-metal ions, micelles impede the reaction. In contrast to SDS, CTAB, and $C_{12}E_7$, Cu(DS)$_2$ micelles catalyze the Diels-Alder reaction with extremely high efficiency, leading to rate enhancements up to 1.8×10^6 compared to

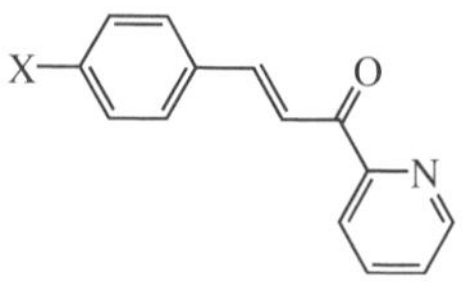

$X = NO_2$, Cl, H, Me, OMe, $CH_2SO_3^-Na^+$, $CH_2N^+(CH_3)_3Br^-$

Figure 12.1   Various 3-(p-substituted phenyl)-1-(2-pyridyl)-2-propen-1-ones.

the noncatalyzed reaction in acetonitrile.[46] This is primarily due to the complete complexation of the dienophiles to the copper ions at the micellar surface. When the dienophile does not bind to the micelle, the reaction is repressed because the uptake of the diene in the micelle lowers its concentration in the aqueous phase. The researchers contend that the retardation of the reaction results from a significant difference in the binding location of the diene and dienophile, with the dienophile preferring the outer regions of the micelle and the diene in the interior.

The use of aqueous surfactant aggregates to control the regiochemistry of Diels-Alder reactions was investigated extensively by Jaeger and co-workers (Eq. 12.10).[47] They have shown that the Diels-Alder reaction of a surfactant 1,3-diene with a surfactant dienophile with a short tether between their functional groups and head groups can proceed with high regioselectivity.

$$(12.10)$$

Under various reaction conditions, the isomer ratio of **A** was consistently higher than **B**. Isomer **A** is the expected regioisomer if the diene and dienophile react in their preferred orientation within a mixed micelle in which the quaternary ammonium groups are at the aggregate-water interface and the rest of the molecule is extended into the micelle interior (Figure 12.2). Isomer **B** comes about from the misalignment of the diene and dienophile within the mixed micelles.

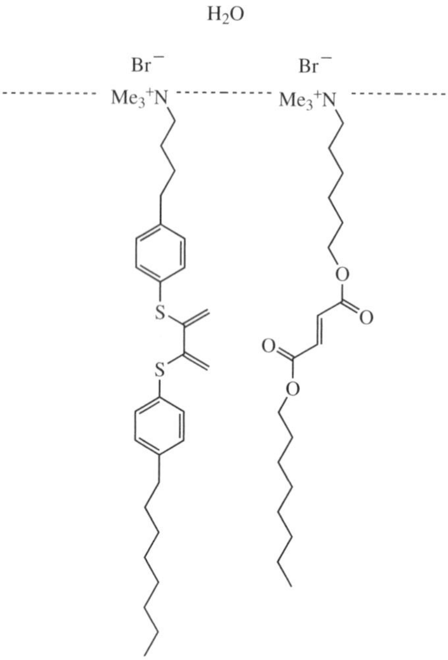

Figure 12.2 Preferred orientation of diene and dienophile at a surfactant aggregate-water interface.

The Diels-Alder reactions of alkyl-substituted benzoquinones with penta-1,3-diene and isoprene were also studied in aqueous cyclodextrin solutions. Highly enhanced *ortho-* and *meta*-regioselectivities were observed (Eq. 12.11).[48]

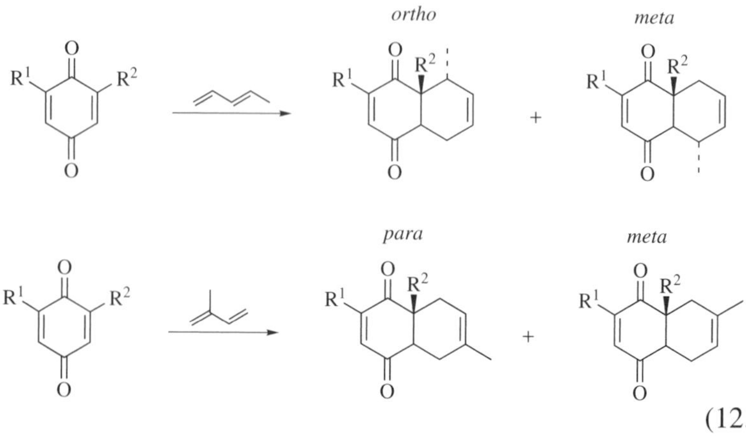

$$(12.11)$$

In addition to Lewis acids, Diels-Alder reactions in aqueous media are also catalyzed by bovine serum albumin,[49] enzymes,[50] antibodies,[51] and amines.[52]

The amine-catalyzed Diels-Alder dimerization reaction of α,β-unsaturated ketones in water was developed by Barbas et al. to form cyclohexanone derivatives (Eq. 12.12). They believe that the reaction proceeds via the in situ formation of 2-amino-1,3-butadiene and iminium-activated enone, as the diene and dienophile, respectively.

Ratio a/b = 6:1

$$(12.12)$$

The antibody-catalyzed Diels-Alder reaction developed by Schultz utilized a "Diel-Alderase" enzyme-like catalyst evolved from an antibody-combining site (Eq. 12.13). The idea is that the generation of antibodies to a structure that mimics the transition state for the Diels-Alder reaction should result in an antibody-combining site that lowers the entropy of activation by binding both the diene and dienophile in a reactive conformation.

$$(12.13)$$

A self-assembled coordination cage[53] and micelles[54] were found to accelerate Diels-Alder reactions in an aqueous media. The catalysis of Diels-Alder reactions via noncovalent binding by synthetic, protein, and nucleic acid hosts has been surveyed and compared to explore the origin of the noncovalent catalysis. These catalysts consist of binding cavities that form complexes containing both the diene with the dienophile and the reaction occurring in the cavity. The binding requires no formation of covalent bonds and is driven principally by the hydrophobic (or solvophobic) effect.[55]

Yoshida and co-workers developed the concept of the "removable hydrophilic group" using 2-pyridyldimethylsilyl (2-PyMe$_2$Si) for aqueous organic reactions including the Diels-Alder reactions, hetero-Diels-Alder reactions, Claisen rearrangement, radical reactions, and transition-metal-catalyzed reactions. Although the low solubility of organic molecules in water has been a bane in aqueous organic reactions, the incorporation of hydrophilic groups into the substrate structure can overcome the solubility problem and at the same time enhance the hydrophobic effect.[56] The Diels-Alder reaction of 2-PyMe$_2$Si-substituted 1,3-dienes with *p*-benzoquinone occurs at room temperature in water (Eq. 12.14). There is a simultaneous desilylation and aromatization to afford naphthoquinones quantitatively, which also means that there is no need for an additional step to remove the 2-PyMe$_2$Si group.

PySi = 2-PyMe$_2$Si

$$(12.14)$$

Diels-Alder reactions in supercritical water have also been investigated.[57] Kolis has shown that Diels-Alder reactions of dienes with various electron-poor dienophiles can be performed in supercritical water with high yields of the desired product without the addition of any catalysts (Eq. 12.15).

$$(12.15)$$

Metal-free, noncovalent catalysis of Diels-Alder reactions by neutral hydrogen bond donors was studied in water.[58] The researchers examined the catalytic activity of a number of substituted thioureas (Figure 12.3) in Diels-Alder and 1,3-dipolar cycloadditions (Eq. 12.16). They concluded from their kinetic data that the observed accelerations in the relative rates are more dependent on the thiourea substituents than on the reactants or solvent. However, even though the catalytic efficiency is highest in a noncoordinating, nonpolar solvent such as cyclohexane, it is also present in a highly coordinating polar solvent such as water, meaning that both hydrophobic and polar interactions can coexist, making the catalyst

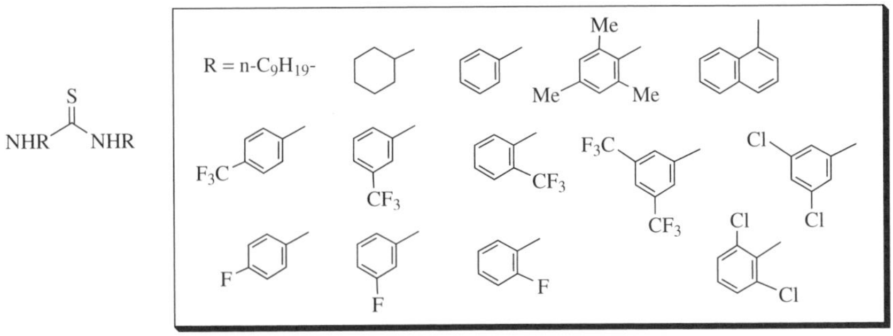

Figure 12.3   Various substituted thioureas.

active, even in highly coordinating solvents. These catalysts increase the reaction rates along with the *endo*-selectivities of Diels-Alder reactions, in a manner similar to weak Lewis acids, but without the associated product inhibition.

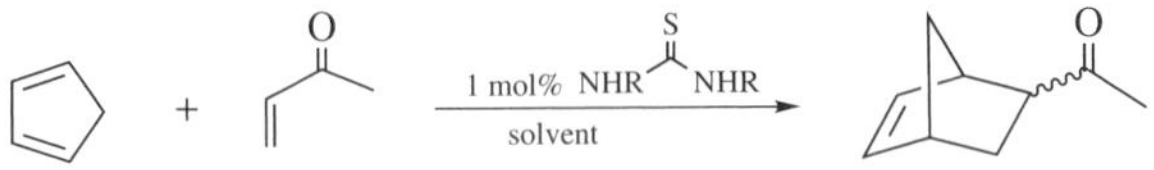

(12.16)

It should be noted, however, that despite many examples of the acceleration of Diels-Alder reactions by the use of aqueous media, Elguero[59] reported that the Diels-Alder reaction between cyclopentadiene and methyl (and benzyl) 2-acetamidoacrylates proceeded better in toluene than in water both in yield and in exo/endo selectivity. Additionally, ultrasonic irradiation did not improve the yield.

12.2.2   Asymmetric Diels-Alder Reactions

The use of aqueous Diels-Alder reactions for generating optically active compounds is one of the efforts recently made on the subject. A diene bearing a chiral water-soluble glyco hydrophilic moiety was studied extensively by Lubineau (Eq. 12.17).[60] The use of water soluble glyco-organic compounds in water achieved higher reagent concentration and resulted in rate enhancement and asymmetric induction. Even though the diastereoselectivity was modest (20% de), separation of the diastereomers led to chiral adducts in pure enantiomeric form after cleavage of the sugar moiety by acidic hydrolysis or by using glycosidase in neutral conditions at room temperature. A variety of substrates bearing other glyco-derivatives was also studied.

$$(12.17)$$

Chiral dienophiles, prepared from an aldehyde and asparagine in water followed by reacting with acryloyl chloride, reacted with cyclopentadiene at room temperature in water or ethanol-water to provide cycloadducts diastereoselectively and chiral products upon separation and hydrolysis (47–64% ee for the endo isomers; endo/exo 82:18) (Eq. 12.18).[61]

$$(12.18)$$

R = i-Pr, t-Bu, Ph

endo/exo = 82/18–95/5
ee(for endo isomer) = 47–64%

Recently, catalytic asymmetric Diels-Alder reactions have been investigated. Yamamoto reported a Brönsted-acid-assisted chiral (BLA) Lewis acid, prepared from (R)-3-(2-hydroxy-3-phenylphenyl)-2,2'-dihydroxy-1,1'-binaphthyl and 3,5-bis(trifluoromethyl)-benzeneboronic acid, that is effective in catalyzing the enantioselective Diels-Alder reaction between α,β-enals and various dienes.[62] The interesting aspect is the role of water, THF, and MS 4A in the preparation of the catalyst (Eq. 12.19). To prevent the trimerization of the boronic acid during the preparation of the catalyst, the chiral triol and the boronic acid were mixed under aqueous conditions and then dried. Using the catalyst prepared in this manner, a 99% ee was obtained in the Diels-Alder reaction of methacrolein and cyclopentadiene; whereas if the catalyst was prepared in the presence of activated MS 4A under anhydrous conditions, the enantioselectivity for the same reaction was reduced to less than 80% ee.

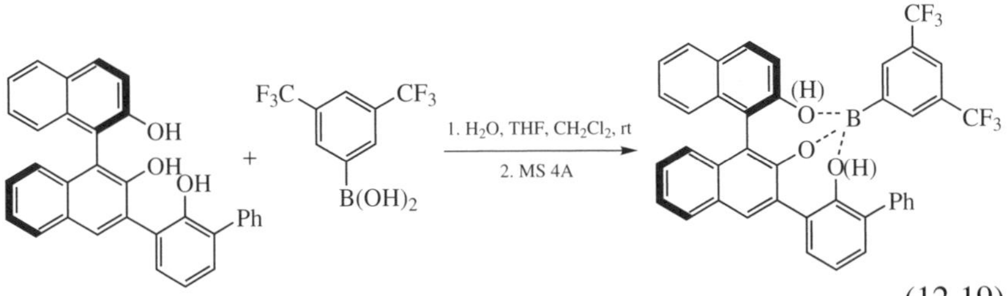

(12.19)

Kanemasa et al.[63] reported that cationic aqua complexes prepared from the *trans*-chelating tridentate ligand (R,R)-dibenzofuran-4,6-diyl-2,2′-*bis*(4-phenyloxazoline) (DBFOX/Ph) and various metal(II) perchlorates are effective catalysts that induce absolute chiral control in the Diels-Alder reactions of 3-alkenoyl-2-oxazolidinone dienophiles (Eq. 12.20). The nickel(II), cobalt(II), copper(II), and zinc(II) complexes are effective in the presence of six equivalents of water for cobalt and nickel and three equivalents of water for copper and zinc.

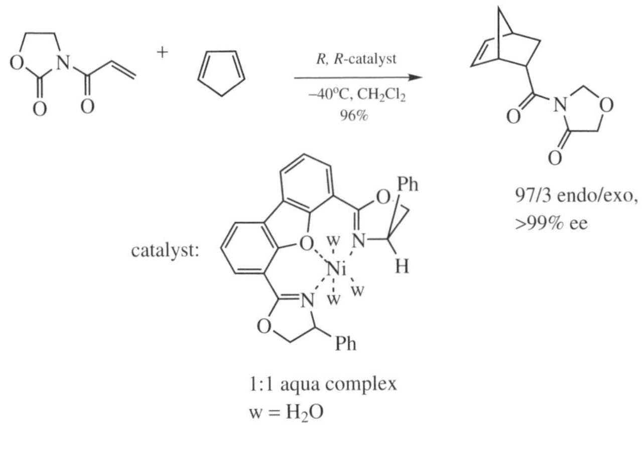

(12.20)

97/3 endo/exo,
>99% ee

catalyst:

1:1 aqua complex
w = H_2O

Desimoni et al. have shown that the use of magnesium perchlorate or magnesium triflate, and three chiral *bis*(oxazolines) and two equivalents of achiral auxiliary ligands such as water or tetramethylurea, induces a strong change of the enantiofacial selectivity with >94% ee in the

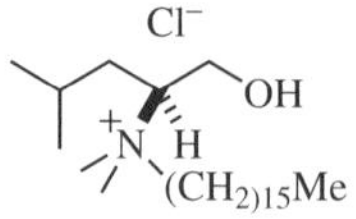

Figure 12.4 (*S*)-Leucine-derived surfactant.

reaction between cyclopentadiene and 3-acryloyl- or (*E*)-3-crotonyl-1,3-oxazolidin-2-ones (Eq. 12.21).[64]

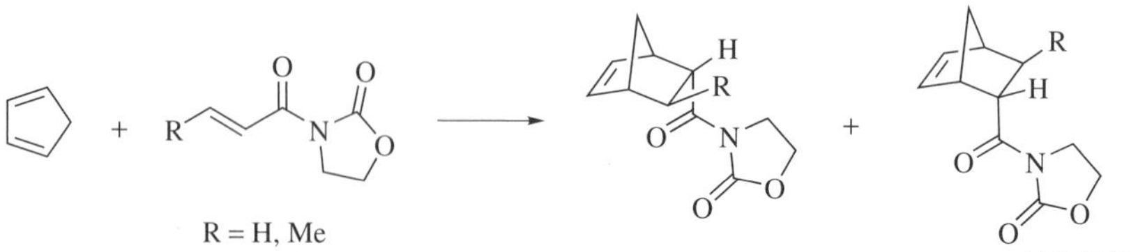

$$(12.21)$$

Chiral surfactants have been used in the aqueous chiral micellar catalysis of a Diels-Alder reaction using an (*S*)-leucine-derived surfactant (Figure 12.4) to catalyze the reaction between cyclopentadiene and nonyl acrylate.[65]

In 1998, Engberts and co-workers reported the first enantioselective Lewis-acid-catalyzed Diels-Alder reaction in pure water.[66] In the presence of 10% of copper(II), complexes of α-amino acids with aromatic side chains such as L-phenylalanine, L-tyrosine, L-tryptophan, and L-abrine as ligands coordinated to copper(II) induced up to 74% ee in the Diels-Alder reaction of 3-phenyl-1-(2-pyridyl)-2-propen-1-one with cyclopentadiene (Eq. 12.22). For the copper-L-abrine-catalyzed reaction, an enantiomeric excess of 74% can be achieved, which is considerably higher than in organic solvents (17–44% ee) and shows that water significantly enhances enantioselectivity. Since significant enantioselectivity was observed exclusively for α-amino acids containing aromatic side groups, the interaction between the aromatic ring of the α-amino acid and the pyridine ring of the dienophile during the activation process was proposed to be responsible for the observed enantioselectivity. In addition, Engberts also investigated the influence of a series of diamine ligands and α-amino acid ligands on the rate and enantioselectivity of the nickel(II)- and copper(II)-catalyzed Diels-Alder reaction between 3-phenyl-1-(2-pyridyl)-2-propen-1-ones and cyclopentadiene in water.[67] However, they found that the diamine ligands did not improve the catalytic efficiency and it was the binding of the aromatic α-amino acid ligands to copper(II) that led to the overall rate increase of the reaction.

$$(12.22)$$

ee up to 74%

Optically active Diels-Alder adducts were also prepared by using a one-pot preparative method and enantioselective formation of inclusion complex with optically active hosts in a water suspension medium.[68] For example, N-ethylmaleimide reacts with 2-methyl-1,3-butadiene in water to give the racemic adduct 1. Racemic **1** and the optically active host **2** form enantioselectively a 1:1 inclusion complex of **2** with (+)-**1** in a water suspension. The inclusion complex can be filtered and heated to release (+)-**1** with 94% ee (Eq. 12.23).

$$(12.23)$$

12.2.3 Theoretical and Mechanistic Studies

The effect of water on Diels-Alder reactions has been studied extensively by various theoretical and experimental methods. It was mentioned previously that Breslow studied the influence of the hydrophobic effect on the aqueous Diels-Alder reactions in detail[8,9] while the volumes of activation for catalyzed Diels-Alder reactions were examined by Isaacs et al.[69]

Recent studies show that the concept of internal pressure cannot be used to explain completely the strong rate enhancement of Diels-Alder reactions when carried out in water with respect to common organic solvents.[70] One factor is solvophobicity; Schneider has reported a quantitative correlation between solvophobicity and the rate enhancement of aqueous Diels-Alder reactions.[10,11] Another factor is hydrogen bonding, and Engberts has suggested that enforced hydrophobic interactions between diene and dienophile and hydrogen bonding can account for the acceleration in water.[12] It was found that an increase in the hydrophobicity close to the reaction center in the diene has a much more pronounced effect on the rate acceleration in water than a comparable increase in hydrophobicity in the dienophile further away from the reaction center. [71]

Density functional theory study of aqueous-phase rate acceleration and endo/exo selectivity of the butadiene and acrolein Diels-Alder reaction[72] shows that approximately 50% of the rate acceleration and endo/exo selectivity is attributed to hydrogen bonding and the remainder to bulk-phase effects, including enforced hydrophobic interactions and cosolvent effects. This appears to be supported by the experimental results of Engberts where a pseudothermodynamic analysis of the rate acceleration in water relative to 1-propanol and 1-propanol-water mixtures indicates that hydrogen-bond stabilization of the polarized activated complex and the decrease of the hydrophobic surface area of the reactants during the activation process are the two main causes of the rate enhancement in water.[13]

Similar studies of Diels-Alder reactions under high pressure by Jenner revealed that water can alter kinetics, chemo-, and enantioselectivity through polarity and hydrophobic effects.[14] The relative importance of these two factors, however, depends on the specific substrates. Another consideration appears to be the concentrations of the reaction substrates. Smith[15] and Griesbeck[16] showed that for Diels-Alder reactions in dilute aqueous ethanol, the rate shows a maximum at 0.5 M. The effect of salt and cosolvent on the stereoselectivity and rate of aqueous Diels-Alder reactions was studied extensively by Kumar.[5,73] The yield of endo products of a Diels-Alder reaction in aqueous LiClO$_4$ can be enhanced by using a simple solvent manipulation.[74] The endo/exo selectivity was found to be mainly dependant on the solvophobic and hydrogen bond donor properties of the solvent, whereas the regioselectivity almost exclusively depends on the hydrogen bond donor ability of the solvent.[75]

An *ab initio* MO calculation by Jorgensen revealed enhanced hydrogen bonding of a water molecule to the transition states for the Diels-Alder reactions of cyclopentadiene with methyl vinyl ketone and acrylonitrile, which indicates that the observed rate accelerations for Diels-Alder reactions in aqueous solution arise from the hydrogen-bonding effect in addition to a relatively constant hydrophobic term.[7,76] *Ab initio* calculation using a self-consistent reaction field continuum model shows that electronic and nuclear polarization effects in solution are crucial to explain the stereoselectivity of nonsymmetrical

Diels-Alder reactions.[77] Using a combined quantum-mechanical and molecular-mechanical (QM/MM) potential, Gao carried out Monte Carlo simulations to investigate the hydrophobic and hydrogen-bonding effects on Diels-Alder reactions in aqueous solution. Enhanced hydrogen-bonding interaction and the hydrophobic effect were found to contribute to the transition-state stabilization.[78] The number of hydrogen bonds was found to cause strong Coulomb interaction and discriminate heats of formation of transition states for exo/endo products.[79]

An interesting mechanistic issue was raised by Firestone on the aqueous Diels-Alder reaction between 2-methylfuran and maleic acid in water, which is found to be 99.9% stereospecific.[80] By adding heavy atom (defined as any below the first complete row of the periodic table) salts to the aqueous media, it was found that addition of heavy but not light atom salts reduced the degree of stereospecificity significantly in the retrodiene reaction. The results suggest that a large portion of the Diels-Alder reaction occurs via diradical intermediates (Scheme 12.2).

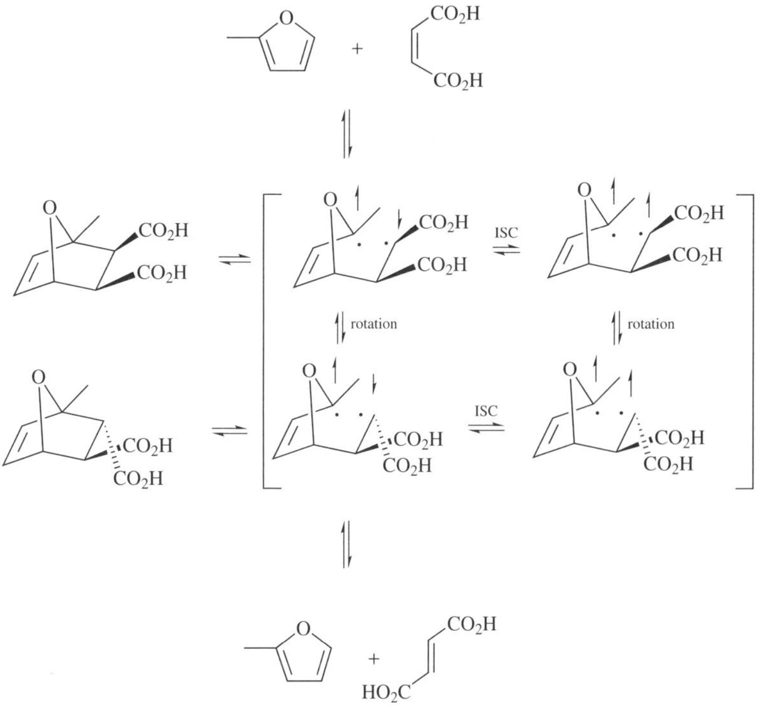

Scheme 12.2

High stereospecificity is observed when the rotation of the diradical intermediate is slow in comparison with cyclization to cycloadduct or reversion to reactants. With the presence of external heavy atoms, it could facilitate the intersystem crossing (ISC) of the first-formed singlet diradical to the longer-lived triplet counterpart. The triplet diradical will have a chance to undergo rotation before it reverts back to singlet and cyclizes or cleaves to reactants. This then accounts for the reduced stereospecificity. The alternative possibility of a zwitterionic intermediate is considered unlikely because there is no interception of zwitterions by water.

12.2.4 Synthetic Applications

Much effort has been directed at developing aqueous Diels-Alder reactions toward the syntheses of a variety of complex natural products. Grieco employed micellar catalysis and pure water as the solvent for the Diels-Alder reaction of dienecarboxylate with a variety of dienophiles. For example, when the Diels-Alder reaction in Scheme 12.3 was carried out in water, a higher reaction rate and reversal of the selectivity were observed, compared with the same reaction in a hydrocarbon solvent (Scheme 12.3).[81] Similarly, the reaction of 2,6-dimethylbenzoquinone with sodium (E)-3,5-hexadienoate (generated in situ by the addition of 0.95 equiv sodium bicarbonate to a suspension of the precursor acid in water) proceeded for 1 hour to give a 77% yield of the adduct

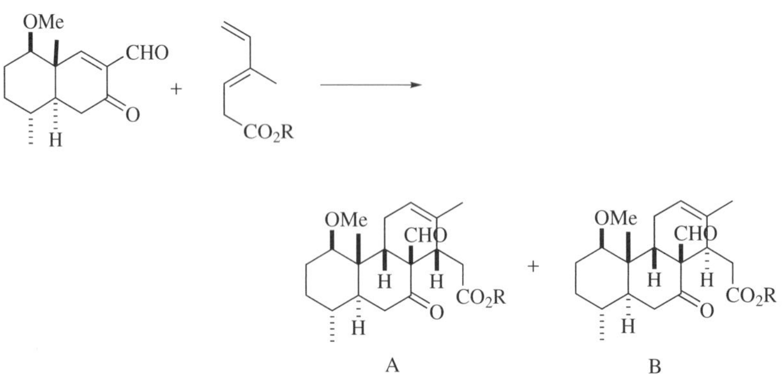

ratio A/B = 3 : 1 (in water)   R = Na
            1 : 0.85 (in benzene)   R = Et

Scheme 12.3

(Eq. 12.24), while for the same reaction performed in toluene there were only trace amounts of the adduct after the reaction was stirred for one week at room temperature. [82]

$$(12.24)$$

However, when 2,6-dimethylbenzoquinone with sodium (E)-3,5-hexadienoate (generated in situ) was reacted in water in the presence of a catalytic amount of sodium hydroxide, pentacyclic adducts were formed via deprotonation of the Diels-Alder adduct followed by tandem Michael-addition reactions with another molecule of 2,6-dimethylbenzoquinone (Eq. 12.25).[83] Similar results were obtained with sodium (E)-4,6-heptadienoate.

$$(12.25)$$

Sensitive dienol ether functionality in the diene carboxylate was shown to be compatible with the conditions of the aqueous Diels-Alder reaction (Eq. 12.26).[84] The dienes in the Diels-Alder reactions can also bear other water-solubilizing groups such as the sodium salt of phosphoric acid and dienyl ammonium chloride (Eq. 12.27).[85] The hydrophilic acid functionality can also be located at the dienophile.[86]

$$(12.26)$$

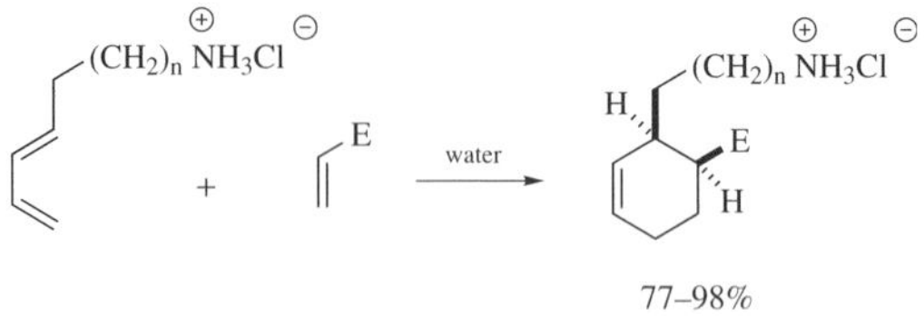

$$ (12.27) $$

77–98%

Grieco utilized an aqueous intermolecular Diels-Alder reaction as the key step in forming the AB ring system of the potent cytotoxic sesquiterpene vernolepin. [87] Cycloaddition of sodium (E)-3,5-hexadienoate with an α-substituted acrolein in water followed by direct reduction of the intermediate Diels-Alder adduct gave the desired product in 91% overall yield (Eq. 12.28).

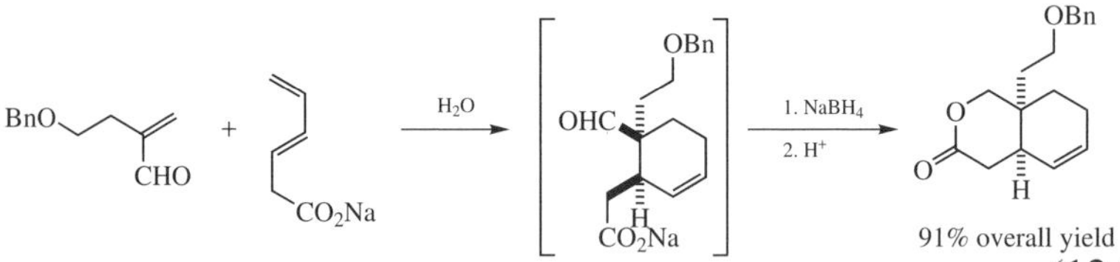

91% overall yield

$$ (12.28) $$

Similar reactions were applied to the syntheses of *dl-epi*-pyroangolensolide, *dl*-pyroangoensolide (Eq. 12.29)[88] and the formal synthesis of the Inhoffen-Lythgoe diol (Eq. 12.30).[89] The key step in the formal synthesis of the Inhoffen-Lythgoe diol is the aqueous Diels-Alder reaction between the sodium salt of the diene and methacrolein to form the cycloadduct, which then undergoes subsequent reactions to form the known hydrindan. Sodium (E)-4, 6, 7-octatrienoate reacted smoothly with a variety of dienophiles to give conjugated diene products.[90]

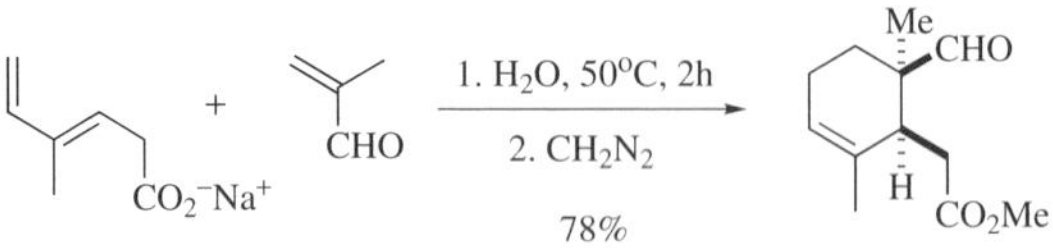

$$ (12.29) $$

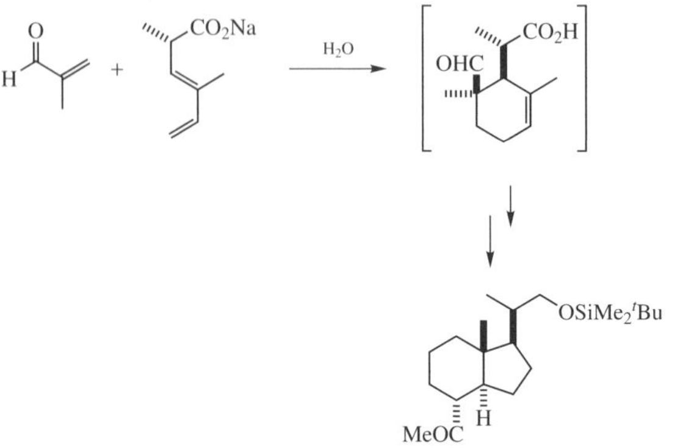

$$ (12.30) $$

An intramolecular version of the Diels-Alder reaction with a diene-carboxylate was used by Williams et al. in the synthetic study of the antibiotic ilicicolin H.[91] The interesting aspect of this work is that they found that under aqueous conditions, there is an observed reversal of regioselectivity (Eq. 12.31). In toluene, there is a 75:25 ratio of **a/b** while in degassed water; the ratio of **a/b** is 40:60.

solvent = toluene, a/b = 75:25
solvent = water, a/b = 40:60

$$(12.31)$$

De Clercq has shown that aqueous Diels-Alder reactions can be used as key steps in the syntheses of $(\pm)$-11-keto-testosterone[92] (Eq. 12.32) and $(\pm)$-gibberellin A_5.[93]

$$(12.32)$$

The reaction of dienes bearing an N-dienyl lactam moiety with activated olefins was examined by Smith.[94] The lactams were excellent enophiles and provided exclusively the ortho regioisomer with good selectivity for the endo (Z) product (Eq. 12.33).

$$(12.33)$$

E/Z = 2:98

In the synthesis of 2,2,5-trisubstituted tetrahydrofurans, a novel class of orally active azole antifungal compounds, Saksena[95] reported that the key step of Diels-Alder reaction in water led to the desired substrate virtually in quantitative yields (Eq. 12.34), while the same reaction in organic solvent resulted in a complicated mixture with only less than 10% of the desired product being isolated. This success made the target compounds readily accessible.

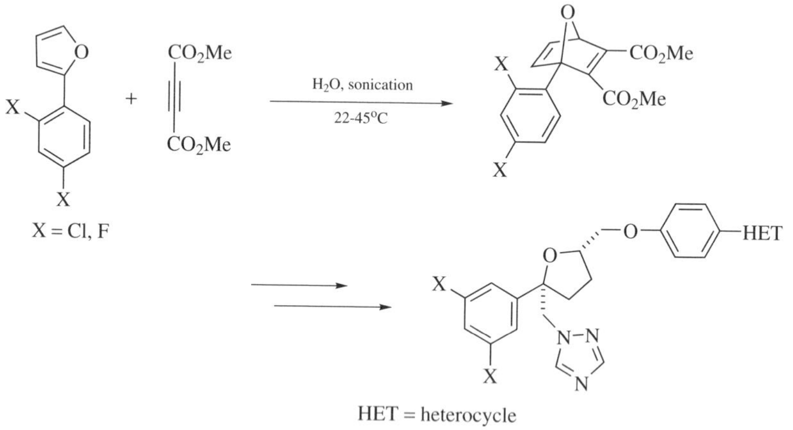

HET = heterocycle

(12.34)

The Diels-Alder reaction between oxazolone and cyclopentadiene in water was investigated by Cativiela (Eq. 12.35).[96] Although the reaction is very slow, the (E)-5(4H)-oxazolone reacted with cyclopentadiene in an aqueous medium for six days at room temperature to form the corresponding sprioxazolones in a 95% yield. The cycloadducts were then readily converted into amino-norbornane carboxylic acids.

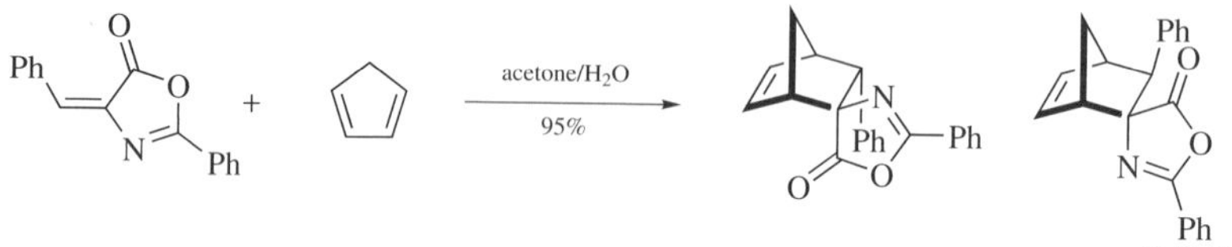

(12.35)

In the synthesis of the tetracyclic intermediates for the synthesis of isoarborinol and its CDE-antipode fernenol, the stereochemistry of the Diels-Alder reaction can be varied using various Lewis-acid catalysts in aqueous media (Eq. 12.36).[97] Their results show that the hydrophobic effects play an important role in enhancing reaction rates and can control product distribution. Novel 2,4-dialkyl-1-alkylideneamino-3-(methoxycarbonylmethyl)azetidines were obtained from aldazines and methyl 3-(alkylidenehydrazono)propionates in an aqueous solution of sodium periodate.[98]

$$(12.36)$$

It was proposed that a Diels-Alder cyclization occurred during a polyketide synthase assembly of the bicyclic core of Lovastatin by *Aspergillus terreus MF 4845*.[99] In vitro Diels-Alder cyclization of the corresponding model compounds generated two analogous diastereomers in each case, under either thermal or Lewis-acid-catalyzed conditions (Eq. 12.37). As expected, the Diels-Alder reaction occurred faster in aqueous media. The cyclization half-life in chloroform at room temperature is 10 days while in aqueous media at either pH 5 or 7, the half-life drops to two days.

$$(12.37)$$

An efficient one-pot synthesis of mikanecic acid derivatives was accomplished from allylic phosphonates, $ClCO_2Et$, and aqueous HCHO (Eq. 12.38).[100] The overall process involves a cascade sequence linking together metalation-alkoxycarbonylation, Horner-Wadsworth-Emmons, and Diels-Alder reactions.

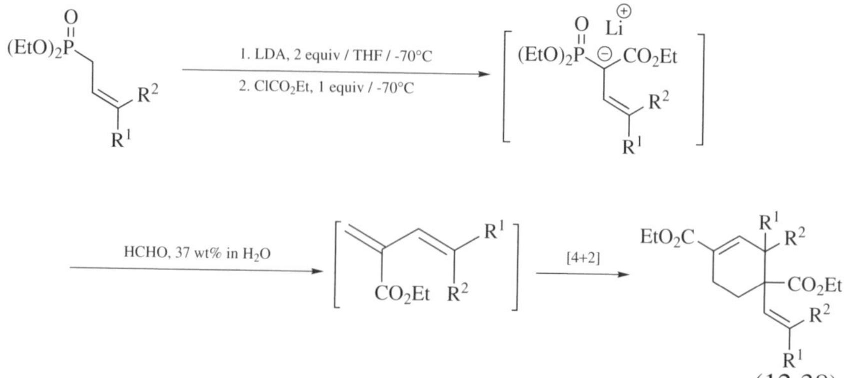

$$(12.38)$$

Cyclopentadienylindium(I) has been shown to be effective in the reaction with aldehydes or electron-deficient alkenes to form highly functionalized cyclopentadienes in aqueous media (See Section 8.4.3).[101] This reaction with the appropriate substrates can be followed by an intramolecular Diels-Alder reaction in the same pot to provide complex tricyclic structures in a synthetically efficient manner (Scheme 12.4).

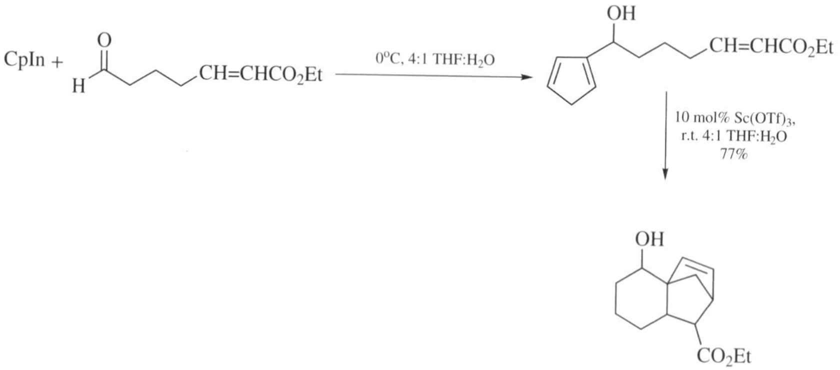

Scheme 12.4

The aqueous Diels-Alder reaction has also been used for bioconjugate studies. A Diels-Alder reaction of diene oligonucleotides with maleimide dieneophiles was used to prepare oligonucleotide conjugates in aqueous media under mild conditions (Eq. 12.39).[102] A Diels-Alder-type cycloaddition of an electronically matched pair of saccharide-linked conjugated dienes and a dienophile-equipped protein was the first method to create a carbon–carbon bond in the bioconjugation step between a saccharide and a protein. These neoglycoproteins were formed at room temperature in pure water and have a reaction half-life of approximately two hours (Eq. 12.40).[103]

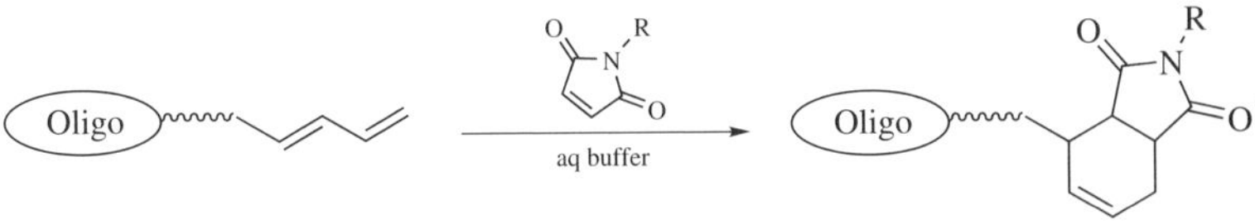

(12.39)

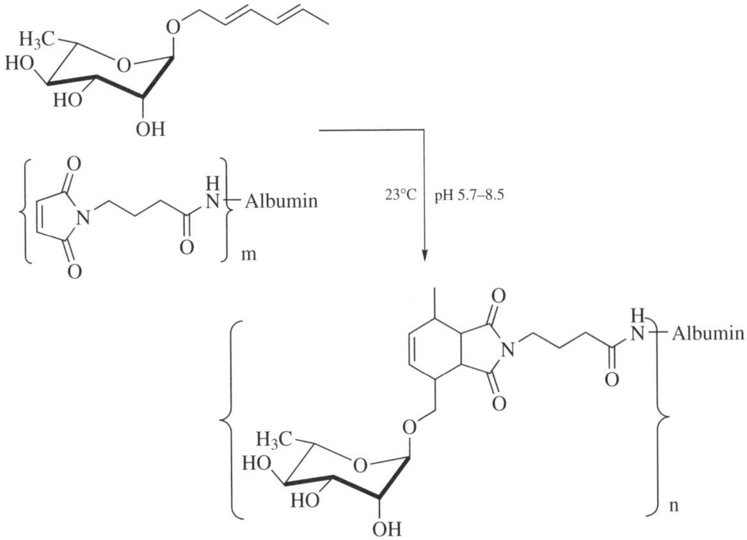

(12.40)

One-pot synthesis of (E)-2-aryl-1-cyano-1-nitroethenes[104] and an approach to the synthesis of nitrotetrahydrobenzo[c]chromenones and dihydrodibenzo[b,d]furans were developed based on aqueous Diels-Alder reactions.[105] Once again, it was found that the reaction occurred faster in water under heterogeneous conditions relative to those performed in toluene and methylene chloride (Eq. 12.41).

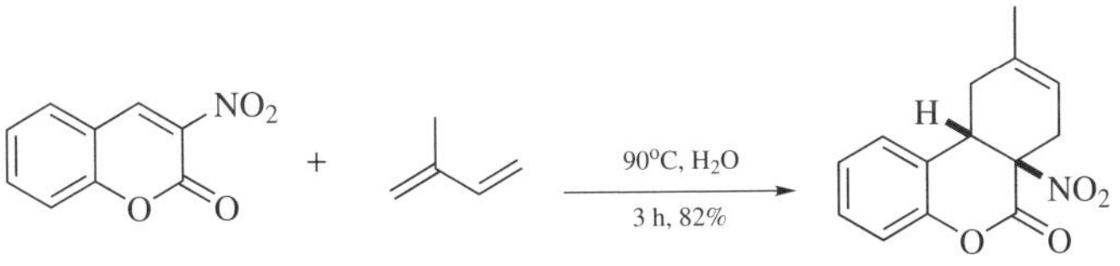

(12.41)

12.2.5 Hetero-Diels-Alder Reactions

For the synthesis of heterocyclic compounds, hetero-Diels-Alder reactions with nitrogen- or oxygen-containing dienophiles are particularly useful. Such reactions have been studied extensively in aqueous media.[106] In 1985, Grieco reported the first example of hetero-Diels-Alder reactions with nitrogen-containing dienophiles in aqueous media (Eq. 12.42).[107] Simple iminium salts, generated in situ under Mannich-like conditions, reacted with the dienes in water to give aza-Diels-Alder reaction products. The use of alcoholic solvents led to a decrease in the reaction rate while the use of THF as a cosolvent did not affect the rate of the reaction.

$$
\text{(diene)} \quad + \quad MeNH_2 \cdot HCl \quad \xrightarrow[\substack{25^\circ C, 3\,h \\ 95\%}]{HCHO,\ H_2O} \quad \text{(bicyclic product, N-Me)} \tag{12.42}
$$

This methodology has the potential to be generally applicable to the synthesis of various alkaloids that have a bridgehead nitrogen via the intramolecular aza-Diels-Alder reaction (Eq. 12.43).[108]

$$
\text{(diene chain)}\ NH_2 \bullet HCl \quad \xrightarrow[\substack{50^\circ C, 48\,h \\ 95\%}]{HCHO,\ H_2O} \quad \text{(bicyclic product)}\ N \bullet Cl \tag{12.43}
$$

Retro aza-Diels-Alder reactions also readily occurred in water.[109] As shown in Eq. 12.44, the 2-azanorbornene undergoes acid-catalyzed retro-Diels-Alder cleavage in water. The produced cyclopentadiene and iminium derivative then reacts with the trapping reagent, N-methylmaleimide, or is hydrolyzed to give primary amines. No reaction was observed in a variety of organic solvents such as benzene, THF, or acetonitrile under similar or more drastic conditions. This means that water accelerates hetero-Diels-Alder reactions in both the forward and reverse directions by lowering the energy of the transition state. This reaction provided a novel method for the N-methylation of dipeptides and amino acid derivatives through reduction of the immonium ion intermediate.[110]

$$
\underset{\substack{N \\ CH_2C_6H_5 \\ \cdot \\ HCl}}{\text{(azanorbornene)}} \quad \xrightarrow[\substack{H_2O,\ 50^\circ C,\ 2\,h}]{N\text{-methylmalemide}} \quad \underset{\substack{OC \\ C \\ O}}{\text{(product)}}{-}NMe \quad + \quad C_6H_5CH_2NH_2 \tag{12.44}
$$

Waldmann used (R) and (S)-aminoacid methyl esters and chiral amines as chiral auxiliaries in analogous aza-Diels-Alder reactions with cyclodienes.[111] The diastereoselectivity of these reactions ranged from moderate to excellent and the open-chain dienes reacted similarly. Recently, the aza-Diels-Alder reaction was used by Waldmann in the asymmetric synthesis of highly functionalized tetracyclic indole derivatives (Eq. 12.45), which is useful for the synthesis of yohimbine- and reserpine-type alkaloids.[112]

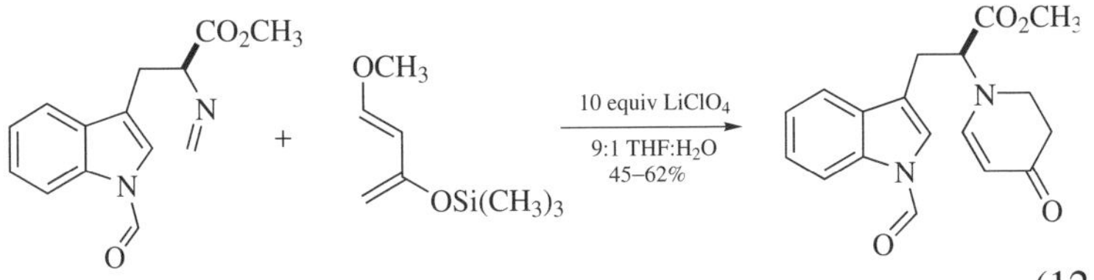

(12.45)

As in the case of Diels-Alder reactions, aqueous aza-Diels-Alder reactions are also catalyzed by various Lewis acids such as lanthanide triflates.[113] Lanthanide triflate–catalyzed imino Diels-Alder reactions of imines with dienes or alkenes were developed. Three-component aza-Diels-Alder reactions, starting from aldehyde, aniline, and Danishefsky's diene, took place smoothly under the influence of HBF_4 in aqueous media to afford dihydro-4-pyridone derivatives in high yields (Eq. 12.46).[114]

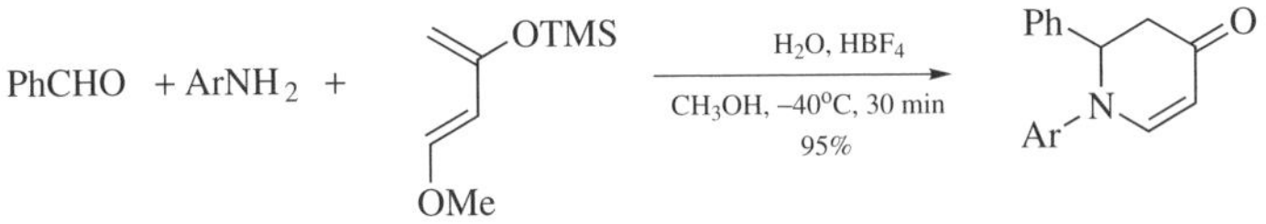

(12.46)

The montmorillonite K10-catalyzed aza-Diels-Alder reaction of Danishefsky's diene with aldimines, generated in situ from aliphatic aldehydes and p-anisidine, proceeded smoothly in H_2O or in aqueous CH_3CN to afford 2-substituted 2,3-dihydro-4-pyridones in excellent yields (Eq. 12.47).[115] Also, complex $[(PPh_3)Ag(CB_{11}H_6Br_6)]$ was shown to be an effective and selective catalyst (0.1 mol% loading) for a hetero-Diels-Alder reaction with Danishefsky's diene and the reaction showed a striking dependence on the presence of trace amounts of water (Eq. 12.48).[116]

$$RCHO + (MeO)C_6H_4p\text{-}(NH_2) + \quad \text{(diene)} \xrightarrow[\substack{\text{Conditions A: } H_2O, 0°C \\ \text{Conditions B: } CH_3CN\text{-}H_2O \ (90:10), -10°C \\ \text{Conditions C: } CH_3CN, -10°C}]{\text{montmorillonite K10}} \quad \text{(product)}$$

$$61\text{-}98\%$$

$$(12.47)$$

$$\text{(diene)} + \text{(imine)} \xrightarrow[CH_2Cl_2]{[(PPh_3)Ag(CB_{11}H_6Br_6)]} \text{(product)} \quad (12.48)$$

$$>99\%$$

Similar aza-Diels-Alder reactions of Danishefsky's diene with imines or aldehydes and amines in water took place smoothly under neutral conditions in the presence of a catalytic amount of an alkaline salt such as sodium triflate or sodium tetraphenylborate to afford dihydro-4-pyridones in high yields (Eq. 12.49).[117] Antibodies have also been found to catalyze hetero-Diels-Alder reactions.[118]

$$\text{(diene)} + \text{(imine)} \xrightarrow[\substack{H_2O, \ r.t. \ 1\text{-}2 \ h \\ 90\%}]{10 \ mol\% \ NaBPh_4} \text{(product)} \quad (12.49)$$

For hetero-Diels-Alder reactions with an oxygen-containing dienophile, cyclopentadiene or cyclohexadiene reacted with an aqueous solution of glyoxylic acid to give α-hydroxyl-γ-lactones arising from the rearrangement of the cyclo-adducts. The reaction was independently studied by Augé[119] and Grieco.[120] Augé showed that using water as the solvent allowed for the direct use of the inexpensive aqueous solution of glyoxylic acid for the Diels-Alder reaction (Eq. 12.50).

$$\underset{H}{\overset{O}{\|}}{C\text{-}CO_2H} + \text{(cyclopentadiene)} \xrightarrow[\text{aqueous solvent}]{pH = 0.9} \underset{a}{\text{(product a)}} \quad \underset{b}{\text{(product b)}}$$

ratio a/b = 73:27

$$(12.50)$$

Using water as the solvent enhanced the rate of the hetero-Diels-Alder reaction relative to the dimerization of cyclopentadiene. In addition, the reaction is much faster at a low pH, which implies that the reaction is acid catalyzed. The 5,5-fused system generated has been used in the total synthesis of several bioactive compounds, including the anti-HIV agent $(-)$-carbovir (Eq. 12.51)[121] and the hydroxylactone moiety of mevinic acids (Eq. 12.52).[122]

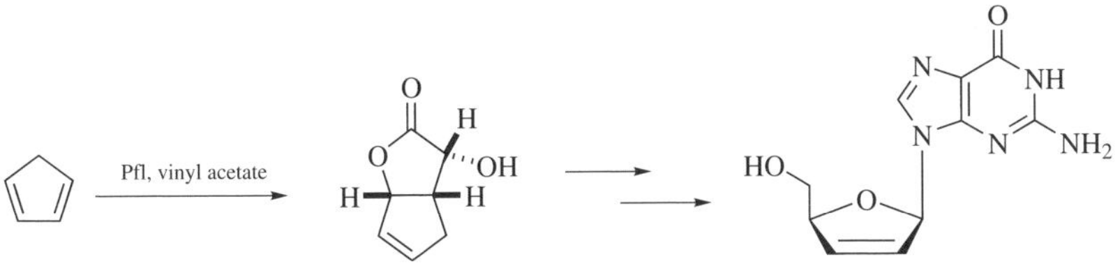

(12.51)

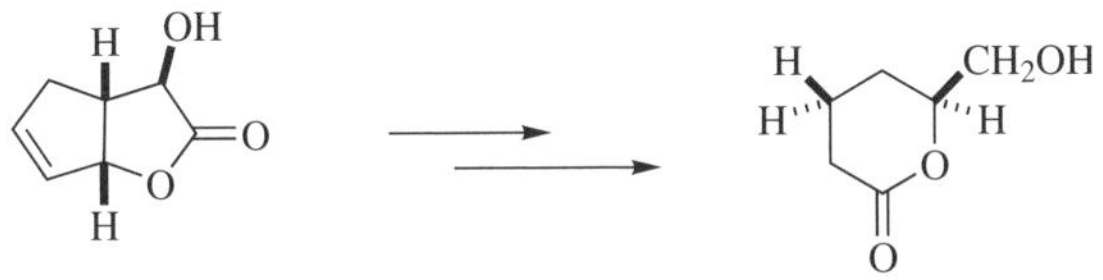

(12.52)

Acyclic dienes react with glyoxylic acid via an oxo-Diels-Alder reaction to give dihydropyran derivatives (Eq. 12.53). An excellent application of the oxo-Diels-Alder reaction is reported by Lubineau et al. in the synthesis of the sialic acids, 3-deoxy-D-manno-2-octulosonic acid (KDO) and 3-deoxy-D-glycero-D-galacto-2-nonulosonic acid (KDN).[123]

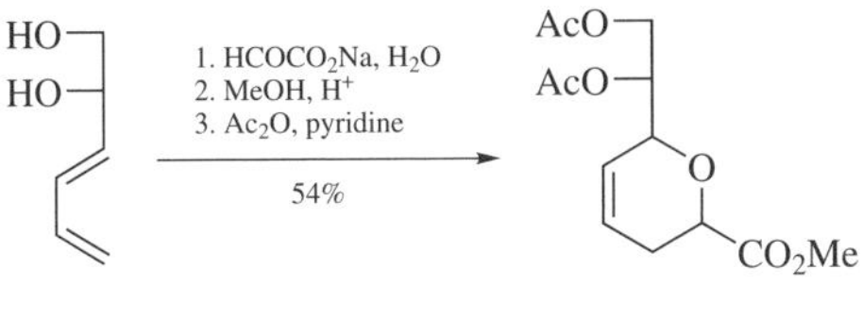

(12.53)

As shown in Eq. 12.54, with glyoxylate as the dienophile, if the attack is on the *si* face of the diene, it would lead to the skeleton of KDO; if the attack is on the *re* face, it would lead to the skeleton of KDN. C-Disaccharide analogs of trehalose were prepared using an aqueous Diels-Alder reaction as a key step.[124]

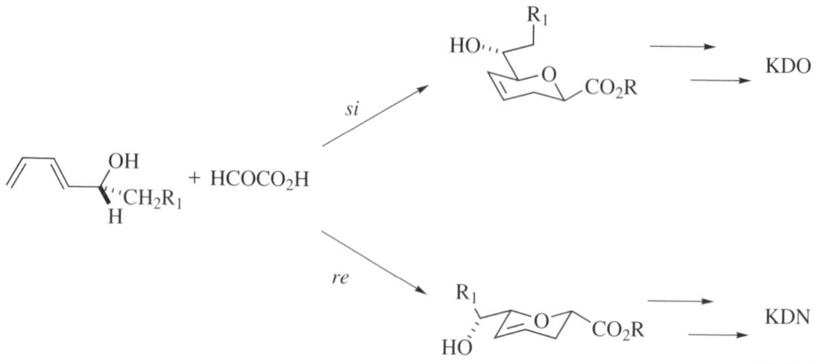

$$(12.54)$$

The hetero-Diels-Alder reaction has also utilized dienophiles in which both reactive centers are heteroatoms. Kibayashi reported that the intramolecular hetero-Diels-Alder cycloaddition of chiral acylnitroso compounds, generated in situ from periodate oxidation of the precursor hydroxamic acid, showed a marked enhancement of the *trans*-selectivity in an aqueous medium compared with the selectivity in nonaqueous conditions (Eq. 12.55).[125] The reaction was readily applied to the total synthesis of (−)-pumiliotoxin C (Figure 12.5).[126]

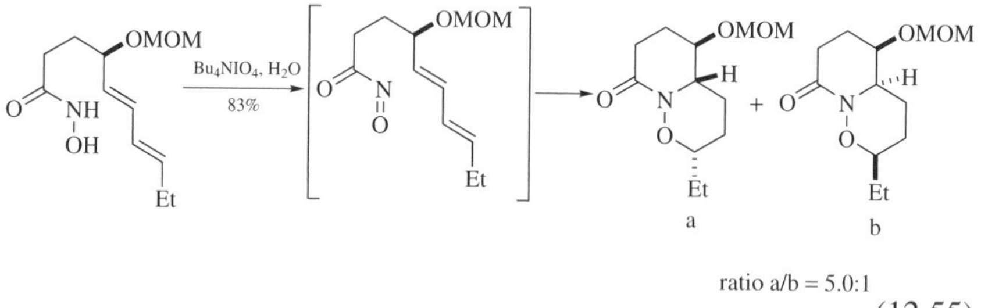

ratio a/b = 5.0:1

$$(12.55)$$

The hetero-Diels-Alder reaction can also employ dienes containing heteroatoms. Cycloaddition of substituted styrenes with di-(2-pyridyl)-1,2,4,5-tetrazine was investigated by Engberts (Eq. 12.56).[127] Again, the rate of the reaction increased dramatically in water-rich media. Through kinetic studies, they showed that the solvent effects on the

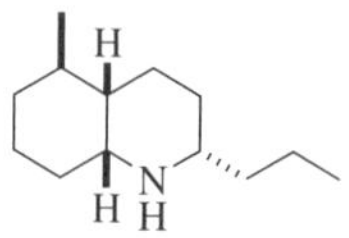

Figure 12.5   (−)-Pumilliotoxin C.

kinetics of hetero-Diels-Alder reactions are very similar to homo-Diels-Alder reactions in aqueous solvent.

$$(12.56)$$

A new noncarbohydrate-based enantioselective approach to (−)-swainsonine was developed in which the key step was an aqueous intramolecular asymmetric hetero-Diels-Alder reaction of an acylnitroso diene (Eq. 12.57).[128] Under aqueous conditions there was significant enhancement of the *trans* stereoselectivity relative to the reaction under conventional nonaqueous conditions.

$$(12.57)$$

Grieco investigated the intramolecular Diels-Alder reaction of iminium ions in polar media such as 5.0 M lithium perchlorate-diethyl ether and in water[129] to form carbocyclic arrays. They showed that water as the solvent provided good-to-excellent yields of tricyclic amines with excellent stereocontrol (Eq. 12.58).

$$(12.58)$$

By reacting aniline with 2,3-dihydrofuran or dihydropyran and a catalytic amount of Lewis acid such as indium chloride in water, various tetrahydroquinoline derivatives were obtained by Li via an in-situ hetero-Diels-Alder reaction.[130] Alternatively, similar compounds were synthesized from anilines and 2-hydroxyltetrahydrofuran or 2-hydroxyltetrahydropyran and a catalytic amount of indium chloride in water (Eq. 12.59).[131]

$$\text{ratio } syn/anti = 24{:}76$$

$$(12.59)$$

One notable result is the treatment of 2-hydroxy cyclic ether analog, 2-deoxy-D-ribose with aniline in water catalyzed by InCl$_3$ to afford the novel tricyclic tetrahydroquinoline compounds (Eq. 12.60). The reaction can also be catalyzed by recoverable cation-exchange resin instead of indium chloride.[132] By using a stoichiometric amount of indium metal, a domino reaction of nitroarenes with 2,3-dihydrofuran generates the same products. [133]

$$(12.60)$$

Yadav et al. explored the reaction of substituted anilines with 3,4,6-tri-*O*-acetyl-D-glucal to offer the tetrahydroquinoline moieties.[134] Most yields are around 80% with excellent distereoselectivity and the reaction was carried out in water (Eq. 12.61). The primary disadvantages are that both CeCl$_3$ and NaI are required in stoichiometric amounts.

$$(12.61)$$

12.2.6 Asymmetric Hetero-Diels-Alder Reactions

The presence of a small amount of water was found beneficial for several asymmetric hetero-Diels-Alder reactions. The asymmetric catalysis of a hetero-Diels-Alder reaction with Danishefsky's diene by chiral lanthanide *bis*(trifluoromethanesulfonyl)amide (*bis*-trifylamide) complexes, showed a significant effect of water as an additive in increasing both the enantioselectivity and the chemical yield (Eq. 12.62).[135] Chiral auxiliaries have also been used in hetero-Diels-Alder reactions for obtaining optically active products.[136] Fringuelli et al. have reacted (E)-2-aryl-1-cyano-1-nitroalkenes with both achiral and enantiopure vinyl ethers in water (Eq. 12.63). In addition, using ($-$)-N, N-dicyclohexyl-(1S)-isoborneol-10-sulfonamide as the chiral auxiliary, asymmetric cycloadditions were observed (Eq. 12.64).

$$ (12.62) $$

yield 88%
ee 66%

$$ (12.63) $$

a/b ratio up to 98:2
yield of a up to 90%

$$ (12.64) $$

endo/exo ratio up to 85:15
de 100%

Aqueous aza-Diels-Alder reactions of chiral aldehydes, prepared from carbohydrates and with benzylamine hydrochloride and cyclopentadiene, were promoted by lanthanide triflates (Eq. 12.65).[137] The nitrogen-containing heterocyclic products were further transformed into aza sugars, which are potential inhibitors against glycoprocessing enzymes.

$$(12.65)$$

12.2.7   Other Cyclization Reactions

12.2.7.1   *Alder-Ene Reactions.*   An ene-iminium one-pot cyclization proceeds smoothly in a mixture of water-THF (Eq. 12.66). [138] The reactivity of the ene-iminium substrates is highly dependent on the substitution pattern of the ethylenic double bond. This methodology can be used to form homochiral pipecolic acid derivatives.

$$(12.66)$$

12.2.7.2   *1,3-Dipolar Cycloaddition Reactions.*   The 1,3-dipolar cyclization of nitrile oxide with dipolarophiles generates structurally important heterocycles. As shown by Lee,[139] the reaction can be carried out in an aqueous-organic biphasic system in which the nitrile oxide substrates can be generated from oximes or hydrazones in situ. The method provides a convenient one-pot procedure for generating a variety of heterocyclic products.

Reaction of preformed aromatic nitrile N-oxides with alkyl disubstituted benzoquinones gives the 1,3-dipolar cyclization product in aqueous ethanol.[140] The effect of the polarity of solvents on the rate of the reaction was investigated. While the reaction is usually slower in more polar solvents than in less polar ones, the use of water as the solvent increases the reactivity. The nitrile oxide reaction has also been catalyzed by baker's yeast[141] and ß-cyclodextrin.[142] Reaction of an azomethine ylide, generated in situ from methyl N-methylglycinate and formaldehyde and with activated olefins, generated the corresponding dipolar products.[143] An elegant application of 1,3-dipolar cyclization of an azide derivative in water was reported by De Clercq in the synthesis of (+)-biotin (Scheme 12.5).[144] Upon thermolysis of the azide compound, a mixture of (+)-biotin and its benzylated derivative was formed directly. The usage of water is necessary as the nucleophile and to accelerate the cyclization via betaine stabilization.

The kinetics of 1,3-dipolar cycloaddition of phenyl azide to norbornene in aqueous solutions was studied (Eq. 12.67).[145] As shown in Table 12.1, when the reaction was performed in organic solvents, the reaction showed very small effects of the solvent, while in highly aqueous media, significant accelerations were observed.

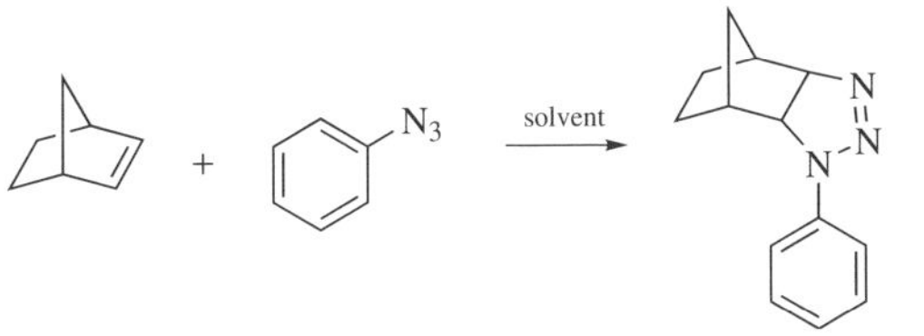

$$(12.67)$$

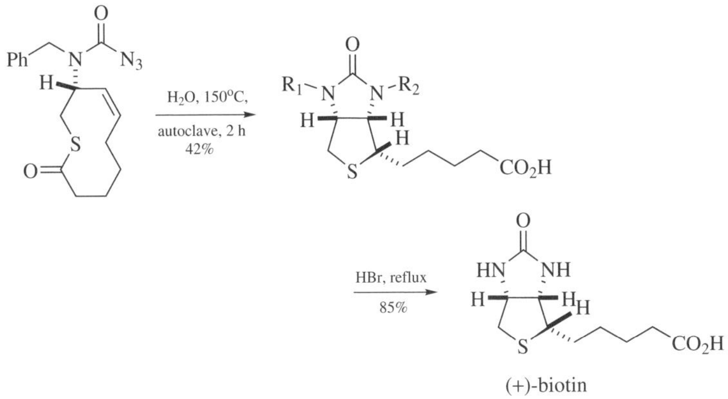

Scheme 12.5

TABLE 12.1 Second-Order Constants for Cycloaddition

Solvent	$10^5 \times k_2$ ($M^{-1}s^{-1}$)
n-hexane	4.7
EtOH	7.4
2-PrOH	8.2
DMSO	17.5
4:1 H_2O/EtOH	37.0
92:8 H_2O/2-PrOH	83.0
99:1 H_2O/NCP	250.0

NCP = 1-cyclohexyl-2-pyrrolidinone.

An intermolecular 1,3-dipolar cycloaddition of diazocarbonyl compounds with alkynes was developed by using an $InCl_3$-catalyzed cycloaddition in water. The reaction was found to proceed by a domino 1,3-dipolar cycloaddition-hydrogen (alkyl or aryl) migration (Eq. 12.68).[146] The reaction is applicable to various α-diazocarbonyl compounds and alkynes with a carbonyl group at the neighboring position, and the success of the reaction was rationalized by decreasing the HOMO-LUMO of the reaction.

$$(12.68)$$

ratio a/b = 91:9
% yield a/b = 82%/7%

12.3 SIGMATROPIC REARRANGEMENTS

12.3.1 Claisen Rearrangements

The enzyme chorismate mutase was found to accelerate the Claisen rearrangement of chorismic acid.[147] For many years, the origin of the acceleration perplexed and intrigued chemists and biochemists. Polar solvents have been known to increase the rate of the Claisen rearrangement reactions.[148] Claisen rearrangement reactions were found to be accelerated going from nonpolar to aqueous solvents.[149] For instance, the rearrangements of chorismic acid and related compounds in water were 100 times faster than in methanol (Eq. 12.69).[150]

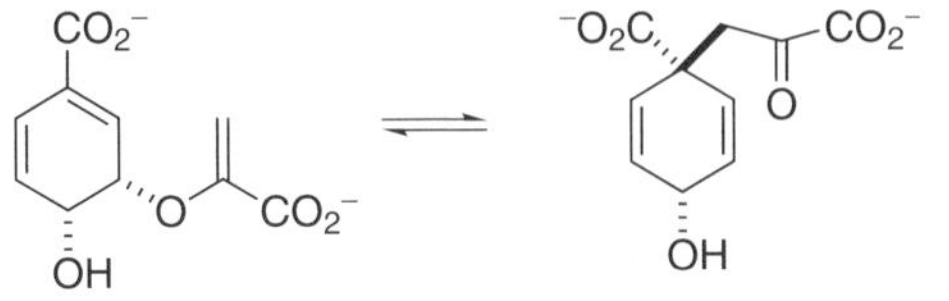

$$\text{(12.69)}$$

Due to the $\Delta V^{\neq}$ (volume change of activation) of Claisen rearrangements having a negative value, as in the Diels-Alder reactions, the Claisen rearrangement reaction is expected to be accelerated by water according to the same effect.[151,152]

Allyl aryl ethers undergo accelerated Claisen and [1,3] rearrangements in the presence of a mixture of trialkylalanes and water or aluminoxanes. The addition of stoichiometric quantities of water accelerates both the trimethylaluminum-mediated aromatic Claisen reaction and the chiral zirconocene-catalyzed asymmetric carboalumination of terminal alkenes. These two reactions occur in tandem and, after oxidative quenching of the intermediate trialkylalane, result in the selective formation of two new C–C bonds and one C–O bond (Eq. 12.70).[153] Antibodies have also been developed to catalyze Claisen[154] and oxy-Cope[155] rearrangements.

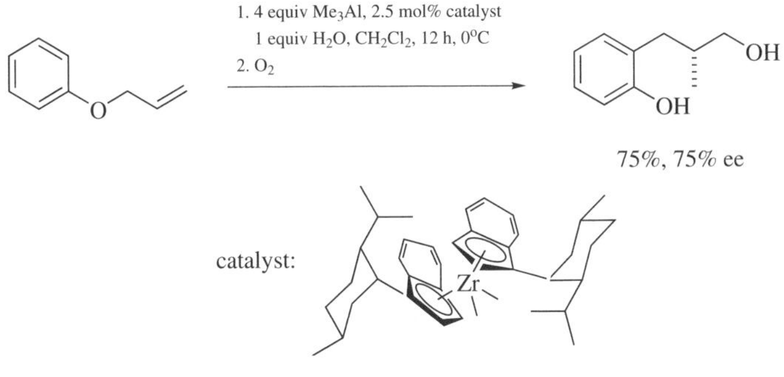

1. 4 equiv Me_3Al, 2.5 mol% catalyst
1 equiv H_2O, CH_2Cl_2, 12 h, 0°C
2. O_2

$$\text{(12.70)}$$

75%, 75% ee

catalyst:

Grieco observed a facile [3,3]-sigmatropic rearrangement of an allyl vinyl ether in water, giving rise to an aldehyde (Eq. 12.71).

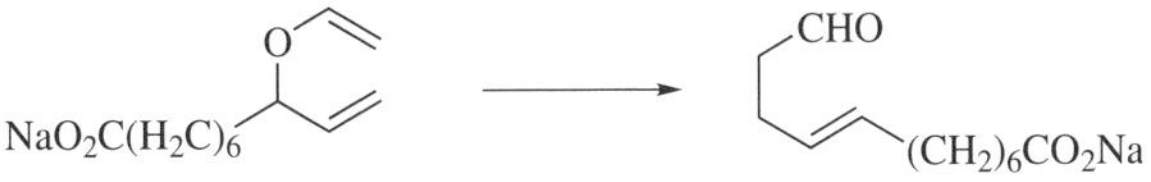

$$\text{(12.71)}$$

The corresponding methyl ester similarly underwent the facile rearrangement. A solvent polarity study on the rearrangement rate of the allyl vinyl ether was conducted in solvent systems ranging from pure methanol to water at 60°C.[156] The first-order rate constant for the rearrangement of the allyl vinyl ether in water was 18×10^{-5} s^{-1}, compared with 0.79×10^{-5} s^{-1} in pure methanol.

The accelerating influence of water as a solvent on the rate of the Claisen rearrangement has also been demonstrated on a number of other substrates. These studies showed that this methodology has potential applications in organic synthesis. In Eq. 12.72, the unprotected vinyl ether in a 2.5:1 water-methanol solvent with an equivalent of sodium hydroxide underwent rearrangement to give the aldehyde in 85% yield.[157]

$$(12.72)$$

The same rearrangement for the protected analog under organic solvent reaction conditions had considerable difficulties and often resulted in the elimination of acetaldehyde.[158]

Water also had an effect on the [3,3]-sigmatropic shift of the allyl vinyl ether, a key intermediate in the synthesis of the Inhoffenn-Lythgoe diol (Eq. 12.73). The rearrangement occurred in only five hours at 95°C in 0.1 N NaOH solution to give the aldehyde in 82% yield, while the corresponding methyl ester led to recovered starting material only upon prolonged heating in decalin.

$$(12.73)$$

It is interesting to note that all previous attempts to utilize the Claisen rearrangement within the carbon framework of the fenestrane system as well as all efforts to prepare a fenestrane in which one of the ring fusions is *trans*, had not been successful. In Eq. 12.74, a facile rearrangement of fenestrene took place in aqueous pyridine to form a fenestrene aldehyde with a *trans* configuration between the two five-membered rings common to the acetaldehyde unit.

$$(12.74)$$

The use of the glucose chiral auxiliary by Lubineau et al. led to moderate asymmetric induction in the Claisen rearrangement (20% de) (Eq. 12.75).[159] Since it could be removed easily, glucose functioned here as a chiral auxiliary. After separation of the diastereomers, enantiomerically pure substances could be obtained.

$$(12.75)$$

The origin of the rate acceleration in Claisen rearrangement has been studied extensively by various methods.[160] A self-consistent–field solvation model was applied to the aqueous medium Claisen rearrangement.[161] The aqueous acceleration of the Claisen rearrangement was suggested to be due to solvent-induced polarization and first-hydration-shell hydrophilic effects.[162] Theoretical studies by Jorgensen[163] and Gajewski[164] suggested that increased hydrogen bonding in the transition state is responsible for the observed acceleration. Isotope effects on the rearrangement of allyl vinyl ether have been carried out.[165] Studies by Gajewski on the secondary deuterium kinetic isotope effects argue against the involvement of an ionic transition state.[166] A combined quantum-mechanical and statistical-mechanical approach used by Gao indicated that different substrates have different degrees of acceleration.[167] The effects of hydration on the rate acceleration of the Claisen rearrangement of allyl vinyl ether were investigated by a hybrid quantum-mechanical and classical Monte Carlo simulation method.[168] A number of continuum models, combined with *ab initio* wave functions, have been used to predict the effect of solvation by water on the Claisen rearrangement of allyl vinyl ether.[169] Monte Carlo simulations have been used by Jorgensen to determine changes in the free

energies of solvation for the rearrangement of chorismate to prephen-
ate in water and methanol.[170] The calculation reproduces the observed
100-fold rate increase in water over methanol. The origin of the rate
difference is traced solely to an enhanced population of the pseudo-
diaxial conformer in water, which arises largely from a unique water
molecule acting as a double hydrogen bond donor to the C_4 hydroxyl
group and the side-chain carboxylate.

The differences in the rate constant for the water reaction and the
catalyzed reactions reside in the mole fraction of substrate present as
near attack conformers (NACs).[171] These results and knowledge of
the importance of transition-state stabilization in other cases support
a proposal that enzymes utilize both NAC and transition-state stabilization in the mix required for the most efficient catalysis. Using a
combined QM/MM Monte Carlo/free-energy perturbation (MC/FEP)
method, 82%, 57%, and 1% of chorismate conformers were found to
be NAC structures (NACs) in water, methanol, and the gas phase,
respectively.[172] The fact that the reaction occurred faster in water than
in methanol was attributed to greater stabilization of the TS in water
by specific interactions with first-shell solvent molecules. The Claisen
rearrangements of chorismate in water and at the active site of *E. coli*
chorismate mutase have been compared.[173] It follows that the efficiency
of formation of NAC (7.8 kcal/mol) at the active site provides approximately 90% of the kinetic advantage of the enzymatic reaction as
compared with the water reaction.

A thio-Claisen rearrangement[174] was used for the regioselective synthesis of thiopyrano[2,3-b]pyran-2-ones and thieno[2,3-b]pyran-2-ones
(Eq. 12.76). A convenient method for the aromatic amino-Claisen rearrangement of N-(1,1-disubstituted-allyl)anilines led to the 2-allylanilines
being produced cleanly and in high yield by using a catalytic amount
of *p*-toluenesulfonic acid in acetonitrile/water (Eq. 12.77).[175]

(12.76)

R= CH$_2$OH, CH$_2$Cl

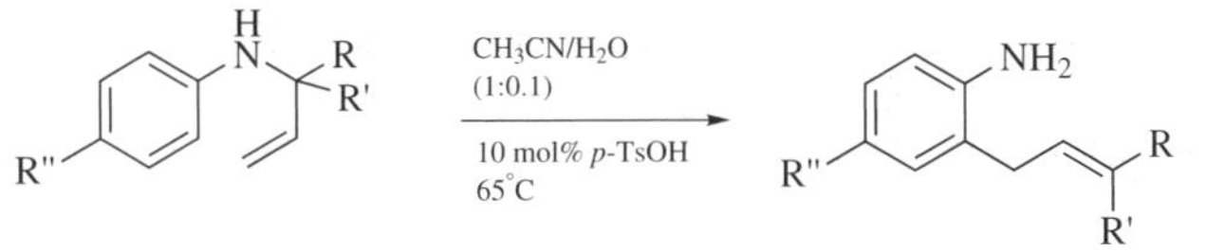

$$(12.77)$$

12.3.2 Cope Rearrangements

Cope rearrangement has also been studied in aqueous media. Both enzymatic and nonenzymatic Cope rearrangements of carbaprephenate to carbachorismate were investigated.[176] Carbaprephenate and its epimer undergo spontaneous acid-catalyzed decarboxylation in aqueous solution. Only at a high pH does the Cope rearrangement compete with the decarboxylation, and at a pH of 12 at 90°C, carbaprephenate slowly rearranges to carbachorismate, which rapidly loses water to give 3-(2-carboxyallyl)benzoic acid as the major product (Scheme 12.6).

A chelation-assisted Pd-catalyzed Cope rearrangement was proposed in the reaction of phenanthroline to generate isoquinolinone derivatives (Eq. 12.78).[177] The use of aqueous media and ligands enables a double-Heck reaction on a substrate favoring alkene insertion over β-hydride elimination.

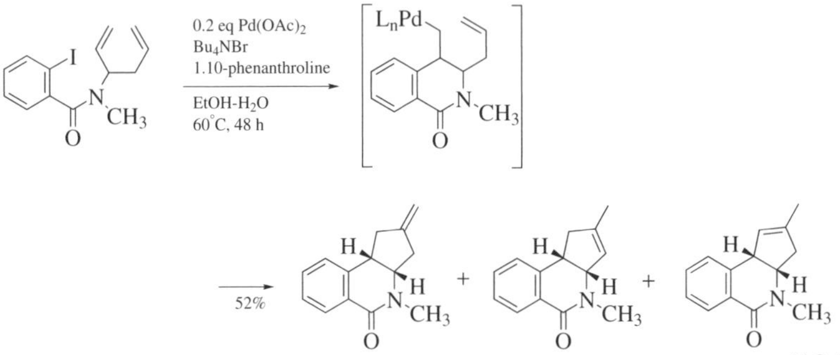

$$(12.78)$$

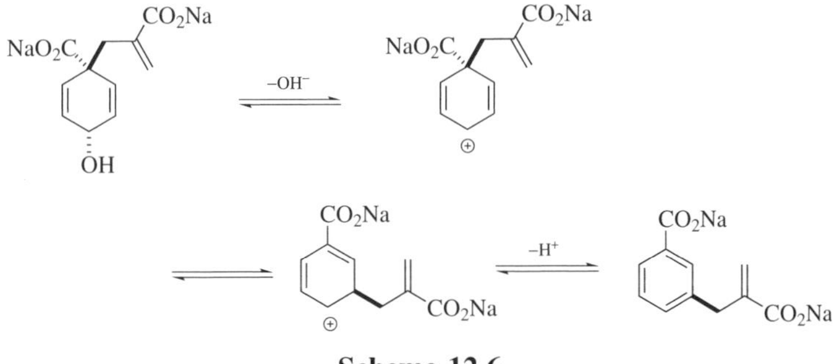

Scheme 12.6

12.4   PHOTOCHEMICAL CYCLOADDITION REACTIONS

An excellent review on organic photochemistry in organized media, including aqueous solvent, has been reported.[178] The quantum efficiency for photodimerization of thymine, uracil, and their derivatives increased considerably in water compared with other organic solvents. The increased quantum efficiency is attributed to the preassociation of the reactants at the ground state.

Organic substrates having poor solubilities in water such as stilbenes or alkyl cinnamates photodimerize efficiently in water. The same reaction in organic solvents such as benzene mainly leads to *cis-trans* isomerizations.[179] As in the case of the Diels-Alder reaction, the addition of LiCl (increasing the hydrophobic effect) increases the yield of dimerizations. On the other hand, the addition of quanidinium chloride (decreasing the hydrophobic effect) lowers the yield of the product. The photodimerization of stilbenes is more efficient in a hydroxylic solvent such as methanol or water rather than in a nonhydroxylic solvent such as hexane, benzene, or acetonitrile. [180] The proposed accelerated photodimerization originates from a formation of a fluorescent solute–solute aggregate. Similarly, coumarin dimerized more efficiently in water than in organic solvents (Eq. 12.79).[181]

$$\tag{12.79}$$

The quantum yield of the dimerization in water is more than 100 times higher than that in benzene and methanol. When a surfactant is added in water, it will aggregate in forming micelles. The formation of such micelles has also been found to have a significant effect on the regio- and stereoselectivity of photochemical reactions. In the micellar case, the hydrophobic interior of micelles provides a hydrophobic pocket within the bulk water solvent. An analogous situation of hydrophobic cage effect is the use of cyclodextrin. Thus, selectivity in product formation could be expected also in this case. Indeed, while four isomers are generated for the photodimerization of anthracene-2-sulfonate, the same reaction gives only one isomer when β-cyclodextrin is present (Eq. 12.80).[182]

$$(12.80)$$

Using an appropriately substituted *o*-hydroxybenzyl alcohol precursor, a photogenerated *o*-quinone methide undergoes efficient intermolecular Diels-Alder cycloaddition in aqueous CH_3CN to generate the hexahydrocannabinol ring system (Eq. 12.81).[183] Laser flash photolysis studies show the intermediacy of an *o*-quinone methide that has a lifetime >2 ms. Quantum yields for the reaction and fluorescence parameters depend strongly on the proportion of water in the H_2O-CH_3CN solvent mixture.

$$(12.81)$$

REFERENCES

1. For general reviews of pericyclic reactions, see: Woodward, R. B., Hoffmann, R., *The Conservation of Orbital Symmetry*, Cambridge University Press, London, 1972; Marchand, A. P., Lehr, R. E., *Pericyclic Reactions*, Academic Press, New York, 1977.

2. Fleming, I., *Frontier Orbitals and Organic Chemical Reactions*, John Wiley & Sons, Chichester, 1976.

3. van Eldik, R., Asano, T., Le Nobel, W. J., *Chem. Rev.* **1989**, *89*, 549.

4. Dauben, W. G., Krabbenhoft, H. O., *J. Am. Chem. Soc.* **1976**, *98*, 1992.

5. Kumar, A., *J. Org. Chem.* **1994**, *59*, 230.

6. Carruthers, W., *Cycloaddition Reactions in Organic Synthesis, Tetrahedron Organic Chemistry Series*, Vol. 8, Baldwin, J. E., and Magnus, P. D., eds., Pergamon Press, 1990.

7. Blake, J. F., Jorgensen, W. L., *J. Am. Chem. Soc.* **1991**, *113*, 7430; Blake, J. F., Lim, D., Jorgensen, W. L., *J. Org. Chem.* **1994**, *59*, 803.

8. Breslow, R., Guo, T., *J. Am. Chem. Soc.* **1988**, *110*, 5613.

9. Breslow, R., Rizzo, C. J., *J. Am. Chem. Soc.* **1991**, *113*, 4340. For recent theoretical studies on the hydrophobic effect, see: Muller, N., *Acc. Chem. Res.* **1990**, *23*, 23.

10. (a) Schneider, H. J., Sangwan, N. K., *J. Chem. Soc., Chem. Commun.* **1986**, 1787, (b) Schneider, H. J., Sangwan, N. K., *Angew. Chem., Int. Ed. Engl.* **1987**, *26*, 896. (c) Hunt, I., Johnson, C. D., *J. Chem. Soc. Perkin Trans. 2*, **1991**, 1051.

11. Sangwan, N. K., Schneider, H. J., *J. Chem. Soc., Perkin Trans. 2*, **1989**, 1223.

12. Blokzijl, W., Blandamer, M. J., Engberts, J. B. F. N., *J. Am. Chem. Soc.* **1991**, *113*, 4241; Blokzijl, W., Blandamer, M. J., Engberts, J. B. F. N., *J. Am. Chem. Soc.* **1992**, *114*, 5440; Otto, S., Blokzijl, W., Engberts, J. B. F. N., *J. Org. Chem.* **1994**, *59*, 5372.

13. Engberts, J. B. F. N., *Pure & Appl. Chem.* **1995**, *67*, 823.

14. Jenner, G., *Tetrahedron Lett.* **1994**, *35*, 1189.

15. Pai, C. K., Smith, M. B., *J. Org. Chem.* **1995**, *60*, 3731; Smith, M. B., Fay, J. N., Son, Y. C., *Chem. Lett.* **1992**, 2451.

16. Griesbeck, A. G., *Tetrahedron Lett.* **1988**, *29*, 3477.

17. Braisted, A. C., Schultz, P. G., *J. Am. Chem. Soc.* **1990**, *112*, 7430; Hilvert, D., Hill, K. W., Nared, K. D., Auditor, M. T. M., *J. Am. Chem. Soc.* **1989**, *111*, 9261.

18. Meekel, A. A. P., Resmini, M., Pandit, U. K., *J. Chem. Soc. Chem. Commun.* **1995**, 571.

19. Jackson, D. Y., Jacobs, J. W., Sugasawara, R., Reich, S. H., Bartlett, P. A., Schultz, P. G., *J. Am. Chem. Soc.* **1988**, *110*, 4841; Hilvert, D., Carpenter, S. H., Nared, K. D., Auditor, M. T. M., *Proc. Natl. Acad. Sci. U.S.A.* **1988**, *85*, 4953.

20. Braisted, A. C., Schultz, P. G., *J. Am. Chem. Soc.* **1994**, *116*, 2211.

21. For reviews, see: Breslow, R., *Acc. Chem. Res.* **1991**, *24*, 159; Grieco, P. A., *Aldrichim. Acta* **1991**, *24*, 59; Otto, S; Engberts, J. B. F. N., *Pure Appl. Chem.* **2000**, *72*, 1365; Wittkopp, A., Schreiner, P. R., *Chem. Dien. Polyen.* **2000**, *2*, 1029; Keay, B. A., Hunt, I. R., *Adv. Cycloadd.* **1999**, *6*, 173.

22. Diels, O., Alder, K., *Liebigs Ann. Chem.* **1931**, *490*, 243; Hopff, H., Rautenstrauch, C. W., U.S. Patent 2,262,002; *Chem. Abstr.*, **1942**, *36*, 1046; Woodward, R. B., Baer, H., *J. Am. Chem. Soc.* **1948**, *70*, 1161; Koch, H., Kotlan, J., Markurt, H., *Monatsh. Chem.* **1965**, *96*, 1646.

23. Rideout, D. C., Breslow, R., *J. Am. Chem. Soc.* **1980**, *102*, 7816.

24. Ben-Naim, A., *Hydrophobic Interactions*, Plenum Press, New York, 1980; Tanford, C., *The Hydrophobic Effect*, 2nd ed., John Wiley, New York, 1980.

25. von Hippel, P. H., Schleich, T., *Acc. Chem. Res.* **1969**, *2*, 257.

26. Breslow, R., Maitra, U., Rideout, D., *Tetrahedron Lett.* **1983**, *24*, 1901; Breslow, R., Maitra, U., *Tetrahedron Lett.* **1984**, *25*, 1239

27. Berson, J. A., Hamlet, Z., Mueller, W. A., *J. Am. Chem. Soc.* **1962**, *84*, 297; Samii, A. A. Z., de Savignac, A., Rico, I., Lattes, A., *Tetrahedron* **1985**, *41*, 3683.

28. Sternbach, D. D., Rossana, D. M., *J. Am. Chem. Soc.* **1982**, *104*, 5853;

29. Wang, W. B., Roskamp, E. J., *Tetrahedron Lett.* **1992**, *33*, 7631.

30. Keay, B. A., *J. Chem. Soc., Chem. Commun.* **1987**, 419.

31. Bourgeois-Guy, A., Gore, J., *Bull. Soc. Chim. Fr.* **1992**, *129*, 490.

32. Wijnen, J. W., Engberts, J. B. F. N., *J. Org. Chem.* **1997**, *62*, 2039.

33. Gonzalez, A., Holt, S. L., *J. Org. Chem.* **1982**, *47*, 3186.

34. Utley, J. H. P., Oguntoye, E., Smith, C. Z., Wyatt, P. B., *Tetrahedron Lett.* **2000**, *41*, 7249.

35. Hollis, T. K., Robinson, N. P., Bosnich, B., *J. Am. Chem. Soc.* **1992**, *114*, 5464.

36. Kobayashi, S., Hachiya, I., Araki, M., Ishitani, H., *Tetrahedron Lett.* **1993**, *34*, 3755.

37. Otto, S., Engberts, J. B. F. N., *Tetrahedron Lett.* **1995**, *36*, 2645; Otto, S., Bertoncin, F., Engberts, J. B. F. N., *J. Am. Chem. Soc.* **1996**, *118*, 7702.

38. Ishihara, K., Hanaki, N., Funahashi, M., Miyata, M., Yamamoto, H., *Bull. Chem. Soc. Jpn* **1995**, *68*, 1721.

39. For a recent review, see: Fringuelli, F., Piermatti, O., Pizzo, F., Vaccaro, L., *Eur. J. Org. Chem.* **2001**, 439.

40. Loh, T.-P., Pei, J., Lin, M., *Chem. Commun.* **1996**, 2315.

41. Zhu, Z., Espenson, J. H., *J. Am. Chem. Soc.* **1997**, *119*, 3507; Bianchini, C., *Chemtracts: Org. Chem.* **1998**, *11*, 99.

42. Prajapati, D., Laskar, D. D., Sandhu, J. S., *Tetrahedron Lett.* **2000**, *41*, 8639.

43. Laurent-Robert, H., Le Roux, C., Dubac, J., *Synlett* **1998**, 1138.

44. Mubofu, E. B., Engberts, J. B. F. N., *J. Phys. Org. Chem.* **2004**, *17*, 180.

45. Tsai, S.-H., Chung, W.-S., Wu, H.-J., *J. Chin. Chem. Soc.* **1996**, *43*, 281.

46. Otto, S., Engberts, J. B. F. N., Kwak, J. C. T., *J. Am. Chem. Soc.* **1998**, *120*, 9517; Manabe, K., Mori, Y., Kobayashi, S., *Tetrahedron* **1999**, *55*, 11203; Rispens, T., Engberts, J. B. F. N., *Org. Lett.* **2001**, *3*, 941.

47. Jaeger, D. A., Wang, J., *J. Org. Chem.* **1993**, *58*, 6745; Jaeger, D. A., Su, D., *Tetrahedron Lett.* **1999**, *40*, 257; Su, D., Jaeger, D. A., *Tetrahedron Lett.* **1999**, *40*, 7871; Jaeger, D. A., Su, D., Zafar, A., Piknova, B., Hall, S. B., *J. Am. Chem. Soc.* **2000**, *122*, 2749.

48. Chung, W.-S., Wang, J.-Y., *J. Chem. Soc., Chem. Commun.* **1995**, 971.

49. Colonna, S., Manfredi, A., Annunziata, R., *Tetrahedron Lett.* **1988**, *29*, 3347.

50. Katayama, K., Kobayashi, T., Oikawa, H., Honma, M., Ichihara, A., *Biochim. Biophys. Act.* **1998**, *1384*, 387.

51. Gouverneur, V. E., Houk, K. N., Pascual-Teresa, B., Beno, B., Janda, K. D., Lerner, R. A., *Science*, **1993**, *262*, 204.

52. Ramachary, D. B., Chowdari, N. S., Barbas, C. F., *Tetrahedron Lett.* **2002**, *43*, 6743.

53. Kusukawa, T., Nakai, T., Okano, T., Fujita, M., *Chem. Lett.* **2003**, *32*, 284.

54. Rispens, T., Engberts, J. B. F. N., *J. Org. Chem.* **2002**, *67*, 7369; Chiba, K., Jinno, M., Nozaki, A., Tada, M., *Chem. Commun.* **1997**, 1403; Diego-Castro, M. J., Hailes, H. C., *Tetrahedron Lett.* **1998**, *39*, 2211.

55. Kim, S. P., Leach, A. G., Houk, K. N., *J. Org. Chem.* **2002**, *67*, 4250.

56. Itami, K., Nokami, T., Yoshida, J.-I., *Angew. Chem., Int. Ed.* **2001**, *40*, 1074; Itami, K., Yoshida, J.-I., *Chem. Rec.* **2002**, *2*, 213; Itami, K., Nokami, T., Yoshida, J.-I., *Adv. Synth. Catal.* **2002**, *344*, 441.

57. Korzenski, M. B., Kolis, J. W., *Tetrahedron Lett.* **1997**, *38*, 5611; Ikushima, Y., *Rec. Res. Dev. Chem. Phys.* **2000**, *1*, 123; Harano, Y., Sato, H., Hirata, F., *Chem. Phys.* **2000**, *258*, 151.

58. Wittkopp, A., Schreiner, P. R., *Chem. Eur. J.* **2003**, *9*, 407.

59. Elguero J., Goya, P., Paez, J. A., Cativiela, C., Mayoral, J. A., *Synth. Commun.* **1989**, *19*, 473.

60. Lubineau, A., Queneau, Y., *Tetrahedron Lett.* **1985**, *26*, 2653; Lubineau, A., Queneau, Y., *J. Org. Chem.* **1987**, *52*, 1001; Lubineau, A., Queneau, Y., *Tetrahedron* **1989**, *45*, 6697; Lubineau, A., Bienayme, H., Queneau, Y., Scherrmann, M. C., *New J. Chem.* **1994**, *18*, 279; Lubineau, A., Bienayme, H., Queneau, Y., *Carbohydr. Res.* **1995**, *270*, 163.

61. Lakner, F. J., Negrete, G. R., *Synlett* **2002**, 643.

62. Ishihara, K., Yamamoto, H., *J. Am. Chem. Soc.* **1994**, *116*, 1561; Ishihara, K., Kurihara, H., Yamamoto, H., *J. Am. Chem. Soc.* **1996**, *118*, 3049; Ishihara, K., Kurihara, H., Matsumoto, M., Yamamoto, H., *J. Am. Chem. Soc.* **1998**, *120*, 6920.

63. Kanemasa, S., Oderaotoshi, Y., Yamamoto, H., Tanaka, J., Wada, E., Curran, D. P., *J. Org. Chem.* **1997**, *62*, 6454.

64. Carbone, P., Desimoni, G., Faita, G., Filippone, S., Righetti, P., *Tetrahedron* **1998**, *54*, 6099.

65. Diego-Castro, M. J., Hailes, H. C., *Chem. Commun.* **1998**, 1549.

66. Otto, S., Boccaletti, G., Engberts, J. B. F. N., *J. Am. Chem. Soc.* **1998**, *120*, 4238.

67. Otto, S., Engberts, J. B. F. N., *J. Am. Chem. Soc.* **1999**, *121*, 6798.

68. Miyamoto, H., Kimura, T., Daikawa, N., Tanaka, K., *Green Chemistry* **2003**, *5*, 57.

69. Isaacs, N. S., Maksimovic, L., Laila, A., *J. Chem. Soc., Perkin Trans. 2*, **1994**, 495.

70. Graziano, G., *J. Phys. Org. Chem.* **2004**, *17*, 100.

71. Meijer, A., Otto, S., Engberts, J. B. F. N., *J. Org. Chem.* **1998**, *63*, 8989.

72. Kong, S., Evanseck, J. D., *J. Am. Chem. Soc.* **2000**, *122*, 10418.

73. Kumar, A., *Chem. Rev.* **2001**, *101*, 1; Kumar, A., Pawar, S. S., *Tetrahedron* **2002**, *58*, 1745; Kumar, A., *J. Phys. Org. Chem.* **1996**, *9*, 287; Pawar, S. S., Phalgune, U., Kumar, A., *J. Org. Chem.* **1999**, *64*, 7055; Kumar, A., Phalgune, U., Pawar, S. S., *J. Phys. Org. Chem.* **2000**, *13*, 555; Kumar, A., Deshpande, S. S., *J. Phys. Org. Chem.* **2002**, *15*, 242.

74. Deshpande, S. S., Pawar, S. S., Phalgune, U., Kumar, A., *J. Phys. Org. Chem.* **2003**, *16*, 633.

75. Cativiela, C., Garcia, J. I., Mayoral, J. A., Salvatella, L., *J. Chem. Soc., Perkin Trans. 2*, **1994**, 847; Cativiela, C., Garcia, J. I., Gil, J., Martinez, R. M., Mayoral, J. A., Salvatella, L., Urieta, J. S., Mainar, A. M., Abraham, M. H., *J. Chem. Soc., Perkin Trans. 2*, **1997**, 653.

76. Chandrasekhar, J., Shariffskul, S., Jorgensen, W. L., *J. Phys. Chem. B* **2002**, *106*, 8078.

77. Assfeld, X., Ruiz-Lopez, M. F., Garcia, J. I., Mayoral, J. A., Salvatella, L., *J. Chem. Soc., Chem. Commun.* **1995**, 1371.

78. Furlani, T. R., Gao, J., *J. Org. Chem.* **1996**, *61*, 5492.

79. Schlachter, I., Mattay, J., Suer, J., Hoeweler, U., Wuerthwein, G., Wuerthwein, E.-U., *Tetrahedron* **1997**, *53*, 119.

80. Telan, L. A., Firestone, R. A., *Tetrahedron* **1999**, *55*, 14269.

81. Grieco, P. A., Garner, P., He, Z. M., *Tetrahedron Lett.* **1983**, *24*, 1897.

82. Grieco, P. A., Yoshida, K., Garner, P., *J. Org. Chem.* **1983**, *48*, 3137.

83. Grieco, P. A., Garner, P., Yoshida, K., Huffmann, J. C., *Tetrahedron Lett.* **1983**, *24*, 3807.

84. Grieco, P. A., Yoshida, K., He, Z. M., *Tetrahedron Lett.* **1984**, *25*, 5715.

85. Grieco, P. A., Galatsis, P., Spohn, R. F., *Tetrahedron* **1986**, *42*, 2847.

86. Proust, S. M., Ridley, D. D., *Aust. J. Chem.* **1984**, *37*, 1677.

87. Yoshida, K., Grieco, P. A., *J. Org. Chem.* **1984**, *49*, 5257.

88. Drewes, S. E., Grieco, P. A., Huffman, J. C., *J. Org. Chem.* **1985**, *50*, 1309.

89. Brandes, E., Grieco, P. A., Garner, P., *J. Chem. Soc., Chem. Commun.* **1988**, 500.

90. Yoshida, K., Grieco, P. A., *Chem. Lett.* **1985**, 155.

91. Williams, D. R., Gaston, R. D., Horton, I. B. III, *Tetrahedron Lett.* **1985**, *26*, 1391.

92. Van Royen, L. A., Mijngheer, R., Declercq, P. J., *Tetrahedron* **1985**, *41*, 4667.

93. Grootaert, W. M., Declercq, P. J., *Tetrahedron Lett.* **1986**, *27*, 1731, Nuyttens, F., Appendino, G., De Clercq, P. J., *Synlett.* **1991**, 526.

94. Zezza, C. A., Smith, M. B., *J. Org. Chem.* **1988**, *53*, 1161.

95. Saksena, A. K., Girijavallabhan, V. M., Chen, Y. T., Jao, E., Pike, R. E., Desai, J. A., Rane, D., Ganguly, A. K., *Heterocycles*, **1993**, *35*, 129.

96. Cativiela, C., Diaz de Villegas, M. D., Mayoral, J. A., Avenoza, A., Peregrina, J. M., *Tetrahedron* **1993**, *49*, 677.

97. Arseniyadis, S., Rodriguez, R., Yashunsky, D. V., Camara, J., Ourisson, G., *Tetrahedron Lett.* **1994**, *35*, 4843.

98. Tsuboi, M., *Chem. Expr.* **1993**, *8*, 441.

99. Witter, D. J., Vederas, J. C., *J. Org. Chem.* **1996**, *61*, 2613.

100. Al-Badri, H., Collignon, N., *Synthesis* **1999**, 282.

101. Yang, Y., Chan, T. H., *J. Am. Chem. Soc.* **2000**, *122*, 402.

102. Hill, K. W., Taunton-Rigby, J., Carter, J. D., Kropp, E., Vagle, K., Pieken, W., McGee, D. P. C., Husar, G. M., Leuck, M., Anziano, D. J., Sebesta, D. P., *J. Org. Chem.* **2001**, *66*, 5352.

103. Pozsgay, V., Vieira, N. E., Yergey, A., *Org. Lett.* **2002**, *4*, 3191.

104. Amantini, D., Fringuelli, F., Piermatti, O., Pizzo, F., Vaccaro, L., *Green Chemistry* **2001**, *3*, 229.

105. Amantini, D., Fringuelli, F., Piermatti, O., Pizzo, F., Vaccaro, L., *J. Org. Chem.* **2003**, *68*, 9263.

106. For reviews, see: Fringuelli, F., Piermatti, O., Pizzo, F., *Targ. Heterocycl. Syst.* **1997**, *1*, 57; Parker, D. T., in *Organic Synthesis in Water*, Grieco, P. A., ed., Blackie, London, UK, 1998, p.47.

107. Larsen, S. D., Grieco, P. A., *J. Am. Chem. Soc.* **1985**, *107*, 1768.

108. Oppolzer, W., *Angew. Chem., Int. Ed. Engl.* **1972**, *11*, 1031.

109. Grieco, P. A., Parker, D. T., Fobare, W. F., Ruckle, R., *J. Am. Chem. Soc.* **1987**, *109*, 5859; Wijnen, J. W., Engberts, J. B. F. N., *Liebigs Ann. Rec.* **1997**, 1085.

110. Grieco, P. A., Bahsas, A., *J. Org. Chem.* **1987**, *52*, 5746.

111. Waldmann, H., *Angew. Chem.* **1988**, *100*, 307; *Angew. Chem. Int. Ed. Engl.* **1988**, *27*, 274; Waldmann, H., *Liebigs Ann. Chem.* **1989**, 231; Waldmann, H., Braun, M., *Liebigs Ann. Chem.* **1991**, 1045.

112. Lock, R., Waldmann, H., *Tetrahedron Lett.* **1996**, *37*, 2753.

113. Kobayashi, S., Ishitani, H., Nagayama, S., *Synthesis* **1995**, 1195; Yu, L., Chen, D., Wang, P. G., *Tetrahedron Lett.* **1996**, *37*, 2169.

114. Akiyama, T., Takaya, J., Kagoshima, H., *Tetrahedron Lett.* **1999**, *40*, 7831.

115. Akiyama, T., Matsuda, K., Fuchibe, K., *Synlett* **2002**, 1898.

116. Hague, C., Patmore, N. J., Frost, C. G., Mahon, M. F., Weller, A. S., *Chem. Commun.* **2001**, 2286.

117. Loncaric, C., Manabe, K., Kobayashi, S., *Chem. Commun.* **2003**, 574.

118. Meekel, A. A. P., Resmini, M., Pandit, U. K., *J. Chem. Soc. Chem. Commun.* **1995**, 571.

119. Lubineau, A., Auge, J., Lubin, N., *Tetrahedron Lett.* **1991**, *32*, 7529.

120. Grieco, P. A., Henry, K. J., Nunes, J. J., Matt, J. E. Jr., *J. Chem. Soc., Chem. Commun.* **1992**, 368.

121. MacKeith, R. A., McCague, R., Olivo, H. F., Palmer, C. F., Roberts, S. M., *J. Chem. Soc., Perkin Trans. 1*, **1993**, 313.

122. McCague, R., Olivo, H. F., Roberts, S. M., *Tetrahedron Lett.* **1993**, *34*, 3785.

123. Lubineau, A., Auge, J., Lubin, N., *Tetrahedron* **1993**, *49*, 4639; Auge, J., Lubin-Germain, N., *J. Chem. Ed.* **1998**, *75*, 1285; Lubineau, A., Queneau, Y., *J. Carbohydrate Chem.* **1995**, *14*, 1295.

124. Lubineau, A., Grand, E., Scherrmann, M.-C., *Carbohydr. Res.* **1997**, *297*, 169.

125. Naruse, M., Aoyagi, S., Kibayashi, C., *Tetrahedron Lett.* **1994**, *35*, 595.

126. Naruse, M., Aoyagi, S., Kibayashi, C., *Tetrahedron Lett.* **1994**, *35*, 9213; Naruse, M., Aoyagi, S., Kibayashi, C., *J. Chem. Soc., Perkin Trans. 1*, **1996**, 1113.

127. Wijnen, J. W., Zavarise, S., Engberts, J. B. F. N., Charton, M., *J. Org. Chem.* **1996**, *61*, 2001.

128. Naruse, M., Aoyagi, S., Kibayashi, C., *J. Org. Chem.* **1994**, *59*, 1358; Kibayashi, C., Aoyagi, S., *Synlett* **1995**, 873.

129. Grieco, P. A., Kaufman, M. D., *J. Org. Chem.* **1999**, *64*, 6041.

130. Zhang, J.-H., Li, C.-J., *J. Org. Chem.* **2002**, *67*, 3969.

131. Li, Z., Zhang, J.-H., Li, C.-J., *Tetrahedron Lett.* **2003**, *44*, 153.

132. Chen, L., Li, C.-J., *Green Chem.* **2003**, *5*, 627.

133. Chen, L., Li, Z., Li, C.-J., *Synlett.* **2003**. 732.

134. Yadav, J. S., Reddy, B. V. S., Srinivas, M., Padmavani, B., *Tetrahedron* **2004**, *60*, 3261.

135. Mikami, K., Kotera, O., Motoyama, Y., Sakaguchi, H., *Synlett* **1995**, 975.

136. Attanasi, O. A., De Crescentini, L., Filippone, P., Fringuelli, F., Mantellini, F., Matteucci, M., Piermatti, O., Pizzo, F., *Helv. Chim. Act.* **2001**, *84*, 513 ; Fringuelli, F., Matteucci, M., Piermatti, O., Pizzo, F., Burla, M. C., *J. Org. Chem.* **2001**, *66*, 4661.

137. Yu, L., Li, J., Ramirez, J., Chen, D., Wang, P. G., *J. Org. Chem.* **1997**, *62*, 903.

138. Agami, C., Couty, F., Poursoulis, M., Vaissermann, J., *Tetrahedron* **1992**, *48*, 431.

139. Lee, G. A., *Synthesis* **1982**, 508.

140. Inoue, Y., Araki, K., Shiraishi, S., *Bull. Chem. Soc. Jpn.* **1991**, *64*, 3079.

141. Rao, K. R., Bhanumathi, N., Srinivasan, T. N., Sattur, P. B., *Tetrahedron Lett.* **1990**, *31*, 899.

142. Rao, K. R., Bhanumathi, N., Sattur, P. B., *Tetrahedron Lett.* **1990**, *31*, 3201.

143. Lubineau, A., Bouchain, G., Queneau, Y., *J. Chem. Soc., Perkin T. 1*, **1995**, 2433.

144. Deroose, F. D., De Clercq, P. J., *Tetrahedron Lett.* **1994**, *35*, 2615.

145. Wijnen, J. W., Steiner, R. A., Engberts, J. B. F. N., *Tetrahedron Lett.* **1995**, *36*, 5389.

146. Jiang, N., Li, C.-J., *Chem. Commun.* **2004**, 394.

147. For a review, see: Ganem, B., *Angew. Chem. Int. Ed. Engl.* **1996**, *35*, 936.

148. White, W. N., Wolfarth, E. F., *J. Org. Chem.* **1970**, *35*, 2196 and 3585.

149. Ponaras, A. A., *J. Org. Chem.* **1983**, *48*, 3866; Coates, R. M., Rogers, B. D., Hobbs, S. J., Peck, D. R., Curran, D. P., *J. Am. Chem. Soc.* **1987**, *109*, 1160; Gajewski, J. J., Jurayj, J., Kimbrough, D. R., Gande, M. E., Ganem, B. Carpenter, B. K., *J. Am. Chem. Soc.* **1987**, *109*, 1170.

150. Copley, S. D., Knowles, J. R., *J. Am. Chem. Soc.* **1987**, *109*, 5008.

151. Brower, K. R., *J. Am. Chem. Soc.* **1961**, *83*, 4370.

152. Walling, C., Naiman, M., *J. Am. Chem. Soc.* **1962**, *84*, 2628.

153. Wipf, P., Ribe, S., *Org. Lett.* **2001**, *3*, 1503; Wipf, P., Rodriguez, S., *Adv. Synth Catal.* **2002**, *344*, 434.

154. Jackson, D. Y., Jacobs, J. W., Sugasawara, R., Reich, S. H., Bartlett, P. A., Schultz, P. G., *J. Am. Chem. Soc.* **1988**, *110*, 4841; Hilvert, D., Carpenter, S. H., Nared, K. D., Auditor, M. T. M., *Proc. Natl. Acad. Sci. U.S.A.* **1988**, *85*, 4953.

155. Braisted, A. C., Schultz, P. G., *J. Am. Chem. Soc.* **1994**, *116*, 2211.

156. Brandes, E., Grieco, P. A., Gajewski, J. J., *J. Org. Chem.* **1989**, *54*, 515.

157. Grieco, P. A., Brandes, E. B., McCann, S., Clark, J. D., *J. Org. Chem.* **1989**, *54*, 5849.

158. McMurry, J. E., Andrus, A., Ksander, G. M., Musser, J. H., Johnson, M. A., *Tetrahedron* **1981**, *37*, 319 (Supplement No. 1).

159. Lubineau, A., Auge, J., Bellanger, N., Caillebourdin, S., *Tetrahedron Lett.* **1990**, *31*, 4147; *J. Chem. Soc. Perkin Trans. 1*, **1992**, 1631.

160. For reviews, see: Houk, K. N., Zipse, H., *Chemtracts: Org. Chem.* **1993**, *6*, 51; Storer, J. W., Giesen, D. J., Hawkins, G. D., Lynch, G. C., Cramer, C. J., Truhlar, D. G., Liotard, D. A., *ACS Symposium Series* (1994), 568 (*Structure and Reactivity in Aqueous Solution*), 24.

161. Cramer, C. J., Truhlar, D. G., *J. Am. Chem. Soc.*, **1992**, *114*, 8794.

162. Gao, J., *J. Am. Chem. Soc.* **1994**, *116*, 1563; Davidson, M. M., Hillier, I. H., Hall, R. J., Burton, N. A., *J. Am. Chem. Soc.* **1994**, *116*, 9294.

163. Severance, D. L., Jorgensen, W. L., *J. Am. Chem. Soc.* **1992**, *114*, 10966.

164. Gajewski, J. J., *J. Org. Chem.* **1992**, *57*, 5500.

165. Gajewski, J. J., *Acc. Chem. Res.* **1997**, *30*, 219.

166. Gajewski, J. J., Brichford, N. L., *J. Am. Chem. Soc.* **1994**, *116*, 3165.

167. Sehgal, A., Shao, L., Gao, J., *J. Am. Chem. Soc.* **1995**, *117*, 11337.

168. Gao, J., Xia, X., *ACS Symposium Series* (**1994**), 568 (*Structure and Reactivity in Aqueous Solution*), p. 212.

169. Davidson, M. M., Hillier, I. H., Hall, R. J., Burton, N. A., *J. Am. Chem. Soc.* **1994**, *116*, 9294; Davidson, M. M., Hillier, I. H., *J. Phys. Chem.* **1995**, *99*, 6748; Davidson, M. M., Hillier, I. H., Vincent, M. A., *Chem. Phys. Lett.* **1995**, *246*, 536.

170. Carlson, H. A., Jorgensen, W. L., *J. Am. Chem. Soc.* **1996**, *118*, 8475.

171. Hur, S., Bruice, T. C., *J. Am. Chem. Soc.* **2003**, *125*, 10540.

172. Repasky, M. P., Guimaraes, C. R. W., Chandrasekhar, J., Tirado-Rives, J., Jorgensen, W. L., *J. Am. Chem. Soc.* **2003**, *125*, 6663.

173. Hur, S., Bruice, T. C., *J. Am. Chem. Soc.* **2003**, *125*, 5964.

174. Majumdar, K. C., Sarkar, S., Ghosh, S., *Synth. Commun.* **2004**, *34*, 1265.

175. Cooper, M. A., Lucas, M. A., Taylor, J. M., Ward, A. D., Williamson, N. M., *Synthesis* **2001**, 621.

176. Aemissegger, A., Jaun, B., Hilvert, D., *J. Org. Chem.* **2002**, *67*, 6725.

177. Grotjahn, D. B., Zhang, X. J., *Mol. Catal. A: Chem.* **1997**, *116*, 99.

178. Ramamurthy, V., *Tetrahedron* **1986**, *42*, 5753.

179. Syamala, M. S., Ramamurthy, V., *J. Org. Chem.* **1986**, *51*, 3712.

180. Ito, Y., Kajita, T., Kunimoto, K., Matsuura, T., *J. Org. Chem.* **1989**, *54*, 587.

181. Muthuramu, K., Ramamurthy, V., *J. Org. Chem.* **1982**, *47*, 3976.

182. Tamaki, T., *Chem. Lett.* **1984**, 53; Tamaki, T., Kokubu, T., *J. Inclusion Phenom.* **1984**, *2*, 815.

183. Diao, L., Yang, C., Wan, P., *J. Am. Chem. Soc.* **1995**, *117*, 5369; Barker, B., Diao, L., Wan, P., *J. Photochem. Photobiol. A: Chem.* **1997**, *104*, 91.

INDEX